Indigenous Ecologies, Sustainability, and Worldviews in Context

Monicca Thulisile Bhuda
https://orcid.org/0000-0002-8506-9562
University of Mpumalanga, South Africa

Tebogo Maria Mothiba
University of Limpopo, South Africa

IGI Global
Scientific Publishing
Publishing Tomorrow's Research Today

Vice President of Editorial Melissa Wagner
Director of Acquisitions Mikaela Felty
Director of Book Development Jocelynn Hessler
Production Manager Mike Brehm
Cover Design Jose Rosado

Published in the United States of America by
IGI Global Scientific Publishing
701 East Chocolate Avenue
Hershey, PA, 17033, USA
Tel: 717-533-8845 | Fax: 717-533-7115
Website: https://www.igi-global.com E-mail: cust@igi-global.com

Library of Congress Cataloging-in-Publication Data

LCCN: 2026032018 (CIP Data Pending)
ISBN13: 9798260028902
Isbn13Softcover: 9798260028919
EISBN13: 9798260028926
British Cataloguing in Publication Data
A Cataloguing in Publication record for this book is available from the British Library.

Table of Contents

Detailed Table of Contents

Indigenous Knowledge Systems (IKS) in India constitute living, community-based epistemologies embedded within Intangible Cultural Heritage (ICH), encompassing oral traditions, ritual practices, craftsmanship, ecological stewardship, and social governance. Despite their historical role in sustaining cultural identity, environmental balance, and social cohesion, these knowledge systems have been systematically marginalized through colonial epistemic domination, modernization, and market-driven cultural homogenization. This chapter conceptualizes Indian IKS as dynamic knowledge systems rather than static traditions and examines their interdependence with ICH. It critically analyzes the structural forces contributing to their erosion while highlighting contemporary community-led revival efforts supported by civil society, educational institutions, and ethical digital interventions. The chapter further explores the integration of IKS with scientific methods, sustainability frameworks, and cultural economies, emphasizing youth engagement and intergenerational transmission.

Indigenous music performed on traditional African instruments has historically been marginalised within formal education and professional performance spaces. Despite their cultural significance, many South African indigenous instruments remain largely confined to rural contexts, where they function more as cultural artefacts than as pedagogical resources. Their limited visibility constrains integration into mainstream education and the broader performing arts sector. Although some performers have introduced these instruments into urban settings, such efforts have not substantially shifted their institutional status. This qualitative study, drawing on interviews with instrument makers and performers, literature review, and performance analysis, examines urban and rural initiatives aimed at revitalising indigenous instruments. Findings indicate that crafting, teaching, and performance are central to sustaining indigenous knowledge within contemporary education. The study advocates for their systematic integration to advance cultural preservation and pedagogical transformation.

This chapter examines Ngwao ya Rona (Our Culture) as expressed through Setswana traditional music, positioning it as a vital Indigenous Knowledge System (IKS) through which cultural consciousness, social values, and historical memory are produced and transmitted among Batswana communities. Setswana traditional music—encompassing song, rhythm, dance, oral poetry, and performance—functions as an embodied and performative knowledge archive, preserving indigenous epistemologies related to identity, morality, spirituality, gender relations, and communal governance. The chapter further situates Setswana traditional music within broader historical processes of colonial disruption, apartheid marginalisation, and contemporary globalisation, highlighting the resilience and adaptability of indigenous musical knowledge. Ultimately, the chapter foregrounds traditional music as central to sustaining ngwao ya rona and advancing indigenous cultural consciousness in contemporary African societies.

This autoethnography traces a personal journey that began as an academic task and evolved into a transformative encounter with the cultural heart of the Sama people—the Tepo of Tawi-Tawi. Initially seen as an ordinary woven mat, the Tepo emerged as a living archive of artistry, memory, and indigenous intelligence. Through exploration, hesitation, and realization, immersion in Tepo deepened cultural empathy, self-discovery, and inspired culturally relevant mathematics teaching. Rather than analyzing weaving techniques, this study centers lived experiences, listening to stories within patterns and honoring relational wisdom across generations. It shows how indigenous knowledge is observed, felt, and negotiated through mindful participation. Curiosity becomes a catalyst for awakening and transformation. This work reimagines indigenous knowledge revitalization, positioning education as a space for honoring culture, community, and self.

This chapter reconstructs the Ramayana as a structured indigenous moral epistemology rather than a merely devotional epic. Through systematic thematic coding of all seven kāṇḍas, the epic's major narrative divisions, using NVivo-assisted analysis, co-occurrence matrices, PCA clustering, and network visualization, it identifies patterns of ethical intensity and maps the relational architecture of dharma (a context-sensitive moral order), leadership, and justice. Situated within the Purushartha framework (the fourfold aims of human life), the study demonstrates that dharma functions as narrative-based moral discernment rather than rigid rule adherence. By critically engaging Aristotelian virtue ethics and Kantian deontology, the chapter establishes the Ramayana's internal normative coherence as a philosophically robust knowledge system. It argues that revitalizing Indigenous Knowledge Systems requires recognizing narrative traditions as norm-generating epistemic structures capable of informing contemporary debates on governance, responsibility, and ethical restraint.

This chapter examines the importance of Indigenous Knowledge Systems (IKS) as key strategies for enhancing cross-cultural communication within the South African National Defence Force (SANDF). The chapter employs Oetzel's (2005) Effective Intercultural Workgroup Communication Theory as the guiding theoretical framework to deepen understanding of the cultural factors that influence workgroup diversity and cross-cultural communication among service members. Employing a qualitative research method through an analysis of documents, the chapter indicates that IKS philosophies can enhance solidarity, reduce conflict and cultural misunderstanding, as well as improve intercultural and cross-cultural interactions within the SANDF. The chapter concludes that integrating IKS as a strategy to cross-cultural communication creates a representative, effective, culturally competent, and united military force.

This chapter examines the role of African Indigenous Knowledge Systems (IKS), with specific reference to Basotho IKS, in response to extreme weather occurrences in Lesotho. Climate change has been reported to affect rain-fed agriculture systems that lack adaptive infrastructure. The research was conducted using the qualitative research approach. Data was collected from 16 purposively selected participants . The data was then analysed using thematic data analysis. From the research findings, it was concluded that Basotho IKS, which relies on ecological observations, rituals, ceremonies, and traditional norms in the communities, plays a crucial role in the response to extreme weather occurrences in Lesotho. The research was theoretically underpinned by the Afrocentric theory, which emphasises the use of African perspectives in knowledge systems. The research findings provide insights that can be used in the response to extreme weather occurrences in Lesotho and beyond.

Food insecurity remains a challenge in many rural African communities despite long histories of self-sustaining food systems (FAO, 2022). Among the Vhavenda of Limpopo Province, indigenous food preservation practices historically supported year-round food availability, nutritional balance, and cultural continuity (Dei, 2018; Chilisa, 2020). This chapter explores food security in Khalavha village, focusing on indigenous science embedded in traditional preservation methods. Drawing on qualitative community-based research, it examines how colonial disruption, modernisation, and socio-economic change have marginalised Indigenous Knowledge Systems, increasing vulnerability to food insecurity. It argues that revitalising indigenous preservation practices can strengthen affordability, resilience, and culturally appropriate food security, particularly in contexts marked by rising food prices, climate variability, and unreliable energy supply in South Africa.

Preface

INTRODUCTION

Indigenous Knowledge Systems (IKS) represent some of humanity's oldest and most enduring ways of understanding the world. Rooted in centuries of observation, experience, spirituality, cultural practice, and community engagement, these knowledge traditions have shaped social organization, environmental stewardship, governance systems, artistic expression, and economic practices across diverse societies. Yet despite their significance, indigenous epistemologies have often been marginalized by colonial legacies, modernization processes, and dominant Western paradigms that have frequently dismissed local knowledge as informal or inferior. In recent decades, however, growing recognition of the limitations of singular approaches to development, sustainability, and knowledge production has renewed scholarly and policy interest in indigenous perspectives.

This volume, *Indigenous Ecologies, Sustainability, and Worldviews in Context*, emerges at a critical moment when global challenges such as climate change, food insecurity, cultural erosion, social inequality, and environmental degradation demand more inclusive and pluralistic approaches to knowledge. Indigenous communities around the world possess valuable insights regarding human relationships with nature, collective responsibility, ethical governance, cultural resilience, and sustainable living. The chapters assembled in this book demonstrate that indigenous knowledge is not merely a relic of the past but rather a dynamic, evolving, and highly relevant resource for addressing contemporary challenges.

The contributions presented in this volume span multiple geographical regions, cultural traditions, and disciplinary perspectives. Authors explore indigenous knowledge systems through lenses that include cultural heritage preservation, education, religion, ethics, environmental sustainability, governance, food security, tourism, economics, communication, and social development. Together, these chapters reveal the diversity and adaptability of indigenous knowledge while highlighting common themes of resilience, interconnectedness, stewardship, and collective well-being. The volume further illustrates how indigenous worldviews continue to inform contemporary practices and contribute to scholarly discourse across multiple fields.

As editors, we hope this collection encourages deeper engagement with indigenous epistemologies and promotes meaningful dialogue between traditional knowledge holders, scholars, policymakers, educators, and practitioners. By bringing together diverse voices and perspectives, this book seeks to contribute to ongoing efforts to recognize, preserve, revitalize, and respectfully integrate Indigenous Knowledge Systems within contemporary global conversations on sustainability, development, and cultural continuity.

OVERVIEW OF THE CHAPTERS

The volume begins by examining the preservation and revitalization of indigenous cultural heritage. Chapter 1 explores the relationship between India's Indigenous Knowledge Systems and intangible cultural heritage, highlighting efforts to revitalize traditional knowledge through education, community engagement, and digital interventions. Chapters 2 and 3 focus on indigenous musical traditions in South Africa, examining the integration of traditional instruments into formal music education and the role of Setswana traditional music in sustaining cultural consciousness and collective memory. Chapter 4 presents an autoethnographic account of the Sama Tepo of the Philippines, illustrating how indigenous cultural practices foster personal transformation, educational innovation, and cultural awareness.

The book then expands into philosophical, ethical, and spiritual dimensions of indigenous knowledge. Chapter 5 reconstructs the *Ramayana* as an indigenous moral epistemology, demonstrating its relevance to contemporary ethical inquiry and governance. Chapter 6 investigates indigenous cosmologies and concepts of non-human personhood across cultures, offering valuable insights into relational ethics and environmental stewardship. Chapter 7 examines ecological ethics through the lens of the Bharatiya concept of *Pañca Rna*, highlighting its contributions to sustainability thinking and contemporary organizational practice.

Questions of governance, colonial influence, and social institutions are addressed in Chapters 8 through 11. Chapter 8 advocates for the revitalization of African kingdoms as custodians of Indigenous Knowledge Systems within South Africa's evolving democratic landscape. Chapter 9 analyzes the impact of British colonial governance on Indigenous Knowledge Systems in India, revealing the lasting consequences of epistemic regulation and knowledge extraction. Chapter 10 explores the experiences of children under extended family systems in Zimbabwe during periods of economic austerity, while Chapter 11 examines how Indigenous Knowledge Systems can enhance cross-cultural communication and cohesion within the South African National Defence Force.

Environmental sustainability, climate adaptation, and food systems form another major theme of the volume. Chapter 12 investigates Basotho Indigenous Knowledge Systems as mechanisms for responding to extreme weather events and climate variability in Lesotho. Chapter 13 examines Vhavenda indigenous food preservation practices and their relevance to food security and resilience in South Africa. Chapter 14 further explores indigenous food systems through a systematic review of the impacts of colonial policies on traditional agricultural and food practices. Complementing these discussions, Chapter 15 reflects on Kwena livestock management traditions in Botswana, demonstrating how indigenous approaches to resource management contribute to environmental sustainability and socio-economic resilience.

The final section of the book explores indigenous knowledge in economics, pilgrimage, and sustainable tourism. Chapter 16 introduces the Vedic-Kautilyan Microfinance Framework, illustrating how ancient Indian economic thought can inform contemporary financial inclusion practices. Chapter 17 investigates visitor motivations and pilgrimage experiences at Sabarimala, one of India's most significant religious destinations, emphasizing the importance of preserving indigenous cultural traditions within modern tourism contexts. Concluding the volume, Chapter 18 examines Chhath Puja in Bihar as a model for sustainable tourism development, demonstrating how indigenous religious festivals can simultaneously support cultural preservation, social cohesion, economic development, and environmental sustainability.

Collectively, these chapters offer a rich and multidimensional exploration of Indigenous Knowledge Systems as living repositories of wisdom that continue to shape societies, inform policy, and contribute to sustainable futures across the globe.

CONCLUSION

The chapters presented in this volume collectively demonstrate that Indigenous Knowledge Systems remain vital sources of intellectual, cultural, ecological, and ethical insight. Far from being confined to historical memory or cultural preservation efforts alone, indigenous knowledge continues to influence contemporary responses to some of the world's most pressing challenges. Whether addressing environmental sustainability, climate adaptation, food security, education, governance, social cohesion, economic development, or cultural resilience, the contributions in this book reveal the enduring relevance of indigenous perspectives in both local and global contexts.

A recurring theme throughout this collection is the importance of relationality—the understanding that humans exist within interconnected networks of relationships involving communities, ancestors, ecosystems, spiritual entities, and future generations. This worldview offers important alternatives to dominant paradigms that often prioritize individualism, extraction, and short-term gains. The studies presented here illustrate how indigenous philosophies foster responsibility, reciprocity, stewardship, and collective well-being, providing valuable frameworks for sustainable and equitable development.

At the same time, many chapters highlight the historical and ongoing challenges facing Indigenous Knowledge Systems, including colonial legacies, globalization, cultural homogenization, environmental pressures, and institutional marginalization. The revitalization of indigenous knowledge therefore requires more than documentation alone; it necessitates meaningful recognition of indigenous voices, equitable partnerships, supportive policy frameworks, culturally responsive education systems, and respectful engagement with indigenous communities as active knowledge holders and contributors.

As editors, we hope this volume serves as both a scholarly resource and a catalyst for continued dialogue and collaboration. We envision this collection contributing to broader efforts aimed at decolonizing knowledge production, strengthening cultural sustainability, and fostering greater appreciation for the diversity of human ways of knowing. Ultimately, Indigenous Knowledge Systems offer not only pathways for preserving cultural heritage but also powerful resources for imagining and building more sustainable, inclusive, and resilient futures for all.

Chapter 1
Revitalizing India's Intangible Cultural Heritage Through Indigenous Knowledge Systems

Priya Chaturvedi
http://orcid.org/0000-0002-9875-661X
Jiwaji University, India

Aditya Ranjan
http://orcid.org/0000-0003-1304-5489
ITM University, India

ABSTRACT

Indigenous Knowledge Systems (IKS) in India constitute living, community-based epistemologies embedded within Intangible Cultural Heritage (ICH), encompassing oral traditions, ritual practices, craftsmanship, ecological stewardship, and social governance. Despite their historical role in sustaining cultural identity, environmental balance, and social cohesion, these knowledge systems have been systematically marginalized through colonial epistemic domination, modernization, and market-driven cultural homogenization. This chapter conceptualizes Indian IKS as dynamic knowledge systems rather than static traditions and examines their interdependence with ICH. It critically analyzes the structural forces contributing to their erosion while highlighting contemporary community-led revival efforts supported by civil society, educational institutions, and ethical digital interventions. The chapter further explores the integration of IKS with scientific methods, sustainability frameworks, and cultural economies, emphasizing youth engagement and intergenerational transmission.

1. INTRODUCTION

Indigenous Knowledge Systems (IKS) in India represent a complex body of knowledge that has developed through long-term interaction between communities and their natural environments, social institutions, and cultural practices (Durie, 2005; Priyadarshini & Abhilash, 2019). These systems include knowledge related to agriculture, biodiversity management, water conservation, health practices, crafts, and community governance, reflecting centuries of experiential learning and adaptation (Orlove et al.,

DOI: 10.4018/979-8-2600-2890-2.ch001

2022; Srivastava, 2012). Researchers increasingly recognize IKS as an authentic knowledge domain that promotes and contributes to sustainability, climate resilience, and social cohesion (Gondo, 2025; Tirivangasi & Kontinen, 2025). These contributions emerge from its foundation in place-based observation and collective responsibility (Oladele & Amara, 2024). Such recognition challenges earlier assumptions that treated Indigenous knowledge as ancient informal or outdated within hegemonic development models (Breidlid, 2009; Gorjestani, 2004).

In India, Indigenous knowledge is closely associated with living cultural heritage (Moitra & Madan, 2025). The living cultural heritage, or intangible cultural heritage includes oral traditions, rituals, performing arts, craftsmanship, and knowledge about nature that are transmitted through participation and practice (Ranjan & Chaturvedi, 2022; UNESCO, 2003). These elements are transmitted intergenerationally through participation, observation, and practice (Smith & Akagawa, 2008; Zabulis et al., 2025). Cultural festivals, storytelling traditions, and apprenticeship-based craft training act as learning environments facilitating the exchange of ecological knowledge, ethical values, and technical skills among generations (Gwerevende & Mthombeni, 2023; Yatim et al., 2025). These practices demonstrate that IKS is not merely symbolic heritage, but a dynamic, living knowledge system embedded in everyday community life (Higgins, 2022; Katelieva et al., 2020). Hence, to comprehend how knowledge is preserved, transmitted, and adapted in different cultural settings, it is crucial to understand the relationship between IKS and ICH (Orlove et al., 2022).

Despite its importance, indigenous knowledge in India has experienced gradual erosion, mainly because colonial interventions, other historical and structural forces disrupted traditional community-based learning systems (Gorjestani, 2004). Colonial policies introduced land management rules, education models, and economic structures that marginalized local knowledge and privileged Western scientific approaches (Breidlid, 2009; Brodt, 1999). Post-independence modernization programs often continued these patterns by promoting industrial agriculture and standardized schooling that overlooked local ecological knowledge and cultural traditions (Knorr & Augustin, 2025; Srivastava, 2012). For example, high-yield crop promotion reduced the use of traditional seed varieties (Brodt, 2001). These changes reduced opportunities for intergenerational transmission and weakened the social institutions that supported knowledge continuity (Gwerevende & Mthombeni, 2023; Malapane et al., 2024).

Contemporary forces such as climate change, migration, and global market expansion have intensified pressures on the transmission and continuity of IKS (Pearson et al., 2023; Shrestha et al., 2025; Tirivangasi & Kontinen, 2025). Climate change is affecting rainfall patterns, crop cycles, and biodiversity, which threatens not only livelihoods but also the knowledge systems tied to specific landscapes and seasonal patterns (Erbach, 2025; Orlove et al., 2022). The erosion of ecological knowledge reduces community resilience to disasters, undermines sustainable resource management, and compromises food security (Oladele & Amara, 2024; Tirivangasi & Kontinen, 2025). Simultaneously, environmental degradation also disturbs traditional farming indicators used by village communities (Erbach, 2025; Orlove et al., 2022). Industrial production and commercialization reduce the economic value of traditional crafts and local food systems, discouraging youth from learning traditional skills (Ma et al., 2022; Shrestha et al., 2025). These pressures highlight the urgent need to revitalize IKS as a resource for sustainable development and cultural resilience.

Additionally, recent researches indicates that revitalization efforts must be grounded in community-centered governance and ethically informed technological integration (Bawack et al., 2025; Chigwada et al., 2025; Eichler, 2021). Moreover, supportive policy frameworks are required to recognize Indigenous knowledge holders as equal partners in development processes (Gorjestani, 2004; Hoppers, 2025). Digital documentation initiatives, participatory research, heritage education programs and tourism initiatives can strengthen knowledge transmission (Ranjan & Chaturvedi, 2023b). However, these processes need to be guided by indigenous data governance principles and grounded in informed community consent (Chigwada et al., 2025; Huang & Xu, 2022). Integrating Indigenous knowledge with scientific methods also improves climate adaptation planning and environmental management because it combines local insights with analytical tools (Durie, 2005; Tirivangasi & Kontinen, 2025).

This chapter therefore examines Indigenous Knowledge Systems in India through the interconnected themes of intangible cultural heritage, sustainability, community development, youth participation, and epistemic justice. By analyzing historical marginalization, contemporary revival efforts, and future pathways for integration with science and technology, the chapter argues that Indigenous knowledge is not a relic of the past but a living resource for addressing global challenges.

2. INDIGENOUS KNOWLEDGE SYSTEMS IN INDIA

Indigenous Knowledge Systems in India have evolved through sustained, historically embedded interactions between communities and their ecological, social, and cultural environments, reflecting context-specific understandings shaped by local environments, customary institutions, and lived experiences (Brodt, 1999). These knowledge traditions are rooted in daily work, ritual life, and community institutions, and they reflect accumulated experience across generations (Durie, 2005; Priyadarshini & Abhilash, 2019). They also show how communities adapt to changing climate, markets, and social pressures by using local understanding of land, water, and biodiversity (Gondo, 2025). Consequently, studying IKS in India requires attention to its conceptual basis, its ways of knowing, and its practical use within plural knowledge traditions (Breidlid, 2009; Hoppers, 2025).

2.1 Defining Indigenous Knowledge Systems in the Indian Context

IKS are commonly described or defined as localized knowledge traditions formed through experience, oral teaching, and close engagement with the environment and community life (Durie, 2005; Gorjestani, 2004). In India, indigenous knowledge exists among tribal, farming, pastoral, coastal, and craft communities, where different practices remain linked through shared ecological and social values (Brodt, 2001; Priyadarshini & Abhilash, 2019). It often relies on memory, practice, and apprenticeship within families and communities (Gwerevende & Mthombeni, 2023; Malapane et al., 2024). Additionally, indigenous knowledge helps communities manage soil, forests, crops, water, and health in ways suited to their local conditions (Oladele & Amara, 2024; Srivastava, 2012). Rather reductively interpreted as folklore or tradition, indigenous knowledge must be regarded as a system built on observation, experimentation, and community review (Breidlid, 2009; Hoppers, 2025). Such knowledge spans agriculture, ethnobotany, architecture, healing, traditions, ritual governance, water management, and artisanal production, demonstrating a comprehensive worldview that integrates material, ethical, and cosmological dimensions (Arslan & Dişli, 2025; Moitra & Madan, 2025). Practices such as local farming methods, seed selection, and water conservation systems show how communities evaluate outcomes and improve techniques over time (Brodt, 2001; Knorr & Augustin, 2025).

Indian indigenous knowledge challenges the dichotomy between "traditional" and "modern" (Brodt, 2001). In contract to western knowledge traditions that privilege abstraction and textual codification,

Indian IKS relies heavily on embodied practice, memory, apprenticeship, and relational learning (Breidlid, 2009; Moitra & Madan, 2025). Indigenous communities routinely innovate, integrate scientific tools, and adapt knowledge practices to environmental and social change (Brodt, 1999; Srivastava, 2012). Communities adopt new tools, crops, and technologies when useful, while keeping core ecological principles (Bawack et al., 2025; Srivastava, 2012). Studies of farming and disaster management in India show how local knowledge works together with scientific information to improve resilience (Erbach, 2025; Orlove et al., 2022). Such examples demonstrate that IKS is dynamic, adaptive, and evolves through cumulative ecological experiences and socio-cultural negotiations. Accordingly, conceptual frameworks need to recognize IKS as context-based expertise that complements modern science, capable of enriching scientific knowledge through local sensitivity, insights, and longitudinal ecological awareness (Orlove et al., 2022).

2.2 The Interconnectedness of IKS and Intangible Cultural Heritage

Indian IKS are associated with practices recognized and categorized as ICH (Moitra & Madan, 2025). Examples include folk songs, rituals, handicrafts, village festivals, and traditional ecological practices (Ranjan & Chaturvedi, 2023a; UNESCO, 2003). These practices transmit ecological knowledge, social values, and collective historical memory (Orlove et al., 2022; Smith & Akagawa, 2008). Such knowledge is conveyed from one generation to another through performance, participation, and embodied cultural practices (Lázaro Ortiz & Jiménez de Madariaga, 2022; UNESCO, 2003). Intangible cultural heritage therefore acts as a living classroom where knowledge is shared and renewed (Katelieva et al., 2020; Zabulis et al., 2025).

In India, weaving traditions, folk songs, pastoral festivals, and ritual healing practices all contain ecological and social knowledge about crops, seasons, plants, community norms and shared responsibilities (Priyadarshini & Abhilash, 2019; Shrestha et al., 2025). These practices guide decision-making about planting time, water use, animal care, and food preparation, showing how culture and environment remain linked (Knorr & Augustin, 2025; Oladele & Amara, 2024). They also help maintain social harmony by teaching shared values and responsibilities (Alam, 2025; Zabulis et al., 2025). Similarly, craft communities exhibit IKS preservation through culture practices. Weavers, potters, metalworkers, and woodcarvers teach young learners through hands-on training that includes knowledge of materials, design, climate effects, and symbolism (Arslan & Dişli, 2025; Ma et al., 2022). Such apprenticeship integrates technical skill with cultural identity, linking economic activity with heritage preservation (Gwerevende & Mthombeni, 2023; Priyadarshini & Abhilash, 2019). Similar patterns are found in ritual healing, where plant knowledge and spiritual belief are taught together (Alum, 2024; Orlove et al., 2022).

IKS and ICH embodies a relational worldview that situates human life within an interconnected network of land, water, animals, and ancestral lineages (Tirivangasi & Kontinen, 2025). Rituals such as seasonal festivals, sacred grove rituals, and pastoral migration reinforce norms of environmental respect and community solidarity (Erbach, 2025). These traditions encode environmental ethics and guide sustainable resource use (Knorr & Augustin, 2025; Orlove et al., 2022). They show that knowledge in Indigenous societies is both practical and moral (Chaudhary, 2025). Language plays an important role in this connection because ecological knowledge, songs, and stories are often preserved in local languages and dialects (Gwerevende & Mthombeni, 2023; Pearson et al., 2023). When languages decline, communities lose important knowledge about plants, animals, and seasons (Alam, 2025; Dembedza et al., 2022). Protecting linguistic diversity therefore supports both cultural heritage and ecological knowledge (Ranjan & Chaturvedi, 2025; Yatim et al., 2025). Thus, intangible culture is integral to indigenous knowledge, serving as a fundamental medium for the creation, dissemination, and revitalization of knowledge (Lázaro Ortiz & Jiménez de Madariaga, 2022; UNESCO, 2003). Researchers and policymakers could be benefited from this connection by better comprehending the importance of preserving cultural traditions for the sustaining ecological wisdom and community resilience (Orlove et al., 2022; Smith & Akagawa, 2008).

2.3 Why Revitalization Is Urgent in the 21st Century?

The need to revitalize IKS in India arises from overlapping ecological, cultural, economic, and intellectual pressures that threaten both community life and environmental balance (Hoppers, 2025; Orlove et al., 2022). Researchers mention that the loss of Indigenous knowledge is linked to land displacement, environmental change, and social transformation that weaken the everyday settings where knowledge is learned and practiced (Malapane et al., 2024; Pearson et al., 2023). In India, mining expansion, forest restrictions, migration to cities, and uniform school systems have reduced the space for local knowledge practices and community learning (Brodt, 2001; Gorjestani, 2004). These changes affect not only livelihoods but also the cultural relations through which ecological knowledge is shared across generations (Durie, 2005; Gwerevende & Mthombeni, 2023).

From an ecological perspective, the decline of Indigenous knowledge weakens climate resilience and environmental management (Oladele & Amara, 2024; Orlove et al., 2022; Sabarivasan et al., 2025). Traditional farming systems such as mixed cropping, seed saving, and soil conservation help maintain biodiversity and reduce climate risks, yet these practices are often replaced by industrial agriculture that depends on external inputs (Knorr & Augustin, 2025; Priyadarshini & Abhilash, 2019). Indigenous knowledge about water storage, rainfall prediction, and local plant cycles also supports adaptation to droughts and floods, especially in rural areas with limited infrastructure (Erbach, 2025; Orlove et al., 2022). When such knowledge disappears, communities lose practical strategies for managing climate uncertainty (Oladele & Amara, 2024; Tirivangasi & Kontinen, 2025). Moreover, cultural pressures also create urgency for revitalization (Puig, 2025; Ranjan & Chaturvedi, 2025). Mass media, migration, and global markets often reshape local traditions into simplified or commercial forms that reduce their deeper social and ecological meaning (Qiu et al., 2022; Smith & Akagawa, 2008). Tourism and market demand sometimes encourage imitation crafts and standardized products that weaken the identity of local artisans and reduce incentives for youth to continue traditional occupations (Ma et al., 2022; Shrestha et al., 2025). Such changes disrupt the link between cultural heritage and knowledge transmission (Katelieva et al., 2020).

Economic challenges further threaten Indigenous knowledge (Shrestha et al., 2025). Many traditional occupations provide unstable income compared with urban employment, which leads young people to leave craft, farming, or pastoral work (Aluko et al., 2025). When knowledge holders cannot sustain livelihoods, apprenticeship systems decline and knowledge transmission weakens (Malapane et al., 2024). Supporting heritage-based livelihoods therefore becomes essential for sustaining knowledge systems (Eichler, 2021). Intellectually, indigenous knowledge continues to face marginalization within academic and policy institutions that often privilege Western scientific models over local experience (Breidlid, 2009; Hoppers, 2025). Researchers argue that true integration requires recognition of knowledge diversity rather than assimilation into dominant frameworks (Durie, 2005; Orlove et al., 2022). When Indigenous perspectives are ignored, development programs may fail because they overlook local realities and cultural values (Gorjestani, 2004; Zabulis et al., 2025).

3. LIVING CULTURAL HERITAGE AS INDIGENOUS KNOWLEDGE SYSTEM

Indian ICH constitute a vase landscape of performative, ritual, artisanal, narrative, and ecological practices embodied in IKS (Orlove et al., 2022; Smith & Akagawa, 2008). These are epistemic systems that generate, encode, interpret, and transfer knowledge about ecological, social norms, and cultural identity (Durie, 2005; Zabulis et al., 2025). Modern researchers recognises that ICH is a living knowledge system rather than a static cultural archive because its vitality depends on social participation, intergenerational learning, and contextual performance (Gwerevende & Mthombeni, 2023; Malapane et al., 2024).

3.1 Overview of UNESCO Intangible Cultural Heritage Categories in India

According to UNESCO (2003), Intangible Cultural Heritage comprises five core domains encompassing oral traditions, performing arts, social practices and festivals, ecological knowledge systems, and traditional craftsmanship (Lázaro Ortiz & Jiménez de Madariaga, 2022). Such categorization demonstrates the breadth of community-based practices that enable the preservation and intergenerational transmission of knowledge (Gwerevende & Mthombeni, 2023; Ranjan & Chaturvedi, 2022). India has several entries on the UNESCO Representative List, including the Vedic chanting tradition, Ramlila performance, Chhau dance, Koodiyattam Sanskrit theatre, the traditional brass and copper craft of Punjab, and Ayurveda and Yoga as systems of traditional knowledge (Moitra & Madan, 2025; Smith & Akagawa, 2008). While these examples have gained global recognition, they represent only a fraction of India's vast intangible cultural landscape. Many local traditions remain undocumented despite their continued role in sustaining community life and ecological understanding (Durie, 2005; Orlove et al., 2022).

UNESCO's framework highlights continuity, community participation, and adaptability as key features of ICH (Lázaro Ortiz & Jiménez de Madariaga, 2022). Such emphasis indicates that heritage remains viable through lived practice and collective participation, rather than through isolated static preservation (Eichler, 2021; Smith & Akagawa, 2008). Research shows that intangible heritage practices embody community norms, values, and ecological knowledge that cannot be separated from the contexts in which they are performed (Shrestha et al., 2025; Zabulis et al., 2025). In India, oral narratives including storytelling often blend with ritual performance (Ranjan & Chaturvedi, 2025). Craft production relies on specialized knowledge of local resources like natural dyes and climatic conditions (Brodt, 2001). Seasonal festivals reflect agricultural cycles as well as cosmological belief systems (Priyadarshini & Abhilash, 2019). The overlapping forms suggest that Indian ICH is intricate and interrelated rather than confined to rigid classifications (Tirivangasi & Kontinen, 2025).

Many ICH in India preserve ecological knowledge related to land stewardship, water management, and biodiversity conservation through ritual practices, oral traditions, and community-based resource governance (Katelieva et al., 2020; Shrestha et al., 2025). Sacred grove rituals reinforce norms of forest conservation, while pastoral traditions preserve knowledge of seasonal mobility patterns (Knorr & Augustin, 2025; Orlove et al., 2022). Agricultural festivals encode information about climate variability and soil management practices (Shrestha et al., 2025). Such examples reflect what Indigenous knowledge scholars describe as the integration of social, spiritual, and ecological dimensions within a unified knowledge framework (Durie, 2005; Pearson et al., 2023). UNESCO classifications therefore provide a useful structure for recognizing heritage elements, but they cannot fully capture the depth of Indigenous knowledge embedded in India's living cultural traditions (Smith & Akagawa, 2008; Zabulis et al., 2025).

3.2 Cultural Expressions as Embedded Knowledge, Not Symbolic Artefacts

A defining feature of Indian ICH is that cultural expressions function as carriers of indigenous knowledge rather than symbolic representations detached from daily life (Breidlid, 2009). Indigenous knowledge is enacted through gesture, rhythm, narrative, craft technique, and spatial practice, all of which encode ecological understanding, technical skill, and social values (Smith & Akagawa, 2008). Studies on Indigenous knowledge emphasize that community practices often integrate observation, memory, and ethical norms into lived traditions rather than written texts (Durie, 2005; Malapane et al., 2024). In India, such practices appear in agricultural rituals, craft traditions, and oral storytelling, where knowledge about seasons, materials, and social conduct is preserved through repeated performance (Orlove et al., 2022; Priyadarshini & Abhilash, 2019).

Artisanal traditions provide illustrative examples of knowledge embedded within cultural practices. Weaving, pottery, metalwork, and dyeing practices involve detailed understanding of raw materials, climate conditions, and design patterns that are passed through apprenticeship and community learning (Arslan & Dişli, 2025; Ma et al., 2022). Research on Indigenous agricultural and forestry systems in India shows that practical techniques often combine ecological observation with ritual practice, demonstrating that technical and cultural knowledge are closely linked (Brodt, 2001; Srivastava, 2012). These examples indicate that craft traditions carry both material skill and ecological awareness. Similarly, ritual acts serve as information systems in addition to being sources of entertainment. Healing rituals, seasonal festivals, and agricultural ceremonies depend on knowledge of plants, weather indicators, and ethical rules that guide community behaviour (Alum, 2024; Orlove et al., 2022). Such practices are learned through participation, observation, and mentorship, which reflects the importance of relational learning within Indigenous knowledge transmission (Gwerevende & Mthombeni, 2023; Malapane et al., 2024).

Researchers describe Indigenous knowledge as relational because it connects people with land, ancestors, and non-human life in shared systems of meaning (Durie, 2005; Tirivangasi & Kontinen, 2025). Understanding cultural expressions as embedded knowledge also challenges heritage approaches that treat traditions as museum objects (Alam, 2025). When crafts are mass-produced or rituals are staged only for tourism, the deeper ecological and ethical meanings within these practices can be weakened (Qiu et al., 2022; Smith & Akagawa, 2008). Researcher discuss that such disconnection from lived context may reduce community control over knowledge and contribute to unequal recognition of Indigenous expertise (Eichler, 2021; Hoppers, 2025). These issues highlight the importance of heritage policies grounded in cultural sensitivity and participatory governance. ICH as embedded knowledge is also connected to specific places. Many indigenous practices depend on local climate, materials, and social relations, and they lose meaning when removed from their original settings (Gondo, 2025; Katelieva et al., 2020; Orlove et al., 2022). Agricultural rituals linked with monsoon cycles or crafts based on specific plant fibres illustrate how knowledge is tied to land and environment (Knorr & Augustin, 2025; Priyadarshini & Abhilash, 2019). These examples show that Indian Indigenous knowledge remains inseparable from lived experience, ecological context, and community relationships.

3.3 Intergenerational Transmission and Community Participation

Intergenerational transmission is the primary process through which IKS in India are transmitted, sustained and adapted across time (Durie, 2005; Malapane et al., 2024). Knowledge is shared through observation, apprenticeship, storytelling, participation in rituals, and everyday work, forming learning pathways that differ from formal classroom-based systems centred on written instruction (Breidlid, 2009; Gwerevende & Mthombeni, 2023). In many Indian communities, learning occurs within families, craft networks, pastoral groups, and farming collectives, where elders guide youth through shared labour, seasonal events, and community festivals (Priyadarshini & Abhilash, 2019; Smith & Akagawa, 2008). These practices show that Indigenous learning is relational and experiential, rooted in social interaction and environmental engagement (Durie, 2005; Orlove et al., 2022).

Community participation is essential because many forms of ICH require collective involvement for meaning and knowledge to be maintained (Eichler, 2021; Zabulis et al., 2025). Rituals, festivals, craft production, and storytelling events often involve entire communities, strengthening social ties while transmitting ecological practices, ethical norms, and local histories (Katelieva et al., 2020). Participation in agricultural festivals, for example, teaches crop cycles, weather knowledge, and shared responsibility toward land and water resources (Knorr & Augustin, 2025; Orlove et al., 2022). Such collective learning supports both cultural identity and environmental stewardship. Intergenerational transmission is increasingly under pressure due to migration, formal education systems, changing occupations, and media influences that reduce opportunities for youth to engage in traditional practices (Breidlid, 2009; Ranjan & Chaturvedi, 2023a). When young people move away from rural or artisanal livelihoods, apprenticeship systems weaken and practical knowledge related to farming, crafts, and local ecology declines (Brodt, 2001; Ngcobo et al., 2025). Research on rural economies in India indicates that alterations in agriculture and employment frequently reduce the motivation to preserve traditional ecological knowledge (Srivastava, 2012). Global media and standardized cultural forms sometimes also reduce youth interest in local traditions, further affecting knowledge continuity (Alam, 2025; Qiu et al., 2022).

Despite these challenges, many Indian communities are developing new ways to sustain knowledge transmission (Ren, 2025). Heritage training workshops, community-run learning centres, digital storytelling projects, and youth apprenticeship programs help adapt traditional learning to contemporary conditions (Huang & Xu, 2022; Ranjan & Chaturvedi, 2025). Educational reforms that include Indigenous knowledge in curricula improve student engagement and strengthen cultural relevance (Moitra & Madan, 2025). These initiatives show that IKS continue to evolve through new forms of participation while maintaining their core principles (Yatim et al., 2025). Intergenerational transmission remains essential because ICH cannot survive without practice and community engagement (Arslan & Dişli, 2025). Knowledge lives through performance, repetition, and shared experience, not only through documentation (Malapane et al., 2024; Smith & Akagawa, 2008). The future of IKS in India therefore depends on creating social, economic, and educational conditions that allow communities to continue practicing and renewing their knowledge traditions within changing social realities (Durie, 2005; Orlove et al., 2022).

4. HISTORICAL AND STRUCTURAL FORCES OF MARGINALIZATION

The marginalization of Indian IKS was shaped by historical processes that changed knowledge hierarchies and weakened the social conditions needed for knowledge transmission (Breidlid, 2009). Colonial rule, administrative reforms, and later development policies often replaced local systems of learning and resource management with centralized institutions that privileged Western science and written knowledge (Gorjestani, 2004). These changes reduced the authority of Indigenous experts and disrupted the cultural settings through which knowledge was practiced and shared (Brodt, 1999).

4.1 Colonial Suppression and Erasure of Indigenous Knowledge

British colonial rule in India reshaped IKS through legal reforms, economic restructuring, and epistemic hierarchies that privileged Western science over local knowledge traditions (Breidlid, 2009; Durie, 2005). Colonial administrators often described Indigenous practices as unscientific, primitive or backward, which reduced the status of herbal medicine, ecological management, and ritual governance in public institutions and policy debates (Gorjestani, 2004; Hoppers, 2025). Such discursive positioning reduced the legitimacy of Indigenous expertise and consolidated the belief that modernity and rational progress were synonymous with Western knowledge systems (Breidlid, 2009; Smith & Akagawa, 2008).

Land, forest, and agriculture policies were among the most powerful colonial interventions affecting Indigenous knowledge (Brodt, 2001). Nineteenth-century Forest regulations restricted shifting cultivation, forest gathering, and community water systems that were based on long-term ecological observation and local resource management (Brodt, 1999; Srivastava, 2012). By limiting access to forests and common lands, colonial policies disrupted the environmental settings where Indigenous ecological knowledge was practiced and transmitted (Priyadarshini & Abhilash, 2019). When communities were separated from forests, agricultural and pastoral lands, and water bodies, they lost opportunities to sustain forestry, agricultural, botanical, and conservation practices built over generations (Pearson et al., 2023). Colonial economic change also affected IKS by redirecting production toward global markets (Priyadarshini & Abhilash, 2019). Industrial imports and centralized trade policies weakened local craft networks and reduced demand for traditional textiles, metalwork, and artisanal goods, which had supported apprenticeship-based knowledge transmission (Arslan & Dişli, 2025). As craft livelihoods declined, youth participation in traditional occupations decreased, leading to erosion of technical and cultural knowledge (Ngcobo et al., 2025).

At the same time, colonial administrative systems replaced local governance structures, reducing the role of community leaders and ritual specialists in managing resources and resolving conflicts (Durie, 2005; Eichler, 2021). Educational reforms further reinforced these changes by promoting English-language instruction, standardized curricula, and Western scientific models that ignored local ecological practices and cultural learning systems (Moitra & Madan, 2025). Schooling often separated young people from traditional learning through apprenticeship, storytelling, and ritual participation, weakening intergenerational transmission (Malapane et al., 2024). Studies of rural knowledge systems in India show that such shifts created generational gaps in environmental knowledge and reduced respect for local expertise (Brodt, 2001; Srivastava, 2012). Additionally, colonial rules also altered the cultural environments that sustained ICH practices by redefining legitimate indigenous knowledge and restricting access to ecological resources (Hoppers, 2025; Smith & Akagawa, 2008). The effects of these policies continue to influence contemporary education, development planning, and cultural life, making the revitalization of Indigenous knowledge in India an important task for achieving cultural continuity and sustainable development (Brodt, 2001; Durie, 2005).

4.2 Effects of Western Educational Models and Modernization

In post-independence India, modernization policies and formal education systems often continued colonial patterns by privileging standardized, text-based knowledge over oral and experiential learning rooted in community life (Brodt, 2001). National education frameworks emphasized scientific rationality, uniform curriculum, and English-language instruction, which reduced the visibility of indigenous ecological practices and local knowledge traditions within schools and universities (Moitra & Madan, 2025). Such institutional bias generated hierarchical knowledge structures privileging formal academic and technical education. Consequently, indigenous youth were encouraged to prioritize institutional learning over community-based transmission through apprenticeship, ritual practice, and oral narratives (Gwerevende & Mthombeni, 2023; Malapane et al., 2024). Such patterns reinforced dual identities in which traditional knowledge was seen as culturally important but economically irrelevant (Eichler, 2021; Priyadarshini & Abhilash, 2019).

Agricultural modernization particularly exemplifies this process (Brodt, 2001). Programs associated with the Green Revolution promoted hybrid seeds, chemical fertilizers, irrigation technologies, and mechanized farming, often replacing traditional systems based on crop diversity, soil regeneration, and local climate knowledge (Sabarivasan et al., 2025; Srivastava, 2012). Indigenous practices such as seed preservation, mixed cropping, and community water management were sometimes discouraged because they did not align with productivity-oriented policies (Brodt, 2001). Studies on Indigenous agricultural knowledge in India show that these changes reduced reliance on local ecological knowledge and weakened intergenerational transmission of farming practices rooted in environmental observation (Brodt, 1999; Orlove et al., 2022). As productivity became the dominant policy goal, youth often perceived Indigenous farming knowledge as outdated, leading to reduced participation in cultural and ecological practices (Knorr & Augustin, 2025; Pearson et al., 2023). Healthcare modernization also marginalized Indigenous medical traditions. Formal health systems prioritized biomedical approaches and centralized institutions, which reduced the social role of local healers and community-based medicinal knowledge (Alum, 2024; Gorjestani, 2004). Although Ayurveda gained some institutional recognition, many tribal and folk healing systems continued to face stigma and limited policy support, which discouraged younger practitioners from learning traditional therapeutic practices (Moitra & Madan, 2025). The loss of medicinal plant knowledge and local healing traditions has been identified as a significant risk for both cultural heritage and community health resilience (Alum, 2024; Pearson et al., 2023).

Urbanization and migration further intensified these pressures by separating younger generations from ritual spaces, agricultural land, and community networks necessary for knowledge transmission (Malapane et al., 2024). As families moved toward wage labour and urban employment, apprenticeship-based craft systems declined, reducing opportunities for artisans to teach traditional skills and weakening the economic base of craft knowledge (Arslan & Dişli, 2025; Brodt, 2001). Research on cultural economies shows that when traditional occupations lose value, participation in associated cultural practices also declines, creating gaps in knowledge continuity (Aluko et al., 2025). Modernization therefore disrupted IKS not only through technological change but also by reinforcing epistemic hierarchies that privileged Western scientific models over community knowledge (Hoppers, 2025). These changes reshaped cultural aspirations, reduced the perceived relevance of Indigenous knowledge in contemporary life, and weakened its role in identity formation among younger generations (Priyadarshini & Abhilash, 2019). The comprehension of these structural authorities is crucial for designing policies that integrate Indigenous knowledge into education, agriculture, health, and development planning in ways that respect its cultural and ecological value.

4.3 Market-Driven Cultural Homogenization and Identity Disruption

Globalization and market-driven cultural homogenization place new pressures on Indigenous communities by reshaping cultural practices into commercial forms, altering identity structures, and changing local economies (Qiu et al., 2022; Smith & Akagawa, 2008). Mass media, global fashion industries, and commercial entertainment promote aspirational lifestyles that often contrast with community-based traditions (Pearson et al., 2023). Such influences reduce youth engagement in local cultural practices and weaken processes of intergenerational knowledge transmission (Puig, 2025; Ranjan & Chaturvedi, 2022). Studies of Indigenous societies show that cultural change linked to globalization can create generational divides in which younger members see local knowledge as incompatible with modern social mobility (Eichler, 2021; Shrestha et al., 2025).

Market forces strongly affect craft-based knowledge systems. Industrial production and imported goods reduce demand for handmade textiles, metalwork, and other traditional crafts, lowering the economic value of artisanal work and weakening apprenticeship systems that transmit technical knowledge (Brodt, 2001; Chaturvedi et al., 2024). Research on cultural economies indicates that when traditional occupations lose income and prestige, youth disengage from hereditary crafts, reducing the social reinforcement that sustains Indigenous skills and identities (Aluko et al., 2025; Ngcobo et al., 2025). Declining craft livelihoods therefore affect not only income but also the transmission of ecological and cultural knowledge embedded in production processes (Arslan & Dişli, 2025). Cultural commodification also reshapes ritual and performative traditions. Festivals, healing ceremonies, and local performances are often adapted for tourism markets, changing their meanings and reducing their role as knowledge systems embedded in community life (Qiu et al., 2022; Tirivangasi & Kontinen, 2025). Researchers address that when heritage is detached from its social context and repackaged for external audiences, it risks losing its relational and educational functions within Indigenous communities (Hoppers, 2025; Zabulis et al., 2025). Such transformations can weaken ethical norms, ecological practices, and community cohesion associated with ritual participation (Orlove et al., 2022; Smith & Akagawa, 2008).

Global markets also affect environmental relationships central to indigenous knowledge. Mining, deforestation, dam construction, and industrial agriculture displace communities from forests, rivers, and grazing lands, disrupting ecological knowledge built through long-term interaction with local environments (Erbach, 2025; Pearson et al., 2023). When access to land and biodiversity is reduced, knowledge related to agriculture, medicinal plants, and resource management cannot be practiced or transmitted effectively, therefore weakening both cultural identity and environmental stewardship (Gondo, 2025; Knorr & Augustin, 2025). Digital globalization adds another layer of change. Youth exposure to global media reshapes aspirations, aesthetics, and value systems, sometimes reducing interest in local cultural practices unless technology is used in culturally grounded ways (Bawack et al., 2025; Lewis et al., 2025). Without community-led digital initiatives, online platforms can amplify dominant cultural narratives and marginalize Indigenous knowledge traditions (Chigwada et al., 2025; Oguamanam, 2025). Market-driven homogenization significantly impacts IKS by undermining economic viability, transforming cultural meanings, and reshaping identity aspirations (Durie, 2005). These processes weaken the social and ecological foundations necessary for sustaining India's ICH (Priyadarshini & Abhilash, 2019).

5. CONTEMPORARY REVIVAL EFFORTS AND COMMUNITY LEADERSHIP

Present initiatives to revitalize Indian IKS are emerging within a historical context shaped by prior marginalization and ongoing economic transformation (Brodt, 2001). They also reflect renewed policy focused on cultural heritage preservation and sustainability objectives. Researchers increasingly recognize Indigenous knowledge as a valid epistemological framework that supports environmental resilience, cultural identity, and inclusive development strategies (Knorr & Augustin, 2025; Orlove et al., 2022).

5.1 Community-Owned Cultural Restoration Initiatives

Community-led initiatives represent some of the most meaningful efforts in revitalizing IKS because they restore authority over knowledge production, interpretation, and transmission to the communities that sustain these traditions (Durie, 2005; Hoppers, 2025). Across India, Indigenous groups have organized youth collectives, women's associations, ritual councils, and craft cooperatives to protect oral histories, ecological practices, and artistic skills that were weakened by earlier marginalization (Priyadarshini & Abhilash, 2019). Such initiatives are effective because Indigenous knowledge survives through lived practice and community engagement (Zabulis et al., 2025). Community-based documentation projects offer clear examples of this approach. In several tribal regions, elders and youth collaborate to record songs, rituals, craft processes, and medicinal knowledge using locally controlled digital tools and archives (Bawack et al., 2025; Huang & Xu, 2022). In addition to preserving the heritage, these events also educate future generations about rich heritage and foster a sense of cultural pride (Malapane et al., 2024; Yatim et al., 2025). Research highlights that locally governed archives protect against cultural appropriation while allowing communities to adapt knowledge to changing ecological and social conditions (Chigwada et al., 2025).

Revival of traditional governance systems is another important community-led strategy. In many tribal areas, councils of elders and customary institutions continue to regulate forest use, manage water resources, and organize rituals based on long-standing ecological ethics (Erbach, 2025; Priyadarshini & Abhilash, 2019). Such governance systems often support biodiversity conservation and sustainable resource use because they integrate social norms with ecological observation (Knorr & Augustin, 2025). Strengthening these institutions therefore contributes both to cultural restoration and environmental resilience (Tirivangasi & Kontinen, 2025). Artisanal revival movements also demonstrate community-led revitalization (Shrestha et al., 2025). Weaver cooperatives, metalwork groups, and pottery collectives across India have formed associations that preserve traditional techniques while finding fair markets for handmade goods (Aluko et al., 2025). Such cooperatives sustain apprenticeship-based learning, encourage youth participation, and protect ecological knowledge embedded in materials and design (Arslan & Dişli, 2025). Research on Indigenous agricultural and craft knowledge shows that stable economic support is necessary for sustaining knowledge systems rooted in everyday practice (Bawack et al., 2025).

Community festivals provide another important space for revitalization (Yatim et al., 2025). Revived harvest festivals and celebrations, initiation rituals, and ancestral ceremonies contribute to the restoration of collective memory (Qiu et al., 2022). Such practices also function as structured spaces for transmitting ecological knowledge, ethical norms, and localized historical narratives across generations (Shrestha et al., 2025). Participation in such events strengthens social cohesion and reinforces cultural identity while transmitting knowledge across generations (Smith & Akagawa, 2008; Zabulis et al., 2025). Such initiatives enable communities to reclaim intellectual sovereignty and reconstruct knowledge systems within culturally grounded frameworks (Priyadarshini & Abhilash, 2019). These processes demonstrate that community-centered action remains indispensable for the long-term sustainability of IKS in present-day India (Durie, 2005).

5.2 Role of Civil Society, Educational Institutions, and Cultural Organisations

Community leadership is central to revitalizing IKS (Eichler, 2021). Partnerships with civil society and academic institutions strengthen revitalization efforts when guided by principles of community ownership, ethical research, and shared decision-making. Civil society organizations, universities, museums, and cultural institutions support these efforts by providing technical expertise, funding, and wider visibility (Zabulis et al., 2025). External institutional involvement can strengthen documentation, training, and advocacy initiatives (Oguamanam, 2025). However, it must follow ethical principles that respect Indigenous autonomy and avoid exploitative research or commercial misuse of cultural knowledge (Hoppers, 2025). Research on Indigenous knowledge governance demonstrates that partnerships with external institutions are most productive when communities maintain epistemological sovereignty (Bawack et al., 2025). However, it works best when communities retain authority over interpretation, ownership, and access to their knowledge (Chigwada et al., 2025; Smith & Akagawa, 2008).

Civil society organizations contribute significantly through capacity building, participatory mapping, community tourism programs, and legal advocacy for land and intellectual property rights (Eichler, 2021; Gorjestani, 2004). Many organizations assist communities in documenting ecological practices, mapping sacred landscapes, and training youth in digital storytelling and archiving using locally managed platforms (Aluko et al., 2025; Gwerevende & Mthombeni, 2023). Such efforts align with research showing that locally governed knowledge systems are essential for safeguarding heritage while allowing communities to adapt to social and environmental change (Tirivangasi & Kontinen, 2025). Support from civil society helps communities access welfare schemes, market opportunities, and legal protection. These mechanisms are crucial for maintaining traditional livelihood systems and ensuring the continuity of cultural practices (Aluko et al., 2025; Ma et al., 2022).

Educational institutions are increasingly recognizing the importance of integrating Indigenous knowledge into curricula and research. Studies on higher education reforms in India show that including Ayurveda, tribal ecological knowledge, and community histories in teaching improves cultural relevance and broadens intellectual perspectives (Breidlid, 2009; Moitra & Madan, 2025). Community–university partnerships, collaborative research, and field-based or experiential learning programs help position Indigenous knowledge holders as collaborators rather than subjects of research (Durie, 2005; Yatim et al., 2025). Such approaches respond to longstanding critiques that formal education has marginalized local knowledge by privileging Western epistemologies (Breidlid, 2009; Smith & Akagawa, 2008).

Museums and cultural organizations are also shifting toward participatory models that involve communities in curating exhibitions, interpreting collections, and presenting living traditions (Zabulis et al., 2025). Community-curated exhibitions and digital archives help reconnect heritage with its social context and support cultural pride among younger generations (Bawack et al., 2025; Ren, 2025). Researchers emphasize that knowledge infrastructures, including museums and archives, should be shaped by Indigenous perspectives to ensure respectful representation and cultural accuracy (Chigwada et al., 2025; Lewis et al., 2025). However, collaboration requires careful attention to consent, benefit-sharing, and cultural sensitivity because external interventions can unintentionally distort meanings or prioritize commercial value over cultural significance (Oguamanam, 2025).

5.3 Barriers: Economic Inequities, Policy Gaps, and Institutional Neglect

Despite growing public recognition of IKS in India, several structural barriers continue to limit their revival. Persistent economic inequality challenges the continuity of IKS by placing knowledge holders in vulnerable livelihood conditions (Priyadarshini & Abhilash, 2019). Key knowledge holders such as artisans, traditional healers, pastoral leaders, and ritual specialists often face insecure livelihoods, restricted market linkages, inadequate infrastructure, and limited access to social protection systems (Aluko et al., 2025). Research on Indigenous livelihoods shows that economic insecurity discourages youth from learning traditional skills and weakens apprenticeship systems essential for knowledge transmission (Ma et al., 2022; Ngcobo et al., 2025). Handmade crafts, community-based tourism, and traditional farming often face competition from mass-produced goods and industrial agriculture, reducing income opportunities for knowledge custodians (Knorr & Augustin, 2025; Qiu et al., 2022).

Additionally, policy gaps are a barrier to IKS revitalization. Although India has laws related to biodiversity conservation, tribal welfare, and heritage protection, many policies focus on technical outputs rather than on the relational and community-based nature of Indigenous knowledge (Brodt, 2001). Health policies, for example, have recognized Ayurveda within national frameworks but have often overlooked tribal and folk healing traditions that remain important for local healthcare and medicinal plant knowledge (Alum, 2024; Moitra & Madan, 2025). Heritage policy frequently prioritizes monuments and archaeological sites rather than living cultural practices that depend on social participation and ecological continuity (Higgins, 2022). This imbalance reduces institutional support for festivals, oral traditions, and craft networks that carry Indigenous knowledge. Moreover, government agencies and research bodies often privilege Western scientific models and standardized data systems, which makes it difficult to integrate experiential and relational learning into development planning (Breidlid, 2009). Studies of Indigenous agricultural knowledge in India show that institutional structures sometimes fail to recognize local expertise in water management, crop diversity, and soil conservation, even when such knowledge supports sustainability (Brodt, 2001). Fragmented governance across ministries of culture, environment, tribal affairs, and tourism also creates policy gaps that overlook the interconnected nature of Indigenous heritage.

Cultural misrepresentation adds another layer of difficulty in IKS revitalization. When festivals, rituals, and crafts are commercialized without community consent, their meanings may be altered or simplified to suit tourist markets, weakening their educational and ecological functions (Qiu et al., 2022; Tirivangasi & Kontinen, 2025). Researchers caution that processes of cultural extraction detach knowledge from its custodial communities (Gorjestani, 2004; Hoppers, 2025). Such practices can undermine cultural dignity and erode intellectual property rights associated with Indigenous knowledge traditions (Oguamanam, 2025). Addressing these barriers requires coordinated reforms that improve economic support for knowledge holders, strengthen community rights, integrate indigenous knowledge into policy and education, and ensure fair representation of cultural practices.

6. INTEGRATIVE PATHWAYS FOR INDIGENOUS KNOWLEDGE SYSTEMS

Indian IKS stand at the convergence of traditional practices and modern global challenges including climate change, cultural loss, and sustainable development (Brodt, 2001; Moitra & Madan, 2025). The twenty-first century presents new opportunities for revitalization through inclusive research, supportive policies, and community-led initiatives (Oguamanam, 2025). Technological innovation, sustainability priorities, heritage policies, and global movements for epistemic justice have created new opportunities (Bawack et al., 2025; Ranjan & Chaturvedi, 2022). These developments support the recognition of Indigenous knowledge as a credible and essential resource for environmental management, community resilience, and culturally rooted development strategies (Tirivangasi & Kontinen, 2025). A conceptual framework shows how IKS integrate sustainability, digital innovation, cultural economies, youth continuity, and epistemic justice (Figure 1).

Figure 1. Integrative Pathways for Indigenous Knowledge Revitalization

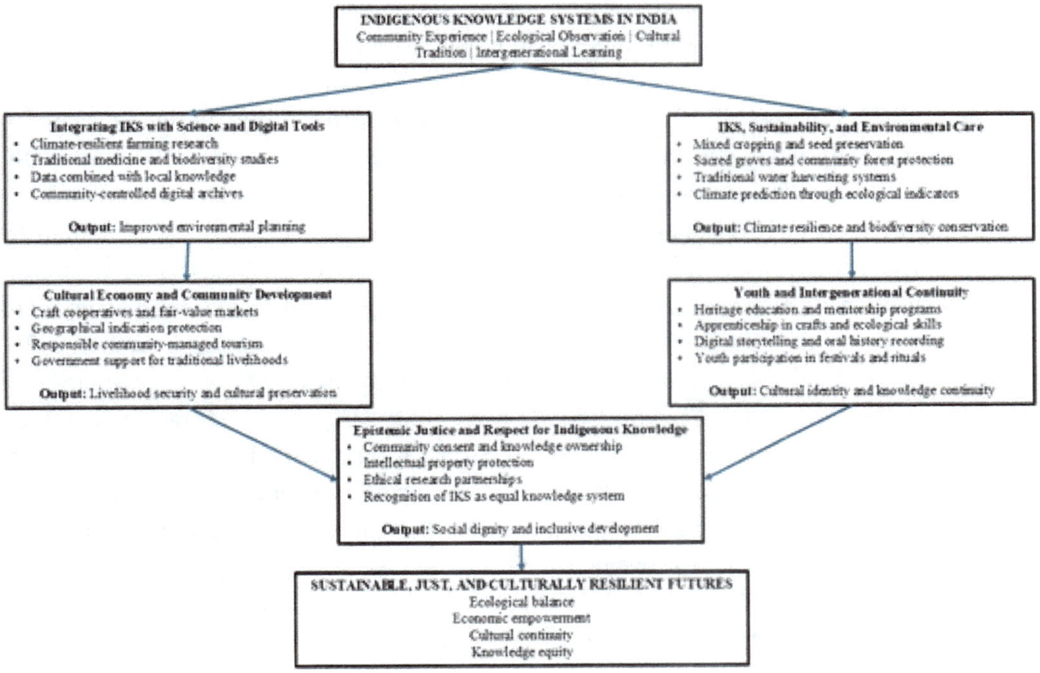

6.1 Integrating IKS with Scientific Methods and Digital Tools

Recent interdisciplinary research recognizes that IKS embodies empirically grounded insights shaped by long-term ecological engagement, repeated observation, and cumulative experimentation across generations (Arslan & Dişli, 2025; Zabulis et al., 2025). Integrating IKS with scientific methods does not require assimilation but rather collaborative engagement that respects the integrity of each knowledge system and values contextual expertise (Hoppers, 2025; Srivastava, 2012). In India, agronomists studying climate-resilient agriculture are examining traditional millet cultivation, mixed-cropping practices, and community water management techniques that demonstrate adaptability to drought and soil stress (Brodt, 2001; Sabarivasan et al., 2025). Ethnobotanical research similarly validates the therapeutic potential of plant-based healing traditions among tribal communities, showing that Indigenous medical knowledge can inform pharmacological innovation and biodiversity conservation (Alum, 2024; Oladele & Amara,

2024). Collaborative research approaches challenge extractive research models and recognize local expertise in areas such as forestry management, biodiversity monitoring, and climate adaptation (Erbach, 2025; Pearson et al., 2023). Indigenous indicators of ecological change such as flowering patterns, animal migrations, soil characteristics, and seasonal water flows are now studied along with satellite data and environmental models to create locally grounded knowledge for conservation planning (Gondo, 2025).

Digital technologies are also expanding opportunities for documenting and transmitting Indigenous knowledge when used under community governance (Ren, 2025; Yatim et al., 2025). Community-curated digital archives, participatory mapping of sacred sites, and audiovisual recording of rituals and craft techniques can support education, intergenerational learning, and cultural continuity (Bawack et al., 2025; Huang & Xu, 2022). Digital storytelling projects and interactive heritage platforms allow youth living in cities to reconnect with community traditions, strengthening identity and participation (Ranjan & Chaturvedi, 2025). Research on Indigenous knowledge and artificial intelligence highlights that technology can support cultural revitalization when guided by Indigenous values and relational ethics (Chigwada et al., 2025; Lewis et al., 2025). These community-governed digital initiatives therefore affirm the complementary relationship between scientific and IKS in responding to complex socio-cultural and environmental challenges (Tirivangasi & Kontinen, 2025).

However, digitization of IKS generates complex ethical concerns concerning custodianship, prior informed consent, and culturally appropriate representation (Bawack et al., 2025). Unregulated documentation and circulation may enable cultural appropriation, commercialization of sacred or context-specific knowledge, and erosion of community-held intellectual property rights (Oguamanam, 2025). Some knowledge related to ritual practice, medicinal plants, or sacred landscapes should remain restricted to custodial communities (Higgins, 2022; Smith & Akagawa, 2008). Effective integration therefore requires community governance, prior informed consent, benefit-sharing mechanisms, and respect for cultural boundaries (Chigwada et al., 2025). When guided by these principles, collaboration among Indigenous communities, scientists, and digital platforms can strengthen knowledge transmission and innovation while preserving cultural integrity (Orlove et al., 2022).

6.2 IKS, Sustainability, and Ecological Stewardship

IKS offer critical insights for responding to ecological degradation and climate change because they are grounded in long-term relationships among land, water, biodiversity, and community life (Gondo, 2025; Oladele & Amara, 2024). Indigenous worldviews often treat humans, non-human beings, and ecosystems as mutually interdependent, promoting resource-use practices grounded in stewardship, reciprocity, and ecological regeneration rather than extractive exploitation (Tirivangasi & Kontinen, 2025). Such perspectives align with research showing that traditional ecological knowledge contributes to sustainable agriculture, biodiversity protection, and climate adaptation in many parts of the world, including India (Pearson et al., 2023; Priyadarshini & Abhilash, 2019).

In India, Indigenous ecological knowledge includes integrated farming systems, crop diversity, seed preservation, forest protection rituals, water-harvesting structures, and rotational grazing practices that reflect long-term adaptation to varied climatic conditions (Brodt, 2001; Srivastava, 2012). Studies of agroecological systems show that traditional mixed-cropping methods improve soil fertility, conserve biodiversity, and reduce climate risk compared with monoculture farming (Sabarivasan et al., 2025). Patterns in plant flowering, animal movement, and wind direction have also been used by native people to predict rain and seasonal changes (Gondo, 2025). Research shows that this kind of knowledge is useful in addition to scientific climate monitoring (Orlove et al., 2022). Sacred groves and community forest traditions across India protect biodiversity through cultural norms and ritual restrictions, demonstrating how ecological conservation is embedded in cultural practice (Chipungu et al., 2005).

Comparisons between Indigenous sustainability practices and industrial development models reveal major differences in values and outcomes. Industrial agriculture and resource extraction often prioritize efficiency and market growth, which can increase environmental stress and reduce ecological resilience (Erbach, 2025; Pearson et al., 2023). In contrast, Indigenous systems emphasize regeneration, cyclical use of resources, and responsibility toward future generations, which supports long-term ecological balance (Katelieva et al., 2020). Research in disaster-prone regions of India shows that combining Indigenous ecological indicators with scientific risk assessment improves community preparedness and recovery planning (Brodt, 1999; Moitra & Madan, 2025).

6.3 Cultural Economies and Community-Centred Development

IKS support cultural economies based on craftsmanship, performance traditions, healing practices, and ecological expertise, offering livelihood opportunities while sustaining cultural continuity and community identity (Aluko et al., 2025; Qiu et al., 2022). Structural inequities in market systems place Indigenous artisans and practitioners in vulnerable economic positions (Gondo, 2025). Market intermediaries, mass-produced goods, and unstable income reduce the social value of traditional livelihoods and discourage youth participation (Ma et al., 2022; Ngcobo et al., 2025). Studies on cultural industries show that without fair pricing, institutional recognition, and market access, knowledge-based livelihoods remain vulnerable despite their cultural importance (Aluko et al., 2025; Qiu et al., 2022). Revitalizing IKS therefore requires restructuring cultural markets to prioritize community ownership, equitable compensation, and respectful representation of heritage practices (Eichler, 2021; Oguamanam, 2025).

Heritage entrepreneurship offers one pathway for community-centred economic development (Aluko et al., 2025). Cooperative models among weavers, potters, and forest-based producers allow communities to manage production, maintain quality, and retain control over cultural knowledge embedded in their crafts (Orlove et al., 2022; Shrestha et al., 2025) Research on Indigenous agricultural and craft knowledge in India shows that traditional skills survive best when supported by stable socioeconomic systems that sustain apprenticeships and protect local resource access (Brodt, 1999; Srivastava, 2012). Community-led enterprises also help maintain ecological ethics by encouraging the use of local materials and sustainable production practices (Aluko et al., 2025). Responsible cultural tourism supports Indigenous knowledge when governed by community protocols (Shrestha et al., 2025). When communities decide which practices may be shared publicly, how rituals are presented, and how benefits are distributed, tourism can strengthen cultural pride and economic resilience (Katelieva et al., 2020; Qiu et al., 2022). Conversely, externally controlled tourism often simplifies traditions into performances for visitors, which can distort meaning and weaken knowledge transmission (Jing et al., 2021; Shrestha et al., 2025). Ethical tourism frameworks therefore stress community consent, fair revenue sharing, and respect for cultural boundaries (Eichler, 2021).

Policy reforms are also essential for strengthening cultural economies rooted in Indigenous knowledge (Aluko et al., 2025). Legal recognition of traditional livelihoods and geographical indication protections can improve economic security (Gorjestani, 2004). Public procurement policies and funding for community cultural centres also support artisans and knowledge holders (Ma et al., 2022). Integrating Indigenous economic practices into sustainable development planning supports both environmental conservation and cultural resilience (Erbach, 2025). Cultural economies grounded in IKS therefore represent more than sources of income. They sustain identity, protect epistemic diversity, and strengthen community autonomy by linking livelihood opportunities with cultural practice and ecological stewardship (Durie, 2005; Orlove et al., 2022).

6.4 Youth, Identity, and Intergenerational Continuity

The long-term survival of indigenous knowledge depends on active youth participation (Ngcobo et al., 2025). However, globalization, standardized education, and digital media exposure often distance young people from community-based knowledge traditions (Malapane et al., 2024). Many youths associate success with urban lifestyles and formal careers, which can lead them to view traditional practices as outdated or economically unviable (Alam, 2025; Ngcobo et al., 2025). This shift creates a transmission gap in which apprenticeship systems weaken and cultural knowledge embedded in crafts, rituals, and ecological practices is not passed to the next generation (Brodt, 2001; Gwerevende & Mthombeni, 2023). In response, indigenous communities across India are developing new learning spaces such as heritage schools, youth craft workshops, and mentoring networks where elders teach ecological knowledge, ritual practices, and artisanal skills through hands-on participation (Yatim et al., 2025). These programs strengthen identity and cultural pride while showing that education rooted in Indigenous epistemologies can remain socially relevant and intellectually meaningful (Durie, 2005; Smith & Akagawa, 2008). Research on Indigenous education indicates that experiential learning and community participation improve both knowledge retention and cultural engagement among young learners (Breidlid, 2009; Malapane et al., 2024).

Digital platforms also provide opportunities for youth engagement when managed within community frameworks (Ren, 2025). Indigenous youth increasingly use social media, podcasts, and video documentation to record oral histories, promote crafts, and share environmental knowledge, helping counter stereotypes and expand visibility (Huang & Xu, 2022). Digital technologies, when guided by Indigenous values and ethical governance frameworks, can enhance knowledge transmission processes (Chigwada et al., 2025). Such approaches enable youth to participate actively in knowledge creation rather than passively consuming dominant cultural narratives (Lewis et al., 2025). Participation in festivals, craft apprenticeships, and ecological stewardship programs further helps young people recognize Indigenous knowledge as a source of pride and belonging (Katelieva et al., 2020; Shrestha et al., 2025) (Shrestha et al., 2025; Katelieva et al., 2020). Engagement in such activities strengthens intergenerational relationships and reinforces ethical values linked to land and community life (Zabulis et al., 2025). When youth reinterpret traditions within contemporary contexts, they help knowledge systems evolve while maintaining continuity, ensuring that IKS remain relevant within modern Indian society.

6.5 Toward Epistemic Justice and Cultural Re-Humanization

The revitalization of IKS is closely linked to the pursuit of epistemic justice, which addresses the systematic devaluation or exclusion of Indigenous ways of knowing within dominant academic, policy, and scientific institutions (Durie, 2005; Hoppers, 2025). Colonial legacies and modernization policies created knowledge hierarchies that privileged Western scientific rationality while marginalizing experiential, relational, and spiritual forms of knowledge rooted in community life (Breidlid, 2009). Restoring balance requires recognizing Indigenous knowledge as a reasonable epistemological tradition rather than a supplementary cultural resource (Smith & Akagawa, 2008). Epistemic justice also involves recognizing Indigenous communities as intellectual authorities with the right to control knowledge governance, interpretation, and transmission (Chigwada et al., 2025; Oguamanam, 2025). Integrating Indigenous knowledge into education, research, and development planning must therefore occur through participatory and consent-based approaches that avoid token inclusion or extractive research practices (Durie, 2005; Eichler, 2021).

Studies on community-based knowledge systems show that inclusive and pluralistic knowledge frameworks improve sustainability outcomes and strengthen social trust (Tirivangasi & Kontinen, 2025). Such pluralism recognizes epistemological diversity as a source of innovation rather than conflict (Knorr & Augustin, 2025; Lewis et al., 2025). Cultural re-humanization complements epistemic justice by restoring dignity to indigenous identities, histories, and worldviews that were long portrayed through deficit narratives (Hoppers, 2025). Research on heritage and community resilience shows that recognizing Indigenous cultural contributions strengthens social cohesion and supports community well-being (Alam, 2025). Revitalizing IKS is therefore not only a cultural project but also a political and ethical process that affirms intellectual equality, protects cultural rights, and supports more inclusive and sustainable knowledge systems in contemporary India (Durie, 2005; Zabulis et al., 2025).

7. CONCLUSION

The revitalization of IKS in India is essential for achieving cultural sustainability, ecological resilience, economic inclusion, and epistemic justice in a rapidly changing world (Durie, 2005). Rooted in place-based experience and relational ethics, Indigenous knowledge offers adaptive strategies for agriculture, resource management, healing, and social organization that have supported community survival across generations (Knorr & Augustin, 2025; Orlove et al., 2022). In the face of climate change, biodiversity loss, and cultural homogenization, these systems provide locally grounded solutions that complement scientific research and technological innovation rather than oppose them (Pearson et al., 2023). Sustainable revitalization of IKS necessitates a shift away from extractive documentation models and static museum preservation. Instead, participatory governance structures must acknowledge Indigenous communities as epistemic custodians empowered to control knowledge documentation, interpretation, and transmission (Hoppers, 2025).

Ethical integration of digital tools and scientific methods can strengthen preservation and learning when guided by consent, intellectual property protection, and cultural sovereignty (Oguamanam, 2025). Educational reforms that include Indigenous knowledge in curricula and research are equally important for restoring dignity, improving relevance, and strengthening intergenerational continuity (Breidlid, 2009; Yatim et al., 2025). Revitalizing IKS is therefore not about preserving the past in static form but about enabling living traditions to evolve within contemporary society. By supporting plural knowledge systems and recognizing Indigenous intellectual authority, India can build inclusive development pathways that respect cultural diversity and ecological balance. Such efforts can help create resilient cultural futures in which Indigenous knowledge and scientific innovation work together to address global challenges while sustaining community well-being for generations to come.

REFERENCES

Alam, M. (2025). Preserving cultural heritage and empowering indigenous communities for sustainable development in Fiji. *Social Sciences and Humanities Open, 12*(September 2024). DOI: 10.1016/j.ssaho.2025.101760

Aluko, H. A., Watson, A., Islam, N., Fernandes, F. A. P., Aluko, A. O., Ayertey, S., & Baig, M. M. (2025). Rethinking Business Practices: Harnessing Indigenous Knowledge Systems and Sustainable Strategies for Resilient Entrepreneurial Success. *Business Strategy and the Environment*, •••, 1–18. DOI: 10.1002/bse.70306

Alum, E. U. (2024). The role of indigenous knowledge in advancing the therapeutic use of medicinal plants: Challenges and opportunities. *Plant Signaling & Behavior*, *19*(1), 20–23. DOI: 10.1080/15592324.2024.2439255 PMID: 39652401

Arslan, E. N., & Dişli, G. (2025). Architectural heritage and traditional knowledge systems: Insights from the ancient settlement of Kilistra, Türkiye #. *Sustainable Communities*, *2*(1), 2477145. Advance online publication. DOI: 10.1080/29931282.2025.2477145

Bawack, R., Roderick, S., Badhrus, A., Dennehy, D., & Corbett, J. (2025). Indigenous knowledge and information technology for sustainable development. *Information Technology for Development*, *31*(2), 233–250. DOI: 10.1080/02681102.2025.2472495

Breidlid, A. (2009). Culture, indigenous knowledge systems and sustainable development: A critical view of education in an African context. *International Journal of Educational Development*, *29*(2), 140–148. DOI: 10.1016/j.ijedudev.2008.09.009

Brodt, S. B. (1999). Interactions of formal and informal knowledge systems in village-based tree management in central India. *Agriculture and Human Values*, *16*(4), 355–363. DOI: 10.1023/A:1007537809389

Brodt, S. B. (2001). A systems perspective on the conservation and erosion of indigenous agricultural knowledge in Central India. *Human Ecology: an Interdisciplinary Journal*, *29*(1), 99–120. DOI: 10.1023/A:1007147806213

Chaturvedi, P., Sharma, J., & Ranjan, A. (2024). The Role of Souvenir Marketing in Sustaining Handmade Craftsmanship. *ShodhKosh: Journal of Visual and Performing Arts*, *5*(3), 278–294. DOI: 10.29121/shodhkosh.v5.i3.2024.1401

Chaudhary, N. (2025). Transcending Time and Space in the Search for Synthesis: A Commentary on Paranjpe's Understanding Yoga Psychology: Indigenous Knowledge with Global Relevance. *Integrative Psychological & Behavioral Science*, *59*(1), 15. DOI: 10.1007/s12124-024-09875-3 PMID: 39779637

Chigwada, J., Mapara, J., Ngulube, P., & Chabvutagondo, T. (2025). Influence of Indigenous data governance principles on Indigenous knowledge management: Lessons from the Southern African Intangible Cultural Heritage Project. *IFLA Journal*, •••, 1–10. DOI: 10.1177/03400352251331471

Chipungu, O., Mamimine, P. W., Chipotereke, C. C., Kwinje, P. C., Mumbengegwi, P., & Mirimi, K. (2005).. . *Convergence of Indigenous Knowledge Systems and Contemporary Approaches in Cultural Heritage Management.*, *2294*, 33–55.

Dembedza, V. P., Chopera, P., Mapara, J., & Macheka, L. (2022). Impact of climate change-induced natural disasters on intangible cultural heritage related to food: A review. *Journal of Ethnic Foods*, *9*(1), 32. Advance online publication. DOI: 10.1186/s42779-022-00147-2

Durie, M. (2005). Indigenous knowledge within a global knowledge system. *Higher Education Policy*, *18*(3), 301–312. DOI: 10.1057/palgrave.hep.8300092

Eichler, J. (2021). Intangible cultural heritage, inequalities and participation: Who decides on heritage? *International Journal of Human Rights*, *25*(5), 793–814. DOI: 10.1080/13642987.2020.1822821

Erbach, T. (2025). Using Cultural Heritage in Climate Adaptation: Fields of Application and Functions. *Wiley Interdisciplinary Reviews: Climate Change*, *16*(4), 1–15. DOI: 10.1002/wcc.70011

Gondo, R. (2025). Bibliometric analysis of Indigenous knowledge systems and climate change adaptation literature, 1993–2023. *IFLA Journal*, •••, 1–14. DOI: 10.1177/03400352251317716

Gorjestani, N. (2004). *Indigenous knowledge for development. Protecting and promoting traditional knowledge: Systems, national experiences and international dimensions.*

Gwerevende, S., & Mthombeni, Z. M. (2023). Safeguarding intangible cultural heritage: Exploring the synergies in the transmission of Indigenous languages, dance and music practices in Southern Africa. *International Journal of Heritage Studies*, *29*(5), 398–412. DOI: 10.1080/13527258.2023.2193902

Higgins, N. (2022). indigenous Intangible Cultural Heritage. *Laws*, *11*(47), 2–15. https://uwspace.uwaterloo.ca/bitstream/handle/10012/11668/Stevens_Julia.pdf?sequence=1&isAllowed=y

Hoppers, C. O. (2025). Research on Indigenous Knowledge Systems: The Search for Cognitive Justice. In *Indigenous Voices and Decolonising Lifelong Education.*

Huang, Z., & Xu, T. (2022). Research on Knowledge Management of Intangible Cultural Heritage Based on Linked Data. *Mobile Information Systems*, *2022*, 1–14. Advance online publication. DOI: 10.1155/2022/3384391

Jing, X., Tan, F., & Zhang, M. (2021). Digital Application of Intangible Cultural Heritage from the Perspective of Cultural Ecology. *Journal of Smart Tourism*, *1*(1), 41–52. DOI: 10.52255/smarttourism.2021.1.1.6

Katelieva, M., Muhar, A., & Penker, M. (2020). Nature-related knowledge as intangible cultural heritage: Safeguarding and tourism utilisation in Austria. *Journal of Tourism and Cultural Change*, *18*(6), 673–689. DOI: 10.1080/14766825.2019.1693581

Knorr, D., & Augustin, M. A. (2025). Towards resilient food systems: Interactions with indigenous knowledge. *Trends in Food Science and Technology, 156*(December 2024), 104875. DOI: 10.1016/j.tifs.2025.104875

Lázaro Ortiz, S., & Jiménez de Madariaga, C. (2022). The UNESCO convention for the safeguarding of the intangible cultural heritage: A critical analysis. *International Journal of Cultural Policy*, *28*(3), 327–341. DOI: 10.1080/10286632.2021.1941914

Lewis, J. E., Whaanga, H., & Yolgörmez, C. (2025). Abundant intelligences: Placing AI within Indigenous knowledge frameworks. *AI & Society*, *40*(4), 2141–2157. DOI: 10.1007/s00146-024-02099-4

Ma, T. F., Chai, C. W., & Chao, T. W. (2022). On the Study of the Sustainable Development of Intangible Cultural Heritage of Indigenous Peoples' Diets—Take the Protection of Geographical Indications as an Example. *Sustainability (Basel)*, *14*(19), 12803. Advance online publication. DOI: 10.3390/su141912803

Malapane, O. L., Chanza, N., & Musakwa, W. (2024). Transmission of indigenous knowledge systems under changing landscapes within the vhavenda community, South Africa. *Environmental Science & Policy*, *161*(August), 103861. DOI: 10.1016/j.envsci.2024.103861

Moitra, P., & Madan, J. (2025). Stakeholder perspectives on integrating Ayurveda and Indian Indigenous Knowledge Systems into higher education: An exploratory study. *Social Sciences & Humanities Open*, *11*(March), 101453. DOI: 10.1016/j.ssaho.2025.101453

Ngcobo, A. S., Khalema, E. N., Masuku, M. M., Mlondo, N. M., Mncube, T., Ramphabana, L. B., & Matse, M. P. (2025). Role of Indigenous Knowledge Systems and Ubuntu in Addressing Gender-Based Violence Against Young Women in Rural KwaZulu-Natal. *Journal of Social Service Research*, *51*(5–6), 1203–1216. DOI: 10.1080/01488376.2025.2515420

Oguamanam, C. (2025). Information systems and digitization of traditional knowledge: Trends in cultural heritage and memory institutions and the WIPO Genetic Resources Treaty*. *The Journal of World Intellectual Property*, (August), 1–32. DOI: 10.1111/jwip.70005

Oladele, O. I., & Amara, A. (2024). Farmers' use of Indigenous knowledge on climate change adaptation across farming systems and agroecological zones of Sierra Leone. *Alternative*, *20*(3), 473–483. DOI: 10.1177/11771801241251862

Orlove, B., Dawson, N., Sherpa, P., Adelekan, I., Alangui, W., Carmona, R., Coen, D., Nelson, M., Reyes-García, V., Wilson, A., Rubis, J., & Sanago, G. (2022). *Intangible Cultural Heritage*. Diverse Knowledge Systems and Climate Change.

Pearson, J., Jackson, G., & McNamara, K. E. (2023). Climate-driven losses to knowledge systems and cultural heritage: A literature review exploring the impacts on Indigenous and local cultures. *The Anthropocene Review*, *10*(2), 343–366. DOI: 10.1177/20530196211005482

Priyadarshini, P., & Abhilash, P. C. (2019). Promoting tribal communities and indigenous knowledge as potential solutions for the sustainable development of India. *Environmental Development*, *32*(September), 100459. Advance online publication. DOI: 10.1016/j.envdev.2019.100459

Puig, D. (2025). Social limits to adaptation in the context of intangible cultural heritage. *Current Opinion in Environmental Sustainability*, *77*, 101569. DOI: 10.1016/j.cosust.2025.101569

Qiu, Q., Zuo, Y., & Zhang, M. (2022). Intangible Cultural Heritage in Tourism: Research Review and Investigation of Future Agenda. *Land (Basel)*, *11*(1), 139. Advance online publication. DOI: 10.3390/land11010139

Ranjan, A., & Chaturvedi, P. (2022). Digital Technologies and the Intangible Cultural Heritage of the Rural Destination. In *Disruptive Innovation and Emerging Technologies for Business Excellence in the Service Sector* (pp. 196–218). IGI Global., DOI: 10.4018/978-1-7998-9194-9.ch010

Ranjan, A., & Chaturvedi, P. (2023a). Digitally Sustaining: The Rural Intangible Cultural Heritage. In *Embracing Business Sustainability Through Innovation and Creativity in the Service Sector* (pp. 14–30). DOI: 10.4018/978-1-6684-6732-9.ch002

Ranjan, A., & Chaturvedi, P. (2023b). Unleashing Digital Innovation for Sustainable Rural Intangible Cultural Heritage Marketing. In *Sustainable Development Goal Advancement Through Digital Innovation in the Service Sector* (pp. 95–113). IGI Global., DOI: 10.4018/979-8-3693-0650-5.ch007

Ranjan, A., & Chaturvedi, P. (2025). Creating Meaningful Tourism Experiences With Digital Storytelling. In *Meaningful Tourism* (pp. 127–144). Emerald Publishing Limited., DOI: 10.1108/978-1-83797-837-320251019

Ren, J. (2025). Inheritance and protection of intangible cultural heritage in drama category based on AI human–computer interaction and digital technology. *Discover Artificial Intelligence*, *5*(1), 67. Advance online publication. DOI: 10.1007/s44163-025-00268-y

Sabarivasan, R., Gokul, S., & Infanta, S. C. (2025). Mapping the global scientific landscape of indigenous knowledge in agroforestry systems: A bibliometric perspective (2000–2024). *Agroforestry Systems*, *99*(8), 232. DOI: 10.1007/s10457-025-01337-y

Shrestha, R. K., L'Espoir Decosta, J. N. P., & Whitford, M. (2025). Indigenous knowledge systems and socio-cultural values for sustainable tourism development: Insights from Indigenous Newars of Nepal. *Journal of Sustainable Tourism*, *33*(1), 143–167. DOI: 10.1080/09669582.2024.2316298

Smith, L., & Akagawa, N. (2008). Intangible Heritage. In Smith, L., & Akagawa, N. (Eds.), *Intangible Heritage*. Routledge., DOI: 10.4324/9780203884973

Srivastava, S. K. (2012). Managing indigenous and scientific knowledge for resilience building: Case studies from disaster‐prone regions of India. *Journal of Advances in Management Research*, *9*(1), 45–63. DOI: 10.1108/09727981211225644

Tirivangasi, H. M., & Kontinen, T. (2025). Reclaiming Indigenous knowledge for sustainability: Afrocentric perspectives on Indigenous intangible cultural heritage and climate adaptation in rural Zimbabwe. *International Journal of Heritage Studies*, •••, 1–28. DOI: 10.1080/13527258.2025.2595080

UNESCO. (2003). *Intangible Heritage - 2003 Convention*. https://ich.unesco.org/en/intangible-heritage -domains-00052

Yatim, H., Jayadi, K., Jamilah, J., Abduh, A., & Prusdianto, P. (2025). Integrating cultural heritage into education: The supportive role of schools in traditional festivals in West Sulawesi. *Discover Education*, *4*(1), 379. Advance online publication. DOI: 10.1007/s44217-025-00847-2

Zabulis, X., Partarakis, N., Zidianakis, E., & Kaplanidi, D. (2025). A Critical Review of the Function of Intangible Cultural Heritage as a Driver for Social Resilience and Cohesion. *Encyclopedia*, *5*(4), 189. DOI: 10.3390/encyclopedia5040189

Chapter 2
Repositioning Indigenous Musical Heritage Within Formal Music Education Frameworks

Evans N. Netshivhambe
http://orcid.org/0000-0003-0362-4110
University of South Africa, South Africa

ABSTRACT

Indigenous music performed on traditional African instruments has historically been marginalised within formal education and professional performance spaces. Despite their cultural significance, many South African indigenous instruments remain largely confined to rural contexts, where they function more as cultural artefacts than as pedagogical resources. Their limited visibility constrains integration into mainstream education and the broader performing arts sector. Although some performers have introduced these instruments into urban settings, such efforts have not substantially shifted their institutional status. This qualitative study, drawing on interviews with instrument makers and performers, literature review, and performance analysis, examines urban and rural initiatives aimed at revitalising indigenous instruments. Findings indicate that crafting, teaching, and performance are central to sustaining indigenous knowledge within contemporary education. The study advocates for their systematic integration to advance cultural preservation and pedagogical transformation.

INTRODUCTION

Education has increasingly been positioned as a critical pathway to socio-economic mobility, particularly for individuals and communities historically excluded from formal education systems (Omoniyi, 2013; Paraschiv, 2017). For many young people from economically marginalised and previously disadvantaged backgrounds, access to education has facilitated a transition from poverty to greater economic stability and social agency (Reddick et al., 2011). Education has long been viewed as an investment that empowers people to become agents of change within their families and communities (Fleming & Grace, 2016). Because education is widely perceived as a mechanism for interrupting cycles of generational poverty, policy priorities and societal expectations have tended to favour disciplines associated with high employability and economic growth (Donald, Ashleigh & Baruch, 2018). Fields such as en-

DOI: 10.4018/979-8-2600-2890-2.ch002

gineering, mathematics, science, and technology are frequently foregrounded as reliable drivers of job creation and national development (Karim & Maat, 2019). Muriithi & Musili (2021, p. 3) argue that, for instance, in Kenya, education investment has increased to support STEM and technology fields of study at all educational levels and institutions because they are linked to youth employability. Within this framework, arts education is often marginalized, framed as a niche domain with limited economic prospects and comparatively uncertain pathways to professional success. This perception continues to influence funding investments, allocations, curriculum design, and student enrolment patterns, thereby shaping the broader hierarchy of disciplines within higher education.

This perception consequently weakens the appeal of arts disciplines to young people who are under sustained social and economic pressure to enrol in programmes perceived to offer clear and immediate employment prospects (Brown, 2007). He (2023, p. 364) indicates that, for instance, in China, technology and artificial intelligence have become the leading career choices for youth because of the employability factor. In contexts marked by high unemployment and economic precarity, students and their families often prioritise qualifications with measurable labour-market returns (Wang, Lim & Ng, 2024). As a result, disciplines positioned outside dominant economic narratives, particularly the arts, are frequently regarded as risky or secondary options (Kovačević, Dekker & van der Velden, 2024). The early twenty-first century has further intensified these dynamics with advances in technology and AI (He, 2023). Against this backdrop, indigenous knowledge education, such as musical instruments, has received the least attention within the education system (Netshivhambe, 2024a). Each graduating cohort in all fields of study enters an environment characterized by heightened competition for limited employment opportunities, reinforcing the emphasis on fields associated with high employability (Donald, Ashleigh & Baruch, 2018). Within this landscape, students passionate about pursuing arts disciplines, especially music, often encounter structural constraints imposed by institutional offerings. Many university music programmes remain anchored in Western art music or jazz traditions, reflecting long-standing curricular models (Scott, 2015). This research focuses on the marginalisation of indigenous musical heritage and how such knowledge can be reposited within institutions of higher learning to foreground African music as a deliberate alternative, aligning with broader calls for curriculum transformation and the recognition of indigenous knowledge systems.

The limited scope of music curricula in many music education departments has further motivated this research, opening the door to indigenous instrument education as a possible career path for students. Perceiving gaps in the representation of indigenous African musical knowledge, some have undertaken field-based research in local communities with the explicit intention of reintroducing and repositioning such knowledge within formal educational contexts (Euba, 2014). This shift reflects both a critique of dominant curricular models and a proactive commitment to epistemic inclusion. This study investigates two distinct yet complementary approaches to the modernisation and crafting of indigenous African instruments, examined from both urban and rural perspectives. By analysing these models comparatively, the research seeks to argue that indigenous instruments can be preserved, adapted, and integrated into contemporary educational frameworks without compromising their cultural integrity by looking at their construction models.

RESEARCH METHODOLOGY

A qualitative research design was employed, incorporating in-depth interviews with indigenous instrument players and builders, participant observation during performances, and engagement in instrument-crafting workshops. These primary data sources were complemented by a comprehensive review of relevant literature. Interviews with six instrument makers, with four from the rural and two from the urban centers, were conducted. The rural instrument makers were multi-instrument makers who produced a range of string, wind, and percussion instruments. While urban instrument makers were very particular about the instruments they made, driven by a profit-driven business model, they predominantly focused on percussion instruments such as marimba and mbira. Recordings were taken from each instrument maker for analysis, and instrument-specific questions were used in an open-ended interview. The findings underscore the importance of intentional, structured strategies to promote and integrate indigenous instruments into formal education from an instrument-construction perspective. The study argues that such integration is essential not only for cultural preservation but also for pedagogical diversification and the sustained advancement of African musical heritage within the broader educational system.

Urbanisation, Decolonial Reform, and the Repositioning of Indigenous Knowledge in Higher Education

Urban centres have historically served as centres of economic concentration and social mobility, attracting individuals seeking employment, education, and improved living conditions. Migration to cities is often driven by the perception that urban environments offer greater access to formal labour markets, professional advancement, and institutional resources (Yang, Shu & Yu, 2022). As a consequence, many rural and hometown communities experience demographic decline, reduced economic activity, and limited prospects for sustainable development (Gheitarani et al., 2020). The concentration of corporations, industries, and universities within metropolitan areas reinforces the city's status as a focal point of opportunity (Lagakos, 2020). However, this aggregation of resources also generates complex social dynamics, and survival aligns with what is in demand. Rapid urbanisation can intensify competition for employment, strain infrastructure, and contribute to socio-economic inequalities, which, in some contexts, manifest as heightened instability and crime (Donald, Ashleigh & Baruch, 2018). Despite these challenges, cities continue to draw people from diverse backgrounds, functioning as spaces where aspirations for mobility and transformation are negotiated. Of particular relevance to this study is the manner in which indigenous knowledge systems have increasingly entered urban environments (Prajapati & Chatwani, 2025). As individuals migrate, they bring cultural practices, musical traditions, indigenous musical instruments, and artisanal skills, leading to new forms of heritage production, circulation, and consumption within metropolitan contexts (Netshivhambe, 2023c; Turino, 2010). The urban space thus becomes not only an economic hub but also a site where indigenous musical heritage is recontextualised, adapted, and, in some instances, commodified within contemporary cultural and educational frameworks (Prajapati & Chatwani, 2025).

Indigenous knowledge systems have historically sustained rural communities, shaping social organisation, cultural expression, ecological stewardship, and systems of intergenerational transmission, but have largely existed only within the rural space (Gómez-Baggethun, 2022; Hlalele, 2019; Netshivhambe, 2023). These knowledge systems continue to play a vital role in preserving traditional heritage, not merely as remnants of the past, but as living epistemologies embedded in everyday practice (Baporikar, 2022; Luca, 2024). Over time, such knowledge has attracted the attention of scholars, cultural institutions such as UNESCO, government structures, ethnographers, musicians, and cultural practitioners (Cavalcanti, 2019; Lázaro Ortiz & Jiménez de Madariaga, 2021). More recently, however, it has also generated renewed interest among broader publics, particularly those who perceive rapid urbanisation and globalisation as forces contributing to the erosion of identity, language, and cultural continuity (Ajani et al., 2024). The turn of the twenty-first century has been marked by intensified debates over the epistemological foundations of education and its role in shaping people's lives. Calls for decolonial reform have challenged the dominance of Eurocentric knowledge paradigms within formal schooling and higher education (Adefila, 2022; Du Plessis, 2021). These debates have foregrounded the Africanisation of curricula, the recognition of indigenous epistemologies as legitimate knowledge systems, and the promotion of African languages as mediums of instruction and scholarship (Le Grange, 2023; Chimbunde & Kgari-Masondo, 2022; Manyike & Shava, 2018). Within the framework of the decolonial agenda, Keebine-Sibanda (2021:183) advances the notion of "epistemological and ontological access," arguing that access to knowledge must extend beyond formal inclusion within institutional spaces. Epistemological access concerns the legitimacy and recognition of diverse knowledge systems, while ontological access foregrounds the modes of being, identity, and lived realities through which such knowledge is embodied and enacted (Keebine-Sibanda, 2021). In this view, meaningful access requires not only exposure to content but also the capacity for learners to engage with knowledge in ways that affirm their own ways of knowing and being.

Knowledge should enable individuals to apply, reinterpret, and mobilise it in socially responsive ways, thereby contributing substantively to society through epistemic frameworks grounded in their own cultural and intellectual traditions (Schieman & Plickert, 2008). Past development trajectories have frequently marginalised local and indigenous epistemologies, privileging models of urban improvement grounded in technocratic and modernist paradigms. Kohsaka and Rogel (2021) characterise this orientation as a top-down approach in which development is largely equated with modernisation. Within such frameworks, progress is measured through infrastructural expansion, economic growth, and institutional standardisation, often without meaningful engagement with community-based knowledge systems. This conceptualisation of development implicitly positions indigenous ways of knowing as secondary or peripheral, treating them as residual cultural practices rather than as dynamic and contextually responsive knowledge systems (Jakes, 2024). As a result of this knowledge privileging approach, indigenous epistemologies are excluded from policy formulation and planning processes, reinforcing hierarchical distinctions between "modern" and "traditional" forms of knowledge (Prajapati & Chatwani, 2025; Sharma et al., 2024). Such epistemic marginalisation not only limits the diversity of perspectives informing development but also undermines the potential contributions of indigenous knowledge to sustainable, locally grounded social transformation.

Collectively, efforts and movements that seek to bring about the inclusion of knowledge diversity within the university system are increasingly widely embraced, as they represent a broader effort to rethink the meaning of Africanness within institutions historically structured by Western intellectual traditions (Nkoane, 2006; Osei-Tutu, 2023; Netshivhambe, 2024b). In this context, indigenous knowledge is increasingly positioned not as peripheral or supplementary, but as central to the reconfiguration of educational thought and practice (Mawere & Mukonza, 2025). Making indigenous knowledge attractive depends on strategically packaging and repackaging it as a legitimate intellectual domain and a viable professional pathway, which play a significant role in attracting young people to pursue it at the university level (Harvey &

Russell-Mundine, 2019). When indigenous knowledge systems are framed not merely as cultural artefacts or objects of admiration in museums, such as indigenous instruments, but as rigorous fields of scholarly inquiry with practical and economic relevance, they gain enhanced legitimacy within academic and career-oriented discourses (Agawu, 2016, p. 79). Presenting indigenous knowledge as both intellectually robust and economically viable shifts its perception from marginal or supplementary content to a credible area of specialisation capable of sustaining livelihoods (Oberhofer, 2020; Magni, 2017). In contexts where students are increasingly concerned with employability and financial stability, demonstrating that engagement with indigenous knowledge can lead to tangible income-generating opportunities across education, performance, research, cultural entrepreneurship, and the creative industries serves as a powerful motivator (Attas & Walker, 2019). The rearticulation of indigenous knowledge within academic and professional frameworks should affirm its epistemic value and enhance its appeal among youth seeking educational pathways that combine cultural relevance with economic sustainability as a form of heritage resilience.

Urbanisation, Market Dynamics, and the Uneven Integration of Indigenous Instruments into Formal Music Education

Many young people are increasingly attracted to the opportunities afforded by urbanisation, including improved access to formal employment, higher education, and broader cultural and social networks (Osutei & Kim, 2023; Sun, Zhao & Zhou, 2025). Urban centres are often perceived as spaces of economic mobility and professional advancement, drawing ambitious youth seeking to secure livelihoods and develop skills that may be unavailable in their rural hometowns (Pedersen & Gram, 2018; Alarima, 2018). According to Pedersen & Gram (2018, p. 624), "education is highlighted as the initial impetus for youth to move away, whereas a lack of professional employment opportunities is said to be the key factor mitigating return migration." Education occupies a pivotal role in shaping life trajectories, particularly for individuals from socio-economically disadvantaged backgrounds. For many, the city represents a space of possibility, a context in which access to formal education is perceived as the primary pathway toward social mobility, economic advancement, and improved living conditions (Kumah-Abiwu, 2022). Education becomes a vehicle of hope, offering the potential not only for individual transformation but also for the upliftment of families and communities. Alongside these opportunities, urban life frequently produces a sense of cultural displacement for others (Demir, Yaman & Koçak, 2025). Individuals experience a growing distance from traditional social structures, communal norms, and inherited cultural heritage, leading to feelings of disconnection or alienation from their cultural roots (Abbas, Ullah & Ullah, 2025; Demir, Yaman & Koçak, 2025). This tension between opportunity and cultural detachment has stimulated a renewed interest in identity and heritage among urban populations. Many young people actively seek ways to reconnect with their ancestral knowledge, practices, and values, viewing these as critical anchors of personal and collective identity (Netshivhambe, 2023c; Lee, 2008).

The urban environment, with its exposure to globalised cultural forms, simultaneously challenges and reshapes these traditional practices, creating a dynamic interplay between modernity and heritage (Pedersen & Gram, 2018). As a result, urban dwellers such as musicians, researchers, and cultural advocates are increasingly exploring strategies to preserve and engage with their cultural roots, whether through participation in music, performance, craft, or other heritage-based activities, even as they navigate the practical demands and social pressures of metropolitan life (Hudson, 2006). This renewed engagement reflects a broader cultural phenomenon in which urban spaces serve as sites of both heritage consumption and innovation. Niglio, Singh, and Gireesh Kumar (2026) examine the *Reconnecting With Your Culture* awakening programme, an initiative developed in partnership with UNESCO that seeks to foster heritage consciousness among young people. The programme is designed to cultivate critical awareness of cultural identity and to encourage participants to recognise the value of their own cultural inheritance (Niglio, Singh & Gireesh Kumar, 2026). By promoting active engagement with heritage practices and knowledge systems, the initiative aims to empower youth to assume responsibility as informed custodians and advocates of their cultural traditions, thereby strengthening intergenerational continuity and sustainable stewardship of heritage (Levstek & Banerjee, 2021; Ferrer-Fons, Rovira-Martínez & Soler-i-Martí, 2022). Urban dwellers are not merely passive recipients of tradition; they actively reinterpret and adapt cultural practices to fit contemporary contexts, demonstrating that cultural preservation and modernisation are not mutually exclusive but can coexist as mutually reinforcing processes (Netshivhambe, 2023a, 2025; Mowakadi, 2025). In this way, urbanisation has become both a challenge and an opportunity for sustaining indigenous cultural practices, providing a platform for young people to negotiate identity, creativity, and economic possibility in ways that honour tradition while responding to contemporary urban realities.

In response, most governments in various contexts have begun to reconceptualise urbanisation not solely as an economic process, but as a cultural and social project. A place where people can call home, with remnants of traditional values that connect diverse people. Efforts to establish and revitalise heritage sites, cultural precincts, and community arts initiatives reflect broader attempts to ensure cultural continuity, promote social cohesion, improve quality of life, and advance inclusive development (Tahseen & Al-Jumaily, 2020; Apaydin, 2020). Such initiatives recognise that sustainable urban growth must engage with historical memory and indigenous cultural expression rather than displace them (Farhan, Merie & Nasar, 2024). Within this framework, education emerges as a central mechanism for heritage preservation and transmission. Formal educational institutions have the capacity to legitimise and disseminate indigenous knowledge, particularly in environments where such knowledge may be perceived as marginal or unfashionable (Mazzocchi et al., 2018). However, access to indigenous music practices and instruments in urban contexts often remains limited because integrating them into the educational system poses serious challenges for a system that has largely promoted the status quo, while those who do so often do so just to tick boxes of transformation (Bohensky & Maru, 2011). Many of these traditions are rooted in rural communities, where they continue to be practiced as part of everyday life; in metropolitan settings, their presence is often mediated, fragmented, or commodified for profit with very little educational focus (Su, 2019).

FINDINGS

Custodianship, Commercialisation, and Curriculum: Urban–Rural Dynamics in the Production and Educational Integration of Indigenous Instruments

Advocating for the inclusion of indigenous musical heritage in professional performance should be centered on the development of indigenous instruments for educational purposes. That means a concerted effort to collaborate with indigenous instrument makers, as the custodians of specialized craftsmanship and material knowledge, who are caretakers of such rich knowledge and producers of these essential instruments, must be a key component of developing musical traditions, performance, studies, formulation, and documentation of transmission skills (Netshivhambe, 2024b;

Smith, 2018). They function as primary enablers within the broader cultural ecosystem, supplying performers, educators, researchers, and the wider public with the material means necessary for artistic and scholarly engagement (Dawe, 2015; Netshivhambe, 2024a). The challenge is that many indigenous instrument makers and practitioners continue to reside in rural communities, where access to their craft and opportunities to sustain a livelihood through their craft remain limited (Onyeji, 2006). Although these artisans possess highly specialised knowledge and technical skill, the economic and infrastructural constraints of rural contexts often restrict their access to markets, formal institutions, and professional networks (Bates, 2012). Their craft skills in making and playing these indigenous instruments cannot be realized in a classroom setting because this knowledge is not widely accessible in educational spaces.

Practitioners who have migrated to urban centres are frequently better positioned to leverage their proximity to universities, cultural institutions, and performance venues to promote and commercialise indigenous musical heritage. One of the indigenous instrument makers who specializes in wood-based instruments such as mbira, tshihwana, and dende, explained that "we are not able to make money by producing these instruments because there is no demand, unless there is a special order" (personal communication, Philemon Nemukula, 2024). The disparity between rural and urban contexts is further compounded by differences in access to tools, materials, and technical resources. For those in urban areas, their only concern is access to traditional materials or materials known to be strong in creating indigenous African instruments (personal communication, Farai Matake, 2024). For someone like Mafemani Kubai, a specialist in woodcraft and an ingenious instrument teacher and player, the problem for many practitioners in rural areas is access to funding (personal communication, Mafemani Kubai, 2024).

Instrument makers in rural areas often work with limited equipment, which can affect consistency in tuning and standardisation, particularly when instruments are intended for ensemble performance. Even when crafted by the same maker, variations in pitch and construction may hinder their collective use within formal school settings, where ensemble compatibility and tuning precision are often required (Netshivhambe, 2023c, personal communication, Maluta Matsheka, 2023). Structural barriers within rural education systems contribute to limited uptake of locally produced instruments, which affects the transmission of their knowledge. Many schools remain unaware of the existence or pedagogical potential of indigenous instruments within their own communities. Others face financial constraints that make acquisition impractical, even when awareness exists (personal communication, Mafemani Kubai, 2024). These intersecting challenges, economic, infrastructural, and informational, collectively inhibit the integration of indigenous instruments into rural school curricula, despite their geographic and cultural proximity, while urban spaces have choices to integrate these instruments into their curricula or not, since certain indigenous instruments, such as modern marimba, are accessible (Miao & Kaewkulthorn, 2024; Gordon, 2020).

One of the most widely promoted indigenous instruments in urban contexts in South Africa and other places is the diatonic marimba, a tuned set of wooden bars arranged sequentially to form a musical scale, commonly classified as a type of xylophone (Masamba & Maguraushe, 2025; Zhang, 2024). In city spaces, makers of this instrument have developed commercially viable enterprises, supplying schools and educational institutions with marimbas for teaching and performance purposes. According to one of the makers, marimba instrument making brings good profit to their business. These urban-based instrument makers typically employ an equal-tempered tuning system, which ensures that the instruments are standardized, replicable, easy to tune, and compatible for ensemble use, thereby enhancing their usability for learners in schools and musicians alike (Miao & Kaewkulthorn, 2024). Urban indigenous African instrument makers also participate in some form of pedagogical knowledge-sharing in schools, such as how to play, handle, and care for the instrument. In contrast, marimba makers in rural areas often face significant barriers to market access (Netshivhambe, 2023a). The tuning systems they traditionally use do not conform to standardized pitch conventions, making their instruments difficult to integrate into school programs or ensemble settings. As a result, despite their technical skill and cultural authenticity, rural makers are often unable to commercialize their instruments effectively, limiting both their economic opportunities and the broader dissemination of indigenous musical knowledge (Post, 2024; Netshivhambe, 2023c).

The mbira is also another indigenous instrument that has garnered considerable attention from scholars, learners, and musicians alike. It is predominantly found in rural communities, where makers are readily available in many villages, unlike the marimba, which is less commonly produced there (personal communication, Farai Matake, 2024). Its introduction to the music education curriculum is not popular and is largely used by performers. The tuning of the mbira remains highly individualised, with each maker crafting instruments according to personal or local conventions rather than adhering to any standardised system (Netshivhambe, 2023c). However, because the mbira is often performed as a solo instrument, it is generally well-received in performance contexts, as it does not need to conform to the tuning of other instruments. Its compatibility with the human voice, which is naturally multi-tonal, further enhances its versatility and cultural resonance (Kashyap, 2025). This accessibility and adaptability make the mbira one of the most prominent and widely used instruments in rural areas. In contrast, it is relatively rare in urban spaces, partly because it lacks standardised tuning and partly because instrument makers in cities do not prioritise its production, given its lower profit margins compared to the marimba, something many urban instrument makers have highlighted. For urban-based makers, investing in mbira production is often seen as commercially unviable, as the instrument does not generate the same economic returns as standardised ensemble instruments like the marimba. Consequently, despite its cultural significance and popularity in rural contexts, the mbira remains underrepresented in formal urban music education and professional performance circuits (personal communication, Philemon Nemukula, 2024).

Most instrument makers operating in urban contexts are primarily driven by commercial considerations, carefully weighing production costs against potential returns on investment. This business-oriented model or approach tends to favour instruments such as the marimba, which are in high demand in urban spaces and are designed for ensemble performance with standard tuning (personal communication,

Gareth Walwyn, 2025). Urban schools frequently purchase multiple marimbas, often in pairs or sets of three to four, to create complete ensembles, making these instruments a lucrative venture for urban makers. By contrast, attempts to commercialise the mbira in city markets have generally proven less profitable. The demand for mbiras is limited in urban contexts, and their individualised tuning and solo performance orientation reduce their appeal for ensemble-based teaching or institutional purchases. Marimbas have gained particular appeal within institutional and educational contexts, in part due to the proliferation of marimba competitions held across various regions (Miao & Kaewkulthorn, 2024; Scott, 2015). These competitions function as significant promotional platforms, enhancing the instrument's visibility, standardisation, and prestige. By providing structured performance opportunities for school ensembles and community groups, such events reinforce the marimba's suitability for group pedagogy and formal assessment. Competitive circuits have stimulated demand for the marimba instrument within schools and universities. Rural marimba production rarely yields returns comparable to those of the makers due to the scarcity of buyers, the logistical challenges of ensemble-standard tuning, and limited institutional demand. This dynamic illustrates the complex interplay between cultural value, market forces, and geographical context in shaping the production and dissemination of indigenous instruments for educational purposes.

Other traditional instruments, such as flutes and one-string bowed instruments, remain largely absent from urban spaces and educational settings, primarily due to the scarcity of suitable raw materials for their manufacture. While these instruments are commonly found within rural communities, their presence in city contexts is limited. Many of these instruments, such as the *uhadi, dende, tshihwana, and umakweyana*, rely on natural resources, including specific tree species, which are rarely available in urban environments. Manufacturers of these instruments often collaborate closely with universities or schools that demonstrate a deliberate interest in incorporating indigenous instruments into their curricula or employ teachers proficient in their use. Another instrument with considerable potential for educational applications is the African drum, commonly known as the djembe (Nzewi, 2007). Although the djembe has been widely incorporated into diverse performance contexts, ranging from traditional ensembles to contemporary and intercultural music settings, its integration into formal education has not been accompanied by a correspondingly robust body of pedagogical frameworks (Nzewi, 2007). Structured teaching methodologies, graded curricula, and standardised instructional materials for systematic classroom use remain comparatively limited. As a result, despite its popularity and adaptability, the instrument's educational potential has yet to be fully realised within institutional music programmes.

Despite the abundance of many of these indigenous instruments in rural areas, they are rarely integrated into formal educational settings. Their teaching methods, such as playing techniques, are not found in the classroom. This is largely because schools are hesitant to engage in the exploratory and experimental processes required to teach and learn indigenous instruments, which are perceived as outside conventional pedagogical frameworks (personal communication, Maluta Matsheka, 2024). Many of these culturally significant instruments remain underutilised in the education sector, highlighting a gap between rural production and urban educational adoption. According to Matsheka, who has tried to teach some of these instruments in schools around his rural home, it is very difficult to get the learners' interest because many are unfamiliar with the instruments (personal communication, Maluta, Matsheka, 2024). Another significant challenge is the lack of trained teachers for many of these indigenous instruments.

The majority of self-taught teachers of these indigenous instruments are elderly and found in rural spaces, which can create perceptions that they are less suited to teaching in formal school settings (Netshivhambe, 2023a). Nemukula also highlights that the pedagogical methods of teaching indigenous instrument knowledge are based on observation, as that is how they learned to play the instrument (personal communication, Philemon Nemukula, 2024). Many of these practitioners from rural communities, who are great players of indigenous instruments, are not fluent in English, which limits their ability to teach learners who may not speak the practitioner's language. Their teaching methods are also based on observation and intuitive response, which is how they have learned to play the instrument (Lebaka, 2022; Urrieta Jr, 2015). Indigenous instruments crafted in rural communities often remain inaccessible for formal educational use. Other contributing limitations include cultural beliefs surrounding the instruments alongside the restricted availability and incompatibility in tuning systems (Isabirye, 2021; Berlina, 1993; Netshivhambe, 2023c). Furthermore, many of these instruments are traditionally designed as solo instruments and are typically performed alongside the voice, rather than in an ensemble such as a marimba. In such contexts, precise tuning is often considered less critical, further complicating their integration into school-based music programs that rely on standardised tuning for group performance.

DISCUSSION

Urban Advantages and the Professional Visibility of Indigenous Instrument Makers: Implications for Education, Career Pathways, and Cultural Sustainability

The efforts of indigenous African instrument makers and performers in urban spaces symbolise the adaptive capacity of indigenous music heritage to assume new forms of preservation and institutional relevance. Their presence and activity within city contexts demonstrate how indigenous musical practices can be recontextualised, professionalised, and integrated into contemporary cultural economies. Many of these urban indigenous instrument players have become guest speakers or lecturers to teach students how to play certain indigenous instruments, at best, but the challenge is that such institutions don't invest in purchasing those instruments for teaching and learning. In doing so, they provide tangible models for how indigenous music heritage can inform and shape curriculum development at the university level, contributing to more inclusive and contextually grounded approaches to music education. Urban-based instrument manufacturers operate within a context that offers distinct structural advantages. Their proximity to established wood-processing and manufacturing industries provides easier access to raw materials, including affordable off-cuts, which are particularly valuable for emerging indigenous instrument makers seeking to minimise production costs (Dawe, 2015). This logistical advantage significantly lowers barriers to entry and enhances the scalability of their operations. In addition to material access, urban manufacturers benefit from their closeness to schools, universities, and cultural institutions, which constitute a reliable and concentrated customer base (personal communication, Farai Matake, 2024). Through direct marketing, demonstrations, and institutional partnerships, these makers can position their instruments within formal education systems, thereby strengthening demand and sustaining their enterprises. Many indigenous instrument makers have initiated enrichment programmes designed to empower students by sharing technical knowledge about the construction of indigenous instruments. These initiatives not only expand market visibility but also contribute to skills development and the pedagogical integration of indigenous instruments within urban educational contexts.

Urban-based instrument manufacturers are frequently invited to schools and universities to serve as guest speakers, where they present instrument playing and construction as viable professional pathways for students. In addition to these engagements, they offer workshops and practical training sessions addressing technical aspects such as tuning systems, maintenance and repair procedures, and the safe handling and transportation of large or delicate instruments, including marimba ensembles. They also teach students short, easy music pieces that can be learned during the presentation sessions (personal communication, Maluta Matsheka, 2024). Activities such as these enhance their visibility and credibility within urban cultural and educational networks, positioning them as accessible specialists who can respond directly to institutional needs. Despite their growing presence, the visibility of these instrument makers and their instruments is not necessarily sufficient to motivate young people to pursue music as a professional career (Netshivhambe, 2023a). Students are generally drawn to fields of study that they associate with tangible employment prospects and long-term economic stability. Students are easily motivated when they see real indigenous music practitioners teaching or performing for them, which inspires them to play or learn indigenous instruments.

Marketing indigenous instruments directly to students and providing first-hand experience with practitioners can also help students see indigenous musical heritage as a viable career path. In contexts characterised by high unemployment and socio-economic uncertainty, career choice is frequently guided by perceived labour-market viability rather than cultural or artistic interest alone. For indigenous music heritage to attract sustained student engagement, it must be positioned not only as culturally meaningful but also as professionally viable. Students require assurance that studying indigenous musical traditions can lead to concrete employment pathways, including performance, teaching, instrument manufacturing, cultural entrepreneurship, and related creative industries (personal communication, Philemon Nemukula, 2024). In this regard, urban musicians and instrument makers who have established sustainable careers can serve as visible models of possibility, embodying the potential for indigenous music to function as both a cultural vocation and a viable livelihood (Netshivhambe, 2023a). The viability of indigenous instruments within educational contexts is significantly strengthened by performers who actively incorporate them into contemporary and popular music practices. By foregrounding these instruments in commercially visible platforms, such artists demonstrate their adaptability and relevance beyond traditional or ceremonial settings. This public use not only reaffirms cultural continuity but also demonstrates that indigenous instruments can meaningfully participate in modern music production economies.

When many people observe indigenous instruments integrated into professional recordings, live performances, and commercially successful projects, the perceived distance between heritage and livelihood diminishes. The instruments become associated not only with cultural identity but also with creative entrepreneurship and income generation. Musicians in both rural and urban contexts have contributed to this shift by producing albums and performances that intentionally celebrate their cultural heritage through the incorporation of indigenous instruments. In doing so, they expand the instruments' symbolic and economic value, reinforcing their legitimacy within both educational and professional spheres. Increased visibility of indigenous instruments in performance spaces, including university settings, has led to a discernible rise in student interest in pursuing them as viable career pathways. Exposure to indigenous instrument manufacturers and performers who facilitate workshops, as well as to performers who have established local and international reputations through indigenous music as a core artistic identity, contributes significantly to this shift in perception. Such encounters provide students with tangible examples of professional success grounded in cultural heritage. As indigenous instruments become more accessible and publicly visible, the likelihood increases that students will select them as primary areas of study or professional focus. Expanded access not only broadens artistic choice but also enhances students' perceptions of future opportunities in the performance, education, and instrument manufacturing sectors.

LIMITATION

This study is limited by the relatively small number of indigenous instrument makers available in both urban and rural settings, which restricted the breadth of participant engagement and the diversity of perspectives represented in the research. Additionally, there is a limited body of existing scholarship specifically focused on indigenous musical instrument construction within higher education, which constrained the availability of comparative academic literature and theoretical frameworks. Another important limitation is the study's geographical scope, which primarily focuses on Southern African practices. As a result, the findings cannot be generalised as a comprehensive representation of indigenous musical instrument construction and educational integration across the entire African continent, since practices, cultural contexts, and institutional approaches differ significantly from one region to another.

FUTURE RESEARCH DIRECTION

Future research should expand beyond the Southern African context to examine how indigenous instrument production, tuning systems, pedagogical practices, and commercialisation strategies operate across different African regions. Comparative studies between rural and urban contexts could further explore how standardisation, technology, and institutional support influence the educational integration and sustainability of indigenous instruments. Additional research is also needed on the development of pedagogical frameworks, curriculum models, and teacher-training programmes for indigenous instruments such as the *mbira, dende, djembe, uhadi*, and *tshihwana* within formal education systems. Future studies may further investigate how digital technologies, community partnerships, and creative industries can support the preservation, transmission, and economic viability of indigenous musical heritage for younger generations. Further research should also be informed by the integration of both rural and urban conceptualisations of indigenous music performance and instrument craftsmanship to advance indigenous instrument development pedagogies. Rural indigenous instrument makers and performers possess extensive cultural knowledge, practical expertise, and deeply rooted understandings of indigenous musical traditions. However, they often have limited access to formalised systems of tuning, standardisation, and large-scale production processes. In contrast, urban-based instrument makers and performers are generally better positioned in terms of access to institutional networks, commercial markets, technological resources, and standardisation practices. Future research should therefore explore how urban-centred indigenous music industries and practitioners can incorporate and promote less commercially viable indigenous instruments, rather than focusing primarily on instruments driven by profitability and market demand.

CONCLUSION

The activities of indigenous African instrument makers and performers in urban contexts demonstrate that indigenous music heritage is a viable career choice as it is neither static nor confined to rural spaces. Rather, it is capable of adaptation, professionalisation, and meaningful integration into contemporary cultural and educational systems. Urban environments provide structural advantages, access to factory-cut and ready-to-use materials, institutional networks, and concentrated markets that enable instrument makers to scale production, enhance standardisation, and position their craft within formal education and expose it to potential students. Through workshops, enrichment initiatives, and public engagements, these practitioners extend their influence beyond performance into pedagogy, skills development, and curriculum transformation. Visibility alone is insufficient to guarantee sustained student engagement. In socio-economic contexts marked by uncertainty and high unemployment, students are guided by considerations of employability and long-term stability. Indigenous music heritage must therefore be framed not only as culturally significant but also as economically viable. When performers successfully integrate indigenous instruments into commercially recognised platforms, they reduce the perceived divide between heritage and livelihood, demonstrating that cultural practice can coexist with professional sustainability. The growing presence of indigenous instruments in public performance, recordings, and university spaces contributes to shifting perceptions among students, who increasingly recognise these traditions as legitimate areas of specialisation, while, in turn, repositioning indigenous musical heritage within educational spaces. Indigenous instrument makers and performers emerge not merely as cultural practitioners but as key agents in reshaping curriculum, expanding opportunity, and ensuring the long-term sustainability of indigenous music heritage within modern academic and economic landscapes.

REFERENCES

Abbas, H., Ullah, I., & Ullah, W. (2025). Displacement and urban alienation: An ecocritical reading of The Reluctant Fundamentalist. *ACADEMIA International Journal for Social Sciences*, *4*(1), 677–688. DOI: 10.63056/ACAD.004.01.0122

Adefila, A., Teixeira, R. V., Morini, L., Garcia, M. L. T., Delboni, T. M. Z. G. F., Spolander, G., & Khalil-Babatunde, M. (2022). Higher education decolonisation:# Whose voices and their geographical locations? *Globalisation, Societies and Education*, *20*(3), 262–276. DOI: 10.1080/14767724.2021.1887724

Ajani, Y. A., Oladokun, B. D., Olarongbe, S. A., Amaechi, M. N., Rabiu, N., & Bashorun, M. T. (2024). Revitalizing indigenous knowledge systems via digital media technologies for sustainability of indigenous languages. *Preservation, digital technology & culture, 53*(1), 35-44.

Alarima, C. I. (2018). *Factors influencing rural-urban migration of youths in Osun State*.

Apaydin, V. (2020). Heritage, memory and social justice: reclaiming space and identity. *Critical perspectives on cultural memory and heritage*, 84-97.

Attas, R., & Walker, M. E. (2019). Exploring decolonization, music, and pedagogy. *Intersections*, *39*(1), 3–20. DOI: 10.7202/1075339ar

Baporikar, N. (2022). Strategies to manage and preserve indigenous knowledge. In *Handbook of research on protecting and managing global indigenous knowledge systems* (pp. 207–222). IGI Global Scientific Publishing. DOI: 10.4018/978-1-7998-7492-8.ch012

Bates, E. (2012). The social life of musical instruments. *Ethnomusicology*, *56*(3), 363–395. DOI: 10.5406/ethnomusicology.56.3.0363

Bohensky, E. L., & Maru, Y. (2011). Indigenous knowledge, science, and resilience: What have we learned from a decade of international literature on "integration"? *Ecology and Society*, *16*(4), art6. DOI: 10.5751/ES-04342-160406

Brown, R. (2007). Enhancing student employability? Current practice and student experiences in HE performing arts. *Arts and Humanities in Higher Education*, *6*(1), 28–49. DOI: 10.1177/1474022207072198

Cavalcanti, M. L. V. D. C. (2019). The Legal Protection of the Intangible Cultural Heritage in Brazil. In *The Legal Protection of the Intangible Cultural Heritage: A Comparative Perspective* (pp. 19–35). Springer International Publishing. DOI: 10.1007/978-3-319-72983-1_2

Chimbunde, P., & Kgari-Masondo, M. C. (2022). The decolonisation of African languages: Insights from Southern Africa. Language Matters. *Studies in the Languages of Southern Africa*, *53*(3), 5–24.

Dawe, K. (2015). Materials matter: Towards a political ecology of musical instrument making. In *Current directions in ecomusicology* (pp. 109–121). Routledge.

Demir, K. A., Yaman, F., & Koçak, B. (2025). Social Alienation with International Students in New York City: A Phenomenological Research. *Siyasal: Journal of Political Sciences*, *34*(1), 134–151. DOI: 10.26650/siyasal.2025.34.1478556

Donald, W. E., Ashleigh, M. J., & Baruch, Y. (2018). Students' perceptions of education and employability: Facilitating career transition from higher education into the labor market. *Career Development International*, *23*(5), 513–540. DOI: 10.1108/CDI-09-2017-0171

Du Plessis, P. (2021). Decolonisation of education in South Africa: Challenges to decolonise the university curriculum. *South African Journal of Higher Education*, *35*(1), 54–69. DOI: 10.20853/35-1-4426

Euba, A. (2014). *JH Kwabena Nketia: Bridging Musicology and Composition: A Study in Creative Musicology*. MRI Press.

Farhan, S. L., Merie, U. A. A. K., & Nasar, Z. (2024). Revitalizing historic city center a comparative methodology of current approaches and alternatives. *Journal of Cultural Heritage Management and Sustainable Development*.

Ferrer-Fons, M., Rovira-Martínez, M., & Soler-i-Martí, R. (2022). Youth empowerment through arts education: A case study of a non-formal education arts centre in Barcelona. *Social Inclusion (Lisboa)*, *10*(2), 85–94. DOI: 10.17645/si.v10i2.4923

Fleming, M. J., & Grace, D. M. (2016). Agents of aspiration: The (often unintended) benefits to university students working in outreach programmes. *Educational Research*, *58*(3), 300–318. DOI: 10.1080/00131881.2016.1185316

Gheitarani, N., El-Sayed, S., Cloutier, S., Budruk, M., Gibbons, L., & Khanian, M. (2020). Investigating the mechanism of place and community impact on quality of life of rural-urban migrants. *International Journal of Community Well-being*, *3*(1), 21–38. DOI: 10.1007/s42413-019-00052-8

Gómez-Baggethun, E. (2022). Is there a future for indigenous and local knowledge? *The Journal of Peasant Studies*, *49*(6), 1139–1157. DOI: 10.1080/03066150.2021.1926994

Gordon, G. A. (2020). *Marimbas in South African schools: gateway instruments for the indigenous African music curriculum* (Doctoral dissertation, Stellenbosch: Stellenbosch University).

Harvey, A., & Russell-Mundine, G. (2019). Decolonising the curriculum: Using graduate qualities to embed Indigenous knowledges at the academic cultural interface. *Teaching in Higher Education*, *24*(6), 789–808. DOI: 10.1080/13562517.2018.1508131

He, S. (2023). Liberal arts education and influence on employability: A case study of a Sino-foreign university. *Chinese Education & Society*, *56*(5-6), 350–368. DOI: 10.1080/10611932.2024.2303918

Hlalele, D. J. (2019). Indigenous knowledge systems and sustainable learning in rural South Africa. *Australian and international journal of rural education, 29*(1), 88-100.

Hudson, R. (2006). Regions and place: Music, identity and place. *Progress in Human Geography*, *30*(5), 626–634. DOI: 10.1177/0309132506070177

Isabirye, J. (2021). Can indigenous music learning processes inform contemporary schooling? *International Journal of Music Education*, *39*(2), 151–166. DOI: 10.1177/0255761421996373

Jakes, V. (2024). The role of traditional knowledge in sustainable development. *International Journal of Humanities and Social Science*, *3*(2), 40–55.

Karim, Z. I. A., & Maat, S. M. (2019). Employability skills model for engineering technology students. *Journal of Technical Education and Training, 11*(2).

Kashyap, A. (2025). Significance of Folk Musical Instruments in reinforcing Cultural Identity and Belonging: Discerning the dialogue between Soundscape Ecology-Ethnobiology-Ethnomusicology. *JOURNAL OF ADVANCE AND FUTURE RESEARCH, 3*(7), 28–53.

Keebine-Sibanda, M. J. (2021). Academics' conceptions of higher education decolonisation. *South African Journal of Higher Education, 35*(3), 182–199.

Kohsaka, R., & Rogel, M. (2021). Traditional and local knowledge for sustainable development: Empowering the indigenous and local communities of the world. In *Partnerships for the Goals* (pp. 1261–1273). Springer International Publishing. DOI: 10.1007/978-3-319-95963-4_17

Kovačević, M., Dekker, T. J., & van der Velden, R. (2024). Employability development in undergraduate programmes: How different is liberal arts education? *Teaching in Higher Education, 29*(8), 2184–2204. DOI: 10.1080/13562517.2023.2212602

Kumah-Abiwu, F. (2022). Urban education and academic success: The case of higher achieving Black males. *Urban Education, 57*(9), 1565–1591. DOI: 10.1177/0042085919835284

Lagakos, D. (2020). Urban-rural gaps in the developing world: Does internal migration offer opportunities? *The Journal of Economic Perspectives, 34*(3), 174–192. DOI: 10.1257/jep.34.3.174

Lázaro Ortiz, S., & Jiménez de Madariaga, C. (2021). The UNESCO convention for the safeguarding of the intangible cultural heritage: A critical analysis. *International Journal of Cultural Policy, 28*(3), 327–341. DOI: 10.1080/10286632.2021.1941914

Le Grange, L. (2023). Decolonisation and anti-racism: Challenges and opportunities for (teacher) education. *Curriculum Journal, 34*(1), 8–21. DOI: 10.1002/curj.193

Lebaka, M. E. K. (2022). Informal learning of indigenous music and dance through observation and imitation: The case of Bapedi children's musical arts. *European Journal of Language and Literature Studies, 8*(1), 55–67. DOI: 10.26417/658qju31

Lee, D. H. (2008). Re-imagining urban space: Mobility, connectivity, and a sense of place. *Mobile technologies: From telecommunications to media* (pp. 235-251)https://doi.org/DOI: 10.4324/9780203884317

Levstek, M., & Banerjee, R. (2021). A model of psychological mechanisms of inclusive music-making: Empowerment of marginalized young people. *Music & Science, 4*, 20592043211059752. DOI: 10.1177/20592043211059752

Luca, C. (2024). The Effect of Globalization on Preserving Indigenous Knowledge in Education. *Research Gate, May.*

Magni, G. (2017). Indigenous knowledge and implications for the sustainable development agenda. *European Journal of Education, 52*(4), 437–447. DOI: 10.1111/ejed.12238

Manyike, T. V., & Shava, S. (2018). The decolonial role of African indigenous languages and indigenous knowledges in formal education processes. *Indilinga, 17*(1), 36–52.

Masamba, G., & Maguraushe, W. (2025). The Uptake of Chromatic Marimbas by Learning Institutions in Zimbabwe. *African Musicology Online*, *14*(2), 69–81. DOI: 10.58721/amo.v14i2.1208

Mawere, J., & Mukonza, R. M. (2025). Empowering marginalised communities: Leveraging indigenous knowledge for sustainable energy development in South Africa. *Sustainable Development (Bradford)*, *33*(2), 2440–2448. DOI: 10.1002/sd.3243

Mazzocchi, F., Simandan, D., Demneh, M. T., Morgan, D. R., Ghazinoory, S., Saghafi, F., & Mirzaei, M. (2018). Why 'Integrating'Western science and Indigenous knowledge is not an easy task: What lessons could be learned for the future of knowledge. *Journal of Futures Studies*, *22*(3), 19–34.

Miao, K., & Kaewkulthorn, P. (2024). Historical Development and Learning of Marimba in Chinese Educational Institutions. *International Journal of Sociologies and Anthropologies Science Reviews*, *4*(6), 553–562. DOI: 10.60027/ijsasr.2024.5129

Mowakadi, J. (2025). Urbanization and Indigenous Cultural Adaptation in Lagos: A Multidisciplinary Analysis of Urban Cultural Coexistence. *Global Interdisciplinary Perspectives*, *1*(2), 20–27. DOI: 10.64229/b6jas597

Muriithi, G. M., & Musili, W. M. (2021). *The Effect of Science, Technology, Engineering and Mathematics Skills on Youth Employment in Kenya*. Kenya Institute for Public Research and Analysis.

Netshivhambe, E. N. (2023a). Indigenous African Music Performance Re-imagined in Public Spaces. In Mpofu, P., Fadipe, I. A., & Tshabangu, T. (Eds.), *African Language Media* (1st ed., pp. 243–257). Routledge., DOI: 10.4324/9781003350194-23

Netshivhambe, E. N. (2023c). *Ignored or Invisible: The status of indigenous South African instruments In the Venda society*. UNISA Press.

Netshivhambe, E. N. (2023d). Living the heritage through Indigenous music competitions. In Kumar, S., & Lakshminarayana, M. B. (Eds.), *Indigenous People - Traditional Practices and Modern Development*. IntechOpen., DOI: 10.5772/intechopen.1003226

Netshivhambe, E. N. (2024a). De-neocolonizing the South African Music Education: Selective Transformation. In Akpan, U. S. (Ed.), *De-neocolonizing Africa: Harnessing the Digital Frontier* (pp. 393–416). Palgrave Macmillan Cham., DOI: 10.1007/978-3-031-66304-8_22

Netshivhambe, N. E. (2024b). Towards Sustainable Preservation: Revitalising Indigenous African Instruments and Craftsmanship. *E-Journal of Humanities. Arts and Social Sciences*, *5*(15), 91–107.

Netshivhambe, N. E. (2025). Promoting Wellness in the Aged through Indigenous Instrument. *Social Determinants of Active Ageing-Building Resilience and Empowerment: Building Resilience and Empowerment*, 21.

Niglio, O., Singh, R. P., & Gireesh Kumar, T. K. (2026). RECONNECTING WITH YOUR CULTURE (RWYC): TOWARDS CULTURAL AWAKENING, PROSPECTS AND VISION. *EDA, ESEMPI DI ARCHITETTURA*, *13*(2), 317–338.

Nkoane, M. M. (2006). The Africanisation of the university in Africa. *Alternation (Durban)*, *13*(1), 49–69.

Nzewi, M. (2007). Learning the Musical Arts in Contemporary Africa. In *Informed by Indigenous Knowledge Systems* (Vol. 1). CIIMDA Publishing.

Oberhofer, C. (2020). Decolonization and Indigenization in music education. *Canadian Music Educator, 62*(1), 48–53.

Omoniyi, M. B. I. (2013). The role of education in poverty alleviation and economic development: A theoretical perspective and counselling implications. British Journal of Arts and Social Sciences, 15(2), 176-185.Benešová, A., & Tupa, J. (2017). Requirements for Education and Qualification of People in Industry 4.0. *Procedia Manufacturing, 11*, 2195–2202. DOI: 10.1016/j.promfg.2017.07.366

Osei-Tutu, A. A. (2023). African-Centered hybridity: A Reconceptualization of Africanness in this colonially guised globalized era. *Alliance for African Partnership Perspectives, 2*(1), 61–73.

Osutei, N., & Kim, J. H. (2023). Examining talent attraction and retention in small and medium-sized cities: Where do universities fit in? *Growth and Change, 54*(3), 736–753. DOI: 10.1111/grow.12672

Paraschiv, C. I. (2017). The role of education in poverty alleviation. *Theoretical & Applied Economics, 24.*Zhong, R. Y., Xu, X., Klotz, E., & Newman, S. T. (2017). Intelligent manufacturing in the context of industry 4.0: A review. *Engineering, 3*(5), 616–630. https://doi.org/https://doi.org/10.1016/J.ENG .2017.05.015

Pedersen, H. D., & Gram, M. (2018). 'The brainy ones are leaving': The subtlety of (un) cool places through the eyes of rural youth. *Journal of Youth Studies, 21*(5), 620–635. DOI: 10.1080/13676261.2017.1406071

Post, J. (2024). Ecosystems and Sounding Lives: Musical Instrument Makers in Greater Central Asia. *Journal of Musicological Research, 43*(2-3), 175–196. DOI: 10.1080/01411896.2024.2376352

Prajapati, M. B., & Chatwani, M. J. (2025). Indigenous Knowledge Systems and Sustainable Urban Futures: Lessons from Traditional Cities and Planning Practices. In *International Conference on 'Indigenous Knowledge Systems and Sustainable Development* (Vol. 9, p. 10).

Reddick, R. J., Welton, A. D., Alsandor, D. J., Denyszyn, J. L., & Platt, C. S. (2011). Stories of success: High minority, high poverty public school graduate narratives on accessing higher education. *Journal of Advanced Academics, 22*(4), 594–618. DOI: 10.1177/1932202X11414133

Schieman, S., & Plickert, G. (2008). How knowledge is power: Education and the sense of control. *Social Forces, 87*(1), 153–183. DOI: 10.1353/sof.0.0065

Scott, R. C. (2015). *The art of marimba in Tokyo: emergence in the twentieth century*. University of Toronto.

Sharma, Y., Tochhawng, L., Kumar, V., Nikam, R. J., Deka, P., & Baruah, C. J. (2024). Cultural heritage conservation in urban areas: Integrating traditional knowledge and modern planning practices. *Journal of Applied Bioanalysis, 10*, 260–265.

Smith, A. (2018). Reconnecting the music-making experience through musician efforts in instrument craft. *International Journal of Music Education, 36*(4), 560–573. DOI: 10.1177/0255761418771993

Su, J. (2019). Understanding the changing intangible cultural heritage in tourism commodification: The music players' perspective from Lijiang, China. *Journal of Tourism and Cultural Change*, *17*(3), 247–268. DOI: 10.1080/14766825.2018.1427102

Sun, M., Zhao, M., & Zhou, W. (2025). Assessment and Model Construction of Talent Attraction in the Creative Industry: A Case Study of Underdeveloped Cities and Regions in China. *Mediterranean Archaeology & Archaeometry. International Journal*, *25*(1).

Tahseen, E., & Al-Jumaily, S. (2020). Mechanisms for reviving the intangible cultural heritage to revitalize urban spaces. *International Journal of Environment. Engineering and Education*, *2*(3), 31–42.

Turino, T. (2010). *Moving away from silence: music of the Peruvian altiplano and the experience of urban migration*. University of Chicago Press.

Urrieta, L.Jr. (2015). Learning by observing and pitching in and the connections to native and indigenous knowledge systems. [). JAI.]. *Advances in Child Development and Behavior*, *49*, 357–379. DOI: 10.1016/bs.acdb.2015.08.003 PMID: 26955937

Wang, J., Lim, X. J., & Ng, S. I. (2024). Why can't I get a job? Exploring the drivers of perceived employability among fine art graduates. *Current Psychology (New Brunswick, N.J.)*, *43*(4), 3002–3020. DOI: 10.1007/s12144-023-04561-w

Yang, S., Shu, T., & Yu, T. (2022). Migration for better jobs or better living: Shifts in China. *International Journal of Environmental Research and Public Health*, *19*(21), 14576. DOI: 10.3390/ijerph192114576 PMID: 36361454

Zhang, W. (2024). Teaching Design of High School Music Class: Taking Marimba piece Bamboo Forest" by Keiko Abe as Extended Example. *Arts, Culture and Language, 1*(6).

ADDITIONAL READINGS

Addaquay, A. P. (2026). Reclaiming the African voice beyond Eurocentric instrumental taxonomies. *African Identities*, •••, 1–21. DOI: 10.1080/14725843.2026.2623958

Aderogba, J. A. (2025). Fusing Traditions: Exploring Selected Western Vocal Music Through African Instruments In a Nigerian Recital Experience. AE-FUNAI JOURNAL OF MUSIC AND THE ARTS (AFUJMA), 2.

Aidoo, S. (2024). Towards the Standardisation of African Musical Instruments: The At? nt? b? n in Perspective. *JOURNAL OF AFRICAN ARTS AND CULTURE*, *7*(3), 59–82.

Akere, A. O. (2023). Technological Innovation of Agidigbo Traditional Musical Instrument: Conceptual and Theoretical Perspectives. International Journal of Operational Research in Management [IJORMSSE]. *Social Sciences & Education*, *9*(2), 280–296.

Baumann, M. P. (2000). The local and the global: Traditional musical instruments and modernization. the world of music, 121-144.

Bleibinger, B., & Ncozana, J. (2024). The Constructive and Connecting Side of Indigenous African Music and Instruments: Examples from the Eastern Cape in South Africa and Germany. *Musicology Australia*, *46*(2), 168–195. DOI: 10.1080/08145857.2025.2485716

Brauer-Benke, J. (2009). The Kora: The History of a West African Instrument. *Hungarian J. Afr. Stud.*, *3*, 104.

Cole, J. Constructing Racial Identity and Power in Music, Ceremonial Practices, and Indigenous Instruments in Early Modern African Kingdoms. In The Routledge Companion to Race in Early Modern Artistic, Material, and Visual Production (pp. 161-185). Routledge.

KEY TERMS AND DEFINITIONS

Africanisation of Curriculum: An educational approach that prioritises African histories, cultures, languages, philosophies, and knowledge systems within formal education, particularly higher education institutions.

Commercialisation of Indigenous Instruments: The process through which indigenous instruments are produced, marketed, and sold within formal economic and educational systems for financial sustainability and broader circulation.

Creative Industries: Economic sectors based on cultural creativity and artistic production, including music performance, instrument manufacturing, recording, cultural tourism, and arts education.

Cultural Commodification: The transformation of cultural practices and artifacts into marketable products for economic purposes, often within urban and commercial contexts.

Cultural Entrepreneurship: Economic activities based on cultural production and heritage practices, including indigenous instrument manufacturing, performance, workshops, and arts-based enterprises.

Cultural Sustainability: The preservation, transmission, and continued relevance of cultural practices, knowledge, and artistic traditions across generations within changing social contexts.

Curriculum Transformation: The process of restructuring educational curricula to include diverse epistemologies, particularly African and indigenous knowledge systems, in order to challenge Eurocentric dominance and promote inclusive education.

Decolonial Reform: Efforts aimed at dismantling colonial and Eurocentric structures within education and knowledge production by recognising and legitimising indigenous epistemologies and cultural practices.

Equal-Tempered Tuning System: A standardised tuning method commonly used in Western music where pitch intervals are equally spaced, enabling compatibility among ensemble instruments such as the marimba.

Heritage Resilience: The ability of indigenous cultural traditions and practices to adapt, survive, and remain relevant despite pressures from modernisation, urbanisation, and globalisation.

Indigenous Instrument Makers: Practitioners who craft indigenous African musical instruments using culturally informed techniques, traditional materials, and inherited artisanal knowledge.

Indigenous Knowledge Systems (IKS): Refers to culturally grounded systems of knowledge, practices, values, and skills developed and transmitted within local communities across generations.

Indigenous Music Pedagogy: Teaching and learning approaches grounded in indigenous musical traditions, often based on observation, imitation, participation, oral transmission, and experiential learning.

Indigenous Musical Heritage: The collective musical traditions, instruments, performance practices, craftsmanship, cultural meanings, and transmission methods associated with indigenous African communities. It encompasses both tangible and intangible cultural expressions linked to music.

Marginalisation of Indigenous Music: The exclusion or limited representation of indigenous musical traditions within formal education systems, institutional policies, and mainstream cultural industries.

Pedagogical Integration: The incorporation of indigenous instruments, teaching methods, and musical knowledge into formal educational curricula and classroom practices.

Professionalisation of Indigenous Music: The process through which indigenous music practices become recognised as viable professional and academic fields through institutional support, commercial participation, and curriculum inclusion.

Standardisation: The adaptation of indigenous instruments to consistent tuning systems, measurements, and construction methods to facilitate ensemble performance, institutional use, and formal music education.

Transmission of Indigenous Knowledge: The intergenerational passing down of cultural knowledge, skills, and practices through observation, participation, oral instruction, and apprenticeship.

Urbanisation: The social and economic process through which populations migrate from rural to urban areas in search of employment, education, and improved living conditions, often reshaping cultural practices and heritage.

Urban–Rural Dynamics: The social, economic, cultural, and infrastructural differences between urban and rural contexts that shape the production, accessibility, commercialisation, and educational integration of indigenous instruments.

Chapter 3
Ngwao ya Rona:
Setswana Traditional Music and Cultural Consciousness

Mojuta Steven Motlhamme
http://orcid.org/0000-0001-7613-7220
University of the Witwatersrand, South Africa

ABSTRACT

This chapter examines Ngwao ya Rona (Our Culture) as expressed through Setswana traditional music, positioning it as a vital Indigenous Knowledge System (IKS) through which cultural consciousness, social values, and historical memory are produced and transmitted among Batswana communities. Setswana traditional music—encompassing song, rhythm, dance, oral poetry, and performance—functions as an embodied and performative knowledge archive, preserving indigenous epistemologies related to identity, morality, spirituality, gender relations, and communal governance. The chapter further situates Setswana traditional music within broader historical processes of colonial disruption, apartheid marginalisation, and contemporary globalisation, highlighting the resilience and adaptability of indigenous musical knowledge. Ultimately, the chapter foregrounds traditional music as central to sustaining ngwao ya rona and advancing indigenous cultural consciousness in contemporary African societies.

INTRODUCTION

Ngwao ya Rona in the Setswana language means our culture and heritage. It is an evocative cry for the preservation, celebration and commemoration of Setswana culture and identity. This chapter seeks to demonstrate and illustrate how Setswana traditional music is part of the culture and heritage of the Batswana speaking people (Molato & Dube, 2025). Music is one of the ways Batswana have transmitted, preserved, and showcased their cultural identity and consciousness (Wood: 1985). A lot of people are introduced to Setswana culture and identity through music as entry point into the culture. Many people who do not speak Setswana can instantly tell you about the music of Mma Ausi (Galaletsang Lekoma) and Culture Spears (headlined by Kabelo Mogwe and Magdeline Lesolebe). This chapter uses traditional

DOI: 10.4018/979-8-2600-2890-2.ch003

music as a discourse of indigenous knowledge and cultural consciousness to argue for the salience of music in understanding Setswana culture and heritage.

Setswana traditional music has long served as a repository of indigenous knowledge, social regulation, and aesthetic expression. However, forces such as colonisation, missionary activity, and contemporary globalisation continue to influence how traditional music is practiced and preserved. At the same time, a resurgence of interest in indigenous arts among youth and cultural institutions points to a renewed effort to reclaim identity and heritage (Molefe, 2019). Despite this, limited scholarship directly connects Setswana music to larger debates around decoloniality, identity politics, and cultural memory—an academic gap this chapter seeks to fill.

Setswana traditional music has played a major role in the proliferation of Setswana cultural consciousness. Setswana cultural consciousness is defined as the ability to understand and become proud of one's identity as a Motswana. Motswana in this context is not limited to the national boundary of Botswana but encompasses Batswana in different countries of Southern Africa, such as Namibia, Zimbabwe, and South Africa. Through music, Setswana musicians are transmitting cultural consciousness to Batswana and other people in the world about being Motswana. Through music, one is transplanted into the Setswana speaking world. Through case studies of different songs and musicians, we learn and understand the Setswana speaking world. This paper has looked at different songs and came with the following conclusions: (1) Songs can be used to understand Setswana culture, (2) Music is a universal language that allows people to overcome cultural differences, (3) Batswana have many different merafe (totemic clusters), (4) we can learn about Setswana culture through songs, (5) songs carry different thematic meanings and speak about the rich tapestry of Setswana history, traditions, and cultural events.

RESEARCH METHODOLOGY

This study adopts a qualitative, interpretive research design grounded in Indigenous Knowledge Systems (IKS) and decolonial epistemologies. The methodological approach recognises Setswana traditional music not merely as cultural artefact but as a living knowledge system embedded within community practices, oral traditions, and embodied performance (Hoppers, 2025). Emphasis is placed on privileging indigenous ways of knowing, being, and transmitting knowledge. The research is located within an indigenous research paradigm, which values relationality, communal ethics, oral knowledge, and lived experience. This paradigm challenges Eurocentric hierarchies of knowledge by foregrounding indigenous epistemologies and methodologies as legitimate and authoritative sources of scholarship. A decolonial lens further informs the study by interrogating the historical marginalisation of Setswana cultural practices under colonialism and apartheid (Shillington, 1985). Selected musical forms, including *dikoma*, *setapa*, *phathisi*, and *mmino wa setso*, were analysed as pedagogical practices (Phibion, 2013). Analysis focused on lyrics, rhythms, movement, and call-and-response structures to understand how cultural values and indigenous knowledge are encoded and communicated. audio recordings, and heritage documents were consulted to contextualise contemporary practices within broader historical trajectories.

THEORETICAL FRAMEWORK

This study is grounded in Indigenous Knowledge Systems (IKS) theory, which conceptualises Setswana traditional music as a legitimate and authoritative knowledge system rather than a peripheral cultural practice (Khan, 2025). Within this framework, *ngwao ya rona* (our culture) is understood as embodied, oral, communal, and performative knowledge transmitted through music, dance, and oral poetry (Johnston, 1973). Musical forms such as *dikoma*, *setapa*, and *phathisi* function as indigenous pedagogies that encode social values, moral norms, spirituality, and historical memory, with elders and cultural practitioners recognised as primary knowledge holders.

The framework is further informed by decolonial theory, which interrogates the historical marginalisation of Setswana cultural practices under colonial and apartheid epistemologies (Mignolo & Walsh, 2018). Drawing on decolonial scholarship, the study positions Setswana traditional music as a site of epistemic resistance and cultural re-existence, challenging Western musicological traditions that privilege written notation, individual authorship, and textual knowledge. Decoloniality enables a re-centering of indigenous epistemologies and affirms oral, collective, and performative modes of knowledge production (Gatsheni, 2018).

LITERATURE REVIEW

The study of Setswana traditional music occupies an important yet uneven position within African humanities scholarship. While African music has attracted sustained academic attention since the mid-twentieth century, much of this scholarship has been shaped by Western epistemological frameworks that prioritise documentation, classification, and formal musical analysis (Agawu: 2014; Bebey, 1999). In contrast, indigenous African perspectives have long understood music as inseparable from culture, morality, spirituality, and social organisation (Kipyegon, Wanyama & Mapaya, 2025). This literature review critically examines scholarship relevant to *Ngwao ya Rona* (Our Culture), with a focus on Setswana traditional music as a carrier of Indigenous Knowledge Systems (IKS) and cultural consciousness.

John Blacking's seminal work *How Musical Is Man?* (Blacking, 1973). marked a critical shift by emphasising music as humanly organised sound that reflects social relationships and cultural values. His ethnographic research among the Venda demonstrated that music cannot be understood outside its cultural context. Although Blacking did not work extensively on Setswana music, his theoretical insights profoundly influenced subsequent studies of southern African musical traditions, including those of Batswana communities. It is important for us to understand that music is carried and sustained by the community where it originates. Many South African musicians such as Mirriam Makeba and Hugh Masekela who went into exile often created contextual struggle music that shows their South African origins (Sizemore-Barber, 2012).

Cultural studies approaches have expanded this analysis by situating Setswana cultural practices within broader historical and political processes (Molato & Dube, 2025). Scholars examining colonialism and apartheid in southern Africa have documented how indigenous cultural expressions were marginalised, regulated, or appropriated (Magubane, 1996). Missionary education and colonial governance frequently viewed traditional music as pagan or disruptive, contributing to its exclusion from formal institutions of knowledge (Coplan, 2007).

Indigenous Knowledge Systems scholarship provides a critical framework for reinterpreting Setswana traditional music as an epistemological practice. IKS theorists argue that indigenous knowledge is holistic, relational, and embedded in everyday cultural practices rather than formalised texts (Khan, 2025). Music is part of the ritual of everyday life and its performance. Knowledge is transmitted orally, performatively, and intergenerationally, often through storytelling, ritual, and music. The music is communally owned and thus does not belong to a particular individual. Within this framework, music is understood as an archive of indigenous knowledge that encodes cosmology, ethics, history, and social organisation. African scholars have emphasised that song and dance are pedagogical tools that teach communities how to live, relate, and remember (Kwabena Nketia: 1974). Setswana traditional music, when viewed through an IKS lens, emerges as a system through which *ngwao ya rona* is sustained and renewed.

IKS literature also challenges hierarchical distinctions between "knowledge" and "culture," arguing that such separations reflect Western epistemologies rather than indigenous worldviews (Bhuda & Marumo, 2022). By recognising music as knowledge, IKS scholarship legitimises the authority of elders, performers, and cultural practitioners as knowledge holders. This perspective is particularly relevant in contemporary debates about curriculum transformation and the decolonisation of knowledge in African universities.

Decolonial theory provides an important analytical lens for understanding the historical marginalisation of Setswana traditional music. Scholars such as Ngũgĩ wa Thiong'o, Walter Mignolo, and Sabelo Ndlovu-Gatsheni have argued that colonialism functioned not only through political and economic domination but also through epistemic violence (Wa Thiong'o, 1986, Mignolo & Walsh: 2018, Ndlovu Gatsheni, 2018) . Indigenous languages, cultural practices, and knowledge systems were systematically devalued in favour of Western epistemologies.

In the context of southern Africa, colonial and apartheid education systems privileged Western music theory and Christian hymnody, relegating indigenous music to informal or "traditional" spaces (Coplan, 2007). This epistemic hierarchy persists in many contemporary institutions, where African music is often taught as ethnographic content rather than as theory-producing practice. Decolonial scholars argue for epistemic disobedience and the re-centering of indigenous knowledge systems. From this perspective, Setswana traditional music can be understood as a site of cultural resistance and re-existence. Its continued performance affirms indigenous identity and challenges the dominance of Western cultural forms. Decolonial theory thus complements IKS by providing a critical historical and political analysis of why indigenous music has been marginalised and why its recovery is necessary.

African cultural theory further enriches the analysis of Setswana traditional music by foregrounding communalism, relationality, and ethical responsibility. Concepts such as *Botho* (Ubuntu) emphasise interconnectedness, collective identity, and moral accountability (Bhuda & Marumo, 2022). Music, within this framework, is a communal practice that reinforces social cohesion and shared values. Scholars of African philosophy argue that cultural consciousness is cultivated through participation in communal practices rather than individual consumption. Setswana traditional music exemplifies this principle through its emphasis on collective performance, audience participation, and call-and-response structures. Meaning is produced relationally, through shared experience and embodied engagement.

Struggle songs are part of literature on South Africa music. Struggle songs were an important part of the fight against apartheid in the collective memory of the oppressed. Struggle songs played a major part in conscientizing people to fight against the apartheid system. Struggle songs were the backbone of the struggle and most of them come from Indigenous Knowledge systems (Ramoupi: 2025). Ramoupi argues that the liberation struggle was not only fought on a political terrain but other sites such as culture played an important role in making the struggle to be what it is.

The concept of *ngwao yarona* refers to the collective customs, values, beliefs, and practices that shape the identity and social organisation of Setswana-speaking communities. Foundational work by Kgomotso Mogapi defines *ngwao* as an all-encompassing system that includes moral codes, social norms, and cultural practices transmitted across generations. Importantly, *ngwao* is closely linked to *thuto* (education), as it functions as an indigenous system of teaching and learning embedded in everyday life. Cultural practices such as initiation (bogwera and bojale), marriage rituals, and community gatherings serve as important sites of knowledge transmission. In this sense, *ngwao yarona* is not static tradition but a dynamic framework through which identity, continuity, and social cohesion are maintained (Mogapi, 1991).

Scholars of Indigenous Knowledge Systems (IKS) further emphasise that *ngwao* constitutes a holistic body of knowledge that integrates social, political, economic, and spiritual dimensions of life. Oral traditions—including folktales, proverbs, and praise poetry—play a central role in preserving and transmitting this knowledge, reinforcing values and guiding behaviour within the community. As noted in broader African scholarship, knowledge in indigenous societies is relational and communal, rather than individualistic, and is embedded in lived experience. Institutions such as the *kgotla* exemplify how governance, education, and cultural expression intersect within *ngwao*, illustrating its continued relevance as both a cultural and political resource in contemporary African societies (Mapara, 2009; Ngulube, 2002; Dei, 2000).

In the context of colonialism and its aftermath, *ngwao yarona* has been both marginalised and revitalised. Colonial and missionary interventions often dismissed African cultural systems as inferior, contributing to the erosion of indigenous practices and knowledge. However, postcolonial scholarship highlights a renewed emphasis on reclaiming *ngwao* as a vital component of cultural identity and heritage education. This revival underscores the importance of *ngwao* in addressing issues of identity, belonging, and epistemic justice in African societies. As contemporary research suggests, reconnecting with indigenous knowledge systems is essential not only for cultural preservation but also for developing more inclusive and contextually relevant educational frameworks in Africa (Smith, 1999; Hountondji, 1997; Higgs, 2012).

This literature highlights the role of performance spaces—rituals, ceremonies, initiation schools, and community gatherings—as sites of cultural education. Music functions as a living text through which communities narrate their histories, affirm their identities, and negotiate change. African cultural consciousness theory thus situates *ngwao ya rona* as an active, dynamic process rather than a static inheritance. Despite extensive scholarship on African music, Indigenous Knowledge Systems, and decoloniality, limited research integrates these perspectives specifically in relation to Setswana traditional music. Much existing work either documents musical forms without engaging their epistemological significance or theorises IKS without detailed engagement with musical practice. There is a particular gap in studies that conceptualise Setswana traditional music as a coherent indigenous knowledge system that actively produces cultural consciousness. Furthermore, few studies centre *ngwao ya rona* as an analytical concept rooted in indigenous language and worldview. This study addresses these gaps by foregrounding Setswana traditional music as a living epistemology that sustains indigenous cultural consciousness in contemporary society.

THEMATIC ANALYSIS OF SETSWANA TRADITIONAL MUSIC AND CULTURAL CONSCIOUSNESS

Theme 1: Lerato (Love)

In Setswana culture, lerato (love) is not merely an individual emotion but a socially embedded value that shapes relationships, moral conduct, and communal harmony (Schapera: 1965). Lerato songs form an important category within Setswana traditional music, articulating romantic desire, courtship, marriage, longing, fidelity, and emotional pain. These songs operate as cultural texts through which communities' express acceptable forms of intimacy, negotiate gender relations, and transmit social expectations surrounding love and partnership. This analysis examines lerato songs as cultural expressions and Indigenous Knowledge Systems that encode emotional, ethical, and social knowledge. Within an Indigenous Knowledge Systems (IKS) framework, lerato songs function as repositories of experiential knowledge about relationships. They transmit lessons about patience, commitment, respect, and emotional restraint, often through metaphor, symbolism, and poetic language. Rather than offering abstract instruction, lerato songs teach through narrative and performance, allowing listeners to learn relational ethics by engaging emotionally with the song. The oral and performative nature of these songs ensures that knowledge about love is passed intergenerationally, particularly among young people navigating courtship and marriage. Elders and experienced performers play a key role in shaping lyrical content, ensuring that expressions of love align with communal values rather than purely individual desire.

The song *Nchandinyana* by culture spears is an example of how the Setswana cultural expression *pelo e ja serati* (the heart love whom it loves) plays itself out. The song is about a young man who is in love with a woman called Nchadinyana. The song starts by describing Nchadinyana as a woman who does not fit the gendered notion of womanhood in traditional societies. The lyrics illustrate the dichotomy between love and approval from families:

Nchadinyana ga se mosadi (Ke batla nchadinyana) Nchadinyana is not a woman (I want her)
Hoo a se mosadi (Ke batla nchadinyana) She is not a woman (I want her)
Ga nka mo fitlhela mo lapeng lame If I find her in my house
Nka mo tshuba ka metsi I will burn her with water
Ga nka mo fitlhela mo ntlong ya If I find her in the house
Ngwanake nka mo bolaisa diatla of my son I will beat her up (Culture Spears: 2006)

The lyrics speak to the tension between finding love and your family members approving of that love. In the olden days in African communities, men did not court women as parents were the ones responsible for the Magadi (bride price) and eventual marriage ceremony (Schapera, 1965). With the advent of colonialism, the spirit of communalism started being eroded and many families stopped interfering in their children's love life.

Didimala Moratiwa is a Setswana traditional love song by Culture spears. The song evokes love and marriage, especially one where material things are not in abundance. The song literally says keep quiet my love, our love will get us through the adversity. The lyrics of the song speak to a couple defeating adversity and getting married:

Thula babe(dont cry babe)
Chorus Sweetie ke bone sebaka sa go bua le wena.(sweetie i got the chance to speak to you)
Ka tsatsi la lenyalo(On a wedding day)
Ga kena gauta le selefera(i have no silver nor Gold)

Ema re tsamaye ka bitso la Jeso(let us walk through this journey of marriage in the name of Jesus)
O sethunya saka(You are my Flower)
Naledi yaaka(My Star) Mma lebopo waaka (My Miss Universe)
Pelo molotsana boela mannong,
Mmata saagwe o se bone. (Proverb meaning be at easy, may your heart be merry, you got what you wanted. (Culture Spears, 2019).

The song speaks to the adversity couples face today in getting married. This is due to the fact that magadi (bride price) and lenyalo (marriage) no longer follow the indigenous knowledge way of life but have become modernised due to colonialism (Schapera, 1965). Magadi was initially created as a way to bring families together and pay tribute to the family of the women, it was not an exercise in showing who can pay the highest price for a woman. The man expresses his love for his woman and sees her as a beautiful flower. The proverb at the end cautions the heart to be content with the person who has been chosen as the wife. This is a very evocative love song. The song is about a man fulfilling a romantic promise to his wife as he says keep quite my love.

Tshinada is a song by Master KG featuring Maxy Khoisan and Makhadzi. The song is about a woman being taken by a *Senatla* (a brave man). The song is about a couple understanding that love is about two people and not inviting infidelity. The song evokes a lot of Setswana indigenous knowledge that couples get when being given rules by their families on the wedding day (go laiwa). The song speaks to a couple about to married to make sure their love is protected:

Makoti we selo sena lerato ke lababedi fela Bride love is for two people
Wa boraro motho ke masenyeletsa the third person destroys
khwenyana selo sena lerato ke lababedi fela Groom love is for two people
Omo tshabe motho leoto la boraro be careful of a third wheel
Omo tshabe motho ke modubamotshelo be careful of a pot stirrer (Master KG et al: 2020).

The song speaks to the fact that infidelity is on the rise and couples need to be warned about the sanctity of marriage by respecting your home. Divorce is on the rise in many communities due to the erosion of cultural values and the adoption of colonial cultural values that undermine indigenous knowledge.

The song *lerato ke lona leo* (here is love) is a song by culture spears that focuses on love. The song laments the way love has been commodified in the current age and how people are no longer in relationships for love but for their own reasons (Schapera, 1965). The lyrics lament the way love has lost its core values:

Ke tla nyala ngwana ga mang? Whose child will I Marry?
Ke tla tshola ngwana le mang? Who will I have a baby with?
Tota lerato la matsatsi a The love of today
Leng ke tla go nyala I will marry you
Bogologolo ne go tshelwa monate In the past life was nice
Lerato le ne le tlotlwa le na le Seriti love was respected
Re ne re tlotla malapa a rona we respected our families
Re ne re tshela ka Boitumelo go sena dintwa we lived in peace, no fights
Lerato ne le tlotlwa le na le Seriti love was respected (Culture Spears: 2006).

The song speaks to the ways people are breaking up and killing each other. It also contrasts the love of today with the love of yesterday. Colonialism changed the way Africans viewed love from a communal perspective to a commodified love as exemplified by days like Valentine's day. Today Batswana communities, like most African communities also face passion killings where a lover kills their partner if the partner threatens to break up with them (Modongo, 2021).

Theme 2: Bogadi (Weddings And Marriage)

In Setswana culture, *bogadi*—the institution of marriage—is not only a private arrangement but a community-embedded social covenant. Traditional music plays a central role in wedding ceremonies, functioning as both celebration and pedagogy (Schapera & Comaroff, 2015). Songs associated with bogadi encode cultural values, moral expectations, and relational norms, teaching participants about courtship, fidelity, social responsibility, and communal identity. Music at weddings is an example of how Setswana Indigenous Knowledge Systems (IKS) use performance to transmit social knowledge across generations.

The first song chosen is *Magadigadi (*Bride price*)* from Mma Ausi and Culture Spears. The song is about paying bride price by a man for a wife. The song says:

Re ntshitse magadi we paid dowry
Re ntshitse dikgomo tsa rona we took out our cows
Ke kgomo tsa magadi the cows of bridal price (Mma Ausi & Kulenyane: 2022)

The song speaks to the centrality of cattle when it comes to a marriage in Setswana culture. This song celebrates the enduring nature of marriage and its cultural customs among Batswana. Magadi (bride price) has always been the first part of a Setswana traditional wedding. This song highlights the resilience of African culture and indigenous knowledge systems pertaining to marriage in the modern era where the coloniality of power often erodes the cultural values of African communities.

Jeremane is a song by culture spears that focuses on a woman being chosen by the king of a village. In the olden days in Setswana communities, kings would create ceremonial events where they would choose a wife from different girls in the village (Schapera & Comaroff, 2015). This song highlights how love can conquer and overcome challenges of suffering. The story can be compared to that of Cinderella who was chosen by the prince despite the challenges they had to overcome. The lyrics focus on a woman's struggle in getting Jeremane (Setswana wedding cloth). The song says:

Moletlo o ke o laleditsweng I was invited to this ceremony
Ke wa ngwana wa Kgosi which is for the son of a king
Ngwana we kgosi wa maitibolo the last born son of the king
O batlelwa mosadi wee a wife is being sought for him
Basadi weee ka re nkadimang Jeremane wee (Jeremane! Jeremane!) women borrow me a wedding shawl (Culture Spears, 2006).

The lyrics highlight the importance of the garment Jeremane. Jeremane as a wedding cloth has its origins in German influence. This shows how Setswana culture was also influenced by missionaries and colonialism through cloth and fabric (Comaroff, 2002). By wearing Jeremane, the woman is able to fit in and be eligible to be chosen by the prince. The song highlights the Setswana proverb of ditaba dimatlhong (events happen through the eyes) as the prince expected to meet these women face to face and choose the one, he was going to marry.

Setimela (train) is a popular Setswana marriage song by Mma Ausi. The song urges women to climb a mountain and exclaim that they are married. The song is a celebration of Setswana marriage culture and the pride of place a woman is given in Setswana culture. The lyrics go:

Mosadi palama thaba o kuwe o re Woman climb a mountain and scream
Lenna ke nyetse I am married!

Rre palama thaba o kuwe o re Man climb a mountain and shout

Le nna ke nyetse I am married!

Ba bo ntsheng terene Show them the train

Go re le dirile I do I do That you did the I do do (Mma Ausi, 2022)

The song compares marriage to a train as indicated by the title. Setimela urges us to understand that Setswana traditional marriage is a communal affair. The usage of the train is also an allusion to the fact that men used trains to go to urban areas and leave their women at home due to the Migrant labour system (Coplan, 2007). The song shows how marriage values are still kept up in traditional Setswana communities.

Theme 3: Death (Loso)

In Setswana culture, death (*loso*) is not merely a biological event but a communal, spiritual, and ritual process. Music is central to death-related practices, functioning as a vehicle for mourning, remembrance, social instruction, and spiritual mediation. Setswana death music is embedded in Indigenous Knowledge Systems (IKS), transmitting cultural norms about grief, ancestor veneration, and social responsibility. Unlike Western traditions that often separate music from ritual, Setswana death music integrates song, rhythm, dance, and orality to shape communal experience of loss (Bammann, 2016). Through these performances, Setswana communities reaffirm cultural identity and continuity, demonstrating that music is an essential component of cultural resilience and Indigenous Knowledge Systems surrounding life and death.

The song *Loso la Monnamogolo* (the death of the old man) is a song by Shumba Ratshega, one of the most talented Setswana musical artists in this generation. The song is about a man who is sleeping with someone's wife and is grateful for his death as that means he will be free to commit adultery without consequences. The song alludes to how death has become a different context for some people. The lyrics of the song go like this:

Leso la monnamogolo le mpereketse the old man's death worked for me

Le mpolokile It saved me

Le mpereketse It worked for me

Ka kgwetsa nkuku gareng ga masigo I made love to Nkuku in the night (Ratshega: 2016).

The song shows how life and death are often connected in intimate levels. This song speaks to how Batswana communities today are facing societal challenges due to death. People engage in passion killings, killing partners due to infidelity and killing them to cash in life policies. Songs such as Loso la Monnamogolo, if listened to carefully, warn against modern day killings and death.

Mmamarutle is a song that was made by Dikakapa in honour of Kgosi Linchwe II of the Bakgatla ba Kgafela Nation based in Mochudi Botswana (Makgala, 2009). The song honours the death of this great king and pays homage to him and his leadership. The lyrics of the song say:

Re swetse ke Kgosi Linchwe Kgosi Linchwe II has passed away

Re swetse ke morwa Kgahela Kgahela's son has passed away

Gomotsegang tlhe Bakgatla be comforted Bakgatla

Robala morwa Lentswe Sleep son of Linchwe (Dikakapa, 2008).

The song is a mourning song meant to honour Kgosi Linchwe II and help Bakgatla ba Kgafela through their grief. The song also includes audio of Kgosikgolo Kgafela Kgafela II being told that he is the successor to Kgosi Linchwe II. Kgosi Linchwe II was a Kgosi from 1963 until 2007 of the Bakgatla ba Kgafela in Botswana and South Africa (Makgala, 2009). The song acknowledges his importance and shows how music is part of archival indigenous knowledge of mourning and grieving a public figure.

Phala (the whistle) is a song by culture spears that focuses on death and love. The song is set in a sombre tone acknowledging the deathly context that the song is set in. The lyrics say:

Ontshiya le mang ha o tsamaya Who are you leaving me with when you die
Ke utlwile Phala e lela I heard the whistle
E latola morati wame confirming the death of my partner
Nka swa nka kepela I can die and bury myself
Ke latela morati wame following my lover
Ke a tsamaya kea mo latela I am going following him to death
Kea legodimo kea mo latela I am going to heaven following him (Culture Spears, 2006).

The song Phala speaks about a woman whose husband has just passed on and her wish to join him in heaven. The song highlights the ways in which indigenous knowledge is used to show the importance of the Phala (whistle horn) to announce a death in a Setswana village (Bammann, 2016). The woman speaks about dying and joining her husband in heaven showing the link between lovers in the current life and the afterlife. Death in Setswana communities is a life changing event for married women as they face challenges of conflict with the husband's family and having to raise their children alone.

Theme 4: Bogosi (Chieftancy)

In Setswana culture, *bogosi* refers to the institution of chiefdom and traditional leadership, encompassing political, social, and spiritual authority. Traditional music plays a central role in rituals, ceremonies, and community events associated with bogosi. Songs and performance practices are used to praise chiefs, educate communities about leadership values, and reinforce social hierarchies and cultural norms (Molato & Dube, 2025). Music is thus both expressive and pedagogical, serving as an Indigenous Knowledge System (IKS) through which leadership principles, moral conduct, and communal identity are transmitted. Through these practices, Setswana traditional music functions as a living Indigenous Knowledge System, integrating leadership, history, ethics, and communal values. Music ensures that bogosi is experienced as both a political and cultural institution, not merely a formal office.

Kgori Mpele Bana is a song by Maletangwao Cultural Troupe. The song focuses on alerting Kgosi Montshiwa that Maletangwao have entered the village. The song is a deeply cultural song that evokes the chieftancy of the Barolong people of Mahikeng. The lyrics go:

O nkele Kgotleng Please go to the court for me
O fitlhe o re Kgosi Montshiwa dumela greet Kgosi Montshiwa for me
Maletangwao a tsene ka motse Maletangwao have entered the village
(Maletangwao, 2022).

The song speaks to how Maletangwao respect their king and honour him in their musicology as Barolong Boora Tshidi (Gape, 2023). In Setswana indigenous knowledge epistemology, when a person enters a royal village, the first person they go and greet is the Kgosi who welcomes them to their land. The song shows how respected the Barolong Kgosi Montshiwa is respected by those under his rule. The mention of the Kgotla (court) shows the importance of this space in the cultural knowledge of Setswana music.

Dikgafela is a song by Shumba Ratshega, a highly respected Setswana musician. The song focuses on the importantance of preserving Setswana cultural traditions through Dikgafela (tribute). Dikgafela is an old age tradition where communities came together to honour dikgosi by giving them food from their agricultural harvest (Schapera & Comaroff, 2015). The lyrics go:

Tlotlang Kgosi tsa Botswana respect chiefs of Botswana
Tlotlang Kgosi tsa lehatshe banna respect chiefs of the land
Re a bidiwa ko Kgotleng we are being called at the royal court
Kgosi wa lebitsa Batswana the chief is calling us Batswana
Kgafela tsa ngwaga wa pula tributes for the year of rain
Ke sedile ngwao ya Batswana I picked up the culture of Batswana
Ke sedile le latlhile I picked it up while it was thrown away
Seepapitso Bangwaketse kwena kgari Sechele names of Batswana chiefs
(Ratshega, 2010).

In this song Ratshega honours the age-old tradition of paying tribute to chiefs by giving them a share of the harvest made by the community as a sign of respect and appreciation for their leadership (Bammann, 2016). Ratshega includes different magosi in the song such as Kgosi Seepapitso of the Bangwaketse in Kanye, Kgosi Kgari Sechele III of the Bakwena ba Sechele, Kgosi Mmirwa Malema of the Babirwa and Kgosi Kgafela Kgafela of the Bakgatla. The song centres Setswana court life as the invocation is made that that the Kgosi is calling the community.

Magosi (chiefs) is a song by the late Mr Laaities of Ngwao ya Tshona from Kuruman in the Northern Cape. The song is about the anti-colonial struggle that dikgosi (chiefs) in the North West and Northern Cape fought against both Dutch and British rule (Shillington, 1981). The song focuses mainly on the 1897 Langeberg rebellion which saw the Batlharo, Batlhaping and Barolong fight against the British at Langeberg. The lyrics go:

Ka ngwaga wa 1885 lehatse in the year 1885 the land
La Batlhaping, Batlharo le Barolong of the Batlhaping, Batlharo and Barolong
La tsena mo mushong wa mohumagadi came under the rule of Queen Victoria
La bidiwa British Bechuanaland It was called British Bechuanaland
Magosi a Tshona The chiefs of our people
Kgosi toto o la tshwara a iswa Robben Island Kgosi Toto was sent to Robben Island
Kgosi Phetlhu a tshwara a kgaolwa tlhogo Kgosi Phetlhu's head was cut off
Mafatshe a Batlharo le Batlhaping the land of Batlhaping and Batlharo
a itselwa Ke makgowa ka 1897 were taken by the whites in 1897
(Ngwao ya Tshona, 2021).

The song speaks to how Magosi (chiefs) fought a struggle to protect their people and their land and paid a heavy price for their bravery. Kgosi Jantjie of the Batlhaping and Kgosi Phetlhu of the Batlhaping had their heads cut off, while Kgosi Toto and Kgosi Galeshewe were imprisoned on Robben Island (Shillington, 1981). The song shows how Setswana traditional music can be used to preserve history and tell stories of communities whose history has been marginalised and not archived as indigenous knowledge.

Dithubaruba (a cultural festival) is a song by Mmausi Lekoma. The song is an ode to the Battle of Dithubaruba led by Kgosi Sechele in 1852 (Ramsay, 2021). Bakwena people of Molepolole in Botswana host an annual festival called Dithubaruba in September to honour the heroes who saved Botswana from the Voortrekkers who wanted to annex Botswana into South Africa in the 1850s. The lyrics of the song go:

Dithubaruba wee!
Di bonwa ka ko ntsweng They are seen in Ntsweng
Ka ko ga Mmakgosi At the place of the Queen Mother

Re epa didiba tse di thata we are digging big wells
Go iwa dithubarubeng Bakwena we are going to Dithubaruba Bakwena
Kgari wa boraro ema ka dinao King Kgari please stand up
Rwala tlhako sa borrago mogolo wear the shoes of your grandfathers
Bakwena tla le boneng Molepolole Bakwena come see Molepolole
Petleke wa lebitsa The king is calling you
Lefoko la Kgosi le agelwa mosako The word of the king is respected
(Mmausi: 2022).

The song honours Dithubaruba as a living heritage of the Bakwena ba Sechele and encourages them to safeguard and support this cultural initiative that is comparable to other heritage commemorations like the battle of Isandlwana for the AmaZulu kingdom (Ramsay, 2021). Kgosi (chief) Kgari Sechele is encouraged to safeguard Dithubaruba as a heritage of his people and forebears who came before him.

Dumela Mmabatho (Greetings my chief) is a song by Mma laloo Morwadiakgosi. This song praises the leader of the Bakwena ba ga Sechele Kgari Sechele III (Ramsay, 2021). Mmabatho is a Setswana word meaning Kgosi. The term Mmabatho is often translated as mother of the people, the Kgosi is considered the mother of his people. The lyrics go:

Dumela Mmabatho Greeting Chief
Kgosi kgari wa boraro Mmabatho Chief Kgari III
Ke wele ka lengole My knee is bent
Dumelang Bakwena botlhe Greeting to all Bakwena
Mokwena otlhe dumelang All of Mokwena greetings
Petleke o rebiditse The chief is calling us
(Mmalaloo: 2024).

The song greets Kgosi Kgari Sechele III. This song pays homage to chieftancy among the Bakwena people (Ramsay, 2021). It is also an exhortation for people to heed the call regarding the Dithubaruba cultural festival. In Setswana culture the term Mmabatho is similar to the proverb Kgosi Thotobolo e olela matlakala (The Kgosi is responsible for everyone).

Theme 5: Ngwao (Culture Songs)

Setswana cultural songs (*dipina tsa Setswana*) are more than musical entertainment; they are vehicles of social education, moral instruction, and cultural continuity. Embedded within Indigenous Knowledge Systems (IKS), these songs transmit communal values, historical memory, spiritual beliefs, and ethical norms across generations (Bammann, 2016). They are performed during life-cycle rituals, initiation schools, weddings, funerals, harvest celebrations, and political gatherings, reflecting the holistic integration of music in daily and ceremonial life (Molefe, 2019). Through these songs, Setswana communities maintain cultural consciousness, relational ethics, and continuity, demonstrating that music is a living, functional, and educational practice.

Setswana is a song by Mma Ausi of Mafikeng. The song is one of the best songs to document and capture the culture of Batswana people. The song encourages Batswana to hang on to their culture in the face of a rampant modernity. The lyrics go:

Setswana wee Setswana praises
moretlwa o lele wild raisin is crying
Thetang ka Setswana banyana dance rhythmically with Setswana girls

Kgwathisa banna le basadi Setswana punish men and women
Moropa o lele Setswana let the drum cry Setswana
(Mmaausi, 2021)

Moretlwa is a wild raisin eaten and fermented by Batswana communities. Setswana is given qualities of getting both men and women into line and respecting the heritage and culture of Batswana. Setswana culture is personified into a drum. Drums play an important part in Setswana cultural heritage and indigenous knowledge (Setlhabi, 2014).

Nkwe is a song by Maletangwao Cultural Group. The song focuses on the leopard as a symbol of Batswana royal heritage and culture. The leopard, unlike the lion is preferred as a symbol of royalty and kingship in many African communities (Schapera & Comaroff, 2015). The lyrics go:

Ke nkwe tilodi ya magaga the black and white spotted leopard of the caves
Seapara ke makgosi wee worn by chiefs
Kana ke raya ya dikgaga I am talking of the wild one
Kana ke kobo ya marena the blanket of royalty
Gaena Botshabelo It has no hiding place
(Maletangwao, 2022)

The leopard prefers to live in caves. Many Dikgosi (Kings) hunt for a leopard and wear the skin on their coronation day. The leopard and leopard skin are revered and respected in Setswana culture (Bammann, 2016). The Bophuthatswana homeland flag was adorned with a leopard as a symbol of unity in the homeland. In the Setswana cultural identity there is no known community who have the leopard as their totem.

Lerole is a song by Mma Ausi. The song focuses on Setswana cultural heritage and is considered a war cry. The song centres lerole which means dust as a major part of Setswana culture and initiation. The lyrics go:

Lerole le le la bo Mallelee The dust of crying mothers
Lerole lele ke la setso the dust of traditional
Ba tsile go nwa metsi they came to drink water
Ba bolotsa tlhabano they are initiating the weapons
Kgala kgala dibini dietla dry up the singers are coming
Kgalagadi wee Kgalagadi region praise on you
(Mmaausi, 2022)

The song acknowledges the role of dust in Setswana military culture. Lerole symbolises the marching by Mephato (military regiments) and how they would go about performing their songs (Ramsay, 2021). Batswana totemic groups who were known for military prowess include Bakgatla under Pilane, Bangwato under Khama III and Bangwaketse of Makaba, Bakwena under Sechele and Barolong under Montshiwa just to name a few groups.

Makgona Ngwao (Culture defeaters) is a Setswana traditional music group from Namibia. Namibia has a Batswana population who settled there during precolonial times in the 1800s. The Batswana people in Botswana are called Batlhaping and have kept their own distinct culture alive (Shillington: 1981). The lyrics of the song:

Ke makgona ngwao these are cultural bearers
Banna ba ngwao re ipela ka ngwao the children of culture we pride ourselves in culture
Rona re ikgantsha ka ngwao yeh we are proud of this culture
Ba re Botswana rea tlhaga from Botswana we came

Aforika Borwa rea tlhaga South Africa we came
Lesotho rea tlhaga Lesotho we are from there
Namibia ka Bophara Namibia as well
(Makgona Ngwao: 2015)

The song focuses on cultural preservation by the Batswana people of Namibia where the Setswana language and culture is a minority culture. The song urges Batswana from the three different countries to be proud of their heritage and culture.

CONSOLIDATION OF THEMES AND ARGUMENT

The themes for this chapter have focused on how traditional music transmit cultural consciousness. Ngwao Ya Rona and Setswana traditional music function as cultural archives that preserve identity, transmit moral values, strengthen communal unity, celebrate social life, and maintain spiritual and historical continuity within Batswana society. Through themes such as love, marriage, death, chieftancy and culture, the chapter emphasises how Setswana traditional music uses songs to educate, share the culture and show how cultural processes function. The themes support the statement that in African culture, everything starts with a song. Setswana traditional music is not merely a form of artistic expression but a multifunctional cultural institution that embodies and sustains the social, moral, and spiritual fabric of Batswana society. Rooted in ngwao (culture), it serves as a repository of indigenous knowledge, preserving language, history, and collective memory across generations through oral transmission (Coplan: 2008). Primarily, Setswana traditional music functions as a tool for cultural preservation and identity formation. Through songs, rhythms, and performance practices, communities reinforce a shared sense of belonging and continuity, especially in the face of modernization and cultural change. This positions traditional music as a form of cultural resistance, safeguarding indigenous values and practices.

In addition, Setswana traditional music plays a didactic role, communicating moral values such as respect, discipline, and social responsibility. Songs are often embedded with teachings that regulate behavior and guide individuals within the community, thereby acting as an informal system of education. Furthermore, the music promotes social cohesion and communal participation. It is performed collectively during ceremonies, rituals, and daily activities, strengthening interpersonal relationships and reinforcing communal unity (Molefe: 2019). The participatory nature of call-and-response patterns reflects the collective ethos of Batswana society. Spiritually, Setswana traditional music acts as a medium of connection between the physical and ancestral worlds. It is integral to rituals that honor ancestors and express beliefs about life, death, and continuity, thus maintaining the spiritual worldview of the community. Finally, Setswana traditional music reflects everyday lived experiences, including agriculture, gender roles, and social structures. It documents and interprets the realities of community life, making it both a cultural record and a means of expression.

FUTURE RESEARCH DIRECTIONS

Future research will have to look deeper into Setswana traditional music and how it relates to the culture of Batswana in South Africa, Namibia, Zimbabwe, and Botswana. This music is connected to the history and indigenous knowledge of Batswana and needs to be analysed as such. Future research will also have to also compare Setswana music in its different cultural categories. There is music for children, youth, adults, and elders depending on the cultural category. Future research will also have to look at how Setswana Traditional Music is appreciated by different Setswana language dialects such as Sepetori in Pretoria and Qwathi in Kimberley. Setswana Traditional Music is also linked to the history, culture and community of Batswana such as National Anthems of Botswana and the former Bophuthatswana.

CHAPTER CONCLUSION: NGWAO YA RONA AND INDIGENOUS KNOWLEDGE SYSTEMS

This chapter has explored *Ngwao ya Rona*—Setswana traditional music—through the lens of Indigenous Knowledge Systems (IKS), highlighting its multifaceted role in transmitting cultural values, ethical norms, and communal identity. The analysis of initiation, love (lerato), bogadi (weddings and marriage), death (loso) and bogosi (chiefship and leadership) demonstrates that Setswana musical practices are not merely artistic expressions but dynamic instruments for social, moral, and spiritual education.

Initiation songs (*bogwera* for boys and *bojale* for girls) exemplify the pedagogical and ritual function of music in preparing youth for adulthood. Through structured call-and-response singing, drumming, and metaphorical lyrics, initiates are taught courage, patience, respect for elders, and gendered social responsibilities, illustrating how music encodes complex Indigenous Knowledge (Coplan, 1985; Vansina, 1985). Similarly, *lerato* songs transmit relational and emotional wisdom, teaching patience, fidelity, and ethical conduct in romantic and communal contexts (Agawu, 2003; Mapaya, 2014). These songs provide both emotional expression and moral guidance, demonstrating the intricate intertwining of affective and cognitive learning in Setswana musical culture.

Wedding and marriage songs (*bogadi*) extend this pedagogical role, emphasizing not only the celebration of love but also communal cohesion, lineage recognition, and ethical obligations within matrimonial unions (Coplan, 2008; Ndlela, 2009). The music performed during bride fetching, ceremonies, and post-wedding celebrations reinforces social norms, gendered expectations, and ancestral blessings, ensuring that the union is integrated into broader societal and spiritual structures. Death-related music (*loso*) similarly functions as both emotional expression and social pedagogy, facilitating mourning, commemorating ancestors, and reinforcing communal solidarity (Erlmann, 1996; Ogunyemi, 2015). Music at funerals embodies collective memory, spiritual cosmology, and moral reflection, highlighting the enduring role of song in negotiating life, death, and communal ethics.

Bogosi songs celebrating chiefs and leadership illustrate the political and historical dimensions of Setswana music. Praise songs and ritual performances reinforce hierarchical structures, transmit leadership values, and connect the living community to ancestral guidance (Herbst et al., 2003; Agawu, 2016). These performances ensure that music continues to mediate social order, historical memory, and spiritual legitimacy, bridging the individual and the collective in culturally sanctioned ways.

Across all thematic areas, several key characteristics emerge: the use of call-and-response structures, metaphoric language, participatory performance, and integration of spiritual and ethical knowledge. Collectively, these features underscore the role of Setswana traditional music as a living Indigenous Knowledge System, transmitting complex social, moral, and emotional lessons while fostering communal identity and continuity (Msila, 2017; Molefe, 2019). Contemporary adaptations—through urban music, digital platforms, and cross-cultural fusion—demonstrate the resilience and adaptability of *Ngwao ya Rona*, ensuring the ongoing relevance of cultural knowledge in modern African societies.

In conclusion, the study of Setswana traditional music affirms that *Ngwao ya Rona* is both an artistic and educational system, where sound, rhythm, and lyrics function as conduits for moral instruction, social cohesion, and spiritual engagement. Love, marriage, death and leadership songs collectively reveal the depth of Indigenous Knowledge embedded in performance, illustrating how music sustains cultural memory, teaches ethical conduct, and strengthens community across generations. This chapter underscores the importance of recognizing Setswana musical traditions as dynamic repositories of Indigenous Knowledge, capable of informing contemporary social, ethical, and educational discourses (UNESCO, 2003; Scherzinger, 2019).

REFERENCES

Agawu, K. (2014). *Representing African music: Postcolonial notes, queries, positions*. Routledge. DOI: 10.4324/9781315810942

Agawu, K. (2016). *African Rhythm: A Northern Perspective*. Cambridge University Press.

Ausi, M. (2022). Lerole. Tshimo ea Tlholwa Album. URL: https://www.youtube.com/watch?v=P6ATHdMJS10&list=RDP6ATHdMJS10&start_radio=1

Ausi, M. (2022). Setimela. Tshimo ea Tlholwa Album. URL: https://www.youtube.com/watch?v=feb7Xk9xm-A&list=OLAK5uy_ncrKEvT2Kvj-keE2Hcl8GAyG2MLzFQ4vQ&index=2

Bebey, F. (1999). *African music: A people's art*. Lawrence Hill Books.

Bhuda, M. T., & Marumo, P. (2022). Ubuntu Philosophy and African Indigenous Knowledge Systems: Insights from decolonization and indigenization of research. *Gender & Behaviour*, *20*(1), 19133–19151.

Blacking, J. (1973). *How musical is man?* University of Washington Press.

Comaroff, J. (2002). The Empire's old clothes: Fashioning the colonial subject. In *Cross-cultural consumption* (pp. 29–48). Routledge.

Coplan, D. B. (2007). *In Township Tonight! South Africa's Black City Music and Theatre*. Ravan Press.

Coplan, D. B. (2008). *Zulu Praise Poetry and Social Identity*. Wits University Press.

Dei, G. J. S. (2000). Rethinking the role of indigenous knowledges in the academy. *International Journal of Inclusive Education*, *4*(2), 111–132. DOI: 10.1080/136031100284849

Dikakapa. (2008). *Mmamarutle*. URL: https://www.youtube.com/watch?v=neFnRNd-P0c&list=RDneFnRNd-P0c&start_radio=1

Erlmann, V. (1996). The Aesthetics of the Global Imagination: Reflections on World Music in the 1990s. *Public Culture*, *8*(1), 1–19. DOI: 10.1215/08992363-8-3-467

Gape, H. (2023). *A Way of Life: Barolong Boora Tshidi. Mmabatho*. Bokone Bophirima Publishing House.

Herbst, J., Mills, G., & Nicolson, G. (2003). *Botswana: A Political History*. Macmillan Botswana.

Higgs, P. (2012). African philosophy and the decolonisation of education in Africa. *Educational Philosophy and Theory*, *44*(sup2, s2), 37–55. DOI: 10.1111/j.1469-5812.2011.00794.x

Hoppers, C. O. (2025). Research on Indigenous knowledge systems: The search for cognitive justice. In *Indigenous Voices and Decolonising Lifelong Education* (pp. 7–24). Routledge. DOI: 10.4324/9781032627465-2

Hountondji, P. J. (1997). *Endogenous knowledge: Research trails*. CODESRIA.

Johnston, T. F. (1973). Aspects of Tswana Music. *Anthropos*, (H. 5./6), 889-896.

Khan, K. B. (2025). Driving the Indigenous Knowledge System Quest Through Music and Dance: Promoting Zimbabwe's Heritage Through Indigenous Songs and Dance. In *Harnessing Indigenous Epistemologies for Sustainable Progress* (pp. 253-272). IGI Global Scientific Publishing.

Kipyegon, K. A., Wanyama, M. N., & Mapaya, M. G. (2025). Epistemics of African Music Performance: Navigating Knowledge, Tradition, and the Digital Frontier. *African Musicology Online*, *14*(2), 14–24. DOI: 10.58721/amo.v14i2.1107

Kwabena-Nketia. (1974). *The Music of Africa.* W W Norton & Company.

Magubane, B. (1996). *The making of a racist state: British imperialism and the Union of South Africa, 1875-1910.* Africa World Press.

Magubane, B. (1996). *The Making of a Racist State: British Imperialism and the Union of South Africa, 1875-1910.* Africa World Press.

Makgala, J. (2009). *History of the Bakgatla baga Kgafela in Botswana and South Africa.* Crink Publishers.

Maletangwao Cultural Troupe. (2022). Nkwe. Thipa fa e se bogale ya lwediwa album. URL: https://www.youtube.com/watch?v=-YFDFLlZaE8&list=RD-YFDFLlZaE8&start_radio=1

Maletangwao Cultural Troupe. 2022. *Kgori Mpele Bana.* Thipa fa e se bogale ya lwediwa album. URL: https://www.youtube.com/watch?v=hOfzRhJ6w_w&list=RDhOfzRhJ6w_w&start_radio=1

Mapara, J. (2009). Indigenous knowledge systems in Zimbabwe: Juxtaposing postcolonial theory. *The Journal of Pan African Studies*, *3*(1), 139–155.

Mapaya, I. (2014). *Indigenous Music and Social Pedagogy in Southern Africa.* African Sun Media.

Master, K. G. Maxy Khoisan, Makhadzi. (2020). Tshinada. URL: https://www.youtube.com/watch?v=942XC5uJoB4&list=RD942XC5uJoB4&start_radio=1

Melczer, R. S. (2019). ""The Fat Which Is Begged for Does Not Make the Hair Pretty": K.T. Motsete and the Margins of Self-Determination in the Bechuanaland Protectorate". Unpublished Doctoral Dissertation: University of California, Santa Barbara.

Mignolo, W. D., & Walsh, C. E. (2018). *On decoloniality: Concepts, analytics, praxis.* Duke University Press.

Mma Ausi & Kulenyane. (2022). *Magadigadi.* Dibeisane Album. URL: https://www.youtube.com/watch?v=ronO9ta3H8I&list=RDronO9ta3H8I&start_radio=1

MmaAusi. (2021). *Setswana.* Tshimo ea Tlholwa Album. URL: https://www.youtube.com/watch?v=2oXbSIKckfw&list=RD2oXbSIKckfw&start_radio=1

Mmalaloo. (2024). *Dumela Mmabatho.* Dithubaruba Album. URL: https://www.youtube.com/watch?v=gmGi-hiqxX8&list=RDgmGi-hiqxX8&start_radio=1

Mmausi. (2022). *Dithubaruba.* Ke Bone Molelo O Timile Album. URL: https://www.youtube.com/watch?v=Gz-JtquS6sQ&list=RDGz-JtquS6sQ&start_radio=1

Modongo, L. P. (2021). *When love kills: constructions of masculinities within the intimate femicide crisis in Botswana* (Doctoral dissertation: Stellenbosch University)

Mogapi, K. (1991). *Ngwao ya Setswana*. Longman.

Mohlamme, J. (2000). Blacks in the ex-Boer republics of the Transvaal and Orange Free State in the aftermath of the South African War of 1899-1902. *Southern Journal for Contemporary History*, *25*(2), 270–283. DOI: 10.38140/sjch.v25i2.4089

Molato, K., & Dube, M. W. (2025). Music, Gender and Religion in Botswana: Culture Spears Traditional Group. In *Manyonganise, Molly; Gunda, Masiiwa Ragies; Naicker, Molefe, R., 2019. Traditional Music and Cultural Consciousness in Botswana*. University of Botswana Press. DOI: 10.20378/irb-106054

Morapedi, W. (2009). Botswana, protest and nationalism. In *Immanuel Ness. International Encyclopaedia of Protest and Revolution from the 1500 to the present* (pp. 453–463). Wiley Blackwell.

Msila, V. (2017). *The Role of Indigenous Knowledge Systems in African Education*. Van Schaik Publishers.

National Anthems Info. (2021). "Bophutatswana National Anthem". URL: https://www.nationalanthems.info/bpt.htm

Ndlela, N. (2009). *Marriage Songs and Cultural Transmission in Southern Africa*. University of KwaZulu-Natal Press.

Ndlovu-Gatsheni, S. (2018). *Epistemic freedom in Africa: Deprovincialization and decolonization*. Routledge. DOI: 10.4324/9780429492204

Ngulube, P. (2002). Managing and preserving indigenous knowledge in the knowledge management era. *Information Development*, *18*(2), 95–105. DOI: 10.1177/026666602400842486

Ngwao, M. (2015). Makgona Ngwao. URL: https://www.youtube.com/watch?v=Jq-DGs9rfBQ&list=RDJq-DGs9rfBQ&start_radio=1

Ngwao ya Tshona. (2021). *Magosi a Tshona*. URL: https://www.youtube.com/watch?v=ADuaJpFEm08&list=RDADuaJpFEm08&start_radio=1

Ogunyemi, A. (2015). *Music and Mourning in African Societies: Case Studies from Southern Africa*. University of Lagos Press.

Phibion, O. S. (2013). The ditlhaka music of the Balete of Botswana. *Muziki*, *10*(2), 90–102. DOI: 10.1080/18125980.2013.844986

Ramoupi, N. L. L. (2025). *Robben Island Songs of Struggle: Izingoma Zomzabalazo*. Skotaville Publishers.

Ramsay, J. (2021). The Batswana-Boer War of 1852-53: How the Batswana achieved victory. *Botswana Notes and Records*, *23*(1), 193–207.

Ramsay, J., Morton, F., & Morton, B. (1996). *Historical Dictionary of Botswana*. The Scarecrow Press Inc.

Ratshega, S. (2010). *Dikgafela*. Makirikiri Album. URL: https://www.youtube.com/watch?v=874NSGeNKWQ&list=RD874NSGeNKWQ&start_radio=1

Ratshega, S. (2016). *Leso la Monnamogolo.* Leso la Monnamogolo Album. URL: https://www.youtube.com/watch?v=-TrHXMQmE04&list=RD-TrHXMQmE04&start_radio=1

Schapera, I. (1965). *Married life in an African tribe*. Sheridon House.

Schapera, I., & Comaroff, J. L. (2015). *The Tswana*. Routledge. DOI: 10.4324/9781315684093

Scherzinger, M. (2019). *Ritual and Music in Africa: Performance, Politics, and Cultural Memory*. Oxford University Press.

Setiloane, G. (1976). *The image of God among the Sotho-Tswana*. Balkema.

Setlhabi, K. G. (2014). The politics of culture and the transient culture of Bojale: Bakgatla-Baga-Kgafela women's initiation in Botswana. *Journal of Southern African Studies*, *40*(3), 459–477. DOI: 10.1080/03057070.2014.913424

Shillington, K. (1985). *The colonisation of the Southern Tswana, 1870-1900*. Routledge.

Shillington, K. T. J. (1981). *Land Loss, Labour and Dependence: The Impact of Colonialism on the Southern Tswana: c. 1870-1900*. University of London, School of Oriental and African Studies.

Sizemore-Barber, A. (2012). The Voice of (Which?) Africa: Miriam Makeba in America. *Safundi*, *13*(3-4), 251–276. DOI: 10.1080/17533171.2012.715416

Smith, L. T. (1999). *Decolonizing methodologies: Research and indigenous peoples*. Zed Books.

Spears, C. (2006). *Jeremane*. Kulenyane Album. URL: https://www.youtube.com/watch?v=iRovbnA5lEs&list=RDiRovbnA5lEs&start_radio=1

Spears, C. (2006). Nchandinyana. Kulenyane Album. Kulenyane Studios. URL: https://www.youtube.com/watch?v=TGQa6EqO1yY&list=RDTGQa6EqO1yY&start_radio=1

Spears, C. (2006). *Phala*. Kulenyane Album. URL: https://www.youtube.com/watch?v=-iwSH6X_dwo&list=RD-iwSH6X_dwo&start_radio=1

Spears, C. (2019). Didimala Moratiwa. URL: https://www.youtube.com/watch?v=vDhvKAU3jBI&list=RDvDhvKAU3jBI&start_radio=1

Spears, C. *Lerato*. Kulenyane Album. URL: https://www.youtube.com/watch?v=taxbic-syuY&list=RDtaxbic-syuY&start_radio=1

UNESCO. (2003). *Convention for the Safeguarding of the Intangible Cultural Heritage*. United Nations Educational, Scientific and Cultural Organization.

Vansina, J. (1985). *Oral Tradition as History*. University of Wisconsin Press. DOI: 10.2307/jj.36106057

Wa Thiongo, N. (1986). *Decolonising the Mind: The Politics of Language in African Literature*. James Currey.

Wood, E. N. (1985). Traditional Music in Botswana. *The Black Perspective in Music*, *13*(1), 13–30. DOI: 10.2307/1214791

ADDITIONAL READING

Agordoh, A. A. (2005). *African music: Traditional and contemporary*. Nova Publishers.

Ditsele, T. (2017). The promotion of Setswana through hip hop and motswakolistas. *Journal of the Musical Arts in Africa, 14*(1-2), 1–14. DOI: 10.2989/18121004.2017.1410990

Impey, A. (2009). *The Serankure and Music in Tlôkweng*.

Levine, L. (2005). *The Drumcafe's traditional music of South Africa*. Jacana Media.

Molale, T., & Mgogo, Q. (2023). More than just edutainment: Indigenous South African community radio music programmes as preservers of African traditional cultures. In *100 Years of Radio in South Africa, Volume 2: Community Radio, Digital Radio and the Future of Radio in South Africa* (pp. 87-109). Cham: Springer Nature Switzerland.

Molato, K., & Dube, M. W. (2025). Music, Gender and Religion in Botswana: Culture Spears Traditional Group. *Religion, Gender and Masculinities in Africa: Essays in Honour of Ezra Chitando*.

Phibion, O. S. (2012). Bangwaketse of Botswana Setapa Traditional Music Practices. *International Review of Social Sciences and Humanities*, *4*(1), 1–11.

Phibion, O. S. (2013). The ditlhaka music of the Balete of Botswana. *Muziki*, *10*(2), 90–102. DOI: 10.1080/18125980.2013.844986

Theledi, K. M. A. (2023). Hip Hop Pantsula making Setswana fashionable: impact on Setswana orthography. *South African Journal of African Languages, 43*(sup1), 334-340.

KEY TERMS AND DEFINITIONS

Call and Response: A musical interaction where a leader sings a phrase and a group responds, encouraging participation.

Communal Performance: A form of music-making that involves group participation, emphasising unity and shared identity.

Cultural Function of Music: The role of music in education, rituals, entertainment, and promoting social cohesion within the community.

Dikgafela: A musical and dance celebration associated with the harvest season, expressing gratitude and unity.

Indigenous Instruments: Traditional instruments such as drums (moropa), clapping (seatla), and horns (dinaka) used to create rhythm and accompaniment.

Initiation Songs: Songs performed during initiation ceremonies (bogwera and bojale) to teach cultural values and social responsibilities.

Kgotla: A traditional meeting place led by a kgosi where community discussions and cultural performances, including music, take place.

Mmino wa Setswana wa setso: Refers specifically to indigenous musical forms passed down orally through generations.

Mmino wa Setswana: A general term referring to all forms of Setswana music, including traditional and contemporary styles.

Oral Tradition: The transmission of music and cultural knowledge through word of mouth rather than written form.

Praise Poetry: A rhythmic, chant-like form of expression used to honour individuals, clans, or leaders.

Setswana Traditional Music: Music that originates from the Batswana people, rooted in indigenous cultural practices and used in ceremonies, rituals, and community gatherings.

Chapter 4
Threads of Wonders:
An Autoethnography of Exploration, Curiosity, and Cultural Awakening Through the Sama Tepo

Hana An L. Alvarado
http://orcid.org/0009-0006-6550-616X
Mindanao State University-Tawi-Tawi College of Technology and Oceanography, Philippines

ABSTRACT

This autoethnography traces a personal journey that began as an academic task and evolved into a transformative encounter with the cultural heart of the Sama people—the Tepo of Tawi-Tawi. Initially seen as an ordinary woven mat, the Tepo emerged as a living archive of artistry, memory, and indigenous intelligence. Through exploration, hesitation, and realization, immersion in Tepo deepened cultural empathy, self-discovery, and inspired culturally relevant mathematics teaching. Rather than analyzing weaving techniques, this study centers lived experiences, listening to stories within patterns and honoring relational wisdom across generations. It shows how indigenous knowledge is observed, felt, and negotiated through mindful participation. Curiosity becomes a catalyst for awakening and transformation. This work reimagines indigenous knowledge revitalization, positioning education as a space for honoring culture, community, and self.

INTRODUCTION

What can a woven mat teach a pre-service mathematics teacher about curiosity, knowledge, and the very act of learning? This question, initially incidental, became a critical point of reflection that unsettled my assumptions about mathematics, education, and my own emerging professional identity. Such moments of disruption are central to autoethnographic inquiry. It is where the personal experience becomes an entry point for examining broader cultural and epistemological issues (Adams et al., 2017). In recent years, the autoethnography has been increasingly recognized as a rigorous qualitative approach

DOI: 10.4018/979-8-2600-2890-2.ch004

that connects the self to culture. It allows the researchers to interrogate how lived experiences are shaped by and, in turn, shape the social realities.

This study emerges from an academic engagement in EDM 132: *Problem Solving, Mathematical Investigation and Modelling*. I was tasked to explore the *Sama Tepo*. It is a traditional woven mat from Tawi-Tawi. The activity was initially framed as a mathematical investigation that focused on identifying patterns, symmetry, and spatial relationships. Gradually, it evolved into a deeper encounter with indigenous knowledge systems. This shift reflects broader calls in education research to move beyond purely abstract and decontextualized views of mathematics toward a more culturally situated and meaningful understanding of learning (D'Ambrosio, 2016). Recent scholarship further emphasizes that integrating ethnomathematical practices and Indigenous Knowledge Systems into mathematics education is essential in efforts toward decolonizing educational spaces and making learning more culturally meaningful (Bhuda, 2024). Such perspectives argue that Indigenous cultural practices should not merely function as supplementary classroom examples, but as legitimate epistemological systems capable of reshaping how mathematics is understood, taught, and experienced.

In the southernmost province of Tawi-Tawi, the *Sama Tepo* is not merely an artifact of craftsmanship but a living expression of cultural identity, collective memory, and embodied knowledge. Traditional weaving practices, such as Tepo, are recognized as carriers of intangible knowledge systems where skills and values are transmitted across generations (UNESCO, 2023). In the 21st century, the revitalization of Indigenous Knowledge Systems (IKS) has become an urgent concern in both cultural preservation and educational discourse. Revitalization extends beyond documentation; it involves the active recognition, transmission, and integration of indigenous knowledge into contemporary contexts to ensure its continuity amid rapid modernization. Within this perspective, practices such as *Sama Tepo* weaving are not only cultural artifacts to be preserved but dynamic knowledge systems that must be sustained through meaningful engagement, education, and intergenerational learning. This aligns with recent discussions on the resurgence and revalorization of Indigenous Knowledge Systems in education, which highlight the importance of integrating indigenous ways of knowing into formal schooling to resist epistemic marginalization and sustain cultural continuity (Bhuda & Maditsi, 2025).

My initial engagement with the *Tepo* reflected my training as a pre-service mathematics teacher. Educated within formal academic structures, I approached the woven mat as a mathematical object seeking tessellations, geometric patterns, ratios, and symmetries. However, as highlighted in *ethnomathematics* research, mathematical knowledge is not confined to formal notation or institutional settings but is embedded in cultural practices, daily activities, and community traditions (D'Ambrosio, 2016). Through deeper interaction with the *Sama* weavers, it became evident that the *Tepo* embodies forms of reasoning that are relational, experiential, and historically situated.

This realization aligns with the emerging interdisciplinary perspectives such as EthnoSTEM. It advocates for integrating indigenous knowledge systems into science and mathematics education to make learning more contextualized and culturally responsive. The EthnoSTEM highlights how local practices, such as weaving. It can serve as rich contexts for understanding mathematical and scientific concepts while affirming learners' cultural identities. Despite increasing recognition of indigenous knowledge in education, there remains a gap in research that meaningfully integrates personal engagement, cultural analysis, and theoretical grounding. Many studies either document cultural practices descriptively or analyse them abstractly without engaging the lived experiences of learners and educators. This study responds to this gap by employing autoethnography to examine how direct engagement with the *Sama Tepo* weaving can foster cultural awakening, epistemological reflection, and pedagogical transformation.

More importantly, it explores how such engagement can contribute to the revitalization of Indigenous Knowledge Systems by supporting their recognition, ethical engagement, and integration into mathematics education.

Through this approach, the study situates personal narrative within the broader structural and cultural contexts. Indigenous knowledge systems have historically been marginalized by dominant Western epistemologies that privilege standardization, abstraction, and textualization (Battiste, 2013). As a result, practices like *Sama Tepo* weaving are often undervalued or excluded from formal educational discourse. By critically reflecting on my own assumptions and positionality, this study seeks to contribute to ongoing conversations on decolonizing education and recognizing multiple ways of knowing. Situated within broader efforts to revitalize Indigenous Knowledge Systems, this chapter traces how an initially academic task evolved into a transformative encounter. It asks: *What happens when learning becomes an act of cultural awakening? How does engagement with indigenous practice reshape one's understanding of knowledge, teaching, and identity? And how can educational spaces move toward more inclusive and culturally grounded approaches to mathematics?*

By weaving together personal experience, ethnographic insight, and theoretical perspectives, this study argues that the *Sama Tepo* is not merely an object of analysis but a site of knowledge, reflection, and transformation. It invites a rethinking of mathematics education as relational, culturally embedded, and deeply human—where learning is not only acquired but experienced, negotiated, and lived.

METHODOLOGY

This section discusses the research design, data sources, data analysis, and the positionality and reflexivity of this study. Grounded in qualitative inquiry, the methodology is designed to examine how engagement with *Sama Tepo* weaving shapes meaning-making, professional identity formation, and epistemological transformation.

Research Design. This study utilized a qualitative, reflective autoethnographic design to examine my experience as a pre-service mathematics teacher engaging with *Sama Tepo* weaving, starting from a group activity in EDM 132 *"Problem Solving, Mathematical Investigation & Modelling"*. Rather than measuring variables or outcomes, the methodology is oriented toward understanding meaning-making, professional identity formation, and epistemological transformation as they unfolded through this initial academic task and subsequent immersive engagement with the weavers in Tawi-Tawi. The autoethnographic approach positions my personal journey as both the site and source of inquiry, allowing lived experience to illuminate broader social, cultural, and educational realities (Adams et al., 2017).

The research design of this paper recognized that my engagement was shaped by two intersecting roles. It is as a pre-service mathematics teacher trained to value abstraction and formal curricula and as a participant-observer entering the weaving community. An autoethnography enables the linking of individual experience to the cultural and pedagogical systems. It highlighted how mathematical reasoning, cultural practice, and teacher identity intersect. My initial task in EDM132 was to investigate the patterns in the *Tepo*. It served as both a catalyst and an analytic lens. It situated the study in a context that combines academic expectation, personal curiosity, and ethical engagement with the indigenous knowledge.

Hence, this reflective autoethnographic design was deliberately chosen because becoming a mathematics teacher is not a purely technical process. It is deeply emotional, ethical, and intellectual. The encounter with the *Sama Tepo* generated moments of curiosity, discomfort, humility, and transformation, which are central to understanding how cultural engagement reshapes professional formation. The autoethnography provides the methodological space to trace these tensions, attend to reflexive insight, and examine the evolving meaning of mathematics and teaching within both formal and community contexts.

Throughout the study, I engaged as an active participant-observer and reflective practitioner. Reflexive journals, photographs, field notes, and interviews served as both data sources and analytic tools. Iterative cycles of writing, revisiting, and theorizing allowed patterns, contradictions, and conceptual shifts to surface over time. By foregrounding reflection alongside critical interpretation, the design acknowledges that learning to teach mathematics involves encountering indigenous practice. It is an ongoing, situated process, rather than a fixed achievement.

The rigor of this study is established through the transparency of positionality, sustained reflexivity, and theoretical grounding in autoethnography and *ethnomathematics* (Adams et al., 2017; D'Ambrosio, 2016). The study does not aim for statistical generalizability but for analytic depth, conceptual resonance, and ethical accountability. By centering the *Sama Tepo* as both cultural and pedagogical interlocutor, the methodology contributes to a nuanced understanding of mathematics education as relational, culturally grounded, and transformative, and highlights how a pre-service teacher's engagement with the indigenous knowledge can shape professional identity, pedagogical awareness, and ethical practice.

Data Sources. The data of this study were generated through the multi-layered, immersive engagement, beginning in one of my courses in the College of Education. In the EDM 132 *"Problem Solving, Mathematical Investigation & Modelling"*, where I was first tasked to investigate the *Sama Tepo* as part of a formal course assignment, and extended to direct interaction with the weaving community in Tawi-Tawi. In line with the autoethnographic methodology, data encompassed lived experience, embodied interaction, emotional responses, and reflective interpretation, treating the self as both participant and analytic lens.

The primary data consisted of reflexive journals maintained throughout the course and fieldwork. In the EDM132 group task, journals captured initial observations. It involves the identification of geometric patterns, calculation of symmetries and ratios, documentation of repeated motifs, and reflections on how formal mathematical training shaped my perspective. These early entries reveal my initial analytic lens. This prioritized the abstraction, formulaic interpretation, and classroom-style problem-solving.

As engagement deepened through field immersion, the journals expanded to include thick, sensory-rich descriptions of weaving sessions. The tactile quality of *pandan* strips, the rhythm of hands in motion, the sounds of weaving, and the spatial arrangement of mats within communal spaces. Detailed attention was given to gestures, timing, intergenerational collaboration, and narrative explanations shared by weavers. Each entry combined observation with reflexive interpretation, noting moments of curiosity, ethical awareness, and emotional response, as well as tensions between my academic perspective and the lived knowledge of the community.

Field notes were also collected during the weaving sessions. It focuses on verbal explanations, demonstrations, and subtle decision-making processes within the weaving practice. Special attention was given to implicit mathematical reasoning embedded in cultural practice, such as proportional adjustment, spatial alignment, iteration, and symmetry. Even when these were not formally articulated. Field notes functioned as a bridge. It connects my formal mathematical training to the embodied practices of the weavers.

Material artifacts served as additional data sources. The photographs and sketches of woven patterns were used to document structures and as prompts for iterative reflection. The *Sama Tepo* itself was approached as a living text, with each mat representing both cultural memory and mathematical reasoning.

Visual and tactile engagement with these artifacts enabled recursive examination of earlier interpretations. It reveals the pattern, conceptual insights, and moments of re-evaluation.

Finally, the retrospective narrative reconstruction was integrated as a data source. Autoethnography recognizes memory as interpretive; journal entries and artifacts were revisited critically to reconstruct experiences, link emerging insights, and reflect on professional and ethical growth. This recursive process allowed the temporal evolution of understanding. It is from the initial EDM 132 group task to deeper community immersion to be traced and analysed. Collectively, these data sources constitute a layered, longitudinal, and relational archive of observation, reflection, and engagement. They capture the convergence of classroom investigation, cultural participation, mathematical reasoning, and teacher formation, positioning the *Sama Tepo* not merely as an object of study but as an active interlocutor in shaping my professional, intellectual, and ethical development.

Data analysis. This study employed a reflexive, iterative, and interpretive process consistent with qualitative autoethnography. Analysis was not confined to a post-collection stage. Rather, it emerged simultaneously with participation, observation, and reflective writing, beginning with the initial EDM 132 *Problem Solving, Mathematical Investigation & Modelling* task and continuing through the immersive engagement with the weaving community in Tawi-Tawi.

The analytic process began with the immersive rereading of reflexive journals, field notes, photographs, sketches, and reconstructed narratives from both classroom-based investigation and community engagement. Open coding was employed to identify segments that revealed tensions, moments of curiosity, affective responses, and shifts in professional understanding. Particular attention was paid to passages illustrating how my pre-service mathematics training shaped the interpretation of woven patterns. It highlighted instances where I initially sought patterns or symmetry, and moments where these initial readings were challenged by cultural context, embodied practice, and communal knowledge transmission.

The analysis was both temporal and dialogic. Early classroom-based observations focused on pattern identification and technical mathematical reasoning, reflecting formal training. Later narratives, drawn from immersive fieldwork with weavers, revealed shifts in perception: recognition of community ethics, aesthetic decision-making, and culturally grounded mathematical reasoning. By comparing early and later entries, the analysis traced the evolution of interpretive stance, professional reflexivity, and ethical awareness.

Theoretical engagement strengthened analytic depth. Insights from *ethnomathematics* (D'Ambrosio, 2016) were used to interpret the embedded mathematical structures in weaving, while autoethnographic frameworks (Adams et al., 2017) guided reflection on identity, positionality, and relational knowledge. Rather than imposing abstract frameworks, theory functioned as a lens to critically examine lived experience, connecting my personal journey to broader educational and cultural discourses. Memo-writing and reflective journaling documented analytic decisions, capturing why specific excerpts were grouped, how themes were refined, and where ambiguity remained. The iterative process preserved productive tensions, acknowledging that teacher formation, cultural engagement, and mathematical understanding are dynamic, ongoing, and non-linear.

Ultimately, the analysis does not aim for universal generalization. Its purpose is to illuminate how a pre-service mathematics teacher's encounter with the *Sama Tepo*. It's beginning in EDM132 and extending into the lived cultural engagement. It was a catalysed epistemological transformation, professional growth, and ethical pedagogical reflection. And through the iterative thematic analysis, the woven mat emerges not merely as an object of the study but as a relational interlocutor that actively shapes the understanding of mathematics, teaching, and learning.

Positionality and Reflexivity. In this study, my positionality is central to both the inquiry and its interpretation. I am a graduating mathematics teacher whose professional identity and epistemological assumptions are shaped by formal academic training that privileges abstraction, procedural reasoning, and standardized assessment. These assumptions shaped my engagement with the *Sama Tepo* during EDM 132, where I initially sought to document geometric structures such as symmetry, ratio, and pattern through the familiar mathematical categories. My identity at that moment was that of both student and researcher, located within an academic institution yet interacting with lived cultural practice.

However, my positionality is multifaceted and situated at the intersection of formal mathematics education and indigenous cultural epistemologies. Autoethnographic theory emphasizes that research cannot be understood apart from the researcher's positionality; lived experience becomes both the site and source of analysis (Adams et al., 2017). Reflexivity, defined as the continuous process of examining how one's own assumptions, identities, and cultural positioning shape interpretation and meaning-making, was therefore essential throughout this work. Autoethnography involves critically examining subjective experiences in relation to broader cultural and social contexts, making visible how personal narratives intersect with the structural and cultural factors (Ellis et al., 2011).

My initial analysis of the *Sama Tepo* reflected a formalist orientation to mathematics. It is an orientation common in teacher education, which often treats mathematics as universal and culture-neutral. This orientation was challenged as I deepened my engagement with the weaving practice in Tawi-Tawi. *Ethnomathematics* theory posits that mathematical thought is embedded in culture, arising through the sociocultural activity rather than existing as a universal abstraction (Vandendriessche & Pinxten, 2022). From this perspective, practices such as weaving are not merely cultural expressions but are inherently mathematical, involving spatial reasoning, pattern logic, and iterative problem-solving grounded in local knowledge systems.

The concept of epistemological pluralism further grounds this reflexive stance by valuing multiple ways of knowing rather than privileging one epistemology over others. This aligns with the research that treats indigenous practices as legitimate systems of reasoning that expand rather than subordinate formal mathematical approaches. For example, the edited volume Indigenous Knowledge and *Ethnomathematics* foregrounds how indigenous mathematical practice. For instance, the pattern-making or procedural techniques embody the complex reasoning processes that challenge the dominant Western paradigms of mathematical knowledge (Vandendriessche & Pinxten, 2022).

The reflexivity was sustained through the ongoing journaling, self-analysis, and narrative reconstruction, in which I constantly interrogated how my academic assumptions influenced what I observed and how I interpreted meanings. Early narratives revealed a tendency to prioritize technical mathematical categories, whereas later reflections highlighted relational, ethical, and cultural dimensions of the weaving practice. This shift illustrates the autoethnographic emphasis on the self-interrogation as analytic work where personal discomfort, shifting interpretations, and epistemic tensions become data that deepen understanding (Adams et al., 2017).

By making my positionality and reflexivity explicit, this study acknowledges that knowledge production is situated and relational. My role as both researcher and mathematics teacher shapes not only what is observed, but how meaning is made with and through the *Sama Tepo*. Reflexivity ensured that interpretations were not unexamined projections of training or bias, but critically examined engagement with the cultural knowledge that broadened my understanding of mathematics, pedagogy, and ethical participation. In this way, the study affirms that teacher identity formation is not only technical but also cultural, emotional, and ethical, emerging through ongoing negotiation between formal educational frameworks and indigenous ways of knowing.

FINDINGS

This section presents the findings of the study through a thematic analysis of the researcher's autoethnographic reflections on engagement with *Sama Tepo* weaving in Tawi-Tawi.

Figure 1. Autoethnographic Trajectory of Learning: Exploration, Curiosity, and Cultural Awakening

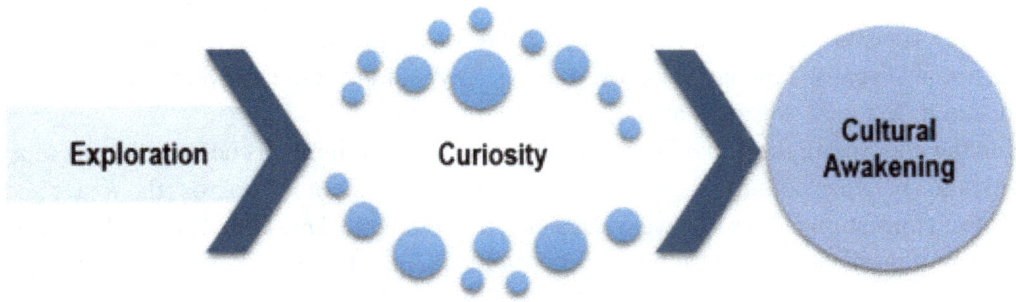

Figure 1 shows the analysis of this study. It is organized into three interconnected themes. It is the exploration, curiosity, and cultural awakening that collectively illustrate the process of epistemological and professional transformation. These themes are not presented as linear stages but as overlapping and recursive dimensions of learning.

Exploration: *From Extraction to Epistemic Humility across Differences*

My first encounter with the *Sama Tepo* did not begin in a weaving house in Tawi-Tawi. It began in a corner of the classroom during my third year in the College of Education. I am completing my undergraduate degree, Bachelor of Secondary Education major in Mathematics. It was in my EDM 132 *"Problem Solving, Mathematical Investigation & Modelling"*. At that time, I did not yet recognize that I was entering the task with a structured epistemological orientation shaped by formal mathematics education and institutional academic training.

I am not *Sama*. I did not grow up in the weaving traditions. I did not inherit the embodied knowledge of *pandan* weaving practices. This positional distance is not merely personal. It is epistemological. It shaped my initial interpretation of the *Sama Tepo* as an object of mathematical analysis rather than a living cultural knowledge system. From an autoethnographic standpoint, this reflects Adams et al. (2017), who emphasize that knowledge production is inseparable from the researcher's positionality.

In EDM 132, our professor tasked us to identify mathematical concepts embedded in the *Sama Tepo*. The task appeared progressive, as it framed culture as a site of mathematical inquiry. However, I now recognize that my initial engagement reproduced what Ubiratan D'Ambrosio critiques as the abstraction of mathematics from cultural life—where structure is privileged over meaning, and pattern over practice.

I approached that task as an outsider trained in formal mathematics. I entered the world of the *Sama* cultural artefact through analytical authority rather than relational accountability. When my group mates and I visited the 2nd *Tepo* Exhibit at Maas Kamlon in MSU-TCTO, my very own university, I remember scanning the hall with anticipation as well as confusion. There were so many mats displayed in the hall. The designs include diagonals as well as colour sequences pulsing with the rhythm. One weave design is longer than the other. Its crossings created a dynamic interplay of repetition and variation.

Without hesitation, I took out my phone and began photographing the *Sama Tepo*. My group mates did the same. There were close-ups of intersections. Angled shots of motifs. I traced imaginary axes with my finger and said to my group mates, *"Guys, there's a symmetry here."* We discussed what were the mathematical concepts can be seen in the mats. There was symmetry, tessellation, and proportion. But what I did not say because I did not yet know it. I was only looking at the weave, not with the community that produced it. My gaze was only analytic, and not relational.

Back in the classroom, the *Sama Tepo* data. We counted strip intervals before patterns repeated. We calculated ratios between coloured segments. I carefully drew straight lines across the weave and labelled geometric transformations. Then, the weave became diagrammatic. I presented our findings with certainty. I put arrows that illustrated rotational symmetry. I explained the proportional relationships between the coloured strips. When our professor nodded approvingly, I felt affirmed and not only intellectually, but institutionally. I believed I had conducted rigorous, culturally responsive exploration. However, the weavers, whose mathematical reasoning animated the *Tepo*, were absent from our analysis. Their voices were not cited. Their intentions were not consulted. And at that moment, the one we thought an exploration meant extraction only.

I searched for familiar mathematical categories within the weave. I did not ask how the *Tepo* functioned within the *Sama* life. I did not ask what meanings, obligations, or histories were interlaced into its patterns. I asked only: *"What mathematical concepts can be found here?"* But months later, when I entered weaving spaces in Tawi-Tawi not as a student completing an assignment or task, but as a researcher. I felt the difference immediately. The *ethnomathematics* reminds us that mathematical reasoning is culturally situated and enacted within lived practice rather than abstracted from it (Vandendriessche & Pinxten, 2022).

I began to understand that my earlier analysis in the *Tepo* in EDM132 was not wrong, but it was only partial. It privileged structure over process, diagram over gesture, and abstraction over embodiment. And perhaps the most transformative realization was the *Sama Tepo*. Beyond serving as an initial point of personal engagement, the process of exploration also functioned as an act of documenting and recognizing the *Sama Tepo* weaving as a living Indigenous Knowledge System. Observing the patterns, techniques, and meanings embedded in the weaving revealed that such practices carry structured, experience-based knowledge that is often overlooked in formal educational contexts. In this sense, exploration becomes more than passive observation; it is a critical first step in the revitalization of Indigenous knowledge, as it allows these practices to be seen, valued, and recorded within academic and pedagogical spaces.

Recognizing this altered my emerging identity as a mathematics teacher. If mathematics exists across epistemological worlds, then my role cannot be extraction or proof. It must be mediation, humility, and reflexive awareness of the hierarchies I inhabit. As a non-*Sama* educator engaging with *Sama* mathematical practice, I must continuously interrogate how my training shapes what I see and what I fail to see. This transformation did not occur in just a single moment of clarity. It unfolded gradually. It is through the discomfort, reflexive questioning, and repeated return. Strip by strip, negotiation by negotiation, just like the weave itself.

Curiosity: From Cognitive Inquiry to Ethical Praxis

Curiosity has long been positioned within education as a positive intellectual disposition associated with inquiry, engagement, and knowledge-seeking. In my own formation as a pre-service mathematics teacher, curiosity was initially framed as a cognitive tool for problem-solving and discovery. In EDM 132 *Problem Solving, Mathematical Investigation and Modelling*, this disposition was activated through the task of identifying mathematical structures within the *Sama Tepo*, particularly symmetry, proportion, and pattern.

However, this initial form of curiosity was not neutral. It was shaped by institutional expectations of mathematics as a discipline grounded in abstraction and formal representation. As a result, my early questions were directed toward extraction: *What mathematical concepts are embedded in the Sama Tepo?* Rather than how *does this practice exist within the lived world of its creators?* This deeper curiosity led me to conduct sustained research on the *Sama Tepo* beyond the classroom requirements. I read an *ethnomathematics* literature, examined cultural accounts of weaving, and drafted reflections identifying geometric transformations, proportional reasoning, symmetry, and spatial structuring embedded in the patterns. But alongside this academic inquiry, another motivation was forming within me. I was not simply studying the *Tepo* to complete an assignment. I was beginning to imagine it as an instructional material for my future classroom. I wanted my students, especially those who share cultural proximity to these traditions and to feel that mathematics is not foreign, not imported, not detached from their lived realities. I wanted them to see that mathematics is already woven into their culture.

I felt productive. I felt inspired. I felt scholarly. Yet my curiosity moved along a particular trajectory. It was still oriented toward mathematics as I had been trained to recognize it. I wanted to know how the *Tepo* could illustrate the geometry, how it could enrich lesson plans, and how it could align with the curricular standards. Even in my desire to culturally ground mathematics instruction, I was still translating the *Tepo* into the language of institutional schooling.

Looking back, I recognize that my curiosity was not neutral. It was institutional. It was shaped by the demands of teacher preparation, by assessment structures, and by curricular frameworks that prioritize demonstrable learning outcomes. My questions were framed within the logic of schooling. As a non-*Sama* mathematics student positioned within the academy, I occupied a space that defines what counts as legitimate knowledge. My curiosity, though genuine, was structured by that institutional authority.

Figure 2. Sama Tepo Weaving Booth

This became more apparent during the 52nd Kamahardikaan Sin Tawi-Tawi celebration. The festival grounds were vibrant. The stalls were lined with different types of woven mats, colours saturated under the heat of the afternoon sun. Deep reds, yellows, and greens interlaced through intricate diagonal patterns. Some weavers were working on-site, their hands moving rhythmically across partially completed mats. I approached them with eagerness.

I held my phone and started to record the explanations of the *Sama* weavers. I felt purposeful. It's like a researcher gathering valuable data. My questions came quickly. It was shaped by the investigative habits I had cultivated in the university. One weaver explains how she begins by folding the base strips before interlacing diagonally. She spoke about keeping the weave balanced, ensuring it does not tilt. It makes sure that the strips remain even. Her explanations were practical, embodied, and relational. She described adjustments through the feeling. The tension of the strips and the firmness of the fold.

As she spoke and demonstrated, something shifted in me. I began to appreciate the *Tepo* beyond its mathematical representations. I was no longer merely identifying geometry. I was witnessing lived knowledge. It was here that I began to see how easily curiosity can slide into epistemic extraction. Even when motivated by admiration and the desire to bring cultural materials into my classroom, my questions risked repositioning *Sama's* knowledge as raw material. Something to be formulated, validated, and circulated through the dominant academic frameworks. As a non-*Sama* researcher, I was not merely asking questions. I was entering a space shaped by history, continuity, and lived cultural memory. My presence carried institutional privilege. My recording device symbolized documentation, not reciprocity.

The tension crystallized later when I encountered the cultural symmetry model articulated by Trinick and Allen (2024), which argues that Indigenous knowledge systems and school mathematics must be positioned symmetrically. Neither the subordinated nor romanticizes. Reading this framework after my field encounters, I recognized how fragile symmetry actually is. Hierarchy does not announce itself loudly. It slips quietly into translation. It hides in well-intentioned clarification. It disguises itself as integration.

With my curiosity, I began to realize, is not inherently ethical. It can reinforce hierarchy when it treats cultural knowledge as something to be mined, repackaged, and aligned with the institutional objectives. It can privilege formal mathematics as the language of legitimacy. It is even while claiming to celebrate cultural practice. Walking from stall to stall during the festival, asking questions, recording responses, I had to confront the possibility that my curiosity. I had initially been more curious about the mathematics in the *Tepo* than about the people who weave it. And admitting that felt uncomfortable. I had equated curiosity with respect and understanding. But these require more than intellectual interest. It requires reflexive awareness of power. It requires accountability.

Through the sustained engagement with the weaving community and continued reflection on my own motivations, curiosity began to shift. It moved from the cognitive inquiry toward ethical praxis. I became more attentive not only to what I was asking, but why I was asking it. My desire to use the *Sama Tepo* as an instructional material remained. But it was now accompanied by a deeper commitment. That its integration into mathematics teaching must honour its relational meanings and cultural integrity. This shift resonates with culturally sustaining pedagogy (Paris, 2021), which insists that educational engagement must sustain and honour cultural practices rather than extract and dilute them for instructional convenience. The danger is not in connecting mathematics to culture. The danger lies in detaching cultural practice from its relational meanings and repositioning it as a pedagogical resource.

Curiosity became slower. Less performative. Less urgent to produce immediate academic output. There were moments when I deliberately chose not to translate explanations into geometric terminology. I allowed descriptions to remain in the language of balance, tilt, tension. I listened without converting. In that restraint, I experienced a different kind of learning—one not centered on mastery, but on relational attentiveness. Curiosity shifted from a tool of academic production to a moral positioning. It required me to recognize that as a non-*Sama* mathematics educator-in-formation, my role is not to validate the *Tepo* through formal mathematics. Rather, my responsibility is to remain accountable to the communities whose knowledge I engage, and to allow that knowledge to reshape my understanding of mathematics itself.

Perhaps the most vulnerable realization in this process was this: I had initially been more invested in demonstrating that mathematics exists in the *Tepo* than in understanding what the *Tepo* means within *Sama* life. Recognizing that imbalance altered my trajectory. Curiosity, when unexamined, can serve the self—expanding one's academic portfolio, strengthening lesson plans, enriching theoretical positioning. Curiosity, when reflexive, can transform the self—disrupting epistemic hierarchies, unsettling inherited assumptions, and reorienting one's ethical commitments. As curiosity deepened, it also required a shift from simply wanting to know toward understanding how to engage responsibly with the knowledge being encountered. Indigenous knowledge systems, such as the *Sama Tepo* weaving, are not merely sources of information to be extracted but are lived practices embedded within cultural identity and community ownership. This realization reframed curiosity as an ethical stance—one that demands respect, permission, and sensitivity to the cultural meanings carried by the practice. In this sense, curiosity becomes a crucial element in the revitalization of Indigenous knowledge, as it encourages not only inquiry but also the protection and honoring of the communities from which such knowledge originates.

Curiosity did not disappear. It deepened. It became both pedagogical and ethical. I still intend to bring the *Sama Tepo* into my mathematics classroom. But nor as proof that mathematics exists in culture, but as a way for students to recognize that mathematics has always been embedded in their cultural worlds. In deepening, curiosity became responsibility. That shaping not only my research trajectory, but my becoming as a mathematics teacher accountable to the communities whose knowledge I seek to engage.

Cultural Awakening: Reconfiguring Teacher Identity through the *Sama Tepo*

Before encountering the *Sama Tepo* beyond the classroom analysis, I had already begun constructing an image of the mathematics teacher I aspired to become. I imagined competence, clarity, and structure. I valued the precision, effectiveness, and efficiency. I believed that good teaching meant explaining concepts coherently, guiding the students toward correct solutions, and ensuring procedural mastery. Rigor, in my imagination, was measured by accuracy and abstraction. My emerging teacher identity had been shaped almost entirely within formal educational institutions. Mathematics appeared stable, universal, and objective. My responsibility, I believed, was to transmit it faithfully and perhaps creatively, engagingly and ultimately, accurately.

As a non-*Sama* pre-service teacher trained within institutional mathematics, I did not initially perceive this vision as culturally situated. It felt neutral. It felt professional. The *Sama Tepo* unsettled that neutrality. The awakening did not arrive as a dramatic revelation. It unfolded gradually through the classroom extraction in EDM132, through questioning the *Sama* weavers, and through the discussion and dialogue with my mentor. It was less an awakening than a slow unravelling of certainty. One evening after interviewing several weavers during the festival, I returned home and re-read my field notes. Patterns in my writing became unmistakable. I had repeatedly described the *Tepo* as something that only demonstrates symmetry, illustrates proportional reasoning, or is something used to teach geometric transformations. Almost every entry framed the *Tepo* in relation to school mathematics. Its value was articulated through pedagogical usefulness.

I paused because educational scholarship describes teacher identity as continuously negotiated across contexts, shaped by experience, discourse, and power relations. Yet until this encounter, I had imagined identity formation primarily as pedagogical refinement—improving strategies, expanding repertoire, deepening content knowledge. The *Sama Tepo* revealed that teacher identity is also epistemological. It reflects what one believes mathematics is, who is authorized to produce it, and how it should circulate within institutions. In this sense, the *Tepo* became an epistemic mirror. It reflected not only geometric structure, but my own assumptions. I began to see how deeply I had internalized a hierarchy: school mathematics as foundational and authoritative; cultural knowledge as illustrative and supplementary. Even when I believed I was *"integrating culture,"* I was positioning it as enrichment rather than as knowledge with its own ontological grounding. Recognizing this was uncomfortable. It required acknowledging that I could unintentionally reproduce epistemic injustice in my future classroom—privileging symbolic abstraction while marginalizing embodied, community-based reasoning.

As a non-*Sama* educator engaging with *Sama* knowledge, the risk of misrepresentation or appropriation was not abstract. It was present in my language, in my framing, in my curricular imagination. *Ethnomathematics* challenges the assumption that Western formal mathematics holds epistemic supremacy, asserting instead that mathematical ideas emerge within diverse cultural practices. Yet reading this theoretically is different from confronting it personally. The *Sama Tepo* did not merely exemplify *ethnomathematics*. It confronted my authority. The *Tepo* was not an example of mathematics. It was a mathematical world. This distinction altered how I began to understand teacher authority. If mathematics exists within weaving practices independent of institutional validation, then my role cannot be that of a validator. Nor can it simply be a translator—converting embodied reasoning into formal terminology. Translation risks hierarchy when one language is positioned as more legitimate.

So what, then, is my role? Am I a mediator? Listener? Facilitator of dialogue between epistemic systems? Learner within spaces I do not inherit? These questions destabilized the image of the teacher as a central knowledge-holder. I began to imagine instead a dialogical classroom—one in which students' cultural experiences are not treated as applications of pre-existing mathematical truth, but as sites of mathematical reasoning in their own right. Similar arguments emerge in recent ethnomathematics scholarship, where Indigenous artistic and cultural practices are positioned as transformative spaces for mathematics teaching and learning. For instance, Motseki et al. (2026) demonstrate how Ndebele art can reshape mathematics instruction by foregrounding cultural reasoning, patterning, and indigenous forms of mathematical expression within classroom practice. This vision aligns with culturally sustaining pedagogy (Paris, 2021), which insists that education must sustain and affirm community practices rather than assimilate them into dominant frameworks. Yet embracing this vision required more than conceptual agreement. It required vulnerability.

Letting go of certainty was not easy. I had been trained to equate rigor with abstraction, generalization, and symbolic proof. *If mathematics is plural and culturally situated, how do I assess it? How do I align it with standardized curricula? How do I navigate institutional accountability without re-centering dominant epistemologies?* These tensions did not disappear with awakening. They intensified. But observing weaving practices began to reshape my understanding of rigor itself. Rigor was present in the repetition of folding and interlacing strips. In the sustained attention required to maintain alignment without measuring tools. In the embodied calibration that corrected the imbalance through touch. In the patience embedded in intergenerational transmission. This was not a romanticization of tradition. It was recognition of discipline—mathematical discipline—manifested differently.

The awakening, then, was not about celebrating cultural artifacts. It was about reconfiguring my professional identity in relation to epistemic plurality. I no longer aspire merely to be a transmitter of abstract content. I aspire to become a mediator of mathematical worlds—creating conditions where formal and cultural knowledge systems can encounter one another symmetrically rather than hierarchically. The cultural symmetry model (Trinick & Allen, 2024) offers conceptual language for this aspiration, arguing that knowledge systems must be positioned in balance rather than subsumed. Yet I remain cautious. Symmetry is not a stable achievement. As a teacher positioned within institutional mathematics, I carry structural authority. The risk of re-centering dominant frameworks never fully disappears. Awakening is not a final state of enlightenment; it is an ongoing ethical commitment.

This cultural awakening extends beyond personal realization and moves toward a sense of responsibility to sustain and transmit Indigenous knowledge systems. Recognizing the richness of the Sama Tepo weaving as a site of mathematical and cultural knowledge highlights the need to integrate such practices into educational spaces. In this way, cultural awakening becomes a transformative force that informs pedagogical choices, encouraging the inclusion of culturally grounded examples that make learning more meaningful and contextually relevant. This shift reflects an important dimension of revitalization, where Indigenous knowledge is not only appreciated but actively sustained through teaching, learning, and continued practice.

Perhaps the most transformative realization was this: becoming a mathematics teacher is not only about learning pedagogy. It is about negotiating epistemology. The *Sama Tepo* did not simply expand my repertoire of culturally relevant examples. It disrupted my assumptions about what counts as mathematics, who produces it, how it circulates, and who holds authority in defining it. Teaching, I now understand, is not the delivery of stable knowledge. It is the careful holding of space between epistemic worlds—without collapsing one into the other. This is my cultural awakening. Not a romantic embrace of tradition. Not a rejection of formal mathematics. But a reconfiguration of identity grounded in relational accountability, reflexive humility, and sustained interrogation of power. Like the weave itself, my teacher identity remains in process—formed through tension, adjustment, and negotiation. Strip by strip. Balance by balance. Pattern emerging, never fully complete.

CONCLUSION

This autoethnography began with a seemingly simple academic task. A task that needed to identify the mathematical concepts embedded within the *Sama Tepo*. As a third-year pre-service mathematics teacher in EDM132, I approached the woven mat as an object of analysis. I traced symmetries, calculated ratios, and mapped geometric transformations. At the time, I believed I was honouring culture by revealing its mathematical sophistication. What I did not yet understand was that my gaze was shaped by a hierarchy I had inherited. Across classroom analysis, the festival encounters during the 52nd Kamahardikaan sin Tawi-Tawi, interviews with the weavers, and sustained reflexive journaling, this study documents a gradual epistemological shift. The *Sama Tepo* moved from being a visual dataset to becoming an epistemic interlocutor. It no longer appeared as an artifact awaiting mathematical validation. It emerged as a living site of mathematical reasoning enacted through the embodied practice, relational knowledge, and intergenerational continuity. This transformation unfolded across three interwoven movements.

First, the exploration shifted from mathematical extraction to relational engagement. *What began as the identification of abstract structures evolved into an attentiveness to how mathematics is lived before it is symbolized?* The *Tepo* revealed mathematics as embodied, adaptive, and communal. It expanded beyond the formal abstractions privileged in institutional schooling. Second, curiosity transformed from the cognitive inquiry to ethical praxis. I came to recognize that curiosity is not neutral. It carries the risk of translation that re-centers dominant frameworks. Through reflexive confrontation with my own tendencies to convert weaving knowledge into geometric terminology, I began to understand curiosity as a moral stance. It requires humility, relational accountability, and attentiveness to epistemic asymmetry. Third, this journey reconfigured my emerging teacher identity. Becoming a mathematics teacher, I now recognize, is not solely about mastering pedagogy. It is about negotiating epistemology. Engagement with the *Sama Tepo* disrupted my assumption that formal mathematics occupies the apex of legitimacy. Instead, I began to envision a pedagogical stance that positions school mathematics and indigenous mathematical worlds symmetrically, neither romanticized nor subordinated.

The conceptual contribution of this study lies in reframing culturally situated mathematical artifacts not merely as pedagogical resources, but as epistemic agents capable of reshaping teacher identity. Rather than asking how indigenous practices can illustrate existing curricula, this autoethnography demonstrates how sustained engagement with such practices can destabilize inherited hierarchies of knowledge and expand what counts as mathematical rigor. Importantly, this awakening is not presented as a resolution. It is an ongoing responsibility. The risk of reinscribing hierarchy remains present in institutional demands, standardized assessments, and curricular structures. Cultural awakening, therefore, is not a completed state but a continuous reflexive practice.

Through the *Sama Tepo*, I learned that mathematics does not require abstraction to be legitimate, nor institutional validation to be rigorous. I learned that precision can reside in fingertips, that symmetry can be felt before it is drawn, and that proportional reasoning can live in memory and movement. Most of all, I learned that teaching mathematics in culturally complex contexts demands more than strategy. It demands epistemological humility. Just like the weave itself, teacher identity forms through tension and adjustment. Each encounter, each reflexive pause, each uncomfortable realization with prior assumptions.

The pattern is never static. I continually negotiated. The threads of wonder, then, are not confined to *pandan* strips. They extend into the ways we see, question, and position knowledge in our classrooms. In learning to approach the *Sama Tepo* not as an example of mathematics but as mathematics itself, I began to reweave my understanding of exploration, curiosity, and cultural responsibility. And in that reweaving, I began to become a different kind of mathematics teacher.

LIMITATION OF THE STUDY

As an autoethnographic inquiry grounded in personal experience, this study is inherently situated and relational. Its strength lies in depth rather than breadth. However, this depth also delineates its boundaries.

First, the findings emerge from a single researcher's lived engagement with the *Sama Tepo* across the specific contexts. These are in the group activity in EDM 132, interactions with actual weavers, and reflexive journaling. While the narrative offers analytic resonance for mathematics teacher education, it does not claim representational authority over the broader *Sama* community or over all the indigenous weaving practices. My interpretations remain shaped by my positionality as a pre-service mathematics teacher navigating institutional training.

Second, the autoethnographic methodology foregrounds reflexivity but relies heavily on memory, field notes, and self-interpretation. Although reflexive writing was iterative and sustained, recollection is never neutral. Experiences are filtered through emotion, hindsight, and theoretical framing. While the analysis reflects my evolving understanding rather than an objective account of events. This is not a methodological flaw but a characteristic of narrative inquiry. Nevertheless, it requires readers to interpret the findings as perspectival rather than definitive.

Third, my engagement with the weaving community was episodic rather than longitudinal. While the interviews and observations during the cultural events and weaving encounters were meaningful, they did not constitute extended ethnographic immersion. A long-term collaborative engagement with the weavers might reveal dimensions of mathematical practice and cultural continuity that exceed the scope of this study. Future research could deepen this relational engagement through the sustained partnership and co-authorship with the community members.

Fourth, the study centers the transformation of a pre-service teacher within a specific institutional and cultural context. Teacher identity formation may unfold differently across regions, teacher education programs, or cultural settings. The analytic transferability of these findings lies not in replication of context but in the resonance of epistemological disruption and reflexive negotiation.

Ultimately, this research engages the *Sama Tepo* as an epistemic interlocutor within mathematics education discourse. However, translating cultural practices into academic scholarship always carries the risk of re-centering institutional frameworks. Despite efforts toward epistemological humility, the act of writing about the *Tepo* within the scholarly conventions may inadvertently reinscribe the hierarchies the study seeks to question. This tension remains unresolved and must be continuously examined.

In acknowledging these limitations, I do not diminish the contribution of the study. Rather, I situate it appropriately. As a reflexive, context-bound account of becoming, a narrative that invites dialogues rather than closure. The purpose is not to generalize across contexts, but to illuminate how sustained engagement with culturally embedded mathematical practices can unsettle inherited assumptions and reshape teacher identity.

The incompleteness of this study mirrors the incompleteness of my own becoming. Like the weave itself, understanding develops through ongoing adjustment. What is offered here is not a final pattern, but a segment that is open to extension, reinterpretation, and further relational inquiry.

IMPLICATIONS

Beyond its contributions to mathematics education, this study situates itself within the broader movement of revitalizing Indigenous Knowledge Systems in the 21st century. The engagement with Sama Tepo weaving demonstrates that revitalization is not limited to cultural preservation or documentation, but also involves recontextualizing Indigenous knowledge within contemporary educational practices. In this sense, education becomes a key site where Indigenous knowledge is not only represented but actively sustained and transmitted to future generations. The journey documented in this autoethnography carries implications that extend beyond my personal transformation. While grounded in a specific engagement with the *Sama Tepo*, the insights generated speak to broader questions in mathematics teacher education, curriculum design, and culturally responsive pedagogy.

First, this study suggests that the teacher education programs must move beyond presenting culturally situated artifacts merely as instructional resources. The teacher education must instead create structured opportunities for epistemological reflection. A space where the pre-service teachers critically examine how their training shapes what they recognize as legitimate mathematics. Engaging deeply with indigenous and community-based practices should not only expand pedagogical repertoire but also destabilize inherited assumptions about the abstraction, rigor, and authority.

The findings further imply that mathematics teachers play a crucial role in the revitalization process. Rather than serving solely as transmitters of formal mathematical knowledge, teachers act as mediators between epistemic systems. By engaging with Indigenous practices such as Sama Tepo weaving, educators participate in the ethical responsibility of sustaining cultural knowledge and ensuring that it is not marginalized within school mathematics. This positions teacher identity formation as inherently linked to cultural continuity and knowledge preservation.

Second, the findings challenge curricular models that treat cultural practices as illustrative examples of pre-determined mathematical content. The *Sama Tepo* demonstrates that culturally embedded mathematical reasoning is not merely an application of school mathematics. It is a parallel epistemic system with its own internal coherence. Therefore, we must move toward epistemic symmetry rather than just assimilation. Instead of just asking, *"How can this practice demonstrate geometry?"* educators might ask, *"What forms of reasoning are already present here,* and *how can dialogue between knowledge systems occur without hierarchy?"* With these, it reframes rigor. Rigor need not be confined to symbolic manipulation. It can reside in embodied precision, iterative adjustment, and relational continuity. Recognizing this broadens students' mathematical identities and legitimizes diverse ways of knowing.

Third, this study challenges curriculum frameworks that treat Indigenous knowledge as supplementary or illustrative. Instead, it advocates for epistemic recognition, where Indigenous mathematical reasoning is acknowledged as a valid knowledge system in its own right. Such a shift contributes to revitalization by resisting assimilationist approaches and promoting dialogue between knowledge systems on more equal terms. This implication resonates with contemporary scholarship advocating for the integration of Indigenous Knowledge Systems into basic education as part of broader decolonization efforts. Bhuda and Maditsi (2025) argue that Indigenous knowledge should not occupy a peripheral position within schooling, but should instead be recognized as a legitimate and valuable knowledge system capable of enriching teaching, learning, and identity formation. Similarly, Bhuda (2024) emphasizes that ethnomathematical activities grounded in Indigenous cultural practices can function as meaningful pathways toward decolonizing mathematics education.

Fourth, this study affirms the value of autoethnography in mathematics education research. Teacher identity formation is often examined through surveys, interviews, or observational studies. Yet, this autoethnography demonstrates how deeply embedded epistemological assumptions can remain invincible without sustained reflexive writing. More importantly, this study suggests that cultural artifacts should not only be studied for the mathematics they contain, but also for the ways they reshape the researcher as well as the teacher.

Ultimately, this journey underscores that the curiosity in culturally complex contexts must be ethically grounded. Educational engagement with indigenous knowledge systems requires relational accountability, humility, and attentiveness to historical marginalization. Without reflexivity, well-intentioned integration can become subtle appropriation.

Hence, this paper invites mathematics education to reconsider a foundational assumption. Assumption that mathematics is universally abstract. The *Sama Tepo* reveals mathematics as lived, embodied, and relational. Recognizing this deepens the mathematics education. If the mathematics teacher education embraces epistemological humility through acknowledging plurality without hierarchy. The future teachers may enter classrooms not only prepared to explain concepts, but prepared to honor the mathematical worlds their students already inhabit. And perhaps that is the most enduring implication of this study. That is to teach mathematics responsibly is to remain open to being transformed by the knowledge one seeks to teach.

Future Research Directions

This study opens several possibilities for future inquiry at the intersection of mathematics education, Indigenous Knowledge Systems (IKS), ethnomathematics, and teacher identity formation. As a reflexive autoethnography grounded in a single researcher's lived engagement with the *Sama Tepo*, the study prioritizes depth and epistemological reflection rather than broad representation. Future research may therefore extend this work through collaborative, comparative, and community-centered approaches.

One important direction involves conducting long-term ethnographic or participatory research with *Sama* weavers and community members themselves. Such studies may further illuminate how mathematical reasoning is transmitted intergenerationally through weaving practices, oral traditions, and embodied apprenticeship. Community-based participatory approaches may also help address concerns regarding epistemic hierarchy by positioning Indigenous knowledge holders not merely as participants, but as co-researchers and co-authors in the production of knowledge.

Future studies may also examine how culturally grounded mathematical practices such as *Sama Tepo* weaving can be integrated into formal mathematics curricula without reducing them into decorative or illustrative examples. Research may explore how culturally sustaining pedagogies can support epistemic symmetry between Indigenous knowledge systems and school mathematics across different educational levels. In particular, classroom-based interventions may investigate how students respond to mathematics lessons grounded in local weaving traditions and whether such approaches influence mathematical engagement, cultural identity, and conceptual understanding.

Another important direction concerns teacher education. Further inquiry may explore how pre-service and in-service mathematics teachers negotiate epistemological tensions when engaging with Indigenous mathematical practices. Comparative studies across teacher education institutions, cultural communities, and regional contexts may reveal how professional identity formation shifts when teachers encounter culturally embedded forms of mathematical reasoning.

Future research may also expand beyond weaving practices and examine other Indigenous cultural activities in Tawi-Tawi and the broader Bangsamoro region as sites of mathematical knowledge. Traditional boat-making, navigation, architecture, embroidery, fishing systems, and community trade practices may offer rich opportunities for ethnomathematical inquiry and culturally grounded mathematics education.

Future studies may also investigate how ethnomathematical activities grounded in Indigenous cultural practices contribute to decolonizing mathematics education across diverse cultural contexts. Comparative studies involving weaving traditions, Indigenous arts, architecture, and local community practices may further illuminate how culturally grounded pedagogies influence teacher identity formation, mathematical engagement, and epistemological awareness. Recent studies involving Indigenous mathematical practices in African educational contexts (Bhuda, 2024; Motseki et al., 2026) suggest promising directions for cross-cultural ethnomathematics research.

Finally, future scholarship may continue interrogating the ethical dimensions of research involving Indigenous Knowledge Systems. Questions surrounding representation, reciprocity, cultural ownership, and epistemic justice remain crucial. As educational institutions increasingly recognize the importance of decolonizing knowledge, future studies must remain attentive to how Indigenous practices are engaged, interpreted, and sustained within academic spaces.

REFERENCES

Adams, T. E., Ellis, C., & Jones, S. H. (2017). Autoethnography. *The international encyclopedia of communication research methods*, 1-11. DOI: 10.1002/9781118901731.iecrm0011

Battiste, M. (2013). *Decolonizing Education: Nourishing the Learning Spirit*. Purich Publishing Ltd.

Bhuda, M. T. (2024). Integrating ethnomathematical activities in the foundation phase as means of decolonisation: Insights from Indigenous Knowledge Systems. *African Journal of Development Studies*, *14*(4), 401. https://hdl.handle.net/10520/ejc-aa_affrika1_v14_n4_a18

Bhuda, M. T., & Maditsi, M. E. (2025). The importance of integrating Indigenous knowledge systems in South Africa's basic education: Perspectives from Indigenous scholars. In *Resurgence and revalorization of Indigenous knowledge systems in the contemporary society* (pp. 43–68). IGI Global Scientific Publishing., DOI: 10.4018/979-8-3373-4124-8.ch003

d'Ambrosio, U. (2016). Ethnomathematics: A response to the changing role of mathematics in society. DOI: 10.1108/978-1-681123-261-420251003

Ellis, C., Adams, T. E., & Bochner, A. P. (2011). Autoethnography: An overview. *Historical social research. Historische Sozialforschung*, •••, 273–290. https://www.jstor.org/stable/23032294

Motseki, P. D., Jojo, Z. M. M., & Gumbo, M. T. (2026). Transforming mathematics teaching through Ndebele art: An ethnomathematics perspective. *Diaspora, Indigenous, and Minority Education*, *20*(1), 1–7. DOI: 10.1080/15595692.2025.2451239

Paris, D. (2021, July). Culturally sustaining pedagogies and our futures. In *The educational forum* (Vol. 85, No. 4, pp. 364-376). Routledge. DOI: 10.1080/00131725.2021.1957634

Trinick, T., & Allen, P. (2024). Wayfinding in an indigenous initial teacher education mathematics programme. *ZDM Mathematics Education*, *56*(3), 485–495. DOI: 10.1007/s11858-024-01589-5

UNESCO. (2023). *What is intangible cultural heritage?* https://ich.unesco.org/en/what-is-intangible-heritage-00003

Vandendriessche, E., & Pinxten, R. (Eds.). (2022). *Indigenous knowledge and ethnomathematics*. Springer., DOI: 10.1007/978-3-030-97482-4

KEY TERMS AND DEFINITIONS

Autoethnography: A qualitative research methodology that uses personal experience as a lens for understanding broader cultural, social, and educational phenomena (Adams et al., 2017). In this study, autoethnography is used to examine the researcher's evolving understanding of mathematics, culture, and teacher identity through engagement with Sama Tepo weaving.

Cultural Awakening: A process of developing critical awareness of one's assumptions, positionality, and relationship to cultural knowledge systems. In this study, cultural awakening refers to the researcher's growing recognition of Indigenous mathematical knowledge as legitimate and epistemologically valuable.

Culturally Sustaining Pedagogy: An educational approach that seeks to sustain, affirm, and value students' cultural practices, languages, and identities rather than assimilating them into dominant frameworks (Paris, 2021).

Epistemological Pluralism: The recognition that multiple ways of knowing and understanding reality can coexist without one being considered universally superior. In this study, epistemological pluralism challenges the dominance of formal Western mathematics as the sole legitimate form of mathematical knowledge.

Ethnomathematics: The study of mathematical ideas and practices embedded within cultural activities, traditions, and daily life (D'Ambrosio, 2016). This study examines the mathematical reasoning present within *Sama Tepo* weaving practices.

Indigenous Knowledge Systems (IKS): Knowledge systems developed and sustained by Indigenous communities through lived experience, cultural practice, oral tradition, and intergenerational transmission. In this study, *Sama Tepo* weaving is approached as a living Indigenous Knowledge System.

Positionality: The recognition that a researcher's identity, experiences, cultural background, and institutional location influence how research is conducted and interpreted.

Reflixivity: The continuous process of critically examining one's assumptions, interpretations, and role within the research process. Reflexivity is central to this autoethnographic inquiry.

Sama Tepo: A traditional woven mat from Tawi-Tawi created by the *Sama* community. Beyond being a cultural artifact, the *Sama Tepo* is understood in this study as a site of mathematical reasoning, cultural memory, and Indigenous knowledge.

Teacher Identity Formation: The ongoing process through which teachers develop professional beliefs, values, pedagogical orientations, and epistemological understandings through experience, reflection, and social interaction.

APPENDIX: ADDITIONAL READING

Readers interested in further exploring the intersections of ethnomathematics, Indigenous Knowledge Systems, autoethnography, culturally sustaining pedagogy, and mathematics teacher education may consult the following works: Adams, T. E., Ellis, C., & Jones, S. H. (2017). *Autoethnography*. This work provides a foundational discussion of autoethnography as a qualitative methodology that connects personal narrative with broader cultural analysis. Battiste, M. (2013). *Decolonizing Education: Nourishing the Learning Spirit*. Battiste examines the marginalization of Indigenous knowledge systems and argues for educational frameworks grounded in epistemic plurality and decolonization. D'Ambrosio, U. (2016). *Ethnomathematics: A Response to the Changing Role of Mathematics in Society*. This text explores how mathematical thinking is embedded in cultural practices and challenges universalist assumptions about mathematics. Paris, D. (2021). *Culturally Sustaining Pedagogies and Our Futures*. Paris discusses the importance of sustaining rather than assimilating cultural practices within educational institutions. Vandendriessche, E., & Pinxten, R. (2022). *Indigenous Knowledge and Ethnomathematics*. This volume examines Indigenous mathematical reasoning across diverse cultural practices and offers important perspectives on epistemological plurality. Trinick, T., & Allen, P. (2024). *Wayfinding in an Indigenous Initial Teacher Education Mathematics Programme*. This study introduces the concept of cultural symmetry and discusses Indigenous-centered approaches in mathematics teacher education.

Readers may also explore literature on culturally responsive pedagogy, decolonizing methodologies, Indigenous epistemologies, and qualitative narrative inquiry to deepen understanding of the broader theoretical foundations informing this study.

Chapter 5
Unravelling the Moral Epistemologies of the Ramayana:
Oral Traditions, Narrative Ethics, and Comparative Philosophical Inquiry

Mani Sachdev
http://orcid.org/0009-0000-8284-9718
Amity Institute of Behavioural and Allied Sciences, Amity University, Jaipur, India

Saad Ullah Khan
http://orcid.org/0000-0002-3392-7844
Centre for Media and Mass Communication Studies, Jamia Hamdard, New Delhi, India

Aahana Saha
http://orcid.org/0009-0004-0172-5932
Department of Applied Psychology, GITAM School of Humanities and Social Sciences, GITAM University, Hyderabad, India

ABSTRACT

This chapter reconstructs the Ramayana as a structured indigenous moral epistemology rather than a merely devotional epic. Through systematic thematic coding of all seven kāṇḍas, the epic's major narrative divisions, using NVivo-assisted analysis, co-occurrence matrices, PCA clustering, and network visualization, it identifies patterns of ethical intensity and maps the relational architecture of dharma (a context-sensitive moral order), leadership, and justice. Situated within the Purushartha framework (the fourfold aims of human life), the study demonstrates that dharma functions as narrative-based moral discernment rather than rigid rule adherence. By critically engaging Aristotelian virtue ethics and Kantian deontology, the chapter establishes the Ramayana's internal normative coherence as a philosophically robust knowledge system. It argues that revitalizing Indigenous Knowledge Systems requires recognizing narrative traditions as norm-generating epistemic structures capable of informing contemporary debates on governance, responsibility, and ethical restraint.

DOI: 10.4018/979-8-2600-2890-2.ch005

INTRODUCTION

The *Ramayana* occupies a foundational position within South Asia's indigenous intellectual traditions, functioning not merely as a literary epic but as a living ethical knowledge system transmitted across generations through oral, performative, and textual practices. Attributed to Sage Valmiki and composed in Sanskrit, its seven Kāṇḍas narrate the life of Lord Rama across domains of kinship, sovereignty, exile, devotion, and war. Yet the epic's enduring significance lies not only in its narrative grandeur but in its capacity to encode and transmit a relational moral philosophy grounded in the Indic concept of dharma. Rather than presenting ethics as abstract doctrine, the *Ramayana* situates moral reasoning within lived contexts, where decisions emerge from the interplay of duty, circumstance, and relational responsibility. Through Rama's journey, ethical principles are not merely stated; they are tested, embodied, and negotiated within complex social and political worlds (Brockington, 1998).

Within the contemporary discourse on revitalizing indigenous knowledge systems, the *Ramayana* may be approached as a dynamic ethical archive that preserves culturally grounded modes of reasoning about governance, justice, ecological balance, and social responsibility. Indigenous knowledge traditions often encode normative frameworks that sustain community cohesion and environmental equilibrium. The *Ramayana*'s emphasis on righteous leadership, restraint in the use of power, respect for ecological habitats during exile, and the interdependence of rulers, citizens, and the natural world offers a model of ethically bounded action that resonates with present-day concerns about sustainability, technological expansion, and social governance. Re-engaging such texts, therefore, constitutes not an antiquarian exercise but a form of critical retrieval—one that repositions indigenous moral systems as viable intellectual resources for addressing contemporary ethical dilemmas. In this respect, the ethical concerns articulated in the epic anticipate modern reflections on responsibility in an age of technological power (Jonas, 1984).

Scholarly engagement with the epic has traditionally focused on its textual formation, religious symbolism, and political imagination (Brockington, 1998; Sharma, 1999). While these approaches have clarified its historical and doctrinal dimensions, another layer emerges when the *Ramayana* is examined as an internally structured moral system whose thematic organization unfolds across its narrative arc. Ethical constructs in the epic are neither incidental nor isolated; they recur with patterned intensity, intersect dynamically, and acquire differentiated significance across contexts. Viewed cumulatively, the narrative reveals a relational ethical architecture rather than a sequence of exemplary moral episodes. This perspective resonates with philosophical accounts that understand moral identity as narratively constituted and historically situated (MacIntyre, 2007; Ricoeur, 1992).

This study advances three interrelated aims. First, it systematically identifies the distribution and interrelation of core ethical categories across all seven Kāṇḍas, thereby illuminating the epic's internal moral structuring as an indigenous ethical knowledge system. Second, it interprets these patterns within the conceptual horizon of Indic philosophical thought, particularly the Purushartha framework and the normative reasoning articulated within Dharmaśāstric traditions (Sharma, 1999; Matilal, 2002). Third, it develops a methodologically rigorous model that combines qualitative hermeneutics with digital analytic tools to examine classical narrative texts without diminishing their philosophical depth. In doing so, the chapter demonstrates how indigenous ethical systems can be critically reinterpreted through contemporary analytical methods, fostering a dialogue between traditional wisdom and modern knowledge practices.

Within the schema of Purushartha—dharma, artha, kama, and moksha—the moral landscape of the *Ramayana* acquires interpretive coherence. Dharma emerges not as a rigid prescription but as a mediating principle that orients personal conduct while sustaining social and cosmic order. Episodes such as

Rama's acceptance of exile, Bharata's renunciation of kingship, Hanuman's disciplined devotion, and the morally contested killing of Vali illustrate ethical judgment under conditions of conflict and uncertainty. Such moments foreground discernment rather than doctrinal certainty, presenting dharma as context-sensitive reasoning embedded in lived relationships. This interpretive orientation aligns with Matilal's (2002) reflections on epic moral reasoning as philosophically substantive rather than merely didactic.

To render these ethical dynamics analytically visible, the study employs qualitative inductive thematic analysis grounded in Braun and Clarke's (2006) six-phase framework, complemented by digital visualization techniques including cluster mapping and network analysis. By identifying ethically charged narrative units—designated as 'ethical shlokas'—and mapping their thematic distribution, the study reconstructs the epic's internal moral ecology. This approach not only preserves the semantic texture of the narrative but also demonstrates how indigenous ethical systems can be systematically examined and recontextualized within contemporary scholarly and technological frameworks. Dialogical engagement with virtue ethics and narrative moral philosophy further situates the findings within cross-cultural ethical discourse (MacIntyre, 2007; Nussbaum, 2001).

Positioned within the broader project of revitalizing indigenous knowledge systems in the twenty-first century, this analysis underscores the *Ramayana*'s continuing relevance as a culturally grounded moral resource. In an era marked by ecological vulnerability, technological acceleration, and governance crises, indigenous ethical traditions offer alternative modes of reasoning that privilege relational balance, responsibility, and restraint. By approaching the *Ramayana* as a structured moral system rather than a static cultural artifact, this chapter contributes to the critical retrieval and contemporary reinterpretation of indigenous knowledge, affirming its potential to inform ethical discourse and sustainable social futures.

OBJECTIVES OF THE STUDY

This study pursues interrelated objectives to illuminate the ethical architecture of the *Ramayana* as an indigenous knowledge system by systematically mapping ethically salient narrative units across all seven Kāṇḍas and examining the relational structure of key thematic domains—Dharma and Moral Duty, Ethical Leadership, Justice and Governance, Compassion and Empathy, Humility and Reverence, Forbearance and Restraint, and Moral Dilemma and Wisdom—situated within the Purushartha framework and classical Dharmic reasoning. By interpreting these thematic interrelations, the study seeks to demonstrate how the epic functions as a culturally grounded moral system that encodes principles of responsible governance, social harmony, and relational balance.

A further objective is to develop a methodologically rigorous model that combines qualitative hermeneutics with digital visualization techniques, including cluster mapping and network analysis, in order to render indigenous ethical patterns analytically visible without diminishing their philosophical depth. Through this integrative approach, the study evaluates the continuing relevance of the *Ramayana*'s ethical reasoning for contemporary debates on governance, sustainability, technological responsibility, and the revitalization of indigenous knowledge systems in the twenty-first century.

CONCEPTUAL AND THEORETICAL FRAMEWORK

This study develops a conceptual framework that situates the *Ramayana* within the broader project of revitalizing indigenous knowledge systems by interpreting its narrative structure as a relational ethical ecology. The framework integrates the classical Purushartha schema with narrative ethics and digital thematic mapping in order to demonstrate how indigenous moral reasoning operates as a dynamic and context-responsive system rather than a static doctrinal code.

The upper tier of the framework is constituted by the four Purushartha principles—Dharma, Artha, Kama, and Moksha—which function as teleological coordinates for ethical judgment. Within Indic intellectual traditions, these principles articulate a holistic vision of human flourishing that balances moral duty, material well-being, desire, and spiritual liberation. Rather than prescribing rigid moral rules, the Purushartha schema provides a flexible normative orientation that allows ethical reasoning to respond to social, political, and ecological circumstances. As such, it represents a foundational component of the subcontinent's indigenous moral philosophy (Sharma, 1999; Matilal, 2002).

The middle tier operationalizes these principles through seven thematic ethical domains: Dharma and Moral Duty, Ethical Leadership, Justice and Governance, Compassion and Empathy, Humility and Reverence, Forbearance and Restraint, and Moral Dilemma and Wisdom. These domains represent the lived enactment of Purushartha ideals within specific social contexts. Their interconnections illustrate the relational character of indigenous ethical reasoning, where moral action emerges through the balancing of multiple values rather than the application of abstract rules. This relational structure reflects narrative conceptions of moral identity, in which ethical understanding develops through action, conflict, and resolution within the continuity of lived experience (MacIntyre, 2007; Ricoeur, 1992).

The bottom tier, designated as Narrative Ethics Enactment, represents ethically charged narrative units or 'ethical shlokas', in which value conflict, moral decision, and transformative action unfold. At this level, ethical reasoning becomes embodied in concrete situations involving kinship, governance, exile, and war. These narrative moments demonstrate how indigenous ethical systems transmit normative knowledge through stories, exempla, and performative traditions rather than through abstract philosophical treatises. In this sense, the *Ramayana* functions as a living repository of culturally embedded moral knowledge.

Methodologically, the framework combines qualitative hermeneutics with digital visualization techniques, including cluster mapping and network analysis, to render ethical patterns analytically visible. Thematic coding grounded in Braun and Clarke's (2006) inductive model enables systematic identification of ethical structures while preserving the semantic richness of the narrative. This integrative approach reflects contemporary efforts to revitalize indigenous knowledge by placing traditional wisdom traditions into dialogue with modern analytical and technological tools.

By positioning the *Ramayana* as a structured indigenous ethical system that addresses questions of governance, responsibility, and relational balance, the framework demonstrates its relevance to contemporary debates on sustainability, technological ethics, and social cohesion. In this sense, the conceptual model does not merely interpret an ancient epic; it participates in the critical retrieval and contemporary rearticulation of indigenous moral knowledge for the twenty-first century.

REVIEW OF LITERATURE

The philosophical status of the *Ramayana* has long remained under-theorized despite sustained textual, philological, and cultural scholarship. Twentieth-century research established the epic's compositional stratification, historical layering, and manuscript plurality (Brockington, 1998; The Rāmāyaṇa of Vālmīki, 1984). These studies decisively displaced devotional essentialism, demonstrating that the text evolved through redactional processes extending across centuries. Yet philological rigor, while indispensable, left largely unaddressed a deeper normative question: does the *Ramayana* merely preserve inherited codes of dharma, or does it articulate a structured mode of moral reasoning capable of philosophical reconstruction?

The pluralist turns in *Ramayana* studies—most prominently articulated by A. K. Ramanujan (1991)—further destabilized the assumption of canonical singularity. By foregrounding the existence of "many *Ramayanas*," Ramanujan emphasized narrative multiplicity, regional reinterpretation, and theological diversity. Paula Richman (1991) extended this approach, demonstrating how vernacular, performative, and gendered retellings recalibrate moral emphasis. Wendy Doniger (2010) situated such plurality within a broader interpretive account of Hindu traditions as internally dialogical. Collectively, this scholarship productively resisted homogenizing readings and exposed the epic as historically dynamic.

Yet the pluralist paradigm generates a philosophical tension. If the *Ramayana* exists only as a network of divergent tellings, can it sustain normative coherence without collapsing into relativism? The persistence of recurrent ethical crises across recensions—filial obedience versus sovereign authority, conjugal fidelity versus public legitimacy, compassion versus juridical responsibility—suggests patterned continuity rather than arbitrary narrative variation. The interpretive problem thus shifts from textual multiplicity to normative architecture: how does ethical coherence persist amid narrative plurality?

A decisive intervention emerges in Bimal Krishna Matilal's *Ethics and Epics: The Collected Essays of Bimal Krishna Matilal Volume II* (2002). Matilal argues that Sanskrit epics should not be reduced to Dharmaśāstric didacticism. Instead, they dramatize moral conflict under conditions of tragic constraint. Dharma, in this account, is neither rigid rule nor metaphysical abstraction; it emerges through context-sensitive deliberation. The killing of Vali, Rama's acceptance of exile etc. are not exemplary affirmations of predetermined codes but instances of moral ambiguity. Matilal's analysis implicitly aligns epic reasoning with Aristotelian phronesis—practical wisdom exercised amid irreducible contingency.

These Aristotelian parallel warrants explicit scrutiny. In the *Nicomachean Ethics* (trans. Irwin, 2000), Aristotle defines virtue as a mean relative to us, determined by reason and guided by the practically wise agent. Moral judgment cannot be mechanically deduced from universal principles; it requires discernment sensitive to particulars. At first glance, the *Ramayana*'s depiction of dharma appears structurally analogous. Rama's actions are repeatedly contextual, shaped by relational obligations and situational constraints rather than by formalizable maxims. Like Aristotle's virtuous agent, Rama navigates ethical life through habituated excellence cultivated within a tradition.

However, the analogy remains incomplete. Aristotelian virtue ethics is teleologically oriented toward *eudaimonia*—human flourishing within the polis and the realization of excellence through civic participation. The ethical horizon of the *Ramayana*, by contrast, extends beyond political and social flourishing to include transcendental aspiration in the form of *moksha* (liberation). Its moral framework is structured through the dynamic relationship among *dharma* (ethical duty), *artha* (political and material order), *kama* (desire), and *moksha* (spiritual release), where worldly responsibilities are not ends in themselves but are situated within a larger metaphysical order. Political responsibility, familial obligation, and personal virtue are thus embedded within a cosmological teleology rather than confined to civic life alone. Where Aristotle's ethical vision culminates in contemplative activity within the framework of the polis, the *Ramayana* locates moral excellence within an ontological horizon that ultimately transcends the political community.

If Aristotelian comparison reveals proximity, Kantian ethics sharpens divergence. In the *Groundwork of the Metaphysics of Morals* (Kant, 1785/1998), Kant articulates morality as grounded in universalizable maxims derived from pure practical reason. The categorical imperative demands that actions be evaluated independently of contingent relationships or particular social roles. Moral worth resides in adherence to duty for its own sake.

At first glance, Rama's unwavering commitment to paternal command appears consonant with Kantian duty. His exile is accepted not because it maximizes happiness but because it fulfils obligation. Yet closer scrutiny reveals decisive differences. Kantian morality abstracts from relational embeddedness; it is grounded in rational autonomy. By contrast, dharma in the *Ramayana* is irreducibly relational. Duties arise not from universalizable maxims but from position within a web of kinship, sovereignty, and cosmic order. Moreover, the epic repeatedly stages conflicts between duties—filial obedience versus protection of subjects, conjugal loyalty versus royal legitimacy. Kantian formalism provides procedural clarity but struggles to accommodate tragic collisions of equally binding obligations without privileging one rule hierarchically. The *Ramayana* instead dramatizes such collisions without resolving them into algorithmic hierarchy.

This divergence exposes a deeper philosophical tension: is morality fundamentally principle-driven or relationally constituted? Alasdair MacIntyre's *After Virtue* (2007) offers a mediating framework. MacIntyre argues that moral rationality is tradition-constituted; virtues are intelligible only within historically extended narratives. Agents inherit practices and goods internal to traditions that structure evaluative reasoning. The *Ramayana* exemplifies precisely such narrative constitution of moral intelligibility. Rama's agency is inseparable from the moral history he inhabits. His actions derive meaning from participation in a shared story extending beyond individual autonomy.

Paul Ricoeur's *Oneself as Another* (1992) deepens this narrative account. For Ricoeur, ethical identity emerges through emplotment across time; responsibility is mediated narratively rather than deduced formally. The *Ramayana*'s seven Kāṇḍas trace precisely such temporal mediation. Exile becomes not mere misfortune but moral apprenticeship. Kingship emerges not as entitlement but as culmination of disciplined selfhood. Narrative continuity thus functions epistemically, shaping moral perception.

Affective theory further complicates the philosophical terrain. Martha Nussbaum's *Upheavals of Thought* (2001) rejects the dichotomy between emotion and reason, arguing that emotions are forms of evaluative judgment. Compassion, grief, anger—these are not irrational residues but cognitive appraisals of value. The epic's pervasive cultivation of karuṇā suggests an affective pedagogy. Ethical knowledge is not transmitted propositionally but enacted through disciplined emotional participation. The audience is invited not merely to observe Rama's choices but to feel their moral gravity. In this respect, the epic challenges rationalist accounts of moral epistemology that privilege formal reasoning over affective formation.

Hans Jonas's *The Imperative of Responsibility* (1984) introduces yet another comparative axis. Writing in response to technological modernity, Jonas argues that ethical reasoning must account for long-term stewardship and vulnerability. Sovereignty entails burdened accountability rather than unrestrained authority. The epic conception of rājyadharma resonates strongly with this responsibility-centred framework. Rama's kingship is portrayed not as dominion but as sacrificial obligation. Political power is legitimized through moral restraint. Such structural resonance indicates that dharmic reasoning engages concerns central to contemporary ethical discourse.

The performative dimension of epic transmission further reinforces its epistemic complexity. Sheldon Pollock's *The Language of the Gods in the World of Men* (2006) situates Sanskrit literary production within broader formations of political authority and cultural self-articulation. Epic narrative functioned not merely as literary artifact but as civilizational discourse. James Fitzgerald (2004) similarly emphasizes the performative shaping of ethical consciousness through epic recitation. The *Ramayana* thus operates as lived moral pedagogy. Ethical knowledge is habituated communally, not merely contemplated individually.

Despite this convergence of scholarship, a systematic reconstruction of the epic's normative architecture remains underdeveloped. Existing literature tends to isolate dilemmas, privilege plurality, or draw comparative parallels without articulating how dharma, virtue, emotion, narrative continuity, and sovereign responsibility interlock as a coherent system. The persistence of structurally homologous ethical tensions across textual strata suggests patterned relationality rather than episodic coincidence. Yet the philosophical articulation of this relational system remains incomplete.

Comparative engagement clarifies what is at stake. Aristotelian virtue ethics illuminates the role of character formation but cannot account fully for the epic's cosmological embedding of moral life. Kantian deontology provides formal universality but abstracts from the relational embeddedness central to dharmic reasoning. MacIntyre and Ricoeur highlight narrative constitution but operate within predominantly Western historical frameworks. Nussbaum foregrounds affective cognition but does not situate it within cosmologically structured obligation. Jonas articulates responsibility but addresses technological modernity rather than epic sovereignty.

The *Ramayana*, by contrast, integrates these dimensions without reducing any single element to primacy. Dharma operates as relational equilibrium among competing goods; character is cultivated narratively; emotions function cognitively; sovereignty is constrained by responsibility; transcendental aspiration frames political order. The epic thereby resists assimilation into any single Western moral paradigm while simultaneously engaging them productively.

For contemporary philosophy of religion and moral theory, this reconstruction carries significant implications. If moral knowledge can be generated through narrative mediation and affective cultivation rather than through purely abstract principle, then prevailing analytic models of moral epistemology require expansion. The epic suggests that ethical rationality may be tradition-embedded, narratively structured, emotionally cultivated, and cosmologically oriented without forfeiting coherence. It challenges the assumption that universality must entail abstraction from relational embeddedness.

The central lacuna in scholarship, therefore, lies not in demonstrating the presence of moral reflection within the *Ramayana*, but in articulating its internal normative architecture as a philosophically intelligible system. Such articulation requires moving beyond thematic description toward structural reconstruction. It demands recognition that dharma is neither rigid deontology nor mere virtue ethics, neither relativistic plurality nor metaphysical absolutism. It is an emergent epistemology enacted through dramatized conflict, temporal formation of character, disciplined affect, and relational accountability.

The *Ramayana* thus enters global moral discourse not as an exotic counterexample but as a substantive interlocutor. Its narrative architecture interrogates Kantian formalism, deepens Aristotelian virtue theory, complements MacIntyrean tradition-constituted rationality, and enriches contemporary debates on moral epistemology. The philosophical task is not to assimilate dharma into Western categories, but to allow its narrative logic to expand the conceptual vocabulary through which moral rationality itself is understood.

RESEARCH METHODOLOGY

This study employs a qualitative, inductive thematic design to reconstruct the ethical architecture of the *Ramayana* through systematic textual analysis. The seven Kāṇḍas of the critical Sanskrit edition (Goldman et al., 1984–2017) constitute the primary corpus. The text was segmented into analytically discrete narrative units, within which ethically salient passages—defined as moments of value conflict, moral deliberation, transformation, or relational responsibility—were identified through close hermeneutic reading.

The analysis followed Braun and Clarke's (2006) six-phase framework for thematic analysis, adapted for classical textual material. Coding proceeded inductively, allowing ethical domains to emerge from sustained engagement with the narrative rather than being imposed a priori. Through iterative refinement and constant comparison, thematic clusters were consolidated into relationally structured domains including Dharma and Moral Duty, Ethical Leadership, Justice and Governance, Compassion and Empathy, Forbearance and Restraint, Humility and Reverence, and Moral Dilemma and Wisdom.

NVivo software facilitated systematic coding and hierarchical organization. Analytical reliability was strengthened through independent coding of selected passages and deliberative reconciliation grounded in textual evidence. To illuminate relational structure, thematic co-occurrence matrices, cluster analysis, Principal Component Analysis (PCA), and network visualization were employed. These digital techniques functioned as heuristic instruments, supporting but not replacing philosophical interpretation.

Throughout, interpretive fidelity was maintained by anchoring analysis within Indic conceptual frameworks and Sanskrit semantic nuance. The study thus integrates hermeneutic rigor with structural mapping to render visible the patterned moral reasoning embedded within the epic narrative.

THEMATIC ANALYSIS & FINDINGS

Thematic Analysis Report: Bal Kāṇḍa

Figure 1. Frequency of key themes extracted from the Bal Kāṇḍa analysis

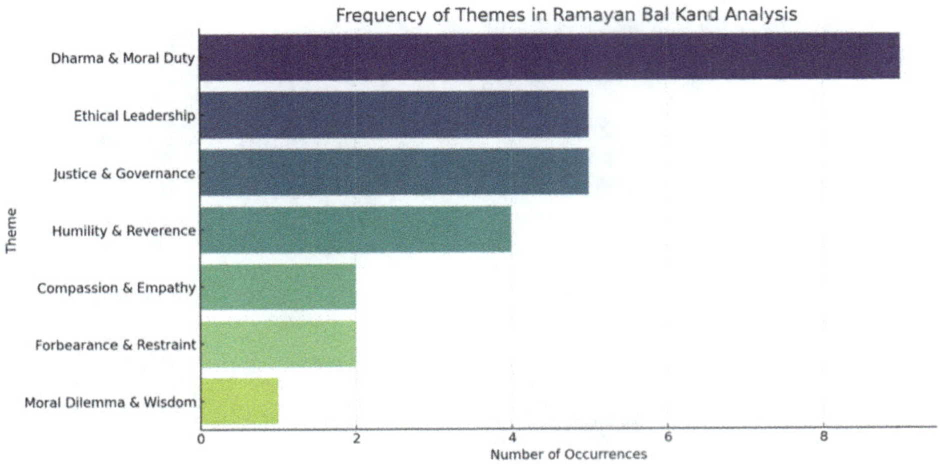

Figure 1 presents a horizontal bar chart illustrating the frequency of seven recurring themes identified in the *Bal Kāṇḍa* of the *Ramayan* through qualitative thematic analysis. Using Braun & Clarke's (2006) six-phase framework and grounded theory principles, themes were inductively coded and refined through constant comparison.

1. **Dharma & Moral Duty (n = 9)** – Most frequent, highlighting the centrality of righteousness and obedience (e.g., Rama's duty to Vishwamitra, Dasharatha's scriptural obligations).
2. **Ethical Leadership (n = 5)** – Reflects righteous governance and spiritual leadership (e.g., Dasharatha, Vishwamitra).
3. **Justice & Governance (n = 5)** – Focuses on law, order, and equitable action, often linked with spiritual authority.
4. **Humility & Reverence (n = 4)** – Embodied through respectful behaviour toward elders, rituals, and traditions.
5. **Compassion & Empathy (n = 2)** – Subtle yet present in responses to suffering or divine care.
6. **Forbearance & Restraint (n = 2)** – Represents emotional control and maturity, particularly in Rama's trials.
7. **Moral Dilemma & Wisdom (n = 1)** – Least frequent but philosophically rich, highlighting ethical discernment in complex situations.

Though frequency does not imply depth, the chart offers a snapshot of thematic emphasis. *Dharma & Moral Duty* dominates, aligning with the *Ramayan*'s role as a moral guide. The analysis reveals *Bal*

Kāṇḍa as a psychologically and ethically rich narrative, blending spiritual ideals with models of character development and governance.

Figure 2. Heatmap showing co-occurrence between major themes

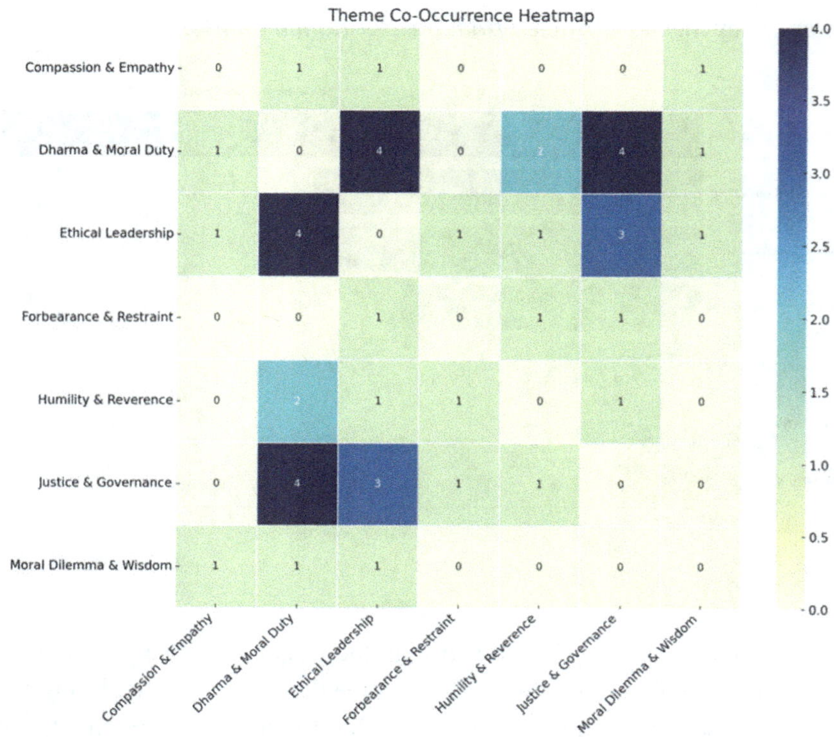

Figure 2 presents a heatmap illustrating the co-occurrence of major themes in the *Bal Kāṇḍa* of the *Ramayana*. The symmetrical matrix shows how often pairs of themes appear together within the same narrative unit (e.g., verse or paragraph). A colour gradient from pale yellow (low) to dark blue (high) indicates the frequency of co-occurrence, with self-pairings excluded.

High co-occurrence values (4) are observed between *Dharma & Moral Duty*, *Ethical Leadership*, and *Justice & Governance*. This triad forms the ethical foundation of the *Bal Kāṇḍa*, particularly in episodes involving Rama and Dasharatha, where moral duty aligns with just leadership and righteous governance.

Moderate co-occurrence (3) between *Ethical Leadership* and *Justice & Governance* reinforces the ideal of the *Rajarshi*—a ruler whose personal virtue ensures fair and ethical rule.

Low to moderate co-occurrence (1–2) is seen with themes like *Humility & Reverence*, which connects most strongly with *Dharma* and lightly with others such as *Justice* and *Forbearance*. Themes like *Compassion*, *Forbearance*, and *Moral Dilemma* appear less frequently alongside others, yet they add emotional nuance and introspective depth to the narrative.

Isolated pairs (0)—such as *Compassion – Forbearance* or *Justice – Moral Dilemma*—indicate themes that tend to function independently, often highlighting singular moments of emotional or philosophical importance.

Overall, the heatmap confirms *Dharma* as the central thematic force, often intersecting with structural and emotional values. While frequently paired themes reflect the integrated moral fabric of the epic, isolated or rare overlaps underscore reflective or transformative moments within the *Bal Kāṇḍa*'s rich ethical universe.

Figure 3. PCA-based cluster visualization of thematic units

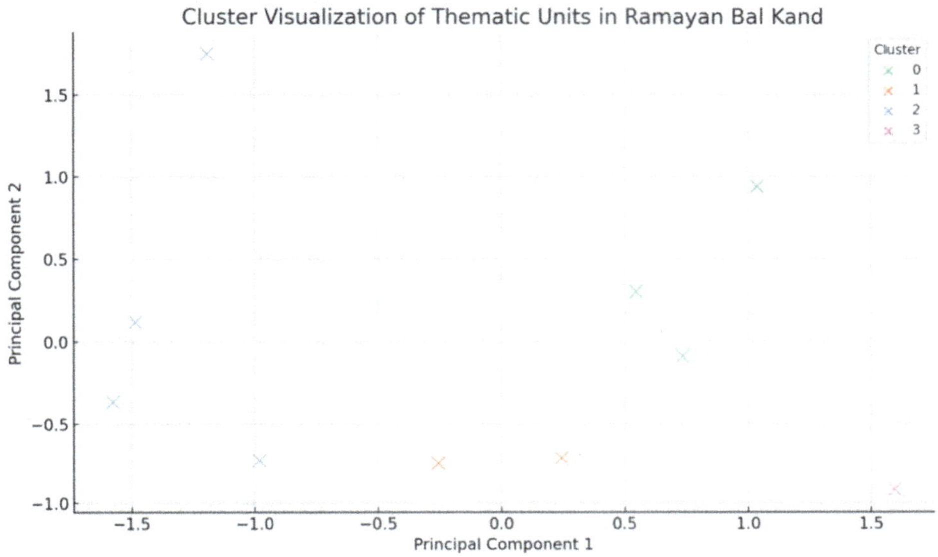

Figure 3 displays a scatterplot based on Principal Component Analysis (PCA) and clustering (e.g., K-means) of thematically coded narrative units from the *Bal Kāṇḍa* of the *Ramayana*. Each 'x' marks a coded segment, grouped into four color-coded clusters. The axes represent two principal components that capture the maximum variance in the data, offering insight into how themes semantically relate across the text.

Cluster 0 (Green): Dharma-Centric Leadership includes segments rich in *Dharma*, *Ethical Leadership*, and *Justice*. Characters like Dasharatha, Vishwamitra, and Rama (as leaders) appear prominently, reflecting the epic's core ideals of moral governance and righteous authority.

Cluster 1 (Orange): Interpersonal Ethics and Emotional Virtues features themes like *Compassion*, *Humility*, and *Forbearance*. These segments, involving Sita, Valmiki, and young Rama, bring emotional depth and emphasize personal virtue, reverence, and social harmony.

Cluster 2 (Blue): Moral Dilemmas and Ethical Tensions centres on *Moral Dilemma & Wisdom*, exploring internal conflicts faced by figures like Rama and various sages. It highlights psychological complexity and ethical ambiguity in decision-making.

Cluster 3 (Purple): Ritual, Renunciation, and Cosmic Order focus on spiritual themes like *tapas*, *tyaga*, and divine rituals. Sages and divine beings dominate this cluster, which underlines the metaphysical foundation of the epic.

Using vectorized thematic codes and unsupervised learning, the analysis identifies coherent patterns. This clustering confirms the *Bal Kāṇḍa*'s multi-layered structure—where political, emotional, ethical, and spiritual dimensions are distinct yet interconnected, reflecting the integrated vision of Indic moral philosophy.

Figure 4. illustrates a mind map of major themes and subthemes from the Bal Kāṇḍa of the Ramayana, using a node-link diagram where central themes connect to related subthemes. Cross-linkages such as Divine Legitimacy and Spiritual Deference appear in multiple clusters, reflecting foundational connections across moral, political, and spiritual domains

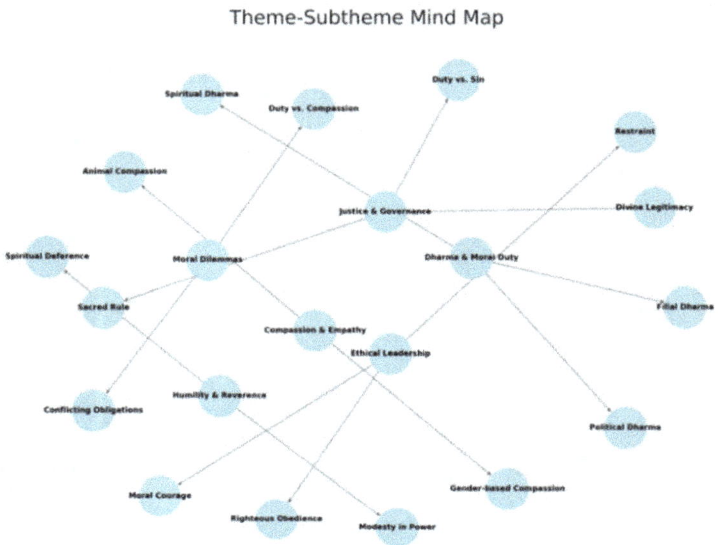

Dharma & Moral Duty branches into *Filial Dharma*, *Political Dharma*, and *Divine Legitimacy*, showing how duty spans personal, social, and divine spheres—aligned with the concept of *svadharma*. *Ethical Leadership* includes *Modesty in Power*, *Gender-Based Compassion*, and *Righteous Obedience*, reflecting the ideal of *rajarshi*—virtuous, humble rulers like Vishwamitra and Dasharatha.

Justice & Governance includes *Duty vs. Sin*, *Sacred Rule*, and shares *Divine Legitimacy*, reinforcing the spiritual basis of just rule. *Moral Dilemmas* involve internal conflict, with subthemes like *Duty vs. Compassion* and *Conflicting Obligations*, emphasizing ethical discernment (*viveka*).

Compassion & Empathy shares subthemes like *Animal Compassion* and *Gender-Based Compassion*, showing that relational sensitivity is integral to dharma. *Humility & Reverence* includes *Moral Courage* and *Spiritual Deference*, positioning reverence as a strength tied to higher wisdom. *Forbearance & Restraint*, though minimal, centers on *Restraint*as a form of inner discipline (*tapas*).

Overall, the map reveals how emotional, ethical, and spiritual values interconnect, reflecting Indian ethical pluralism (*Purushartha*) and the Vedic model of moral cognition—from *manas* (emotion) to *buddhi* (reason) to *atman* (self-realization).

Thematic Analysis Report: Ayodhya Kāṇḍa

This report presents a thematic analysis of selected shlokas from the Ayodhya Kāṇḍa of the Ramayana. The ethical, moral, and philosophical themes are examined through classical Indian ideals and modern parallels.

Figure 5. Frequency of themes in Ayodhya Kāṇḍa

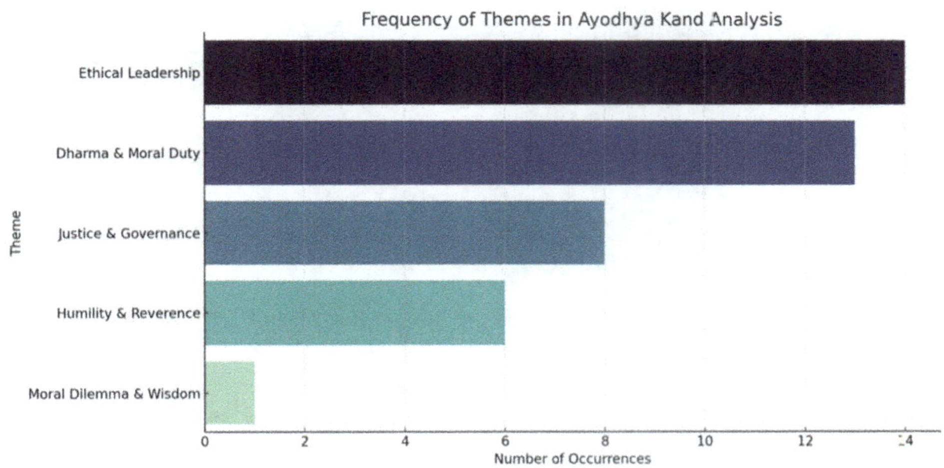

Figure 5 presents a horizontal bar chart representing the frequency of thematic occurrences identified in the *Ayodhya Kāṇḍa* section of the Ramayana. The chart illustrates five primary themes along the y-axis and their corresponding number of occurrences along the x-axis.

The most frequently observed theme is **Ethical Leadership**, which appears 14 times in the analysis. This suggests that the Ayodhya Kāṇḍa places significant emphasis on the moral responsibilities and righteous conduct expected from rulers and leaders—aligning with Lord Rama's portrayal as the ideal prince adhering to dharma. Following closely is **Dharma & Moral Duty**, with 13 occurrences, reinforcing the central motif of dharma (righteous path) in decision-making and personal sacrifice, particularly in the context of Rama's exile and obedience to paternal command. The theme of **Justice & Governance** occurs 8 times, highlighting the narrative's concern with administrative ethics, legal order, and equitable rule. This reflects the social expectations from a king and the importance of institutional fairness. **Humility & Reverence** is represented 6 times, which may relate to the virtues of devotion, respect toward elders and sages, and the submissive acceptance of one's fate as portrayed by characters such as Rama and Lakshmana. The least frequently occurring theme is **Moral Dilemma & Wisdom**, appearing only once. This indicates that while moral complexities are present in the Ayodhya Kāṇḍa, they are not the dominant focus compared to more overt ethical virtues and leadership traits. This thematic frequency distribution highlights that *Ayodhya Kāṇḍa* predominantly emphasizes leadership qualities, moral responsibility, and the embodiment of dharma. The visual data support the interpretation that this section of the Ramayana serves as a narrative foundation for ethical governance and personal sacrifice for collective good.

Figure 6. Co-occurrence of ethical themes

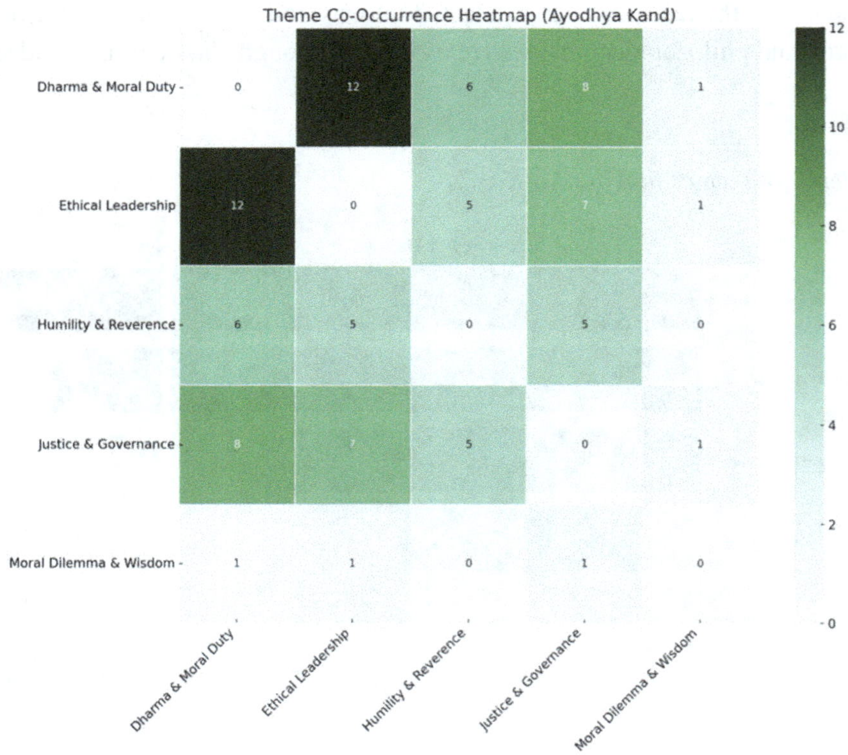

Figure 6 presents a heatmap showing the co-occurrence frequencies of key themes in the *Ayodhya Kāṇḍa* of the *Ramayana*. Each matrix cell indicates how often two themes appear together in the same narrative unit, with darker shades representing higher frequencies. Diagonal values are excluded to avoid self-pairing.

The strongest co-occurrence (n = 12) is between *Dharma & Moral Duty* and *Ethical Leadership*, highlighting that leadership actions in *Ayodhya Kāṇḍa*—such as Rama's exile—are deeply rooted in dharma. Other high co-occurrences include *Dharma* with *Justice & Governance* (n = 8), and *Ethical Leadership* with *Justice & Governance* (n = 7), reinforcing a thematic triad of righteous rule, ethical conduct, and justice.

Moderate co-occurrences are observed between these core themes and *Humility & Reverence* (n = 5–6), reflecting how emotional virtues like humility accompany institutional and moral actions—exemplified in the reverent behaviour of Rama, Bharata, and Lakshmana.

Moral Dilemma & Wisdom shows minimal overlap (n = 1 with a few themes), indicating its marginal role in this section. This supports the view that *Ayodhya Kāṇḍa* emphasizes clarity of righteous action over inner conflict or ethical ambiguity.

Overall, the heatmap underscores the thematic interdependence of *Dharma*, *Leadership*, and *Justice*, portraying an idealized vision of rule and virtue in *Ayodhya Kāṇḍa*.

Figure 7. Cluster visualization of thematically similar shlokas

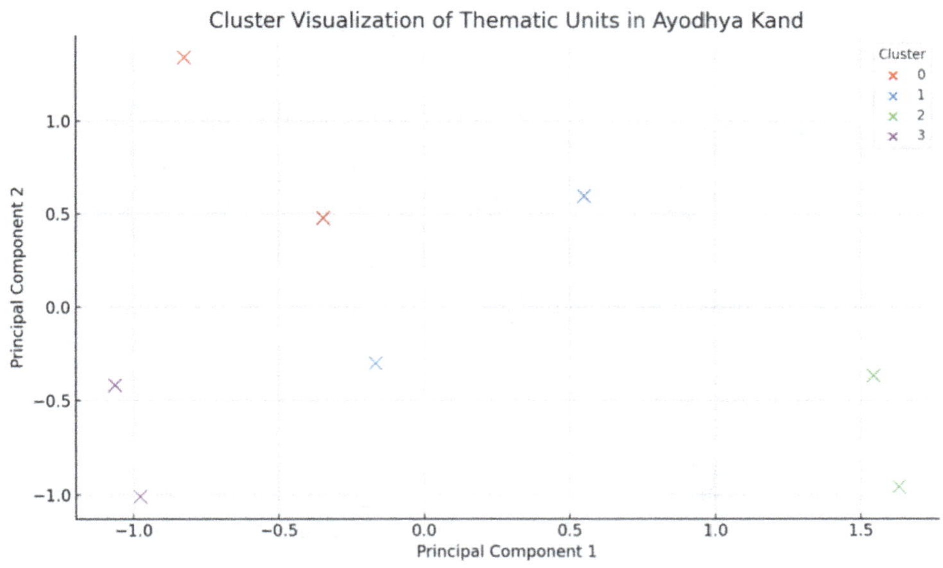

Figure 7 shows a two-dimensional scatter plot based on Principal Component Analysis (PCA), used to visualize thematic clusters in the *Ayodhya Kāṇḍa* of the *Ramayana*. Each point represents a coded thematic unit, and the axes reflect compressed dimensions of multiple themes. Color-coded clusters (0–3), identified through unsupervised learning (e.g., K-means), reveal four distinct groupings, suggesting that certain themes tend to co-occur or share semantic similarities.

Cluster 0 (Red), in the upper-left quadrant, likely centres on moral authority, leadership, and ethical decision-making—possibly linked to figures like Rama or King Dasharatha. Its tight formation suggests strong thematic cohesion.

Cluster 1 (Blue) appears more central, potentially representing transitional themes that blend personal virtues with institutional concerns such as loyalty or justice.

Cluster 2 (Green), isolated in the lower-right, may represent less frequent but introspective themes like moral conflict or philosophical wisdom—possibly involving sages or moral dilemmas.

Cluster 3 (Purple), located in the lower-left, seems to reflect emotional restraint, humility, and reverence—traits often associated with Bharata or Lakshmana. Its density suggests minimal variation among included units.

Overall, the plot reveals a clear thematic segmentation in *Ayodhya Kāṇḍa*, where narrative content clusters around leadership, virtue, emotional depth, and philosophical introspection. These insights can inform deeper qualitative interpretation or quantitative validation of thematic patterns in the epic.

Thematic Analysis Report: Ayodhya Kāṇḍa Uttarardh

Figure 8. Frequency of key ethical themes

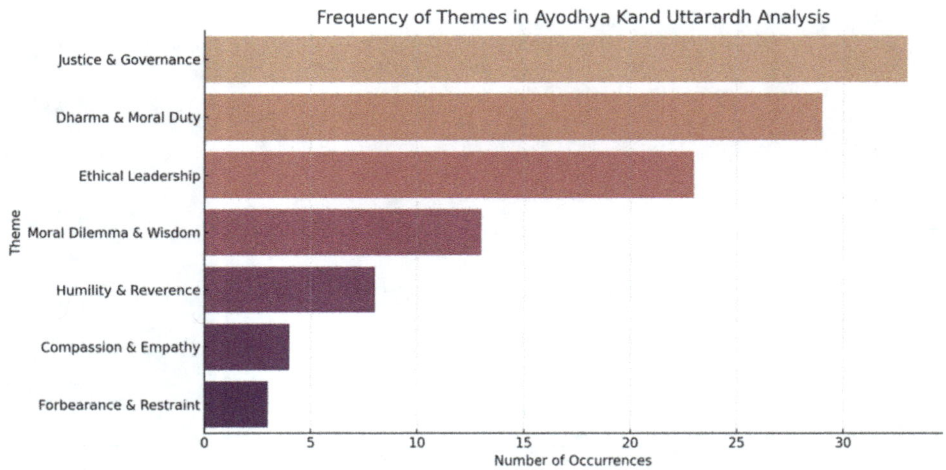

Figure 8 presents a horizontal bar chart showing the frequency of key themes in the *Uttarardh* (latter half) of the *Ayodhya Kāṇḍa*. The most frequent theme is *Justice & Governance* (n = 32), reflecting a strong focus on political responsibility and fair leadership after Rama's exile. *Dharma & Moral Duty* (n = 29) remains central, emphasizing righteous action across roles—be it Bharata, ministers, or citizens.

Ethical Leadership (n = 23) highlights ideals of rulership and self-sacrifice, particularly in Bharata's symbolic preservation of Rama's authority.

Moral Dilemma & Wisdom (n = 13) appears more prominently than in the *Purvardh*, pointing to deeper ethical introspection and conflict. Themes like *Humility & Reverence* (n = 8), *Compassion & Empathy* (n = 4), and *Forbearance & Restraint* (n = 3) add emotional and spiritual depth. Overall, the analysis shows a narrative shift from personal duty to broader institutional and philosophical concerns, affirming the evolving moral complexity of the *Ayodhya Kāṇḍa*.

Figure 9. Heatmap showing theme co-occurrence patterns

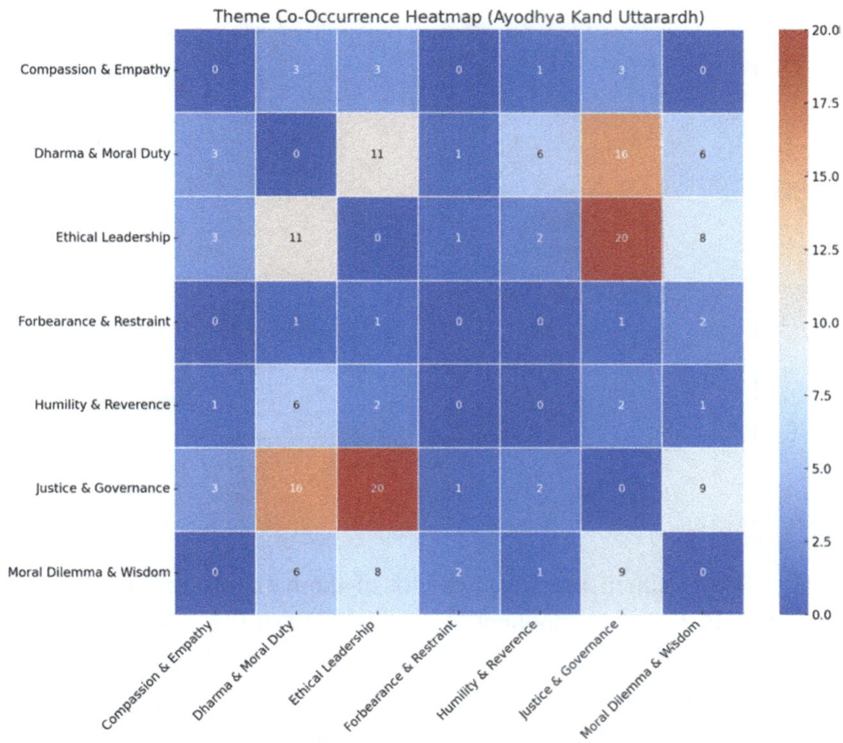

Figure 9 presents a co-occurrence heatmap showing how often thematic pairs appear together in the *Uttarardh* (latter half) of the *Ayodhya Kāṇḍa*. Darker shades indicate higher frequencies, while diagonal values are excluded.

The strongest co-occurrence is between *Justice & Governance* and *Ethical Leadership* (n = 20), reflecting Bharata's morally grounded stewardship during Rama's absence. Other high co-occurrences include *Dharma & Moral Duty* with *Justice & Governance* (n = 16), and with *Ethical Leadership* (n = 11), underscoring a deep link between righteous duty, governance, and ethical rule.

Moral Dilemma & Wisdom shows increased overlap with institutional themes—especially *Ethical Leadership* (n = 8) and *Justice* (n = 9)—highlighting growing introspective tension in leadership decisions. Moderate links with *Humility & Reverence* and *Dharma* (n = 6 each) add emotional and spiritual nuance.

Low co-occurrences with *Forbearance & Restraint* and *Compassion & Empathy* (n ≤ 2) suggest these values remain present but play a quieter, symbolic role.

Overall, the heatmap captures a narrative shift in the *Uttarardh*—from clear-cut dharma to complex intersections of leadership, justice, and ethical introspection.

Figure 10. Clustered thematic units based on co-occurring ethical categories

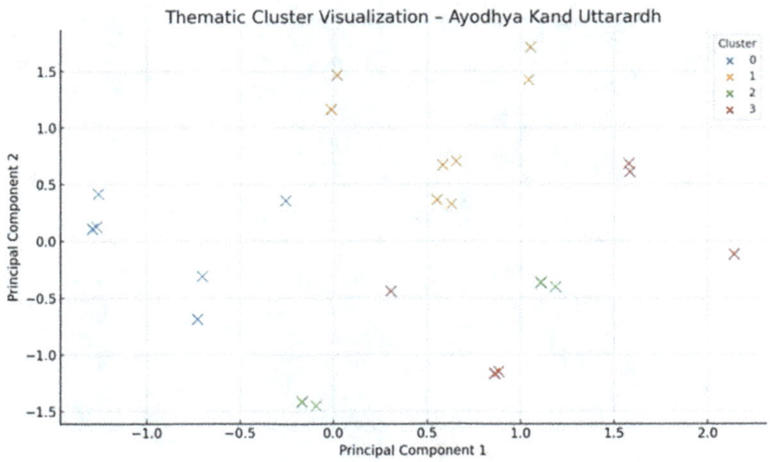

Figure 10 displays a two-dimensional scatter plot of thematic clusters from the *Uttarardh* of the *Ayodhya Kāṇḍa*, based on Principal Component Analysis (PCA) and unsupervised clustering. Each point represents a thematic unit, with four color-coded clusters (0–3) reflecting distinct narrative patterns.

Cluster 0 (Blue), located on the far left, includes emotionally grounded themes like reverence and restraint, with a tight formation indicating thematic consistency.

Cluster 1 (Orange), the largest and most dispersed, spans ethical, dharmic, and governance themes, suggesting it forms the narrative core with high semantic variability.

Cluster 2 (Green) appears in the lower middle-right region, with fewer points possibly tied to philosophical reflections or moral dilemmas.

Cluster 3 (Red), isolated in the lower right, likely represents emotionally intense or morally complex units, showing internal coherence within specialized narrative moments.

Overall, the plot highlights the thematic richness of the *Uttarardh*, with Cluster 1 anchoring central concerns of leadership and duty, while the remaining clusters map emotional and philosophical subcurrents. The PCA-based clustering reveals both dominant structures and nuanced thematic layers within the narrative.

Figure 11. Co-occurrence network of moral themes

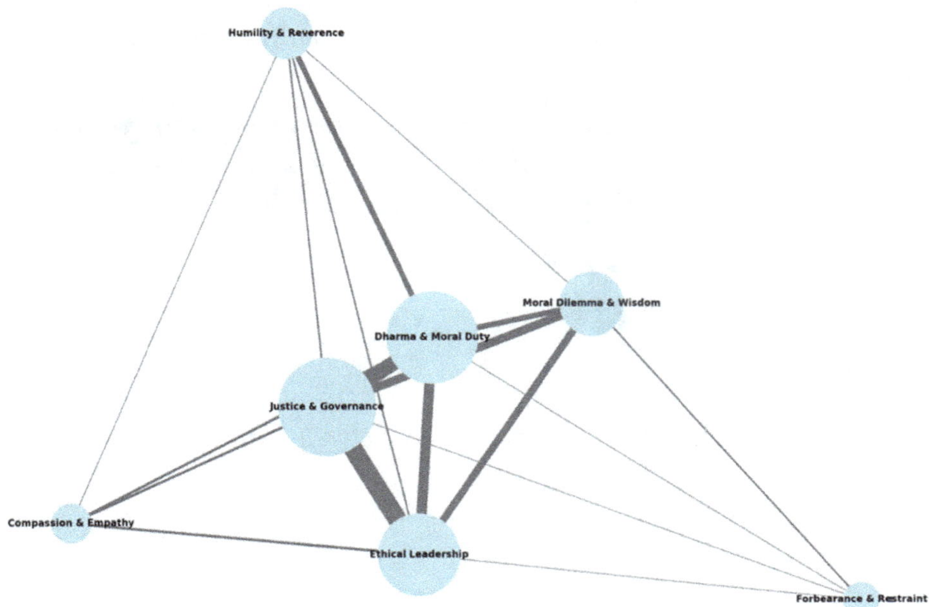

Figure 11 presents a co-occurrence network diagram of themes from the *Uttarardh* of the *Ayodhya Kāṇḍa*. Each node represents a theme, with node size indicating frequency and edge thickness showing co-occurrence strength.

The network reveals a dense thematic structure centred around *Dharma & Moral Duty*, *Ethical Leadership*, and *Justice & Governance*. These core themes form a central triangle, strongly interconnected and linked to others like *Moral Dilemma & Wisdom*, highlighting their foundational role in the narrative. *Moral Dilemma & Wisdom* is closely tied to leadership and dharma, suggesting that ethical ambiguity often arises within these domains.

Humility & Reverence occupies a bridging position between core and peripheral themes, indicating its role in linking emotional and institutional values. In contrast, *Compassion & Empathy* and *Forbearance & Restraint* are peripheral, appearing less frequently and serving more as emotional or contextual enrichments.

Overall, the network underscores the interwoven nature of leadership, justice, and dharma in the *Uttarardh*, while also acknowledging the quieter, supportive presence of emotional and introspective themes.

Figure 12. presents a horizontal bar graph of the 20 most frequent sub-themes in the Uttarardh of the Ayodhya Kāṇḍa, highlighting the detailed moral and emotional texture of the narrative. Leadership-related sub-themes dominate—Moral Leadership (n = 36), Just Leadership (n = 33), and Ethical Leadership (n = 23)—reflecting intense focus on Bharata's ethical struggles and symbolic rulership.

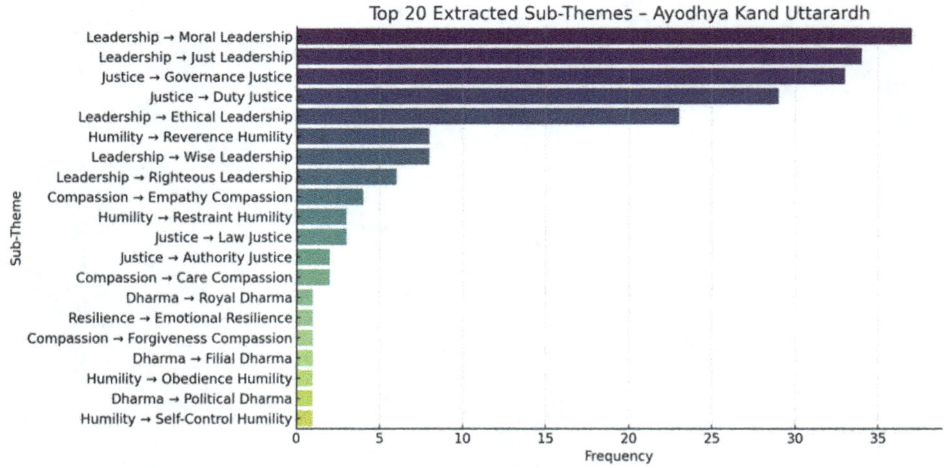

Justice-oriented sub-themes like *Governance Justice* (n = 34) and *Duty Justice* (n = 29) emphasize procedural fairness and legitimacy. Sub-themes under *Humility* (e.g., *Reverence*, *Restraint*, *Obedience*) and *Compassion* (e.g., *Empathy*, *Care*, *Forgiveness*) illustrate the emotional depth and moral restraint embodied by characters like Lakshmana and Bharata.

Dharma appears in forms like *Royal*, *Filial*, and *Political Dharma*, revealing a layered view of duty across personal and political realms. *Emotional Resilience* captures the inner strength shown in moments of moral strain, such as Rama's exile or Bharata's renunciation.

Overall, the analysis shows that the *Uttarardh* is a nuanced ethical reflection on leadership, justice, and emotional intelligence—framing governance not just as power, but as a dharmic, compassionate, and morally anchored practice.

Thematic Analysis Report: Aranya Kāṇḍa

Figure 13. Frequency of key ethical themes in Aranya Kāṇḍa

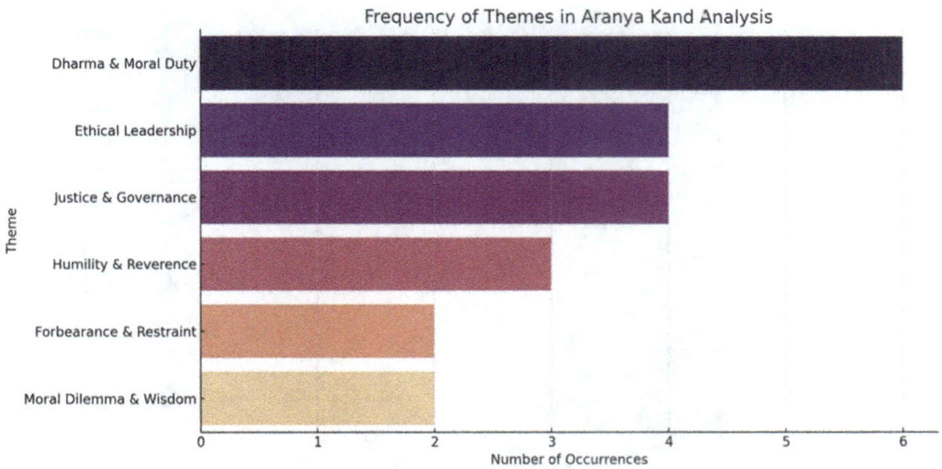

Figure 13 presents a horizontal bar chart showing the frequency of major themes in the *Aranya Kāṇḍa* of the *Ramayana*. The most prominent theme is *Dharma & Moral Duty* (n = 6), highlighting Rama's unwavering commitment to righteousness and moral obligations even in exile. *Ethical Leadership* and *Justice & Governance* (n = 4 each) follow, emphasizing that principles of leadership and justice persist beyond royal settings, especially in Rama's protection of sages and confrontation with demons.

Humility & Reverence (n = 3) appears in Rama's respectful interactions with figures like Shabari, while *Forbearance & Restraint* and *Moral Dilemma & Wisdom* (n = 2 each) reflect internal conflicts and emotional control during key turning points, such as Sita's abduction. Overall, the *Aranya Kāṇḍa* portrays dharma and leadership as enduring moral forces, even amid growing complexity and emotional tension in the forest setting.

Figure 14. Heatmap showing co-occurrence of themes across verses

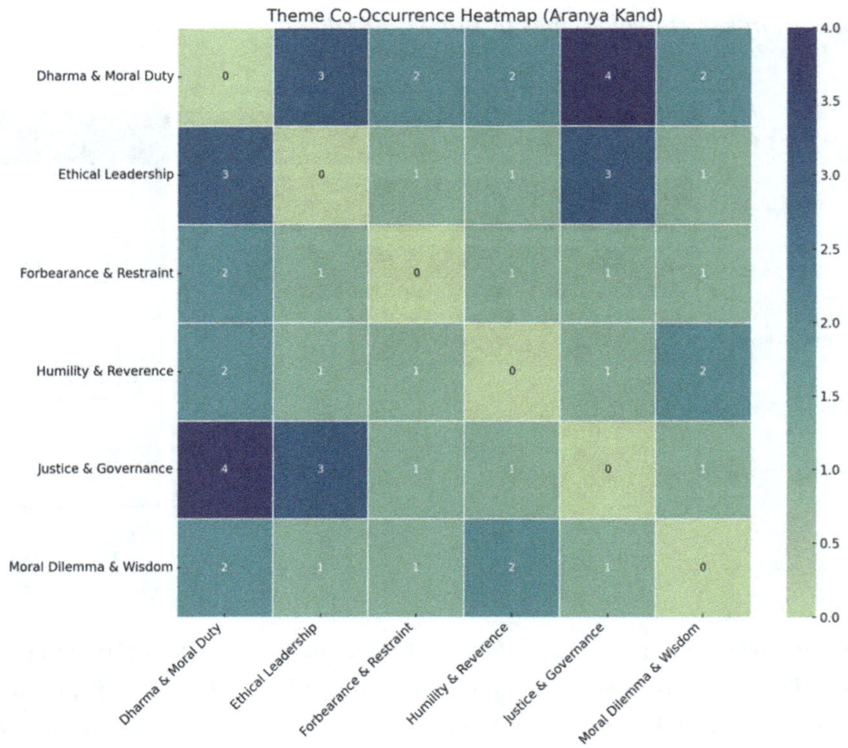

Figure 14 shows a co-occurrence heatmap of thematic pairs in the *Aranya Kāṇḍa* of the *Ramayana*, with darker cells indicating stronger co-occurrence and the diagonal set to zero. The strongest co-occurrence is between *Dharma & Moral Duty* and *Justice & Governance* (n = 4), reflecting Rama's efforts to uphold righteousness in the forest through actions like protecting sages and confronting demons.

Other notable links include *Dharma* with *Ethical Leadership* (n = 3) and with *Moral Dilemma*, *Humility*, and *Forbearance* (n = 2 each), showing how dharma is expressed through virtue, restraint, and ethical choices. *Ethical Leadership* also co-occurs with *Justice & Governance* (n = 3), reinforcing the idea of moral leadership in a non-royal setting.

Lower co-occurrences (n = 1) among themes like *Forbearance*, *Moral Dilemma*, and *Justice* suggest emerging inner tensions and psychological complexity. Overall, the heatmap highlights how dharma, justice, and leadership remain central even in exile, while emotional and ethical challenges begin to surface, foreshadowing the narrative's shift toward deeper conflict.

Figure 15. PCA-based cluster visualization of thematic groupings

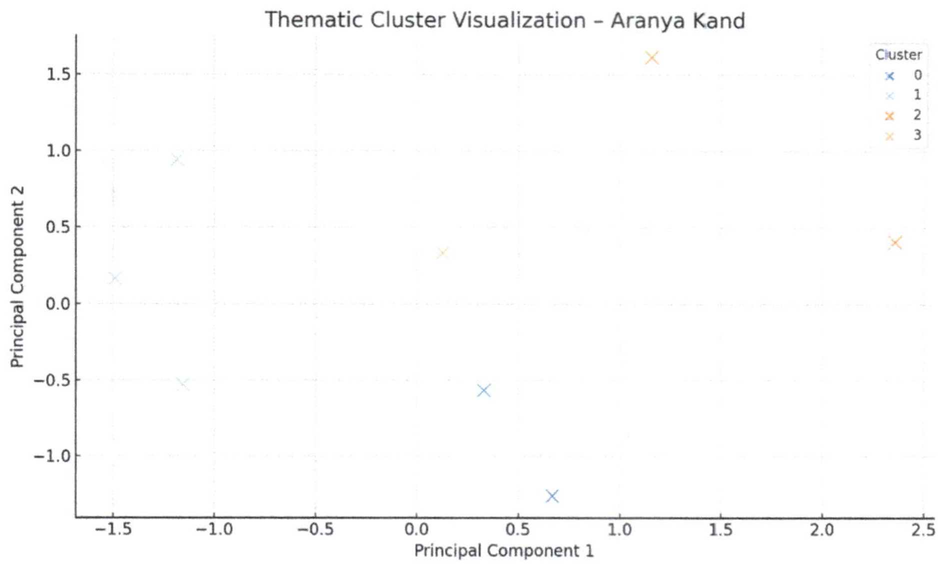

Figure 15 shows a PCA-based scatter plot of thematic units from the *Aranya Kāṇḍa* of the *Ramayana*, grouped into four color-coded clusters (0–3). Each point represents a coded narrative segment, and the axes reflect reduced dimensions that capture thematic variance.

Cluster 0 (light blue) is the largest and spans the left-to-centre area, likely encompassing core themes such as *Dharma & Moral Duty*, *Justice & Governance*, and *Ethical Leadership*—highlighting Rama's continued commitment to righteous conduct even in exile. **Cluster 1 (blue)**, smaller and centrally placed, may include introspective themes like *Forbearance & Restraint* or *Moral Dilemma & Wisdom*, tied to ethical tension and personal decisions.

Clusters 2 and 3 (orange shades) are sparse and peripheral. Cluster 2 possibly captures emotionally charged but infrequent events like Sita's abduction, while Cluster 3 may include spiritual or reflective moments, such as Rama's interactions with sages like Shabari.

Despite fewer narrative units, *Aranya Kāṇḍa* displays clear thematic layering—anchored in action and dharma, yet branching into emotion and spiritual reflection. The visualization affirms the Kāṇḍa's role as a narrative and ethical bridge from royal duty to deeper personal and philosophical challenges.

Figure 16. Network graph visualizing how themes are interlinked

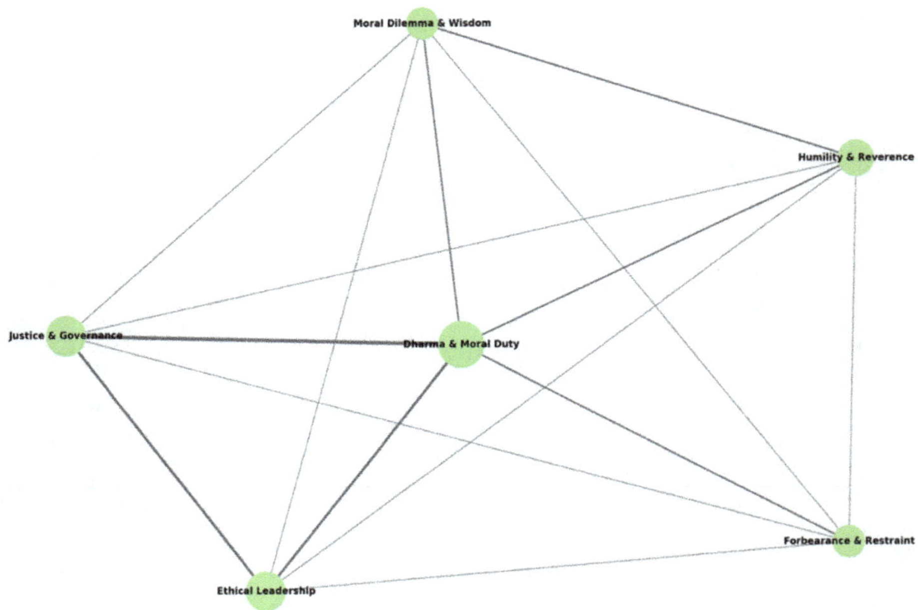

Figure 16 presents a thematic co-occurrence network from the *Aranya Kāṇḍa* of the *Ramayana*, where nodes represent major themes and edges show how often themes appear together. Node size indicates centrality, and edge thickness reflects co-occurrence strength.

At the centre is *Dharma & Moral Duty*, strongly connected to *Justice & Governance*, *Ethical Leadership*, and *Moral Dilemma & Wisdom*. This highlights Rama's continued commitment to righteous action as the driving force behind leadership, justice, and ethical reasoning in the forest.

Justice & Governance and *Ethical Leadership* are closely linked to dharma but more peripheral, reflecting their practical enactment in Rama's actions against injustice. *Moral Dilemma & Wisdom* shows moderate connections—especially with *Dharma* and *Humility*—suggesting reflective moments and moral tension, though not dominant.

Humility & Reverence and *Forbearance & Restraint* are lightly connected, indicating their supportive, context-specific roles in emotional and interpersonal interactions.

Overall, the network reveals a compact yet morally rich structure centred on dharma, with justice, leadership, and wisdom orbiting closely, while emotional virtues add stability and nuance to Rama's ethical conduct in exile.

Thematic Analysis Report: Kishkindha Kāṇḍa

Figure 17. Frequency of dominant ethical themes in Kishkindha Kāṇḍa

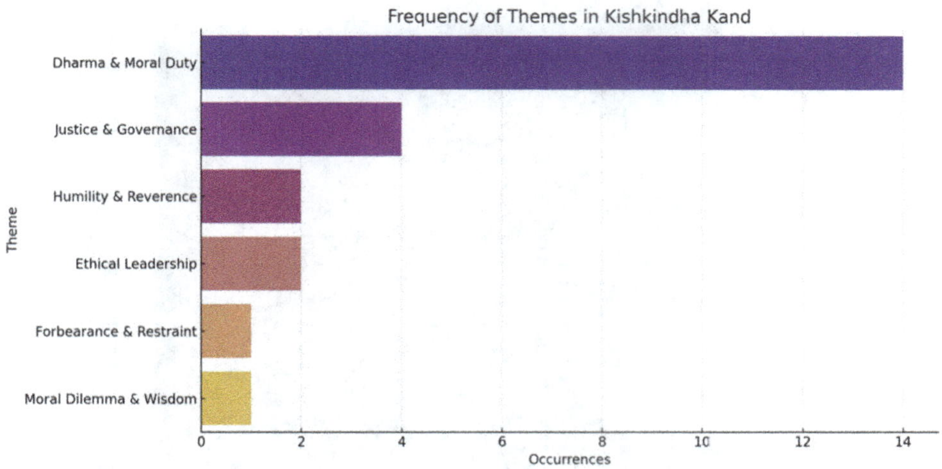

Figure 17 shows a horizontal bar chart of six major themes identified in the *Kishkindha Kāṇḍa* of the *Ramayana*. The most frequent theme is *Dharma & Moral Duty* (n = 14), reflecting Rama's unwavering adherence to ethical conduct, alliances, and vows amid political challenges.

Justice & Governance (n = 4) appears as a secondary theme, particularly in the context of Sugriva's exile and Vali's downfall, reinforcing the idea that rightful leadership must align with dharma. *Ethical Leadership* and *Humility & Reverence* (n = 2 each) emerge in Rama's interactions with Sugriva and Hanuman, highlighting his moral consistency and respectful diplomacy.

Forbearance & Restraint and *Moral Dilemma & Wisdom* (n = 1 each) are minimally present but symbolically important—seen in Rama's restraint and his internal conflict before killing Vali.

Overall, the *Kishkindha Kāṇḍa* emphasizes strategic alignment, moral clarity, and just leadership, with dharma as the overarching guide. Emotional and philosophical introspection plays a limited but meaningful role, marking this Kāṇḍa as a transitional phase focused more on action and alliance than inner conflict.

Figure 18. Heatmap of theme co-occurrences

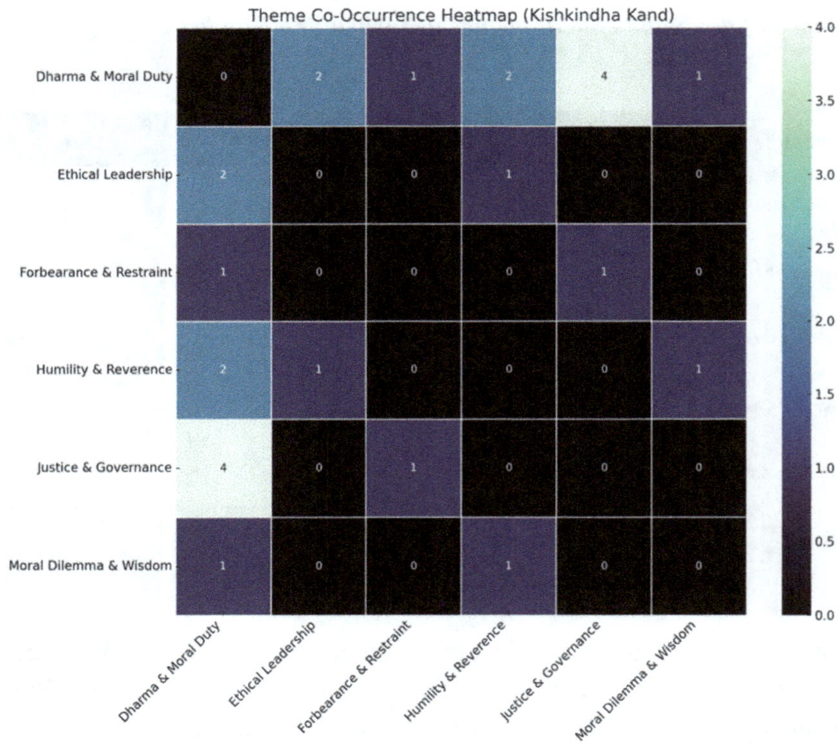

Figure 18 presents a heatmap of thematic co-occurrence frequencies in the *Kishkindha Kāṇḍa* of the *Ramayana*. The strongest co-occurrence is between *Dharma & Moral Duty* and *Justice & Governance* (n = 4), highlighting the integration of ethical obligation and rightful leadership—exemplified by Rama's support of Sugriva and the justified killing of Vali.

Moderate co-occurrences include *Dharma* with *Ethical Leadership* and *Humility & Reverence* (n = 2 each), and *Justice* with *Forbearance* (n = 1), suggesting that moral decisions are reinforced by humility and restraint. *Moral Dilemma & Wisdom* lightly connects with *Dharma* and *Humility* (n = 1 each), reflecting moments of ethical reflection, such as Rama's inner conflict before confronting Vali.

Overall, *Dharma & Moral Duty* emerges as the thematic core, tightly linked with justice, reverence, and ethical conduct. Emotional and philosophical themes, while present, are context-specific and serve to deepen key narrative decisions rather than drive the overall arc—affirming *Kishkindha Kāṇḍa*'s focus on restoring moral and political order.

Figure 19. PCA-based cluster visualization of thematic content

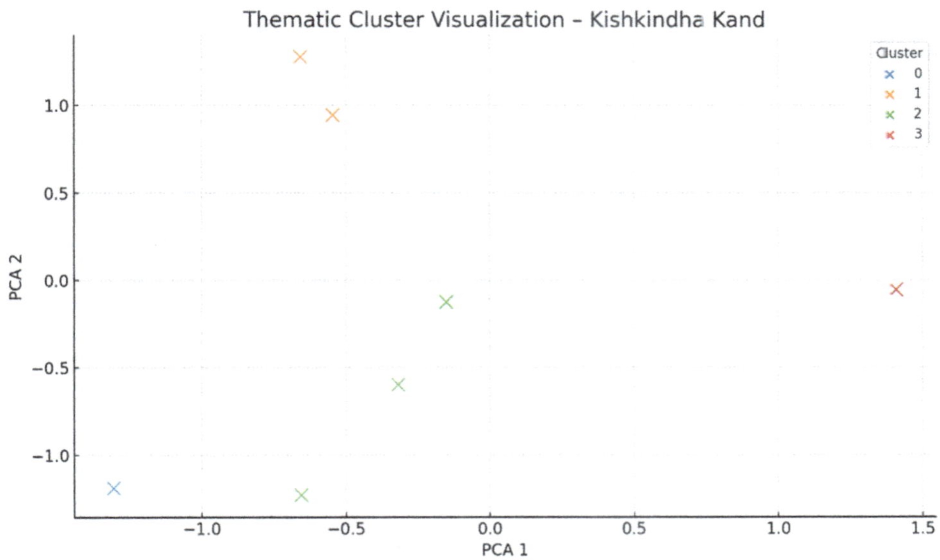

Figure 19 presents a PCA-based scatter plot illustrating thematic clustering in *Kishkindha Kāṇḍa*. Each point represents a coded narrative segment, with color-coded clusters reflecting thematic groupings. The moderate spread of points indicates thematic variety within a focused narrative scope.

Cluster 0 (Blue), in the lower-left quadrant, likely represents dharma-driven segments, especially those involving Rama's principled actions in the Sugriva–Vali conflict. **Cluster 1 (Orange)**, in the upper-left, appears centred on emotionally charged or leadership-focused content, such as Sugriva's hesitation or Hanuman's loyalty, reflecting themes of reverence and ethical leadership.

Cluster 2 (Green) occupies the lower-centre and seems to capture transitional or reflective moments—dialogue-heavy segments balancing action with moral restraint. **Cluster 3 (Red)** is a sparse outlier, possibly representing a singular, ethically complex event—likely Vali's death—marking a philosophical inflection in the narrative.

Overall, the plot reveals that *Kishkindha Kāṇḍa* is structured around dharma and justice, with additional thematic layers of emotional nuance and moral complexity. The isolated position of Cluster 3 suggests a pivotal moment that challenges the dominant ethical framework.

Figure 20. Network graph showing theme interlinkages

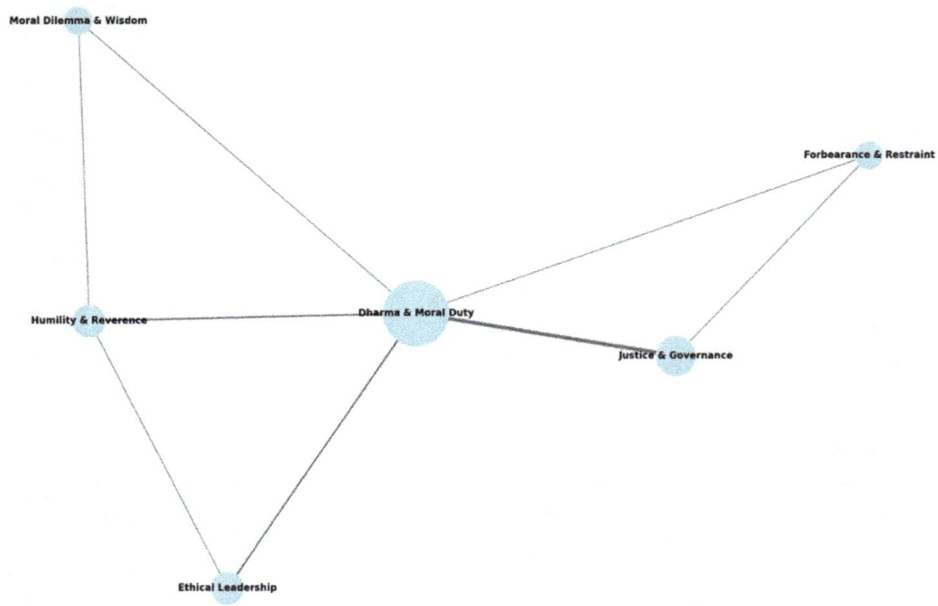

Figure 20 presents a co-occurrence network graph of major themes in *Kishkindha Kāṇḍa*, highlighting the structural prominence and interrelation of ethical motifs within the narrative. Each node represents a theme, with node size proportional to its frequency and edge thickness indicating the strength of co-occurrence between theme pairs. At the center of this network is **Dharma & Moral Duty**, emerging as the dominant thematic hub. It exhibits strong connections with both **Justice & Governance** and **Ethical Leadership**, indicating that most actions within this Kāṇḍa—including Rama's alliance with Sugriva and the ethical dilemma surrounding Vali's death—are anchored in the framework of righteousness and obligation.

The connection between **Dharma** and **Humility & Reverence** as well as **Moral Dilemma & Wisdom** further reflects the introspective and relational depth of Rama's decisions. Meanwhile, peripheral themes such as **Forbearance & Restraint** and **Ethical Leadership** appear as secondary but still significant, providing emotional and governance-related scaffolding to the central dharmic axis. The overall structure of the network confirms that *Kishkindha Kāṇḍa* prioritizes the restoration of moral-political order through dharma-driven decision-making, with other themes acting in service of this central moral imperative.

Thematic Analysis Report: Sundara Kāṇḍa

Figure 21. Frequency of dominant ethical themes in Sundara Kāṇḍa

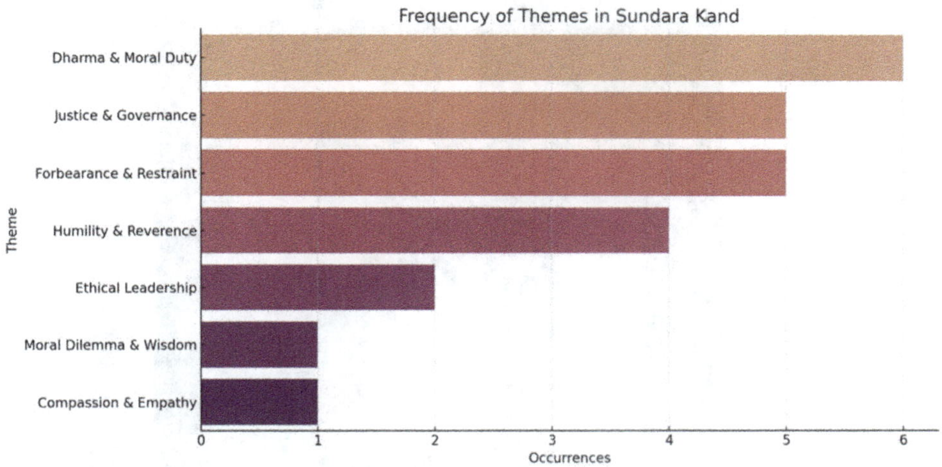

Figure 21 presents a bar chart visualizing the frequency of key thematic elements found in *Sundara Kāṇḍa* of the Ramayana. Among the most dominant themes is **Dharma & Moral Duty** (n = 6), affirming its central role even in this largely action-driven and emotionally resonant episode. Closely following are **Justice & Governance** and **Forbearance & Restraint** (n = 5 each), which reflect Hanuman's measured conduct during his mission to Lanka—where despite immense power, he upholds justice and exercises self-control in service of Rama's cause. The theme of **Humility & Reverence** (n = 4) is significant as well, highlighting the devotion and respect Hanuman displays toward Sita and Rama, as well as his humble approach despite his heroic capabilities.

Ethical Leadership (n = 2) appears moderately, likely referring to Hanuman's responsible leadership in a foreign and hostile environment. Less frequent but notable themes include **Moral Dilemma & Wisdom** and **Compassion & Empathy** (n = 1 each), which may be attributed to Hanuman's moral reflections before confronting the rakshasas, and his empathetic interaction with Sita. Overall, the theme distribution in *Sundara Kāṇḍa* reflects a **narrative balance between action and inner virtue**, where restraint, reverence, and moral wisdom guide heroic deeds rather than aggression or domination.

Figure 22. Theme co-occurrence heatmap showing correlated ethical categories

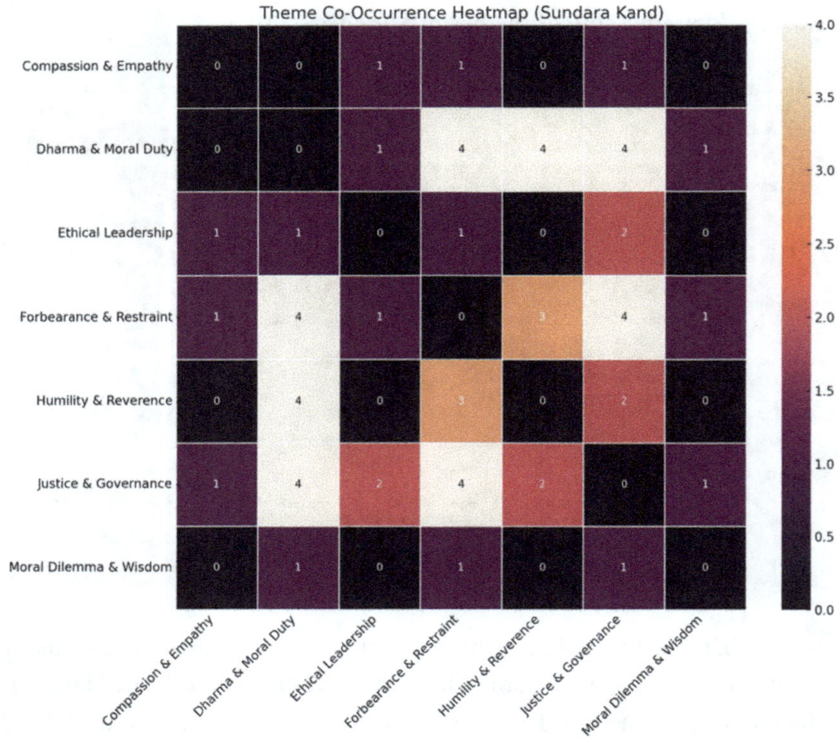

Figure 22 presents a thematic co-occurrence heatmap for *Sundara Kāṇḍa*, capturing the frequency with which themes appear together in the same narrative segments. The most robust co-occurrence patterns (n = 4) emerge between **Dharma & Moral Duty** and three key themes: **Justice & Governance**, **Forbearance & Restraint**, and **Humility & Reverence**. These strong co-activations suggest that Hanuman's mission is simultaneously framed within a dharmic imperative, executed with measured restraint, and infused with humility. Similarly, **Justice & Governance** is centrally co-occurring with multiple themes—most notably **Forbearance & Restraint** (n = 4) and **Ethical Leadership** (n = 2)—highlighting how the pursuit of justice in *Sundara Kāṇḍa* is not marked by domination but by ethical action and control.

The linkage between **Humility & Reverence** and **Forbearance & Restraint** (n = 3) further emphasizes Hanuman's self-regulation, especially in moments of destructive potential, such as in the burning of Lanka. While **Compassion & Empathy** and **Moral Dilemma & Wisdom** co-occur less frequently, their connections (n = 1) with themes like **Justice**, **Restraint**, and **Dharma** illustrate the nuanced emotional and cognitive landscape behind seemingly heroic deeds. Overall, this heatmap underscores the **integrated presence of moral discipline, reverent conduct, and governance ethics**, woven into the unfolding of divine agency.

Figure 23. PCA-based thematic cluster visualization

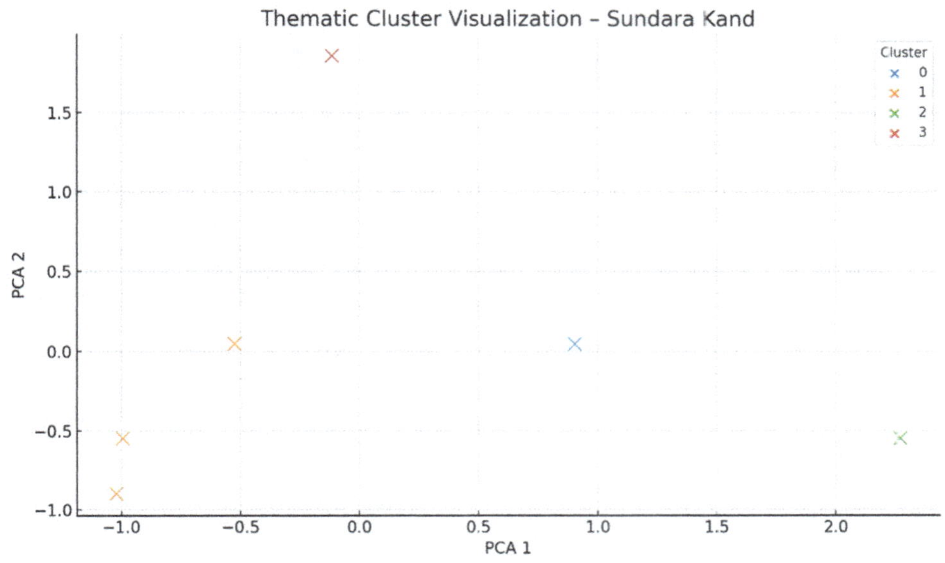

Figure 23 presents a PCA-based scatter plot showing thematic clustering in the *Sundara Kāṇḍa* of the *Ramayana*. Each point represents a narrative unit, grouped into one of four clusters based on reduced thematic dimensions (PCA 1 and PCA 2).

Cluster 1 (Orange) is the most populated, located in the left quadrant, and likely represents core ethical themes such as *Dharma & Moral Duty*, *Justice*, *Humility*, and *Forbearance*, reflecting Hanuman's disciplined and purposeful mission. **Cluster 0 (Blue)**, sparse and on the right, may contain action-oriented or symbolically distinct moments, including Hanuman's assertive acts or significant dialogues.

Cluster 2 (Green) appears as a thematic outlier, likely capturing emotionally nuanced or philosophical content, possibly tied to *Moral Dilemma* or *Compassion*. **Cluster 3 (Red)**, small and positioned centrally, may reflect spiritually charged or introspective episodes, such as Hanuman's devotion or reflections before returning to Rama.

Overall, the scatter plot reveals that *Sundara Kāṇḍa* blends action with moral depth, featuring distinct thematic layers of heroism, reflection, and emotional resonance.

Figure 24. Network graph of theme interrelationships

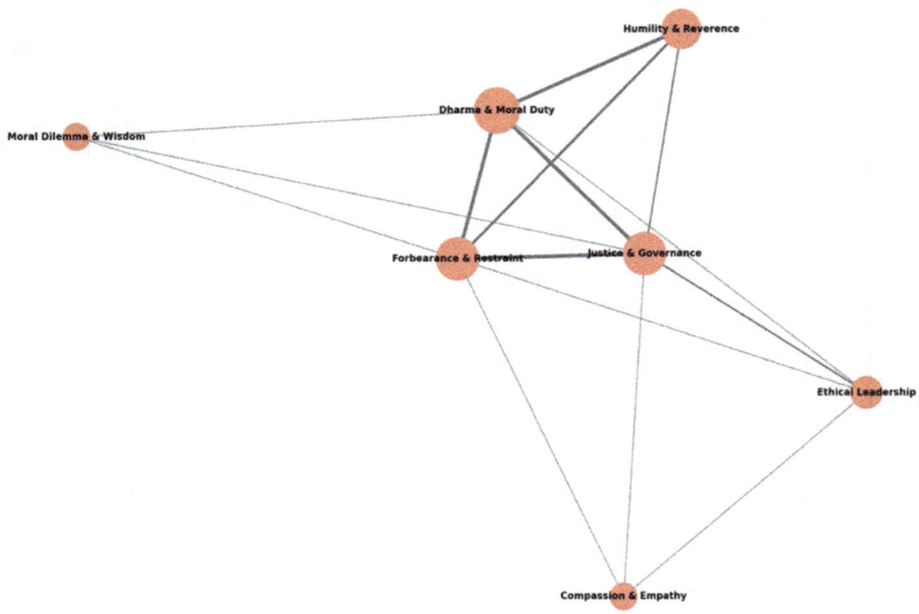

Figure 24 presents a thematic co-occurrence network from *Sundara Kāṇḍa*, highlighting the interconnections among key moral and emotional themes. *Dharma & Moral Duty* is the central node, strongly linked with *Justice & Governance*, *Humility & Reverence*, and *Forbearance & Restraint*. This reflects Hanuman's ethical and respectful conduct throughout his mission to find Sita and serve Rama's cause.

Justice and *Forbearance* also form a tightly knit triad with *Dharma*, emphasizing the role of moral restraint and fairness in action. *Ethical Leadership*, though less frequent, connects to *Justice* and *Compassion*, pointing to Hanuman's service-driven leadership rooted in empathy and discipline.

Peripheral themes like *Moral Dilemma & Wisdom* and *Compassion & Empathy* appear with lighter connections, emerging in emotionally rich moments—such as Hanuman's restraint in burning Lanka or his sensitivity toward Sita.

Overall, the network reveals *Sundara Kāṇḍa* as a morally layered narrative, where heroism is grounded in dharma, restraint, and reverence, and guided by introspection and ethical clarity.

Thematic Analysis Report: Yuddha Kāṇḍa

Figure 25. Frequency of key themes in Yuddha Kāṇḍa

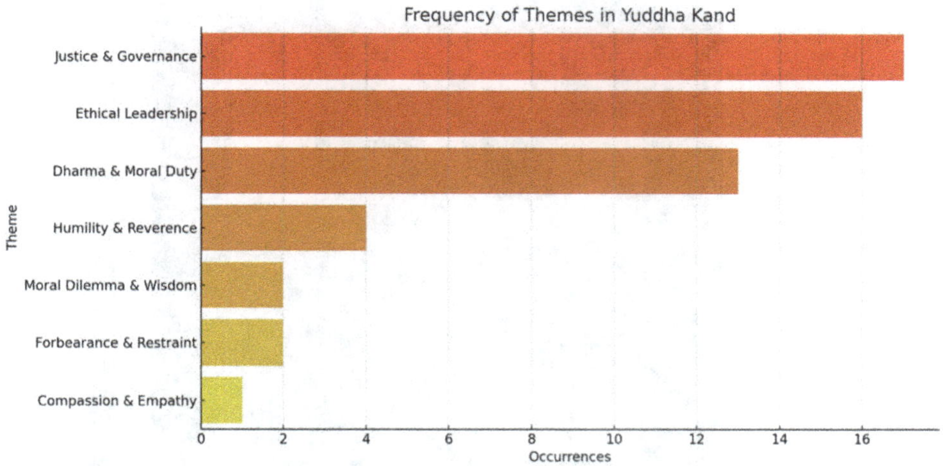

Figure 25 presents a horizontal bar chart showing the frequency of core themes in the *Yuddha Kāṇḍa*, the climactic war section of the *Ramayana*. The most frequent themes are *Justice & Governance* (n = 17), *Ethical Leadership* (n = 16), and *Dharma & Moral Duty* (n = 13), emphasizing the Kāṇḍa's focus on righteous rule, moral warfare, and principled decision-making.

These themes reflect Rama's ethical stance in battle, as well as the conduct of key figures like Lakshmana and Vibheeshana, showcasing leadership grounded in virtue rather than power. *Humility & Reverence* (n = 4) adds moments of spiritual depth, especially in divine interactions and Vanara camaraderie.

Less frequent but meaningful are *Moral Dilemma & Wisdom* and *Forbearance & Restraint* (n = 2 each), which appear in morally charged decisions—such as sparing Ravana or handling Vibheeshana's defection. *Compassion & Empathy* (n = 1), though rare, plays a key emotional role in post-war scenes like Sita's trial and reunion.

Overall, the theme distribution affirms *Yuddha Kāṇḍa* as the epic's ethical high point, where justice, leadership, and dharma overshadow emotional ambiguity, culminating in a victory defined as much by moral clarity as by strategic success.

Figure 26. Theme co-occurrence heatmap reflecting moral complexity

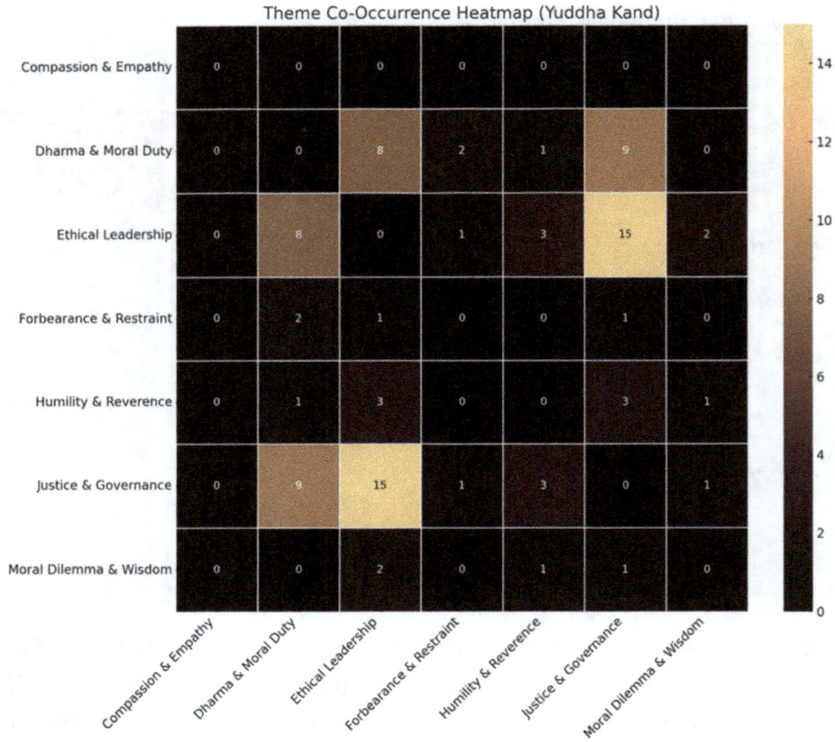

Figure 26 presents a co-occurrence heatmap of major themes in *Yuddha Kāṇḍa*, highlighting how often thematic pairs appear together in the same narrative segments. The most frequent pairing is *Justice & Governance* with *Ethical Leadership* (n = 15), reflecting the Kāṇḍa's central message: that righteous governance must be led by principled individuals, as seen in Rama's battlefield and post-war decisions.

Dharma & Moral Duty also frequently co-occurs with *Justice* (n = 9) and *Ethical Leadership* (n = 8), forming a strong moral triad that guides major actions involving war, mercy, and justice. Moderate co-occurrences with *Humility & Reverence* (n = 3 with both *Justice* and *Leadership*) suggest that even amid conflict, reverence and respect remain ethical ideals.

Themes like *Forbearance & Restraint*, *Moral Dilemma & Wisdom*, and *Compassion & Empathy* show low co-occurrence but appear in crucial scenes—such as Sita's trial or Hanuman's post-war choices—adding emotional and ethical nuance.

Overall, the heatmap confirms *Yuddha Kāṇḍa* as the moral and ethical crescendo of the *Ramayana*, where justice, dharma, and leadership operate in unison, shaping a vision of ideal rulership through ethically grounded action.

Figure 27. PCA-based cluster visualization of thematic groupings

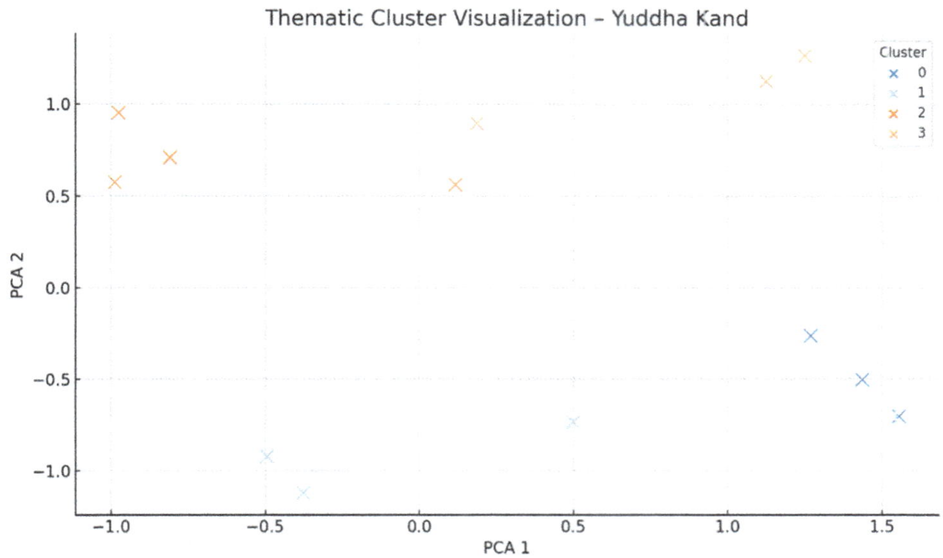

Figure 27 presents a PCA-based scatter plot showing thematic clustering in *Yuddha Kāṇḍa*. Each point represents a narrative unit, and the axes reflect compressed thematic dimensions. The color-coded clusters (0–3) reveal how thematic content naturally groups based on contextual similarity.

Cluster 2 (Orange) is the largest and most widely spread, covering the upper and left areas. It likely includes dominant themes like *Justice & Governance*, *Ethical Leadership*, and *Dharma & Moral Duty*—capturing battlefield decisions, moral rulership, and the enforcement of dharma through action. Its breadth reflects the range of justice-driven content in the Kāṇḍa.

Cluster 0 (Blue) appears tightly grouped in the lower-right quadrant, likely representing emotionally and philosophically reflective segments. This may include themes like *Moral Dilemma & Wisdom*, *Forbearance & Restraint*, and *Humility & Reverence*—found in moments of surrender, moral hesitation, or interpersonal dilemmas.

The separation between Clusters 0 and 2 suggests a contrast between external ethical action and internal moral contemplation. **Clusters 1 and 3**, smaller and more scattered, likely represent hybrid or transitional moments, such as emotionally charged decisions, spiritual symbolism, or unexpected moral turns.

Overall, the visualization reveals *Yuddha Kāṇḍa* as thematically balanced between the pursuit of justice through action and the introspective ethics that temper and validate that action—underscoring the epic's complex vision of righteous warfare.

Figure 28. Inter-theme network graph showing shared moral contexts

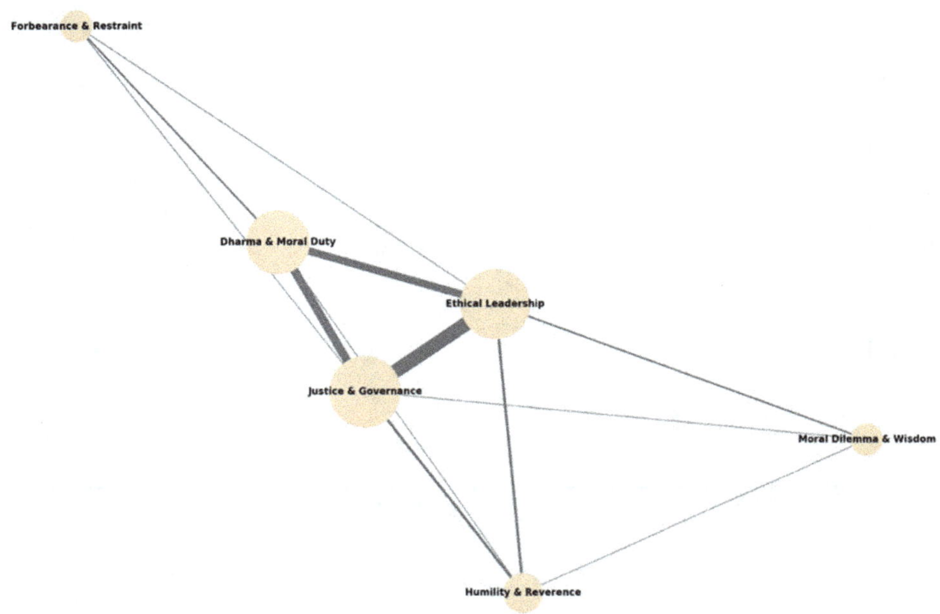

Figure 28 presents a thematic co-occurrence network for *Yuddha Kāṇḍa*, mapping the interconnections between major ethical and moral themes. Node size reflects theme frequency, and edge thickness represents how often two themes co-occur within the narrative.

At the centre are the three dominant themes—*Justice & Governance*, *Ethical Leadership*, and *Dharma & Moral Duty*—forming a tightly linked moral core. The strongest connection is between *Justice* and *Leadership*, highlighting how governance in *Yuddha Kāṇḍa* is inseparable from principled action, exemplified by Rama's conduct during and after the war. Strong links between *Dharma* and both these themes emphasize a triangulated moral framework where all actions are anchored in righteousness.

Humility & Reverence connects moderately with *Justice* and *Leadership*, reflecting moments of respectful dialogue, divine invocation, and allegiance. *Moral Dilemma & Wisdom* appears on the network's periphery but links to key themes, signalling ethical introspection in moments like Sita's trial or Vibheeshana's surrender.

Forbearance & Restraint, though less frequent, connects to the central triad, illustrating the role of self-control and discipline—especially in Rama's and Hanuman's actions.

Overall, the network affirms that *Yuddha Kāṇḍa* presents not just a physical battle, but a morally structured one—where leadership, justice, and dharma are deeply interwoven, and ethical restraint plays a vital role in defining true victory.

Comparative Thematic Analysis Report – Across Seven Kāṇḍas

Figure 29. Heatmap of theme frequencies across all Kāṇḍas

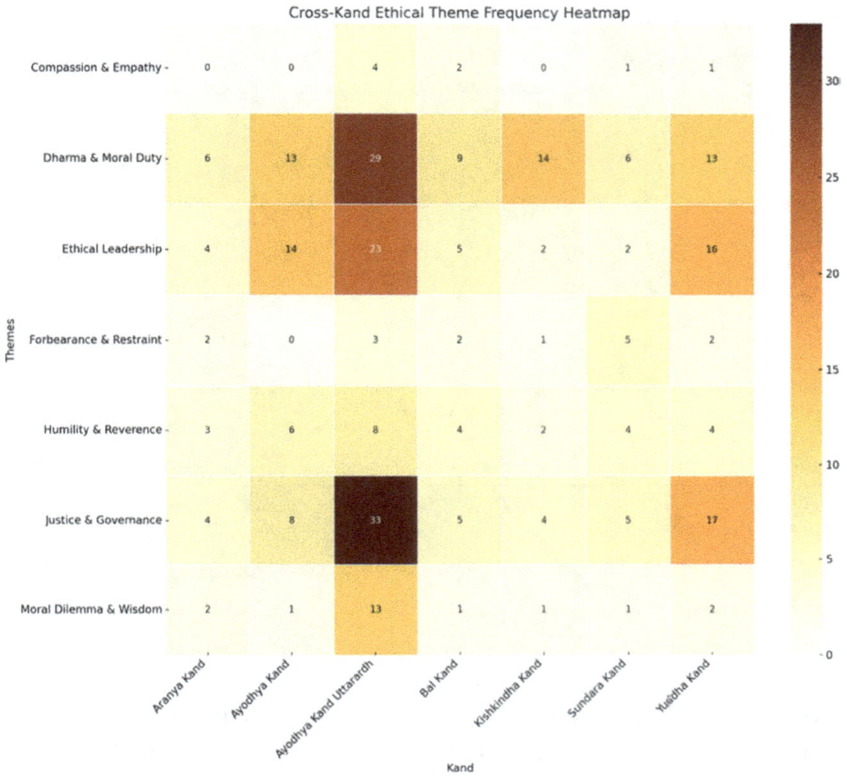

Figure 29 presents a heatmap comparing the distribution of major ethical themes across the *Ramayana*'s Kāṇḍa (books), highlighting shifts in moral focus. *Justice & Governance* and *Dharma & Moral Duty* are the most consistently recurring themes, peaking in *Ayodhya Kāṇḍa Uttararдh* (33 and 29), marking it as a key moral turning point involving exile, duty, and governance.

Ethical Leadership also peaks in *Ayodhya Uttarardh* (23) and remains strong in *Yuddha Kāṇḍa* (16), reflecting principled decision-making in both personal and wartime contexts. In contrast, *Compassion & Empathy* appears infrequently, surfacing modestly in *Bal Kāṇḍa* and occasionally in *Sundara* and *Yuddha Kāṇḍa*, tied to emotional moments.

Forbearance & Restraint is most prominent in *Sundara Kāṇḍa*, aligning with Hanuman's disciplined conduct, while *Humility & Reverence* appears steadily across all Kāṇḍa, underscoring its quiet but constant presence.

The heatmap reveals an ethical arc: early Kāṇḍa build moral tension, *Ayodhya Uttarardh* peaks in complexity, and later books like *Sundara* and *Yuddha* enact dharma through action—charting Rama's journey from moral struggle to righteous resolution.

Figure 30. illustrates the progression of core moral themes across the Ramayana's Kāṇḍas, showing how ethical priorities shift throughout the epic. Leadership peaks in Ayodhya Kāṇḍa Uttarardh (100+ instances), reflecting intense focus on governance and moral decision-making during Rama's exile and Bharata's dilemma. Justice also rises sharply here (≈70), emphasizing the political and ethical crises within the kingdom.

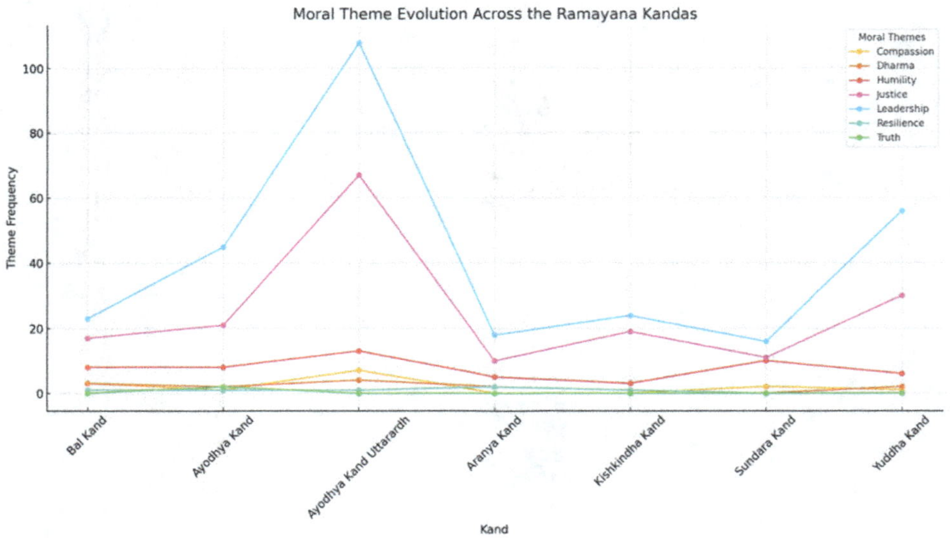

Both themes decline in *Aranya* and *Sundara Kāṇḍa*, signalling a shift from statecraft to personal trials and spiritual action. *Dharma* increases steadily from *Bal Kāṇḍa* to *Yuddha Kāṇḍa*, affirming its role as a constant moral guide. *Humility* rises notably in *Sundara Kāṇḍa*, likely tied to Hanuman's devotion and restraint.

Resilience and *Truth* remain low in explicit frequency but appear consistently, suggesting their subtle but foundational presence. Overall, the chart captures the epic's ethical arc—from divine grounding and leadership struggles to moral action and dharmic resolution—mapping Rama's journey through shifting moral terrains.

Figure 31. presents a heatmap of ethical sub-themes across the Ramayana, showing how moral nuances evolve through its Kāṇḍas. Just Leadership and Moral Leadership dominate, peaking in Ayodhya Kāṇḍa Uttarardh (34 and 37 instances), reflecting the epic's ethical climax during Rama's exile and Bharata's moral conflict. Supporting this are Duty Justice (29) and Governance Justice (33), underscoring the narrative's deep focus on institutional dharma and public duty.

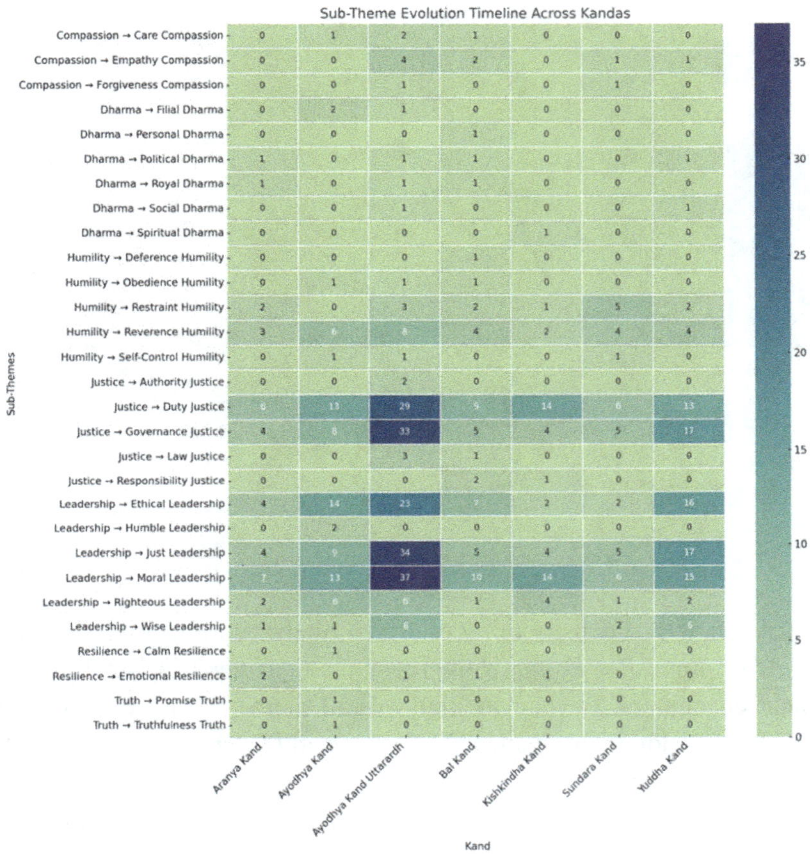

Sub-themes like *Filial*, *Political*, and *Royal Dharma* appear mainly in *Bal* and *Ayodhya Kāṇḍa*, reinforcing familial and political ethics. In contrast, *Spiritual* and *Social Dharma* remain minor, indicating a pragmatic over metaphysical framing of dharma.

Humility sub-themes—especially *Reverence* and *Restraint*—are steady, notably in *Ayodhya* and *Sundara Kāṇḍa*, shaped by Hanuman's devotion and Rama and Sita's restraint. *Compassion* and *Resilience* sub-themes surface selectively, such as *Empathy Compassion* in *Ayodhya* and *Calm Resilience* in *Kishkindha*, marking emotional turning points.

Overall, the heatmap reveals leadership, justice, and humility as the core ethical drivers, with *Ayodhya Kāṇḍa Uttarardh* serving as the moral peak of the epic. The sub-thematic spread highlights the *Ramayana*'s rich philosophical depth and the layered ethical journeys of its characters.

Figure 32. illustrates the thematic density—measured as the number of distinct moral themes per unit of text—across the Ramayana's seven Kāṇḍas. Sundara Kāṇḍa stands out with the highest density (~3.4 themes/unit), reflecting its rich ethical layering in Hanuman's emotionally and spiritually charged journey, where forbearance, dharma, and humility frequently intersect.

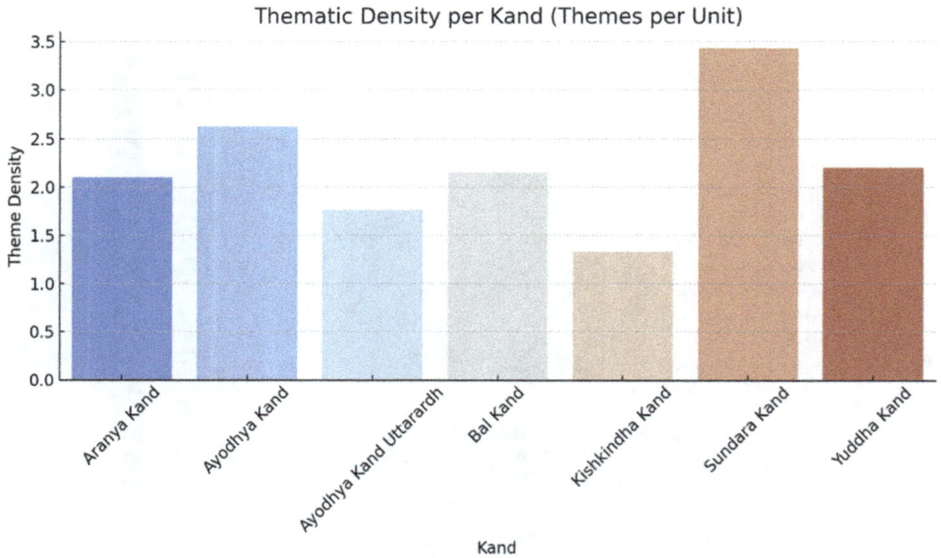

Ayodhya Kāṇḍa follows with a density of ~2.6, driven by the moral complexity of exile, duty, and royal responsibility. *Aranya* and *Bal Kāṇḍa* show moderate densities (~2.1–2.2), containing episodic yet meaningful ethical encounters. Despite its high overall theme count, *Ayodhya Kāṇḍa Uttarardh* has a lower density (~1.8), as dominant themes like justice and leadership extend over longer narrative segments with less thematic variety.

Kishkindha Kāṇḍa records the lowest density (~1.3), focusing more on alliances and strategic preparation than moral diversity. *Yuddha Kāṇḍa* strikes a balance (~2.2), blending action with recurring ethical decisions.

Overall, the chart highlights *Sundara Kāṇḍa* as a philosophical high point, where a concentrated narrative delivers deep moral complexity in a compact form.

Figure 33. presents a PCA-based cluster analysis of thematic units across all Ramayana Kāṇḍas, revealing five emergent clusters (0–4) that group shared moral patterns beyond narrative order.

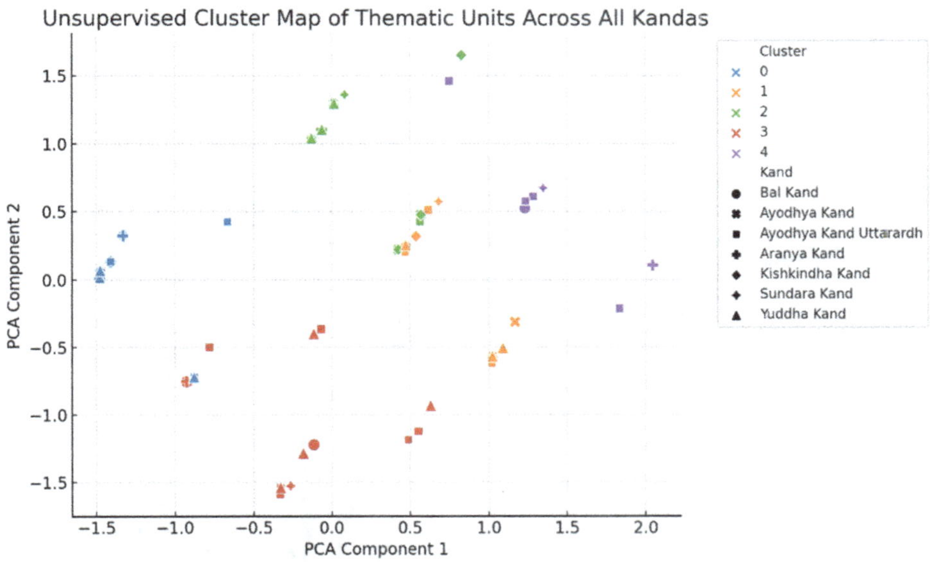

Cluster 0 (light blue) includes units from *Bal*, *Kishkindha*, and *Yuddha Kāṇḍa*, suggesting shared preparatory or strategic ethical themes. **Cluster 1 (orange)** blends *Ayodhya*, *Sundara*, and *Yuddha Kāṇḍa*, reflecting emotional and leadership-related dilemmas. **Cluster 2 (green)** connects *Ayodhya Kāṇḍa Uttarardh* and *Sundara Kāṇḍa*, highlighting philosophical and exile-driven moral parallels.

Cluster 3 (red) draws from earlier Kāṇḍas, pointing to foundational ethical ideas, while **Cluster 4 (violet)**, dominated by *Ayodhya Uttarardh* and *Aranya Kāṇḍa*, centres on justice, duty, and family obligations.

The overlap across Kāṇḍas illustrates that the *Ramayana's* ethical framework is not strictly chronological but thematically interwoven, showing recurring moral concerns across varied contexts.

Figure 34. presents a network graph of key ethical themes across all Ramayana Kāṇḍas, mapping how they co-occur throughout the narrative. Nodes represent themes, with size indicating frequency (centrality) and edge thickness showing co-occurrence strength.

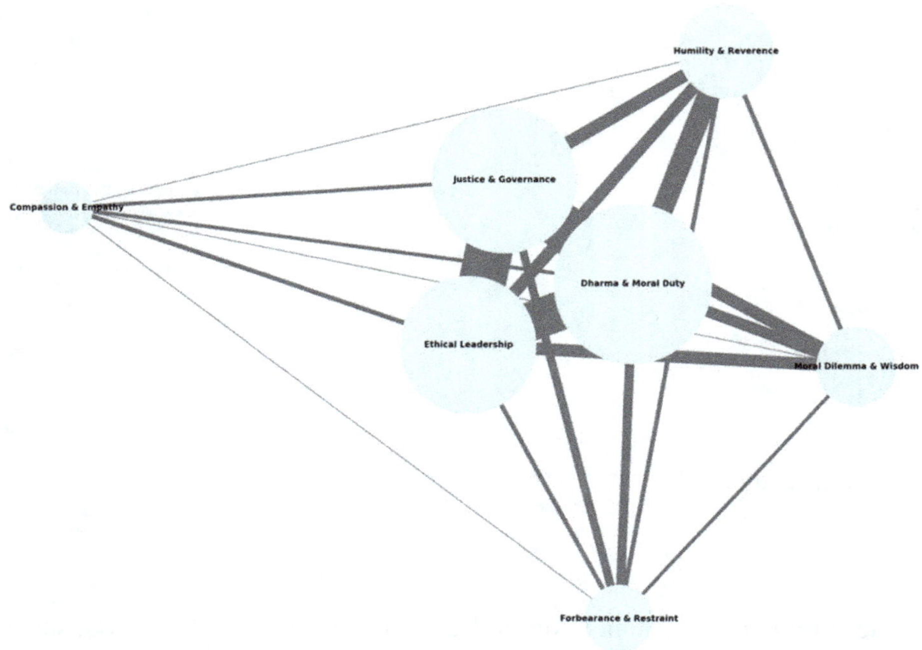

At the centre of the network are the three foundational themes: **Dharma & Moral Duty**, **Justice & Governance**, and **Ethical Leadership**. Their strong, mutual connections reflect the Ramayana's core moral architecture—where governance, principled leadership, and dharma operate in tandem, especially during pivotal events like Rama's exile and Hanuman's mission.

Themes such as **Moral Dilemma & Wisdom** and **Humility & Reverence** also link meaningfully to these core nodes, showing that moments of ethical complexity often emerge through inner conflict, humility, or reverent restraint. Though less central, **Compassion & Empathy** maintains connections across all themes, functioning as an emotional undercurrent.

The present network analysis indicates that the ethical structure of the *Ramayaṇa* is relational rather than episodic. Thematic co-occurrence matrices and cluster visualizations generated in this study reveal that justice, compassion, political responsibility, and restraint consistently operate in conjunction rather than in isolation. Distinct Kāṇḍas display varying concentrations of these ethical domains: the *Bal Kāṇḍa* foregrounds formative restraint and disciplined obedience; the *Ayodhya Kāṇḍa* intensifies tensions between personal loyalty and public duty; the *Araṇya* and *Kiṣkindha Kāṇḍas* emphasize alliance-building, humility, and negotiated authority; the *Sundara Kāṇḍa* exhibits the highest density of affectively charged resilience and devotional steadfastness; and the *Yuddha Kāṇḍa* consolidates questions of justice, statecraft, and post-conflict accountability. These patterns emerge directly from the inductive thematic coding and relational mapping undertaken in this study.

While the specific structural distributions identified here are original to the present analysis, they resonate with established philological and philosophical scholarship that characterizes the epic as thematically coherent and morally complex (Brockington, 1998; Matilal, 2002). Previous scholarship has also emphasized that dharma in the epic functions as situational discernment rather than rigid legal prescription (Fitzgerald, 1983), and that narrative traditions construct moral understanding through historically extended patterns of character formation (MacIntyre, 2007). Furthermore, accounts of moral emotion as cognitively evaluative (Nussbaum, 2001) provide a theoretical lens through which the affective intensity of episodes such as the *Sundara Kāṇḍa* may be interpreted. These scholarly frameworks do not determine the findings presented here; rather, they offer conceptual resources that illuminate the structural patterns identified in the present study.

Taken together, the results support the conclusion that the *Ramayana* develops an internally differentiated yet integrated moral architecture. Ethical reasoning unfolds progressively—from personal discipline to political governance, and from principled duty to negotiated responsibility—without collapsing into static prescription. The epic thus operates not merely as narrative inheritance but as a sustained exploration of context-sensitive moral deliberation embedded within civilizational memory.

DISCUSSION & CONCLUSION

This chapter argues that the *Ramayana* should be understood not as a narrative that merely illustrates dharma, but as a narrative architecture that generates dharma as a mode of moral cognition. Its structure performs epistemic work. Across the seven Kāṇḍas, ethical reasoning is progressively intensified—disciplinary formation, relational rupture, sovereign crisis, and post-conflict accountability unfold in patterned escalation. Dharma emerges neither as codified injunction nor as interior sentiment but as context-sensitive discernment within shifting relational matrices. Such a reconstruction aligns with Alf Hiltebeitel's (2011) demonstration that dharma evolved through narrative, law, and ritual interpenetration rather than abstract systematization.

Situating the epic within Indigenous Knowledge Systems requires epistemological precision. Indigenous knowledge here is not cultural residue but a structured mode of norm production. Sheldon Pollock (2006) has shown that Sanskrit cosmopolitanism fused literary form with normative imagination, embedding ethical reasoning within aesthetic and political discourse. Similarly, Julius Lipner (2010) emphasizes that classical Hindu traditions articulate theological and ethical reflection through narrative embodiment rather than detached theorization. The *Ramayana's* integration of metaphysics, kingship, and moral psychology thus exemplifies a civilizational epistemology in which storytelling constitutes a vehicle of disciplined normative transmission.

Philosophically, the epic resists reduction to dominant modern categories. It does not conform to deontological abstraction, where duty is universalized independent of relational context (Kant, 1998). Nor does it fully align with Aristotelian virtue ethics centred on eudaimonic flourishing. Instead, it articulates a relational conception of personhood akin to what Chadha (2024) identifies in classical Indian thought: agency is constituted through interdependence, and moral legitimacy depends upon self-limitation within inherited normative orders. As Bimal Krishna Matilal (2002) argued, Indian ethical reflection consistently embeds rational deliberation within social and cosmological embeddedness. Justice, sovereignty, and leadership in the epic therefore function as mutually conditioning ethical coordinates rather than isolable virtues.

The contemporary force of this Indigenous epistemology becomes particularly visible under technological modernity. Hans Jonas (1984) famously contended that modern technological power demands an expanded ethics of responsibility oriented toward anticipatory foresight. Shannon Vallor (2016) similarly argues that technological cultures risk erosion of practical wisdom unless virtue is structurally cultivated within socio-technical systems. The *Ramayana's* conception of rājyadharma anticipates such responsibility-centred governance: authority is legitimate only when internally regulated by reflexive accountability and moral foresight. Power without self-limitation destabilizes the ethical field—a narrative principle repeatedly dramatized in the epic's political crises.

Importantly, this interpretive move avoids anachronistic projection. The epic is not retrofitted as proto-digital ethics. Rather, as Jonardon Ganeri (2007) demonstrates, classical Indian philosophy develops sophisticated theories of agency and rational deliberation independent of Western liberal individualism. The *Ramayana* extends this philosophical grammar narratively, modelling systemic moral reasoning in environments of complex causality. Its relational ontology—where actions reverberate across moral networks—anticipates contemporary calls in global ethics for systems-level accountability rather than atomistic blame.

The broader implication for Indigenous Knowledge Systems scholarship is methodological. If ethical theory remains confined to analytic abstraction and individual autonomy, it risks epistemic narrowing. The *Ramayana* demonstrates that narrative can function as a rigorous medium of ethical theorization without sacrificing coherence. Dharma, in this reconstruction, is neither immutable statute nor subjective preference; it is disciplined responsiveness within a relational cosmos that exceeds the individual yet binds the individual to consequence.

The epic's enduring philosophical significance therefore lies in its architectonic coherence. As an Indigenous Knowledge System, it models an alternative configuration of moral knowledge—historically embedded yet normatively generative, culturally situated yet conceptually dialogical. In an era marked by technological amplification of agency and fragmentation of accountability, the *Ramayana* offers not prescriptive closure but an epistemology of responsibility capable of sustaining moral intelligibility amid systemic complexity.

FUTURE RESEARCH DIRECTIONS

Future research must situate the ethical universe of the Ramayana more explicitly within the larger intellectual project of revitalizing Indigenous Knowledge Systems in the twenty-first century by examining how epic traditions function not merely as cultural inheritances, but as living epistemic frameworks capable of addressing contemporary civilizational challenges. While the present study reconstructs the Valmiki Ramayana as a structured moral system grounded in dharma, relational responsibility, and narrative ethics, subsequent scholarship may extend this inquiry by exploring how such ethical knowledge survives, transforms, and gains renewed relevance through oral traditions, performative practices, vernacular retellings, temple recitations, women's narrative traditions, and tribal interpretations across diverse cultural landscapes. This would help reposition indigenous knowledge not as archival residue of the past, but as a dynamic mode of ethical reasoning embedded in collective memory, social practice, and intergenerational transmission. Future studies may also investigate how Ramayana-based ethical frameworks can inform present debates on ecological sustainability, climate responsibility, governance crises, restorative justice, educational philosophy, technological ethics, and leadership accountability, particularly in societies seeking culturally rooted alternatives to purely technocratic or market-driven models of development. Comparative philosophical engagement with African indigenous ethics, Confucian relational morality, Buddhist compassion ethics, and Latin American decolonial thought may further strengthen this discourse by demonstrating that indigenous traditions across civilizations share epistemic commitments to balance, reciprocity, restraint, and community-centred responsibility. Methodologically, the integration of digital humanities, corpus linguistics, network analysis, and AI-assisted interpretive tools offers new possibilities for preserving, mapping, and reactivating indigenous ethical systems without diminishing their philosophical depth. In this sense, future research should move beyond preservation toward epistemic reactivation, where indigenous narratives such as the Ramayana become active intellectual resources in shaping sustainable futures, decolonizing knowledge production, and restoring moral imagination in an age marked by ecological vulnerability, technological acceleration, and fractured social life.

REFERENCES

Aristotle. (2000). Nicomachean ethics (T. Irwin, Trans., 2nd ed.). Hackett Publishing Company. (Original work published ca. 4th century BCE)

Braun, V., & Clarke, V. (2006). Using thematic analysis in psychology. *Qualitative Research in Psychology, 3*(2), 77–101. DOI: 10.1191/1478088706qp063oa

Brockington, J. L. (1998). *The Sanskrit epics.* Brill. DOI: 10.1163/9789004492677

Chadha, M. (2024). Personhood in classical Indian philosophy. In Zalta, E. N., & Nodelman, U. (Eds.), *The Stanford Encyclopaedia of Philosophy* (Winter 2024 ed.)., https://plato.stanford.edu/archives/win2024/entries/personhood-india/

Doniger, W. (2010). *The Hindus: An alternative history.* Oxford University Press.

Fitzgerald, J. L. (2004). Dharma and its translation in the *Mahābhārata. Journal of Indian Philosophy, 32*(5-6), 671–685. DOI: 10.1007/s10781-004-8640-7

Ganeri, J. (2007). *Concealed Art of the Soul: Theories of the Self and Practices of Truth in Indian Ethics and Epistemology.* Oxford University Press. DOI: 10.1093/acprof:oso/9780199202416.001.0001

Hiltebeitel, A. (2011). *Dharma: Its early history in law, religion, and narrative.* Oxford University Press., DOI: 10.1093/acprof:oso/9780195394238.001.0001

Jonas, H. (1984). *The imperative of responsibility: In search of an ethics for the technological age.* University of Chicago Press. DOI: 10.7208/chicago/9780226850337.001.0001

Kant, I. (1998). *Groundwork of the metaphysics of morals* (Gregor, M. J., Trans.). Cambridge University Press. DOI: 10.1017/CBO9780511809590.004

Lipner, J. (2010). *Hindus: Their religious beliefs and practices* (2nd ed.). Routledge.

MacIntyre, A. (2007). *After virtue: A study in moral theory* (3rd ed.). University of Notre Dame Press. (Original work published 1981)

Matilal, B. K. (2002). *Ethics and epics: The collected essays of Bimal Krishna Matilal* (Ganeri, J., Ed.). Vol. 2). Oxford University Press.

Nussbaum, M. C. (2001). *Upheavals of thought: The intelligence of emotions.* Cambridge University Press. DOI: 10.1017/CBO9780511840715

Pollock, S. (2006). *The language of the gods in the world of men: Sanskrit, culture, and power in premodern India.* University of California Press. DOI: 10.1525/9780520932029

Ramanujan, A. K. (1991). Three hundred Rāmāyaṇas: Five examples and three thoughts on translation. In Richman, P. (Ed.), *Many Rāmāya as: The Diversity of a Narrative Tradition in South Asia* (pp. 22–49). University of California Press. DOI: 10.1525/9780520911758-004

Richman, P. (Ed.). (1991). *Many Rāmāya as: The Diversity of a Narrative Tradition in South Asia* (1st ed.). University of California Press., https://www.jstor.org/stable/jj.5973087 DOI: 10.1525/9780520911758

Ricoeur, P. (1992). *Oneself as another* (Blamey, K., Trans.). University of Chicago Press.

Sharma, A. (1999). The Puruṣārthas: An Axiological Exploration of Hinduism. *The Journal of Religious Ethics*, *27*(2), 223–256. https://www.jstor.org/stable/40018229. DOI: 10.1111/0384-9694.00016

Vallor, S. (2016). *Technology and the virtues: A philosophical guide to a future worth wanting*. Oxford University Press. DOI: 10.1093/acprof:oso/9780190498511.001.0001

ADDITIONAL READING

Ganatra, A. (2022). *Ramayana Unravelled.* Bloomsbury Publishing India Pvt Ltd.

Lutgendorf, P. (1991). *The Life of a Text: Performing the Ramcaritmanasof Tulsidas*. University of California Press. https://www.jstor.org/stable/10.1525/j.ctt1ppzkw

Neelakantan, A. (2025). *Many Ramayanas, Many Lessons*. HarperCollins Publishers India.

Richman, P. (Ed.). (1991). *Many Ramayanas: The Diversity of a Narrative Tradition in South Asia.* University of California Press. DOI: 10.1525/9780520911758

Thapar, R. (2013). *The Past Before Us: Historical Traditions of Early North India.* Harvard University Press. DOI: 10.4159/harvard.9780674726512

KEY TERMS AND DEFINITIONS

Araṇya: The forest as both a geographical setting and a symbolic arena of withdrawal, ethical testing, and inner transformation.

Artha: The structured pursuit of material security, governance, and socio-economic stability that enables ordered collective life.

Atman (Ātman): The innermost principle of conscious being that underlies thought and emotion and grounds spiritual identity.

Dharma: A relational principle of rightful conduct that sustains balance across personal, social, and cosmic domains.

Dharmic: Describing actions or dispositions aligned with ethical responsibility and sustaining order.

Divine Legitimacy: The grounding of authority in alignment with transcendent moral or sacred principles.

Filial Dharma: Obligatory responsibility toward parents and elders rooted in gratitude, lineage continuity, and inherited moral order.

Indigenous Knowledge System: A culturally embedded framework of understanding transmitted through lived practice, narrative, and intergenerational continuity.

Kāṇḍa: A structural division within the epic marking a distinct narrative and moral phase.

Moksha (Mokṣa): The culmination of spiritual striving characterized by freedom from existential limitation and realization of one's deepest nature.

Moral Ecology: An interconnected ethical field in which institutions, emotions, duties, and spiritual aims mutually influence one another.

Purushartha: The fourfold schema of human aims—ethical responsibility, material order, desire, and liberation—structuring a balanced life orientation.

Relational Accountability: The principle that moral evaluation occurs within networks of relationship rather than isolated acts.

Renunciation: The deliberate relinquishment of attachment, privilege, or power for a higher moral or spiritual aim.

Sage: A contemplative seer whose authority arises from disciplined insight rather than institutional power.

Shloka: A metrical verse form foundational to classical Sanskrit epic and philosophical expression.

Svadharma: Context-specific obligation emerging from one's position, role, and stage of life.

Tapa: Transformative discipline characterized by sustained effort and self-restraint that generates inner refinement.

Tyaga: Intentional relinquishment undertaken to subordinate personal gain to ethical or spiritual priority.

Viveka: The cultivated capacity for moral discernment when values conflict.

Yuddha: Conflict understood both as physical battle and as ethical confrontation aimed at restoring order.

Chapter 6
Intersubjective Cosmologies and Non-Human Personhood in Indigenous Religions:
A Cross-Cultural Systematic Review

Gedong Maulana Kabir
http://orcid.org/0009-0003-1809-3065
Bina Nusantara University, Indonesia

ABSTRACT

Indigenous cosmologies challenge Western anthropocentrism by extending the concept of personhood to non-humans, such as spirits and other elements of nature. This study examines how Indigenous religions view these non-human entities, interact with them, contribute to conservation efforts, and negotiate legal frameworks. In accordance with the PRISMA protocol, a thematic analysis was performed on 15 Scopus articles published between 2015 and 2025. The findings indicate that Indigenous peoples attribute personhood to relational capacities. This recognition is expressed through rituals, dreams, and shamanic mediation, which establish reciprocal commitments, ethics, and responsibilities between humans and non-humans. These intersubjective relationships influence better approaches to environmental conservation. However, positive legal acceptance of these cosmologies presents challenges. This review concludes that Indigenous cosmologies are sophisticated philosophical systems that offer important insights for the Indigenous religion studies.

INTRODUCTION

In recent decades, the perspective of indigenous peoples has begun to receive serious attention from academics who study religious, environmental, and legal issues. This attention is reasonable. Indigenous peoples have a deep understanding of the interconnectedness between living things that are rooted in cosmology and their worldviews. In contrast to the Western perspective which tends to place humans as the center and ruler of nature, indigenous peoples respect the existence of all creatures equally. They built what researchers call a relational ontology—a way of understanding the world that prioritizes the relationship between humans and non-human, rather than the separation of both (Maarif (2023). Un-

DOI: 10.4018/979-8-2600-2890-2.ch006

fortunately, throughout history, the dominant Western anthropocentric understanding has often reduced indigenous religion to mere cultural practices, thus ignoring the worldviews and ecological wisdom contained in it (M. S. Weatherdon, 2022). In fact, this kind of perspective is very relevant to conservation efforts and sustainability that are now on the global ecological agenda.

To understand why indigenous perspectives are so different, we need to trace the philosophical roots of the Western perspective. Cartesian dualism—the thought that separates mind and body, humans and nature—has shaped the way the West views the environment for centuries (Alves & Vidal, 2024). This separation, according to many researchers, contributed to the ecological destruction we are witnessing today (Keegan, 2022; M. Weatherdon & Schermerhorn, 2022) The indigenous epistemology offers a different path. They do not separate humans from nature, but rather see the two as part of an interconnected life. In this view, non-human entities—whether spirit, animals, plants, or other natural elements—have agency and deserve to be treated with respect (Nurit Bird-David, 1999). It makes indigenous frameworks allow for a more holistic approach to environmental management, especially in addressing ecological imbalances due to industrialization.

How exactly do Indigenous peoples practice this relational perspective? One key concept that helps explain this is intersubjectivity. This concept describes a view in which humans and non-human beings engage in a reciprocal relationship based on respect and care. A real example can be found in the *Suku Anak Dalam* community in Jambi, Indonesia. Najib's research (2020) shows that the intersubjective relationship between these communities and the forests around them is the foundation of their conservation practices. For them, protecting forests is not just an ecological obligation, but a form of respect for the entities they consider to be fellow inhabitants of the world. These practices produce real and measurable conservation impacts, while challenging us to rethink how humans should interact with nature in an ecosystem (M. S. Weatherdon, 2022).

The findings from the *Suku Anak Dalam* community are not a single case. More broadly, indigenous cosmology in various parts of the world shows a similar pattern in perceiving non-human entities. They believe that animals, plants, mountains, rivers, and spirits are part of a network of life in which relationships—not hierarchies—determine the existence and identity of each entity (Maarif, 2019). This kind of perspective encourages more sustainable interactions with nature. A systematic review of the existing literature confirms that recognition of non-human agency is a central element in indigenous belief systems, which directly influence the way indigenous peoples manage their environment (Keegan, 2022).

Although the contribution of indigenous perspectives to conservation has been widely documented, efforts to integrate them into academic discourse face challenges. One of the main problems lies in translation: relational cosmologies are often acknowledged, but they are not easily rendered into the academic, legal, and policy vocabularies commonly used in research and governance. Many studies recognize the existence of this framework but have not fully developed an adequate methodology to assess sustainability practices sourced from indigenous cosmology. This gap opens opportunities for research in this area to explore how relational cosmology can be translated into operational conservation strategies without reducing indigenous categories to metaphor, folklore, or environmental sentiment (Keegan, 2022).

Behind these methodological challenges, there is a more fundamental problem, namely epistemological recognition, both academic and paradigm of law. Indigenous people often experience marginalization because their worldview is considered just a culture, not an equal knowledge system or a 'legitimate religion' (Maarif, 2023; Keegan, 2022), and the lack of formal legal recognition. Changing this perspective requires a paradigm shift that is not easy (Kabir, 2022). However, if we are willing to acknowledge that non-human entities can be treated as a person in a relational sense, our understanding of personhood diversity and human responsibility to the environment will be stronger. By using relational epistemology, we open the possibility of more completely understanding of the relationship between humans and the broader ecological system. Ultimately, fair and sustainable environmental management requires full recognition of indigenous cosmology as a knowledge system with equal value to other scientific traditions and with continuing relevance to current debates on revitalization, governance, and ecological futures.

All the above analysis is the main concern in the issue of indigenous religious knowledge in this paper. This paper will systematically review the issue by following these questions: (RQ1) How do indigenous religions across different cultural contexts conceptualize and attribute personhood to non-human entities?; (RQ2) How do indigenous religions across different cultural contexts engage with non-human personhood?; (RQ3) How do indigenous cosmologies and worldviews contribute to conservation practices and environmental stewardship?; and (RQ4) What is the legal recognition and rights framework for adherents of indigenous religions/communities? By answering these questions, the chapter also clarifies why indigenous cosmologies matter for present-day debates on legal pluralism, environmental policy, and the revitalizing of Indigenous knowledge systems.

THEORETICAL FRAMEWORK

The theoretical landscape underlying the discussion of indigenous religions, and their conceptualization of non-human personhood is a fundamental challenge to modern post-Cartesian ontological assumptions. This section traces the academic debates about the ontological foundations, relational cosmology, and epistemological validity that shape our understanding of intersubjective cosmology in the indigenous religion context. Through critical engagement with anthropological theory, particularly ontological debates, this framework establishes a conceptual basis for analyzing how indigenous religions relate agency, consciousness, and personhood to non-human entities.

Ontological Foundations of Indigenous Religion

The ontological turn in anthropology marks a profound epistemic correction to the dominance of representationalism in indigenous cosmological understandings. Representationalism is rooted in Cartesian ontological assumptions that divide reality into binary oppositions of subject–object, mind–body, and human–nature. Within this dichotomous perspective, the worldview of indigenous peoples is often misunderstood. Contemporary critiques suggest that this dualism is not simply an abstract philosophical choice, but rather an epistemic foundation that systematically limits recognition of non-human agency and reduces relational ethics to instrumental relationships (Keegan, 2022). In this context, the ontological turn serves as a conceptual repositioning by recognize indigenous cosmologies as coherent and equal philosophical systems, while at the same time open space for indigenous religious paradigms that make relationality and intersubjectivity valid ontological foundations (Maarif, 2015).

One of the most operational articulations of the Cartesian ontological dualism critique can be found in the Amerindian perspectivism developed by Eduardo Viveiros de Castro. Perspectivism views the world as a ground inhabited by diverse perspectives of various *persons*—humans, animals, spirits, mountains, seas, and other natural entities—each of which experiences reality from the perspective of an ontologically equal subject (De Castro, 2019). Status as a subject is not only determined by human cognitive capacity, but also by the perspective and affection that shape the way each entity perceives and relates to the world. Within this framework, the shaman acts as an ontological mediator due to his ability to adopt non-human subjectivity, allowing for a perspective shift that shows other entities also view themselves as persons. Thus, nature is not reduced to a passive object of human representation, but rather is understood as a multinaturalism cosmos, where diverse *persons* are simultaneously present through diverse perspectives that share the same world.

In line with these criticisms, Philippe Descola provides an analytical tool that expands the understanding of ontological variation across cultures through the typologies of animism, totemism, analogism, and naturalism. With these typologies, he shows that Western naturalism—which assumes physical continuity but discontinuity of interiority (soul or spirituality) between humans and non-humans—is only one particular historical configuration. Descola emphasized that the separation of humans and nature is not a universal norm (Descola, 2013). In indigenous community context, for instance, they recognize the continuity of interiority in various species, thus extending subjectivity and social capacity to various entities in their cosmology. In short, through these typologies, various indigenous communities understand that various entities also have the capacity to be called subjects, a status that is not exclusively dominated by humans as colonial anthropology tends to suggest.

Colonial anthropology generally reduces the worldviews of indigenous communities that perceive various non-human entities as subjects to mere primitive beliefs. Boughton points out that the so-called "new animism" reveals a complex philosophical framework in understanding non-human agency, relational ethics, and ecological connectedness (Boughton, 2020). However, this ontological recognition at the same time requires critical awareness so that it does not turn into a conceptual appropriation that separates indigenous cosmology from the context of power relations and colonial history that surrounds it. By incorporating this reflection, the new animism does not stop at the affirmation of non-human agency but is positioned in a critical dialogue with the epistemic and political boundaries of academic knowledge itself.

The common thread that unites these approaches lies in the placement of relationality as the initial ontological principle. In many indigenous cosmologies, relationships are not understood as a result of the existence of entities, but rather as conditions that enable entities to emerge, be recognized, and be meaningful (Maarif, 2015). This position fundamentally distinguishes the *indigenous religion paradigm* from the Cartesian ontological dualism. If Cartesian dualism starts from the separation of subject and object and then establishes a relationship between the two, relational ontology starts from relationships as a prerequisite for existence between persons. Consequently, personhood is understood as an actual and ongoing relational capacity, rather than as an exclusive biological attribute of humans, so it can be attached to any entity that participates in social and cosmic life (Weatherdon, 2022).

Departing from this foundation, non-human personhood in indigenous religions should be understood as an ontological recognition of real agents who have consciousness, intentionality, and relational capacity, not as a symbolic metaphor or naive anthropomorphism. This is the main ontological foundation of indigenous religion: the world is understood as an intersubjective network inhabited by various persons who are intertwined in ethical, ecological, and cosmological relationships. This foundation allows for a re-reading of the boundaries of environmental ethics, conservation practices, and the modern legal imagination that has relied on the separation of human and non-human. Thus, the theoretical flow moves argumentatively from a critique of the foundations of Cartesian ontological dualism to an affirmation of relational ontology as a coherent conceptual basis for understanding indigenous religions in cross-cultural studies.

Figure 1. Comparative Ontological Framework

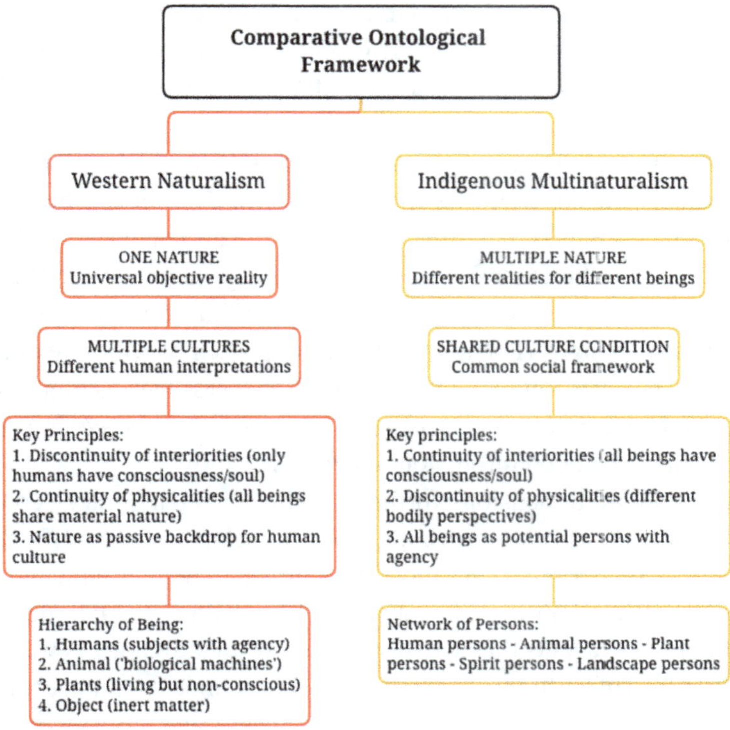

Indigenous Relational Cosmology

As has been emphasized in the discussion of ontological foundations above, the critique of Cartesian ontological dualism opens conceptual space for the recognition of indigenous cosmology as an ontological system based on relationships. The expansion of this criticism does not stop at the philosophical level, but rather obtains an explicit methodological articulation through the work of Shawn Wilson, who proposes indigenous epistemology as an alternative research paradigm. In *Research Is Ceremony*, Wilson asserts that the knowledge production is inseparable from the network of relationships that make up the social world and the cosmos, so research is understood as a ceremonial practice that demands relational involvement and openness. This position directly challenges Western epistemological assumptions regarding objectivity, distance (from the research object), and the neutrality of the researcher, by placing the researcher as part of the same ontological relationship with the research subject, both human and non-human (Wilson, 2020). Thus, relationality serves not only as an ontological claim, but also as an epistemic foundation for the way of knowing the world.

These epistemological foundations gained ontological deepening through Graham Harvey's contribution to the development of the study of contemporary animism. Through the framework of new animism, Harvey shows that indigenous cosmology consistently views non-humans as persons involved in social relations governed by protocols of respect, reciprocity, and ethical obligation. Different from classical readings that reduce animism to primitive beliefs or cognitive illusions, Harvey asserts that new animism represents a sophisticated religious system, in which cross-entities relationships are maintained through ritual practices and everyday ethics. By positioning the new animism as a relational epistemology—not just a collection of doctrines—Harvey makes it clear that knowledge of more-than-human world arises from the practice of living together, not from symbolic representations which separate from actual relationships (Harvey, 2019). At this point, criticism of Western naturalism transforms into a concrete understanding of how cross-entities relations are lived and interpreted.

This relationality then was expanded by Eduardo Kohn by shifting the focus from mere social relations to the semiotic process of life itself. In *How Forests Think*, Kohn develops a bio-semiotic approach that rejects the limitations of meaning and communication in human language by showing that meaning-making takes place throughout biological systems. His ethnography with the Runa people in Ecuador shows that forests are understood as "thinking" entities, inhabited by non-human persons who interpret and respond to environmental signs (Kohn, 2013). This approach provides a strong theoretical justification for indigenous cosmology's claims of non-human consciousness and agency, while expanding the critique of epistemological anthropo-centrism by showing that interpretation and representation are relational processes that are scattered throughout the world of life.

The relational and semiotic dimensions further acquire a contemporary ecological and historical context through the concept of "multispecies collections" developed by Anna Tsing. Through her study of the *matsutake* mushroom economy, Tsing shows how humans and non-humans collectively shape the world through reciprocal interactions that are not always harmonious, but interdependent (Tsing, 2023). Fungi are treated as ecological agents that actively shape forest landscapes as well as human social life, thus revealing the limitations of anthropocentric frameworks in understanding environmental dynamics. By emphasizing collaborative survival in damaged landscapes, Tsing provides an analytical lens to read how indigenous cosmology maintains relational obligations to non-humans while responding to global ecological crises.

Collectively, Wilson, Harvey, Kohn, and Tsing's contributions make it clear that ontological critiques of Western dualism inevitably have implications for the recognition of indigenous cosmology as indigenous relational cosmology. This cosmology can be defined as an ontological system that understands the world as a cross-entities intersubjective network, where existence, knowledge, and ethics are rooted in the ongoing relationship between humans and non-humans. If the practice of knowing is based on ceremonial relationships, relational ethics, and semiotic processes of life, then indigenous epistemology can no longer be treated as inferior local knowledge. Rather, it needs to be recognized as a legitimate way of knowing with its own criteria of validity—namely, relational consistency, relationship sustainability, and ontological responsibility to a world beyond humans. At this point, the discussion of the validity of indigenous epistemology becomes not only relevant, but theoretically inevitable.

Indigenous Epistemological Validity

Contemporary academic discourse shows increasing tensions in the way indigenous cosmologies are interpreted and epistemologically positioned, particularly regarding issues of authenticity, adaptation, and legitimacy of knowledge. The debate over the traditional and contemporary expressions of indigenous religions cannot be separated from the more fundamental issue of cultural continuity amid colonization, globalization, and ongoing environmental change (Weather & Schermerhorn, 2022). Within this framework, the question of the validity of indigenous epistemologies arises as a direct continuation of the ontological critique of Cartesian dualism, as well as a test of the capacity of modern knowledge frameworks to recognize ways of knowing rooted in relationality and intersubjectivity.

One major area of debate centers on the issue of authenticity, specifically whether indigenous practices that adopt non-traditional elements or adapt to contemporary contexts can still be understood as legitimate indigenous cosmologies. Critics from outside, and in some cases from within the community itself, often view syncretism as a deviation from authentic tradition. However, recent studies have shown that indigenous cosmology has since its inception operated as a dynamic and adaptive system, capable of integrating new relationships, entities, and practices without losing its basic ontological principles (Maćkowiak, 2024). In this perspective, so-called syncretism—although it is highly debated—does not signify epistemic degradation, but rather an expression of indigenous agency in navigating complex and often asymmetrical colonial and postcolonial conditions (Zaluchu, 2025).

Tensions surrounding authenticity are further complicated by the issue of translation, which reveals deeper epistemological challenges. Many key concepts in Indigenous cosmologies have no direct equivalents in Western languages or conceptual categories, so the process of translation often results in a narrowing or distortion of meaning. Even the term 'religion' itself becomes problematic when applied to cosmologies that fail to separate the sacred from the profane, the spiritual from the material, or religion from everyday life practices (Mendoza, 2024). These difficulties are not purely linguistic, but rather reflect the fundamental epistemic incompatibility between Western academic categories and the way indigenous organizations experiences, knowledge, and relationships with the superhuman world (Årsheim, 2018).

In addition to conceptual issues, discourse on the validity of indigenous epistemologies also increasingly emphasizes the ethical dimensions of research practice. The long history of Western knowledge production of indigenous peoples has often been marked by a pattern of knowledge extraction and epistemic violence, in which indigenous cosmology is studied without recognition of intellectual sovereignty or mutual benefit to the communities concerned (Weaver, 2024). This situation demands the use of decolonial methodologies that explicitly position indigenous peoples as producers of knowledge, not just objects of study. This approach requires collaboration, relational accountability, and an orientation to concrete benefits for indigenous communities as a prerequisite for the epistemic legitimacy of research.

Overall, this debate asserts that the validity of indigenous epistemologies cannot be understood as an additional claim within the framework of academic knowledge, but rather as an epistemic claim that stands alongside other knowledge traditions. Indigenous epistemologies have their own validity rooted in relational consistency, continuity of practice, and ontological responsibility to the more-than-human world. Thus, academic engagement with indigenous cosmology demands more than theoretical acceptance. It requires an ethical commitment to indigenous sovereignty, self-determination, as well as the right of communities to defend and develop their ontological and epistemological frameworks in their own way. It also requires comparative caution, attention to internal diversity, and sensitivity to indigenous self-representation in the present. In this sense, the theoretical framework functions not only as an instrument of analysis, but as a normative basis for responsible scientific practice in dealing with the way of knowing and the way of being indigenous.

RESEARCH METHOD

This study adopts the Systematic Literature Review (SLR) method by following the guidelines of Preferred Reporting Items for Systematic Reviews and Meta-Analyses (PRISMA) proposed by (Moher et al., 2009). PRISMA is an internationally recognized standard for improving the quality of reporting and systematic review methodologies (Page et al., 2021). This method was chosen because it has proven effective across disciplines for tracing patterns in a body of literature while making the search and screening process transparent. In this chapter, the method is used to identify, screen, and analyze published studies relevant to indigenous religion and non-human personhood. This process utilizes the Watase Uake research collaboration tool, which is divided into three stages: identification, screening, and analysis (Wahyudi, 2024). At the same time, the review remains interpretive rather than exhaustive, because the corpus is intentionally limited to a narrow search strategy and a final set of 15 articles.

Identification

Article searches were conducted using the Scopus database, which was chosen for its excellence in providing high-quality scientific articles through a rigorous indexing process (Oliveira et al., 2018). Scopus was chosen as the primary alternative to Google Scholar because Google Scholar has several shortcomings, such as 1) repetitive search results and duplication of articles from various sources, and 2) inclusion of articles from predatory journals that do not meet academic standards. In addition to Scopus, searches were also conducted using other methods, including the Watase database, to ensure completeness of sources. The initial search was conducted using specific relevant keywords, namely 'indigenous religion' and produced 92 records. This keyword was selected to keep the search closely aligned with the focus of the chapter. However, that choice also creates an important limitation, since relevant studies that use adjacent vocabularies—such as animism, sacred ecology, or more-than-human relations—may not be fully captured by the corpus.

Screening

All the articles obtained then go through an initial screening process with several steps. First are irrelevant articles. A total of 24 articles were removed because they were deemed irrelevant by the automated tool based on the year of publication (2015-2025). The second is an article without an abstract. There are 7 articles that were deleted because they do not have abstracts that can be used for playback. And third, there are 2 other articles that were deleted because they did not meet the quality criteria of the journal. After this initial screening process, there are 59 articles left ready for the next screening stage.

The next stage is content article screening. Conference abstracts, book reviews, and non-academic sources were excluded due to a lack of detailed quality assessment. Studies that treated indigenous beliefs as mere folklore or cultural artifacts, without acknowledging their ontological validity, were also excluded, in accordance with the principles of decolonial methodology (Chambers et al., 2018). Articles that do not have an explicit discussion of non-human agencies or personalities are considered outside the scope of this review. Furthermore, studies that focus only on Western interpretations of animism without indigenous perspective are excluded to maintain epistemological consistency with indigenous knowledge systems. From the 59 articles screened for relevance and validity, 40 did not meet the criteria and were excluded. A further 19 articles were selected for the report development process. Of these, 4 articles cannot be taken, leaving 15 articles to be assessed. All 15 articles were judged to meet the eligibility criteria, and none were deleted. A total of 15 articles were finally included in the systematic review and used as analysis material in this study. However, to support the argument and provide valid academic evidence, this paper also refers to various other relevant articles.

Qualitative Analysis

After obtaining the articles included in the final stage of PRISMA, a qualitative analysis was carried out using thematic analysis. This process follows the PRISMA protocol and organizes the findings according to the four research questions mentioned earlier. Thematic analysis was used to identify recurring patterns, points of convergence, and areas of difference across the reviewed studies. This approach makes it possible to move beyond summary and to interpret how indigenous cosmologies are discussed in relation to personhood, ritual practice, conservation, and legal recognition.

Figure 2. PRISMA Flow Diagram

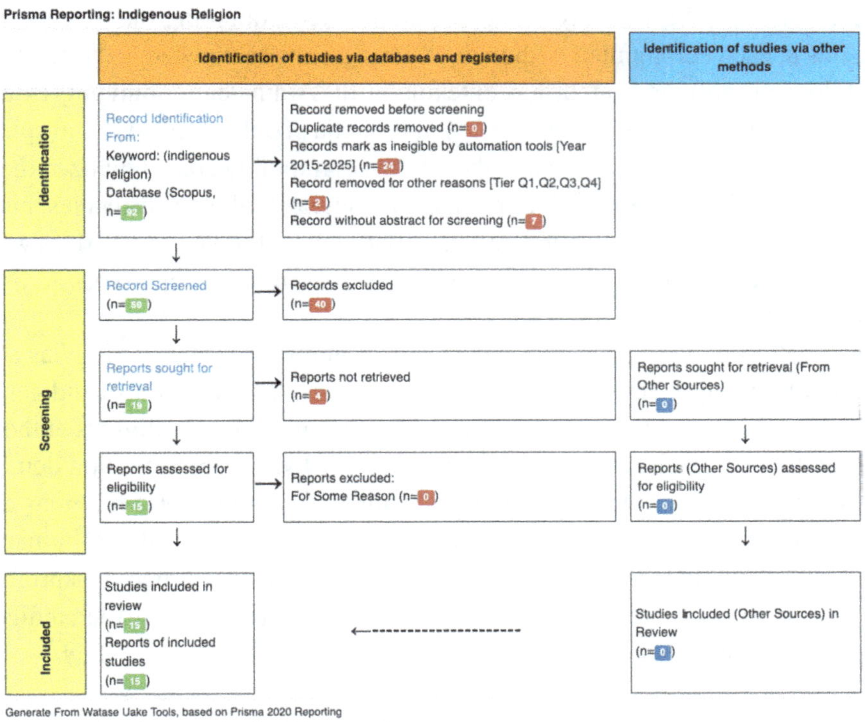

FINDINGS

Ontological Configuration of Non-Human Personality

Analysis of ontological configurations across indigenous communities reveals a prominent consistency in the way indigenous cosmology extends personhood beyond human boundaries, even though expressed through different cultural practices and symbols. A synthesis of data from eight Indigenous groups in Southeast Asia, North America, and Latin America (see Table 1) shows that social consciousness, agency, and capacity are also systematically attached to non-human entities through relational systems that are empirically variable but remain philosophically coherent. These findings provide empirical support for previous theoretical arguments that relational ontologies are not isolated local traits, but rather cross-cultural patterns in Indigenous cosmologies.

If we examine the criteria for personhood attribution more deeply, the basis for personhood recognition is relational, not rational centered. This pattern directly challenges the Western philosophical tradition that privileges rationality, reflective awareness, or individual mental capacity as the primary requirement of personality. The *Suku Anak Dalam*, for example, recognizes forest creatures and *Badewo* spirits as social actors based on their ability to "contribute to the sustainability of life" (Najib, 2020). This reciprocal framework reinforces the theoretical findings of multinaturalism discussed earlier, in which personality is achieved through active participation in relationships, rather than through the possession of predetermined cognitive attributes.

Cross-cultural variations become even more apparent when examining the spectrum of entities recognized as persons in each community. Although various groups acknowledge the existence of spiritual beings, their ontological configurations differ contextually: the Ammatoan community extends personality to land, forests, plants, and animals (Maarif, 2015); the Toraja people place ancestral spirits at the center of their worldview (Maćkowiak, 2024); while the Ohlone community conceptualized the land itself as a relative who had agency. This variation does not indicate ontological inconsistencies, but rather reflects a relational scheme that is in harmony with Descola's ontological typology (2013), as well as indicating that new animism encodes the theory of the complex distribution of consciousness across entities in the cosmos (Gomez, 2024).

The encounter between indigenous religions and emerging world religions in Nias and Toraja communities adds an additional analytical layer to these findings. The survival of indigenous ontologies alongside Christianity challenges assumptions about the articulation of 'religious authenticity.' Rather than signaling the dilution or disappearance of tradition, this hybrid configuration demonstrates indigenous agency in maintaining ontological principles while confronting world religious coercion and its attendant power structures (Zaluchu, 2025). The persistence of the concept of non-human personhood in the Christian context shows that the ontological configuration serves as a structure of profound worldview and is relatively resistant to total religious conversion. Thus, the encounter between indigenous religions and world religions does not always negate each other but can influence each other.

Table 1. Ontological configurations of non-human personhood across indigenous communities

Indigenous Group	Geographic Location	Non-Human Persons Recognized	Personhood Criteria	Relational Practices	Ontological Framework
Suku Anak Dalam (SAD)	Jambi, Indonesia (Najib, 2020)	Forest beings, *Badewo* (forest spirits), animals as social actors	Mutual contribution to life sustainability, perceived as humans with agency	Living together in mutual relationships, reciprocal obligations for forest sustainability	Intersubjective cosmos where humans and non-humans are equal social actors
Ammatoans	South Sulawesi, Indonesia (Maarif, 2015)	Land, forest, plants, animals, spirits	Sharing life with humans, constitutive of intersubjective relationships	Strict forest conservation regulations, punishment systems, religious ecology practices	All beings perceived as persons/subjects in intersubjective relationships
Toraja	Sulawesi, Indonesia (Maćkowiak, 2024)	Ancestral spirits, supernatural beings in *aluk* worldview	Continuation of personhood after death, agency in affecting living world	*Rambu Solo* funeral ceremonies, *gora-gora tongkon* ritual practices	Hybrid ontology combining indigenous *aluk* and Christian cosmologies

continued on following page

Table 1. Continued

Indigenous Group	Geographic Location	Non-Human Persons Recognized	Personhood Criteria	Relational Practices	Ontological Framework
Nias Indigenous Communities	Nias Island, Indonesia (Zaluchu, 2025)	Ancestral spirits, *Lowalangi* (sky god)	Protective agency, spiritual authority, continued presence	Prayer to ancestors, contextualized Christian-indigenous practices	Syncretic framework blending animist-dynamic beliefs with Christianity
Ohlone Tribes	California, USA (Gomez, 2024)	Land as relative, ancestral beings, natural entities	Relational identity, knowledge holders, treaty makers	Research as ceremony, land-based practices, relational protocols	Relational ontology emphasizing kinship with land and more-than-human beings
Indigenous Communities (General)	Indonesia (Maarif, 2023)	All cosmic beings - human and non-human	Inherent relational dignity, capacity for interpersonal relationships	*Adat* (customary) systems, reciprocal dignity exchanges	Inter-relational cosmology with relational dignity as religious norm
Native American Nations	USA & Canada (Weatherdon, 2022)	Animals as persons, relatives, knowledge holders, treaty makers	Agency, consciousness, capacity for knowledge transmission	Ceremonial protocols, kinship practices, treaty-making with animals	Pluralistic cosmology of beings with equal standing
Latin American Indigenous Peoples	Paraguay, Mexico, Colombia (Petri & Klocek, 2025)	Sacred sites, natural entities, spiritual beings	Collective spiritual significance, communal identity markers	Collective religious practices, sacred site protection	Cosmologies under threat from external and internal FoRB violations

Communicative Practices and Rituals with Non-Humans

Continuing previous findings regarding the expansion of non-human personhood in indigenous cosmology, the modality of communication between humans and non-humans emerged as a key mechanism that allowed such relationships to operate practically. In many indigenous contexts, this communication is mediated through rituals and ceremonial protocols that include dreams, prophecies, and the reading of natural signs. These practices mediate intersubjective relationships that enable dialogue across entities and strengthen the understanding of the world as an interconnected network of relationships. Najib's study of the SAD community demonstrates how rituals and dreams play an integral role in establishing dialogue with the forest, emphasizing that communication is not limited to human language but is embedded in relational existence itself (Najib, 2020).

In line with the practice of these rituals, the use of signs and prophecies as a medium of communication indicates the epistemological dimension of human and non-human relations. This communication mode allows individuals to interpret environmental messages within a cosmological framework that views non-humans as active and communicative agents. Weatherdon emphasizes that rituals are often directed at maintaining reciprocal relationships with animals and other natural entities, while also emphasizing the importance of respecting the messages conveyed by the non-human world (Weatherdon, 2022). Thus, knowledge is not produced through mere cognitive abstraction, but through lived experience and continuous interaction with the environment.

For regular and meaningful interspecies communication to occur, many indigenous communities rely on ritual specialists or shamans (*dukun*) as mediators. These actors act as translators between the human and non-human worlds, drawing on their in-depth knowledge of culturally and contextually specific ceremonial protocols. Keegan asserts that such mediators play a crucial role in maintaining the balance of interspecies social relations, as they mediate non-human interests into human understanding and action (Keegan, 2022). This role highlights that ritual communication is not an individual practice, but rather a social institution that sustains ecological harmony. When understood collectively, the practice of ritual communication and the role of *dukun* show a high level of complexity in navigating human and non-human relationships. Through ceremonial practices, indigenous peoples build sustainable relationships with non-human entities based on mutual respect and understanding. These findings deepen our understanding of indigenous knowledge systems as relational epistemologies that serve not only to interpret the cosmos but also to ethically regulate communal life.

This ethical dimension of cross-species communication is clearly articulated in the concept of mutual obligation that rules interspecies relationships. In much indigenous cosmology, reciprocity is not limited to interactions between humans, but extends to relationships with more-than-human entities. The community well-being is understood as a result of the maintenance of healthy reciprocal relationships within the ecosystem, where traditional ceremonies serve as a formal expression to renewal or remind of such responsibility. In this context, rituals are not symbolic artifacts, but rather vital practices that bind human and non-human agents in networks of collective responsibility.

This mutual obligation principle is concretely manifested in ceremonial practices that facilitate dialogue with forest creatures and other natural entities (see table 2). Among Ammatoan, rituals play a role in affirming the commitment to coexistence and caring for the environment in a way that respects the spirit and the power that inhabits it (Maarif, 2015). This practice articulates a relational ontology in which humans are positioned as part of a broader ecological framework, with roles and responsibilities that are not separate from non-humans.

In addition to serving as an ethical mechanism, rituals also have a real ecological impact through so-called ritual efficacy and ceremonial outcomes. Weatherdon notes that indigenous rituals encourage recognition of the role of non-humans, particularly animals and other natural entities, in shaping ecological dynamics (Weatherdon, 2022). These impacts contribute to environmental management and biodiversity conservation practices, as rituals articulate the interconnectedness of life forms and enable collective responses to ecological challenges. This finding is reinforced by Keegan's observation that historical representations of Indigenous peoples often fail to capture how they interact with the environment through ritual practices. Ethnographic studies show that active engagement with non-humans through rituals effectively establishes ecological relationships, which positively impacts social cohesion and environmental health (Keegan, 2022). Thus, rituals not only reflect cosmology, but also produce material meaning in the lives of people who are in harmony with their environment.

Overall, understandings of reciprocal obligations and cross-species ritual communication affirm indigenous cosmologies' commitment to respectful relationships between humans and non-humans. By recognizing the role of non-humans and strengthening shared responsibilities through ceremonial practices, indigenous perspectives offer significant epistemological and ethical contributions to contemporary debates on sustainability and human-environment relations. These findings reinforce previous claims that indigenous relational ontologies are not simply belief systems, but frameworks of living knowledge that have practical implications for managing a world beyond human capabilities.

Table 2. Communication and ritual practices with non-human persons

Indigenous Community	Communication Methods	Ritual Contexts	Mediator Roles	Reciprocal Obligations	Documented Outcomes
Mentawai/ Siberut Communities	Shamanic mobility between home and forest (Hammons, 2016)	Indigenous religious ceremonies, healing rituals	Shaman as mobile mediator between village and forest realms	Maintaining spatial, social, and cosmological continuity	Contested authenticity between mobile and sedentary practitioners
Suku Anak Dalam	Direct interaction with *Badewo*, forest communication (Najib, 2020)	Forest conservation practices, daily interactions	Community members as direct participants	Mutual contribution to forest life sustainability	Successful integration with *Bukit Duabelas* national park zoning system
Ammatoans	Religious perception governing daily behaviors (Maarif, 2015)	Forest conservation rituals, punishment ceremonies	Community leaders enforcing religious ecology	Strict adherence to conservation regulations	Effective forest preservation through religious practices
Toraja	*Rambu Solo* ceremonies, ancestral communication (Maćkowiak, 2024)	Funeral ceremonies, life-cycle rituals	*Gora-gora tongkon* ritual officials, Christian-indigenous mediators	Maintaining relationships with deceased, cultural obligations	Hybrid ritual practices accommodating multiple religions
Nias Communities	Prayer to ancestors, spiritual communication (Zaluchu, 2025)	Protection rituals, community ceremonies	Traditional practitioners and Christian leaders	Honoring ancestral spirits while maintaining Christian faith	Persistent syncretism despite Christianization
Ohlone Tribes	Research as ceremony methodology (Gomez, 2024)	Land-based ceremonies, community gatherings	Community members as knowledge keepers	Strengthening relationships through engaged scholarship	Enhanced community-researcher relationships
Indonesian Indigenous Communities	*Adat*-based communication systems (Maarif, 2023)	Customary ceremonies, dignity exchanges	*Adat* leaders and community elders	Maintaining relational dignity with all beings	Resilience despite government and missionary pressures
Mobile Indigenous Peoples	Movement as communication, walking as knowing (Weatherdon & Schermerhorn, 2022)	Pilgrimage, seasonal movements, ceremonial journeys	Mobile practitioners maintaining networks	Cultivating personhood through movement	Political alliances and religious meaning generation

Environmental Management through Relational Cosmology

Based on the recognition of non-human personhood, ritual communication, and cross-entities mutual obligations, indigenous conservation practices rooted in relational cosmology have been shown to produce measurable ecological impacts, particularly in forest and biodiversity conservation. Najib points out that SAD people in Indonesia utilize intersubjective relationships with non-human entities to build an understanding of ecological balance that underpins effective forest conservation practices (Najib, 2020). This approach positions environmental management as a relational responsibility, where humans and non-humans are understood as part of an interdependent ecological system, thereby strengthening adaptive management strategies and ecosystem sustainability.

The environmental management framework is in line with the broad findings of Indigenous Ecological Knowledge (IEK), which has historically prioritized sustainability and biodiversity conservation. Indigenous conservation practices that recognize non-human personhood emphasize the spiritual and cultural value of the species, which in turn reinforces the collective commitment to environmental management (Keegan, 2022). In this context, indigenous ontology forms a holistic management system that is not only oriented towards ecological outcomes, but also on the maintenance of relational relationships within ecosystems (Maarif, 2015).

When compared with Western conservation approaches, the effectiveness of IEK is clear. Indigenous knowledge systems, which include ontological relationships between humans and non-humans, allow for a more sensitive form of environmental management to the complex interrelationships between species and ecosystems (Weatherdon, 2022). These practices and relationships are not merely technical instrumental relationships but constitute a much more binding spiritual responsibility. In contrast, Western approaches often adopt a reductionist framework that separates humans from their environment, thus risking overlooking the relational dynamics underlying ecological resilience. Findings on the impact of indigenous conservation communities confirm that local understandings rooted in relational cosmologies yield more adaptive strategies in the face of environmental change.

In addition to ecological advantages, the integration of IEK into environmental governance also shows a significant contribution to social well-being. By recognizing the intrinsic value of non-human entities and fostering collaborative relationships across species, indigenous practices broaden the horizons of conservation ethics beyond narrow utilitarianism. Research shows that the adoption of a relational cosmology-based approach contributes to improving ecological outcomes while strengthening the social cohesion of indigenous peoples (Maarif, 2023).

However, the effectiveness of these practices cannot be separated from the challenge of integration into formal policy frameworks. Exploration of indigenous conservation reveals that while IEK offers huge potential for sustainability, recognition of non-human agency and sacred ecology is often not accommodated in state policies. This gap is evident in the context of customary forest management, where local practices based on relational cosmology are confronted with regulations that prioritize a technocratic approach. It is the case of SAD in Jambi, where government policies often fail to integrate indigenous wisdom into forest conservation strategies. Najib noted that the absence of shared management mechanisms and meaningful recognition of indigenous knowledge creates a disconnect between local conservation practices and mainstream environmental policies (Najib, 2020). This condition emphasizes the need for collaboration between indigenous knowledge holders and state institutions so that conservation efforts are in harmony with the cultural and spiritual values of indigenous peoples.

In a cross-regional perspective, relational ontology provides a conceptual framework for understanding the success of indigenous conservation in a variety of contexts, including Indonesia, the Amazon, and North America (see Table 3). Cross-regional studies show that customary practices that prioritize relationships between species consistently produce positive conservation outcomes, as these strategies are rooted in a deep understanding of ecological linkages (Weatherdon, 2022). In the Amazon, indigenous groups practice conservation through rituals and land management that reaffirm relational responsibility to the land and the creatures that inhabit it. Similar patterns are also found in Native American communities, where spiritual beliefs play a central role in shaping environmental management strategies (Keegan, 2022). These examples show that the integration of indigenous cosmology into conservation models is not only normatively important, but also empirically effective in responding to the biodiversity crisis.

However, the interaction between indigenous worldviews and modern conservation policy frameworks is still colored by the marginalization of indigenous epistemology. The systemic failure to recognize IEK as a legitimate source of knowledge often results in policies that are insensitive to local ecological realities and undermines conservation effectiveness (Maarif, 2023). Therefore, fair recognition of indigenous epistemology is a prerequisite for inclusive environmental governance. Overall, overcoming the challenges of implementing customary conservation models requires a commitment to understanding and respecting the relational ontology that underpins the practice. By recognizing the value of indigenous environmental management and the effectiveness of conservation strategies based on relational cosmology, policymakers can build equitable and sustainable partnerships. This approach not only has the potential to improve biodiversity outcomes, but also to maintain the cultural and epistemological integrity of indigenous peoples as key actors in global environmental management.

Table 3. Environmental conservation through indigenous relational cosmologies

Case Study Location	Conservation Practice	Cosmological Foundation	Measurable Impact	Challenges Faced	Policy Integration
Bukit Duabelas National Park (TNBD), Indonesia (Najib, 2020)	Integration of SAD customary values with state zoning	*Badewo* and animals as social actors in forest conservation	Successful forest preservation in TNBD	Dynamics between state zoning and indigenous spaces	Configuration of intersubjective relationship between government ecology and indigenous religion
Tanah Toa, South Sulawesi (Maarif, 2015)	Strict forest regulations with punishment systems	Non-human beings as persons sharing life with humans	Well-preserved forest areas under Ammatoan management	External pressures from development	Religious ecology model for conservation policy
Indigenous territories, USA/ Canada (Weaver, 2024)	Indigenous ecological knowledge (IEK) practices	Knowledge as relational and sacred, not commodity	Not quantified - focus on epistemic resistance	Settler appropriation and bio-colonialism of IEK	Critique of settler-environmental integration attempts
Multiple Indonesian sites (Maarif, 2023)	*Adat*-based environmental management	Relational dignity extending to all beings	Maintained biodiversity despite pressures	Government non-recognition, corporate encroachment	Call for expanding religious freedom to include relational dignity
Siberut Island, Indonesia (Hammons, 2016)	Ancestral land management through mobility	Spatial and cosmological continuity between home and forest	Preservation of clan territories	Government relocation programs disrupting mobility	Resistance to sedentary policies
North American Indigenous lands (Weatherdon, 2022)	Animals as treaty partners in conservation	Animals as persons, relatives, and knowledge holders	Traditional territory preservation	Western conservation models excluding animal personhood	Proposing decolonized conservation approaches
Latin American territories (Petri & Klocek, 2025)	Sacred site protection as conservation	Collective spiritual significance of landscapes	Variable due to FoRB violations	External threats from extractive industries, internal religious conflicts	Need for FoRB protections incorporating indigenous cosmologies
California Ohlone territories (Gomez, 2024)	Land-based practices maintaining relationships	Land as relative requiring reciprocal care	Community resilience and cultural continuity	Ongoing occupation and limited land access	Engaged scholarship as conservation support

Indigenous Community Legal Recognition

Responding to the challenges faced by indigenous communities, both epistemological and environmental conservation practices, recognition through legal frameworks is important in the governance of contemporary life. In recent decades, legal developments concerning rivers, forests, and other natural entities have shown that indigenous relational cosmologies increasingly intersect with debates on legal personhood and Rights of Nature. Recent comparative analysis of 14 Rights of Nature cases worldwide shows that these legal innovations fail through symbolism alone. Their effectiveness depends greatly on questions of design, especially the clarity of guardianship, the allocation of liability, and the extent to which local communities and Indigenous peoples remain central rather than merely consultative actors (Kahui et al., 2024). The recognition of the legal personality of the Whanganui River in New Zealand, for example, reflects the Māori cosmological view that the river is an ancestor with equal rights with humans (Collins & Esterling, 2019). This case serves as an important reference point in expanding the definition of legal personality beyond humans, in line with the principles of indigenous cosmology that emphasizes the interconnectedness of all beings.

In the Indonesian context, customary practices have long articulated intersubjectivity-based relationships that implicitly recognize the rights of non-human entities. Maarif points out that indigenous cosmology places humans and non-humans as relational subjects who have reciprocal rights and obligations in maintaining ecological balance in the Ammatoan society (Maarif, 2015). These findings suggest that modern legal recognition of natural rights is not ahistorical, but rather an institutionalization of ontological principles that have long operated in indigenous practice. A similar normative pattern appears in recent work on water conservation through local wisdom, where ecological protection is framed simultaneously as social justice and ecological justice and where ritualized community ethics mediate the relation between human need, shared water sources, and environmental restraint (Hermanto Candra et al., 2024).

In line with the recognition of non-human personalities, the legal protection of freedom of religion or belief (FoRB) for indigenous peoples further highlights the importance of protecting sacred sites and ritual practices. Legal cases in North America show that the protection of sacred landscapes is not only related to the preservation of culture, but also to maintaining the ecological integrity embodied by these sites (Keegan, 2022). Keegan also notes that throughout history, indigenous beliefs have often been misinterpreted or ignored in the legal system, undermining effective advocacy for the spiritual and ecological rights of indigenous peoples. This tension underscores the need for a more inclusive and sensitive legal framework for customary cosmology.

By placing the recognition of non-human personalities side by side with the protection of religious rights, the contemporary legal landscape shows a shift towards a more holistic approach to governance. At the same time, this shift remains fragile. Ugwu (2025) argues that rights-of-nature discourse may become institutionally thin when it is detached from the religious or ethical worlds that give it social depth. In that sense, the legal field becomes a crucial site where indigenous cosmologies are not only described, but selectively institutionalized. The integration of indigenous worldviews into law not only expands protections for non-human entities, but also strengthens guarantees over sacred practices and spiritual landscapes, thus linking cultural and ecological preservation within a single normative framework.

At the international level, frameworks such as the United Nations Declaration on the Rights of Indigenous Peoples (UNDRIP) and ILO Convention 169 play an important role in protecting indigenous cosmology, particularly spiritual relations with non-humans. These instruments emphasize the interconnectedness between spiritual practice and nature as the basis for the recognition of rights. Najib emphasized that indigenous cosmology fosters intersubjective relationships that support a holistic understanding of the environment, thus demanding legal protection of these traditional practices and knowledge in the national legal system (Najib, 2020).

Nevertheless, the implementation of UNDRIP and ILO 169 often faces structural challenges. Tensions arise when the individualistic and anthropocentric Western legal framework is confronted with a relational indigenous ontology (Petri & Klocek, 2025). The concept of "relational dignity", which recognizes non-humans as social subjects, is often at odds with Western jurisprudence categories that prioritize human interests. Maarif points out that the misrecognition of indigenous religions as mere cultural expressions, rather than legitimate spiritual beliefs, further complicates the enforcement of the collective and ecological rights of indigenous peoples (Maarif, 2023). This tension is also reflected in efforts to recognize the collective rights of indigenous peoples. Individually oriented legal systems are often unable to accommodate claims based on social and ecological relations. Keegan highlights that historical narratives that marginalize indigenous perspectives have complicated legal advocacy for their spiritual and environmental rights (Keegan, 2022). Weatherdon adds that the failure to recognize the relational dimension of indigenous beliefs—where non-humans are understood as active agencies in the socio-ecological order—significantly undermines legal protection and advocacy effectiveness (Weatherdon, 2022).

Table 4. Legal recognition of non-human personhood and indigenous cosmologies

Country/ Region	Legal Framework	Indigenous Group Involved	Non-Human Rights Recognized	Implementation Status	Outcomes/Challenges
Canada (Årsheim, 2018)	Ktunaxa Nation v. British Columbia Supreme Court case	Ktunaxa Nation	Sacred site (*Qat'muk*) spiritual significance	Case decided, limited recognition	Court struggled with "intersectional" vocabulary for indigenous rights claims
Norway (Årsheim, 2018)	Jovsset Ante Iversen Sara v. Norway Supreme Court case	*Sámi* peoples	Reindeer herding as religious practice	Partial recognition of religious dimensions	Challenges in translating indigenous practices into legal categories
Indonesia (Maarif, 2023)	Limited constitutional recognition, advocacy for expanded FoRB	Multiple indigenous communities	Relational dignity concept proposed	Ongoing advocacy, limited implementation	Need to expand definition of religion to include relational dignity
USA (Keegan, 2022)	US Federal Indian policy, religious freedom cases	Federally recognized Native American tribes	Sacred sites, ceremonial practices	Variable by tribe and jurisdiction	Gaps in protection for non-federally recognized tribes
Paraguay (Petri & Klocek, 2025)	National and international FoRB frameworks	Various indigenous groups	Collective religious practices	Significant violations documented	External threats from development, internal from religious conversion
Mexico (Petri & Klocek, 2025)	Constitutional protections, international conventions	Multiple indigenous peoples	Sacred territories and practices	Partial implementation, ongoing violations	Drug cartels and extractive industries threatening sacred sites
Colombia (Petri & Klocek, 2025)	Constitutional Court decisions, FoRB protections	Various indigenous communities	Territorial rights with spiritual dimensions	Progressive legal framework, implementation gaps	Armed conflict impacting religious freedom
USA Museums (Mendoza, 2024)	NAGPRA (Native American Graves Protection and Repatriation Act)	Fort Peck Tribes, various Native nations	Ancestors and sacred objects as relatives requiring care	Ongoing repatriation processes	Museums slowly recognizing relational ontologies in repatriation

Overall, while international frameworks such as UNDRIP and ILO 169 provide an important normative foundation, their effective enforcement requires a deeper understanding of indigenous relational ontologies. A collaborative and sensitive legal approach to indigenous cosmology has the potential to strengthen constitutional recognition, protect collective rights, and build more equitable relationships between indigenous peoples, their spiritual heritage, and the environment they inhabit. These findings confirm that the legal recognition of non-human personality is not simply a juridical innovation, but rather a logical consequence of the recognition of indigenous epistemology and ontology.

DISCUSSION

Relational Cosmology, Revitalization, and the Contemporary Strakes of Indigenous Knowledge

A systematic analysis of non-human personhood in indigenous religions reveals the existence of a consistent and profound ontological framework, which fundamentally challenges Western anthropocentric assumptions while offering important insights for responding to contemporary ecological crises. A synthesis of these 15 studies shows that, despite being expressed through diverse cultural practices and conceptualizations, indigenous communities consistently extend personhood beyond human boundaries. This discussion places the findings in a broader theoretical framework by examining their ontological implications, methodological challenges, practical applications, and contemporary relevance to the revitalizing of Indigenous knowledge systems. At the same time, the chapter does not claim that the reviewed literature represents a single Indigenous worldview or the full diversity of Indigenous traditions.

The most prominent finding in this study is the fundamental relational nature of person attribution in indigenous cosmologies. In contrast to the Western philosophical tradition that associates person with human rationality, indigenous peoples define person through relational capacity—that is, the ability to participate in social relationships and contribute to collective well-being (Maarif, 2015). This foundation describes a broad spectrum of entities recognized as personal, from forest spirits and ancestral creatures to natural landscapes such as rivers, forests, mountains, and so on. The SAD recognition of the *Badewo* spirit as a social actor based on "mutual contribution to the sustainability of life" confirms that personality emerges through relational involvement, not individually inherent attributes (Najib, 2020).

However, this ontological consistency faces tension when explored in contemporary academic contexts, particularly in relation to issues of syncretism, authenticity, and self-representation. The case of the Toraja and Nias communities, where indigenous cosmology coexists with Christianity, challenges the linear narrative of cultural decline or corruption in syncretic religious narratives (Maćkowiak, 2024; Zaluchu, 2025). Rather than exhibiting dilution, such configurations often represent adaptive strategies for maintaining core ontological principles in colonial and postcolonial conditions. However, the literature remains divided on the implications of this adaptation, and some communities themselves question the authenticity of hybrid practices, suggesting that authenticity is a dynamic internal negotiation field, rather than a stable academic category (Hammons, 2016). These tensions also caution against overgeneralization, because internal diversity within communities is part of the contemporary life of Indigenous religions.

This relational dimension becomes even more complex when analyzed through communication practices and rituals. The research documented a variety of species communication modalities—dreams, prophecies, and ceremonial protocols—that uncovered mechanisms for interacting with non-humans (Weatherdon, 2022). However, methodological limitations remain significant. Reliance on ethnographic observation and interviews risks filtering indigenous experiences through Western interpretive frameworks, so that aspects of communication that reject linguistic representation can be reduced or distorted

(Weatherdon, 2022). This challenge confirms the epistemological limits of cross-ontological research and helps explain the persistent translation gap between conceptual recognition and institutional understanding.

In addition, the ecological implications of relational cosmology present both opportunities and risks. Documented indigenous conservation practices—such as the preservation of Ammatoan forests and the integration of SAD indigenous values with state zoning—demonstrate the practical efficacy of non-human personality recognition in environmental management (Maarif, 2015; Najib, 2020). These findings challenge the dominant conservation paradigm that separates human culture from ecosystems. However, Weaver's (2024) critique of IEK appropriation warns of the dangers of reducing complex cosmology to a set of conservation techniques that are detached from their ontological foundations. Revitalization therefore cannot be understood only as preservation of practices; it also concerns the continuing authority of the cosmological principles that make those practices meaningful.

Similar tensions arise in the legal context. Attempts at the recognition of non-human personhood reveal a structural incompatibility between indigenous ontology and Western legal frameworks based on individual rights. While precedents such as the Whanganui River's legal personality appear progressive, broader comparative analysis shows that Rights of Nature regimes remain fragile when guardianship, liability, and community authority are poorly designed (Kahui et al., 2024). Ugwu (2025) sharpens this problem from another angle by arguing that rights-of-nature discourse may remain normatively thin when it is detached from the religious or ethical worlds that sustain it. Edirisinghe and Suchet-Pearson (2025) similarly show that legal personhood and Rights of Nature can help bridge relational Indigenous worldviews and state law, but only when Indigenous communities are treated as active actors in more-than-human governance rather than as symbolic sources of legitimacy. This problem is not merely semantic but reflects fundamental epistemological differences regarding the nature of existence, rights, and legal authority.

The implications of these findings for the methodology of religious studies are fundamental. Indigenous cosmology's rejection of the sacred–profane or spiritual–material dichotomy challenges this basic category of discipline (Keegan, 2022). This requires the development of an analytical framework that can accommodate ontological pluralism without reducing indigenous cosmology to a variant of world religions. The concept of "relational dignity" proposed by Maarif (2023) offers a conceptual path to reformulate religious freedom as a cross-entities relationship. In a more applied sense, Sohad and Mafrolla (2025) argue that integration between Indigenous and scientific knowledge becomes productive only when institutions move beyond extraction and create arrangements for co-production, contextual respect, and reciprocal exchange. This insight helps clarify that the translation gap identified in this chapter is not merely semantic; it is also organizational, political, and epistemic.

Going forward, this review confirms that research must move towards collaborative methodologies that recognize indigenous peoples as producers of epistemic knowledge and paradigms. Policy frameworks need to be restructured to accommodate relational ontologies substantively, rather than symbolically. In addition, environmental conservation and governance initiatives must understand that effectiveness lies not only in the adoption of indigenous practices, but in respect for, and legal recognition of, their cosmological foundations. Recent work on Indigenous Guardians, Elders, harvesters, and biocultural indicators associated with Netukulimk also shows that revitalization takes practical form through community governance, ecological stewardship, and intergenerational transmission rather than through static preservation alone (Lukawiecki et al., 2025). In the context of the global ecological crisis, indigenous ontologies can no longer be treated as ancient beliefs, but rather as sophisticated knowledge systems that offer crucial insights for the sustainability of the planet.

FUTURE RESEARCH DIRECTIONS

Future research needs to develop a collaborative methodology that positions indigenous communities as subjects as well as producers of epistemic knowledge, not just objects of study. This approach should include greater attention to the gender dimension in human and non-human relations, urban and diaspora indigenous communities, and internal diversity within indigenous traditions. In addition, longitudinal studies on the transformation of indigenous religions in the midst of socio-ecological changes are urgently needed. Future research also needs to encourage the reconstruction of legal and policy frameworks that accommodate the cosmological underpinnings of indigenous religions in a substantive, rather than symbolic, manner.

CONCLUSION

A systematic review of the 15 studies in this chapter shows that indigenous religions consistently articulate non-human personhood through relational ontological frameworks that challenge the dominant anthropocentric assumptions in the Western scientific tradition. Answering the research question, the findings of this study confirm that (RQ1) in indigenous religions, non-human personhood are not derived from individual cognitive capacity, but rather from participation in relations and mutual contributions to the sustainability of life; (RQ2) cross-cultural patterns show the recognition of spirits, ancestors, animals, and other natural entities as religious and social actors, although expressed through locally distinctive symbolic and ritual configurations; (RQ3) such intersubjective cosmology directly shapes effective religious and ecological practices, especially through the ethics of mutual obligations that inform indigenous religion-based environmental management and conservation; and (RQ4) the legal recognition of non-human personhood and the spiritual rights of indigenous peoples demands the substantive integration of indigenous religious cosmology into national and international legal frameworks. Theoretically, this chapter contributes to the indigenous religion studies by affirming indigenous cosmology as a coherent and reflective religious-philosophical system, rather than merely a cultural expression, while expanding environmental discourse by showing that conservation success is inseparable from a framework that recognizes non-human agency. In the context of law, these findings highlight the limitations of modern legal categories in understanding indigenous cosmology and religiosity rooted in relationality. However, this synthesis also has limitations, especially the small evidence base, regional imbalance, limited attention to internal diversity, and the lack of longitudinal studies on the transformation of indigenous religions amid socio-ecological changes. Therefore, future research needs to develop collaborative approaches that position indigenous communities as subjects as well as producers of religious knowledge, while encouraging legal and policy frameworks that can accommodate the cosmological foundations of indigenous religions more substantively. Ultimately, indigenous religions must be understood not as relics of the past, but as living traditions that continue to shape ethical relationships between humans, non-human beings, and the world in which they live.

OPEN CONTRIBUTIONS

Gedong Maulana Kabir is the single author responsible for the conceptualization, methodology, data collection and analysis, visualization, and writing the initial draft and final revision of the article.

OPEN DATA

The data is available at: https://doi.org/10.5281/zenodo.19042677

REFERENCES

Alves, F., & Vidal, D. G. (2024). Plural Nature(s): An Overview of Their Sociocultural Construction. In *Encyclopedia* (Vol. 4, Number 1, pp. 1–12). DOI: 10.3390/encyclopedia4010001

Årsheim, H. (2018). Including and Excluding Indigenous Religion through Law. *Numen*, *65*(5–6), 531–561. DOI: 10.1163/15685276-12341511

Boughton, J. (2020). New animism: Relational epistemologies and expanding Western ontologies. *Student Anthropologist*, *7*(1), 56–59. DOI: 10.1002/j.sda2.20200701.0012

Chambers, L. A., Jackson, R., Worthington, C., Wilson, C. L., Tharao, W., Greenspan, N. R., Masching, R., Pierre-Pierre, V., Mbulaheni, T., Amirault, M., & Brownlee, P. (2018). Decolonizing Scoping Review Methodologies for Literature With, for, and by Indigenous Peoples and the African Diaspora: Dialoguing With the Tensions. *Qualitative Health Research*, *28*(2), 175–188. DOI: 10.1177/1049732317743237 PMID: 29182046

Collins, T., & Esterling, S. (2019). Fluid Personality: Indigenous Rights and the Te Awa Tupua (Whanganui River Claims Settlement) Act 2017 in Aotearoa New Zealand. *Melb. J. Int'l L.*, *20*, 197.

De Castro, E. V. (2019). Exchanging perspectives: The transformation of objects into subjects in Amerindian ontologies. *Common Knowledge*, *25*(1–3), 21–42. DOI: 10.1215/0961754X-7299066

Descola, P. (2013). *Beyond Nature and CUlture*. The University of Chicago Press. DOI: 10.7208/chicago/9780226145006.001.0001

Edirisinghe, A., & Suchet-Pearson, S. (2025). Legal Personhood and Rights of Nature: Bridging Relational Vedda Worldviews and Sri Lankan Law. *Transnational Environmental Law*, *14*(3), 630–658. DOI: 10.1017/S2047102525100101

Gomez, A. R. (2024). Religion Scholarship on Indigenous Lands: Research, Relationship, and the Potential of Engaged Scholarship. *Journal for the Study of Religion, Nature and Culture*, *18*(4), 484–501. DOI: 10.1558/jsrnc.25258

Hammons, C. S. (2016). Indigenous Religion, Christianity and the State: Mobility and Nomadic Metaphysics in Siberut, Western Indonesia. *The Asia Pacific Journal of Anthropology*, *17*(5), 399–418. DOI: 10.1080/14442213.2016.1208676

Harvey, G. (2019). Animism and ecology: Participating in the world community. *The Ecological Citizen*, *3*(1), 79–84.

Hermanto Candra, P., Hamim, M., & Maulana Khabir, G. (2024). Water Conservation through Local Wisdom as A Manifestation of Social Justice and Ecological Justice. *BIO Web of Conferences, 96*, 04007. DOI: 10.1051/bioconf/20249604007

Kabir, G. M. (2022). Revisiting Javanese Islam: Towards a New Paradigm in Religious Studies. *Jurnal Dinamika Penelitian: Media Komunikasi Sosial Keagamaan*, *21*(2), 315–330.

Kahui, V., Armstrong, C. W., & Aanesen, M. (2024). Comparative analysis of Rights of Nature (RoN) case studies worldwide: Features of emergence and design. *Ecological Economics*, *221*, 108193. DOI: 10.1016/j.ecolecon.2024.108193

Keegan, B. (2022). Contemporary Native American and Indigenous Religions: State of the field. *Religion Compass*, *16*(9), e12448. Advance online publication. DOI: 10.1111/rec3.12448

Kohn, E. (2013). *How forests think: Toward an anthropology beyond the human*. Univ of California Press.

Lukawiecki, J., Jeddore, G., Roth, R., & Moola, F. (2025). Practicing "Netukulimk": Indigenous governance of culturally significant species in Ktaqamkuk (Newfoundland, Canada). *Facets*, *10*, 1–15. DOI: 10.1139/facets-2025-0152

Maarif, S. (2015). Ammatoan Indigenous Religion and Forest Conservation. *Worldviews: Global Religions, Culture, and Ecology*, *19*(2), 144–160. DOI: 10.1163/15685357-01902005

Maarif, S. (2019). Indigenous Religion Paradigm: Re-interpreting Religious Practices of Indigenous. *Studies in Philosophy*, 103–121. https://doi.org/http://doi.org/10.15068/00155157

Maarif, S. (2023). Human (Relational) Dignity: Perspectives of Followers of Indigenous Religions of Indonesia. *Religions*, *14*(7), 848. DOI: 10.3390/rel14070848

Maćkowiak, A. M. (2024). Christian and Indigenous: Multiple "Religions" in Contemporary Toraja Funerals. *Religions*, *15*(9), 1112. DOI: 10.3390/rel15091112

Mendoza, R. J. (2024). Undisciplining the Museum: Indigenous Relationality as Religion. *Religions*, *15*(11), 1325. DOI: 10.3390/rel15111325

Moher, D., Liberati, A., Tetzlaff, J., & Altman, D. G. (2009). Preferred reporting items for systematic reviews and meta-analyses: The PRISMA statement. *BMJ (Clinical Research Ed.)*, •••, 339. PMID: 19622551

Najib, K. (2020). Government Ecology and the Indigenous Religion of the Suku Anak Dalam: Intersubjective Relations in Forest Conservation in Jambi, Indonesia. [Journal of Tropical Forest Management]. *Jurnal Manajemen Hutan Tropika*, *26*(3), 303–315. DOI: 10.7226/jtfm.26.3.303

Nurit Bird-David. (1999). "Animism" Revisited: Personhood, Environment, and Relational Epistemology. *Current Anthropology Special Issue Culture—A Second Chance? 40*(S1). https://doi.org/https://doi.org/10.1086/200061

Oliveira, A. S., de Barros, M. D., de Carvalho Pereira, F., Gomes, C. F. S., & da Costa, H. G. (2018). Prospective scenarios: A literature review on the Scopus database. *Futures*, *100*, 20–33. DOI: 10.1016/j.futures.2018.03.005

Page, M. J., McKenzie, J. E., Bossuyt, P. M., Boutron, I., Hoffmann, T. C., Mulrow, C. D., Shamseer, L., Tetzlaff, J. M., Akl, E. A., Brennan, S. E., Chou, R., Glanville, J., Grimshaw, J. M., Hróbjartsson, A., Lalu, M. M., Li, T., Loder, E. W., Mayo-Wilson, E., McDonald, S., & Moher, D. (2021). The PRISMA 2020 statement: An updated guideline for reporting systematic reviews. *International Journal of Surgery*, *88*, 105906. https://doi.org/https://doi.org/10.1016/j.ijsu.2021.105906. DOI: 10.1016/j.ijsu.2021.105906 PMID: 33789826

Petri, D. P., & Klocek, J. (2025). External and Internal Threats to the Freedom of Religion or Belief of Indigenous Peoples in Latin America. *Religions*, *16*(2), 209. DOI: 10.3390/rel16020209

Sohad, M. K. N., & Mafrolla, E. (2025). Bridging science and society: The integration of indigenous and scientific knowledge management. *Journal of Knowledge Management*, *29*(7), 2258–2284. DOI: 10.1108/JKM-11-2024-1326

Tsing, A. (2023). A Multispecies Ontological Turn? In *The Routledge International Handbook of More-than-Human Studies* (pp. 116–128). Routledge. DOI: 10.4324/9781003262619-7

Ugwu, I. P. (2025). Environmental Protection, Rights of Nature, and Religious Beliefs in Europe. *International Journal for the Semiotics of Law - Revue Internationale de Sémiotique Juridique, 38*(3), 993–1014. DOI: 10.1007/s11196-024-10193-5

Wahyudi, L. (2024). *Watase Uake: Research Collaboration Tools.*

Weatherdon, M., & Schermerhorn, S. (2022). Movement and Indigenous Religions: A Reconsideration of Mobile Ways of Knowing and Being. *Material Religion : the Journal of Objects, Art and Belief, 18*(1), 3–15. DOI: 10.1080/17432200.2021.2015921

Weatherdon, M. S. (2022). Religion, Animals, and Indigenous Traditions. *Religions*, *13*(7), 654. DOI: 10.3390/rel13070654

Weaver, C. B. (2024). How Not to Undiscipline Religion and Science: Indigenous Ecological Knowledge, Epistemic Resistance, and the Settler Imagination. *Religions*, *15*(11), 1290. DOI: 10.3390/rel15111290

Wilson, S. (2020). *Research is ceremony: Indigenous research methods*. Fernwood publishing.

Zaluchu, S. E. (2025). Pray to the ancestors: A case study of Christianity and Nias indigenous religion. *Cogent Arts & Humanities*, *12*(1), 2492419. Advance online publication. DOI: 10.1080/23311983.2025.2492419

ADDITIONAL READING

Adam, R., & Bagir, Z. A. (2022). The Indigenous Politics of Justice: The Case of the Sedulur Sikep Movement in Central Java. *Jurnal Kawistara*, *12*(2), 181–199. DOI: 10.22146/kawistara.78485

Afiff, S., & Lowe, C. (2007). Claiming Indigenous Community: Political Discourse and Natural Resource Rights in Indonesia. *Alternatives*, *32*(1), 73–97. DOI: 10.1177/030437540703200104

Maarif, S. (2025). Decolonizing Religion: Toward an Engaged "Religious" Methodology. In *Religion, Decolonization, and the Planetary Community: Voices from the Indonesian Archipelago* (pp. 38–57). Routledge. DOI: 10.4324/9781003527398-4

Riyanto, G. (2025). Indigeneity as a Sphere of Differences: State Enclosure and Counter-enclosure of Rural Spaces in Indonesia. *Anthropological Theory*, *25*(4), 440–464. DOI: 10.1177/14634996251313825

KEY TERMS AND DEFINITIONS

Cartesian Dualism: The philosophical separation concept of mind-body and human-nature attributed to the French philosopher, Rene Descartes.

Indigenous Ecological Knowledge: A holistic environmental knowledge system held by indigenous communities, integrating spiritual and relational understandings of ecosystems to guide conservation and sustainable management.

Indigenous Epistemology: A way of knowing rooted in relational engagement, ceremonial practice, and responsibility to the more-than-human world.

Intersubjective Cosmology: A cosmological framework in which humans and non-humans engage in mutual relationships based on responsibility, respect and ethical principles.

Non-Human Personhood: The recognition of spirits, animals, plants, rivers, and other natural entities as real agents with consciousness and relational capacity.

Relational Ontology: A worldview in which relationships between humans and non-humans are the primary condition for existence, meaning, and identity.

Chapter 7
Indigenous Ecological and Environmental Ethical Practices:
Alignment With Pañca Rna From Bharatiya Scriptures

Pallavi Joshi
http://orcid.org/0000-0001-7645-2140
Banasthali Vidyapith, India

Smriti Dixit
http://orcid.org/0000-0002-0515-8371
Jaipuria Institute of Management, Ghaziabad, India

Vaishali Singh
http://orcid.org/0009-0001-6127-6188
Banasthali Vidyapith, India

ABSTRACT

The interaction of humans with nature significantly impacts the environment and unsustainable practices are a major factor. Human actions have roots in the worldview and philosophy. If the worldview positions humans at the centre of nature as master, unsustainable acts will be severe and nature would be only a source for extraction. On the other hand, a nature based worldview would define philosophy and practices to develop a sense of responsibility. This chapter examines how Pañca Ṛṇa from Bharatiya civilizational thought, comprising the five fundamental debts to divine forces (Deva), sages (Ṛṣi), ancestors (Pitṛ), life forms (Bhūta), and humanity (Manuṣya), provides a framework for sustainability. It also captures indigenous ecological knowledge and posits that sustainability could be achieved through adoption of these ideas. The review the concept of Pañca Ṛṇa from scriptures is followed by cases of Tata Power, Gita Press, WWF India, Zoho Corporation and Banasthali Vidyapith for contemporary applications.

DOI: 10.4018/979-8-2600-2890-2.ch007

1. INTRODUCTION

The contemporary world is witnessing environmental challenges which range from climate change and biodiversity loss to ecological degradation and the depletion of natural resources (Hameed et. al., 2025). Although climate change is often presented as the most discussed concern, it also reflects broader patterns of human activity and the cumulative impacts of development choices (Khatib, 2025). Overconsumption, loss of restraint, and development models driven by growth have severely impacted ecological systems and have deepened the imbalance between humans and nature (Richardson, 2025). Indigenous ecological traditions, in this context, offer alternative ethical and cultural viewpoints to understand the association between humans and nature where nature is treated with respect.

Many contemporary responses to environmental problems are a result of the interventions which are largely imposed externally. Such efforts have raised the awareness but they often operate after the occurrence of the substantial damage and consider sustainability as an add on to development (Engerbretsen & Greenhalgh, 2025; Liang & Segalas, 2025). This limitation invites a deeper exploration of ethical foundations that can shape ecological behaviour at the level of everyday life.

Indigenous ecological practices, which this chapter focuses on, provide a different orientation toward nature through holistic frameworks of harmonious living (Gauthier et al., 2025). In these traditions ecological responsibility arises from an internal sense of duty grounded in dharma (Khoshoo, 1999). Human beings are viewed as participants within a larger ecological order (Chakraborty, 2002; Jindal et al., 2025) and nature is approached with reverence, gratitude, and restraint, without an exploitative attitude (Jain, 2024). This is how preservation and the passing on of natural resources to future generations are embedded into routine conduct and everyday life (Saxena, 2025).

Within this worldview, Bharatiya scriptures offer a structured framework for understanding the human nature relationship through the concept of Pañca Ṛṇa, the five obligations with which a person is born (Kumar & Prasad, 2025). These include obligations toward divine forces (Deva Ṛṇa), sages and knowledge traditions (Ṛṣi Ṛṇa), ancestors (Pitṛ Ṛṇa), other living beings (Bhūta Ṛṇa), and humanity at large (Manuṣya Ṛṇa). The framework is based on gratitude, humility, and responsibility and recognizes the sustenance of human life through direct and indirect contributions which requires acknowledgement and reciprocation through conduct (Muniapan, 2023; Saxena, 2025).

Fulfillment of Pañca Ṛṇa provides a basis for ecological responsibility which can strengthen sustainability as an intrinsic way of living. The chapter regards duty and gratitude as the behavioural foundations which can complement contemporary discourses on sustainability.

To show the practical relevance of this framework, the chapter also discusses contemporary examples of selected organizations. Tata Power is discussed for their energy related ecological commitments representing Deva Ṛṇa. Gita Press Gorakhpur is included for their dimension of knowledge transmission which aligns with Ṛṣi Ṛṇa. Banasthali Vidyapith is considered as an educational example of living sustainable practices representing Pitṛ Ṛṇa. Example of WWF India is discussed for conservation of biodiversity linked with Bhūta Ṛṇa. Zoho's example is discussed for their community oriented practices aligning with Manuṣya Ṛṇa. Community led projects in India have contributed to the ecological balance of the planet like Earth5R initiative & JSW's Mangrove Initiative which bridges the gap between nature & community. These examples are used to relate the panch rin framework with visible practices and present day relevance.

1.1 Research Objectives and Methodology

- To explore the concept of Pañca Ṛṇa as a framework of individual obligation in Bharatiya culture and its implications for sustainable and responsible ways of living in contemporary contexts.
- To examine selected contemporary examples of organizations that demonstrate the practical applicability of Pañca Ṛṇa in present day ecological and social practices.

This is a conceptual study. It presents a review of key ideas grounded in ancient wisdom and examines organizational examples whose practices align with this framework. The organizational examples are reviewed using existing case studies and data available from secondary sources. The synthesis across these materials is used to draw conclusions aligned with the stated objectives.

2. ETHICAL AND PHILOSOPHICAL FOUNDATIONS OF INDIGENOUS ECOLOGICAL THOUGHT

This section presents the foundations of indigenous ecological thinking, based on the Bharatiya concept of Pañca Ṛṇa amd its historical evolution, followed by a discussion of indigenous ecological practices, and their mapping.

2.1 The Concept of Pañca Ṛṇa in Bharatiya Scriptures

The value systems and practices that shape any civilization also shape its approach to ecology (Singhal & Sharma 2025). In the Bharatiya worldview, life is understood in interdependence in which human beings are connected with other people, with natural forces, and with other living beings. Through this understanding, the concept of Pañca Ṛṇa offers a foundational framework by identifying the obligations with which a human being is born (Sarkar, 2021; Suresh & Sreejith, 2025). These obligations reflect a deeper understanding that human life is sustained through visible and invisible contributions and therefore must be lived with gratitude, restraint, reciprocity, and responsibility.

Malhotra & Babaji (2020) argue that some crucial words that have cultural significance and are linked to worldview cannot be translated. Similar is the case with Ṛṇa. Although words like obligations and responsibility are used, but these do not capture the essence. The idea of Ṛṇa is very inherent and intrinsic as it reminds the individual that human beings should not consider themselves as the exclusive owner of natural resources. The human being can advance and excel through judicious and responsible usage by following certain protocols. It is not an external imposition of rules and regulations that matter, the important aspect is a genuine feeling of responsibility (Saxena 2025). Scholars have also interpreted the related framework of Pañca Mahāyajña as a fivefold structure of obligation intended to give back to society, nature, the divine, ancestors, and knowledge traditions, thereby extending its significance beyond ritual observance alone (Suresh & Sreejith, 2023).

Thus the idea of Pañca Ṛṇa is closely connected with the tradition of Pañca Mahāyajña through which these obligations are expressed in everyday life. This framework emphasizes that the existence of humans cannot be considered in isolation. Breathing, eating, learning, producing, consuming, and participating in society all take place within an already given order sustained by nature, culture, lineage, and knowledge. Pañca Ṛṇa therefore transforms daily conduct into an ethical response to this received order. The concept is not confined to material offerings alone, but includes non material and spiritual forms of action grounded in sincerity and devotion (Made & Somawati, 2021).

अध्यापनं ब्रह्मयज्ञ:पति यज्ञस्तु तर्पणम्। होमो दैवो बलभौर्तो नृयज्ञोSतथिपूजनम्॥ Manusmriti 3.70

वेद का अध्ययन और अध्यापन करना 'ब्रह्मयज्ञ'(ऋषक्रिण) है, तर्पण करना 'पतृयज्ञ' (पतृ ऋण) है, हवन करना 'देवयज्ञ' (देव ऋण) है, बलविश्वदेव करना 'भूतयज्ञ' (भूत ऋण) है तथा अतथियों का भोजन आदसि सत्कार करना 'नृयज्ञ' (नृ / मनुष्य ऋण) है।

This verse from Manusmṛti explains that Brahma Yajña is related to the study and teaching of sacred knowledge and is linked with Ṛṣi Ṛṇa which is the obligation toward sages and knowledge traditions. Pitṛ Yajña which is expressed through acts such as tarpaṇa, represents gratitude toward ancestors and aligns with Pitṛ Ṛṇa. Deva Yajña which is fulfilled through home and offerings, relates to Deva Ṛṇa representing the obligation toward the life sustaining forces of nature. Bhūta Yajña which is expressed through bali-vaiśvadeva and compassion toward other living beings, reflects Bhūta Ṛṇa. Nṛ Yajña which is often understood in the context of hospitality, service, and care toward fellow human beings, corresponds to Manuṣya Ṛṇa.

2.2 Historical Evolution and Continuity of Pañca Ṛṇa

Through the discussions above it can be seen that the significance of Pañca Ṛṇa remains relevant because of its influence in everyday life, social practices, and ecological behaviour over time (Kumar & Prasad, 2025; Muniapan, 2023). The idea sustains because it was framed as a continuing moral orientation regulating the relations between the individual, society, and the larger natural order. In this sense, Pañca Ṛṇa may be understood as a doctrinal idea as well as a civilizational ethic shaping everyday practice. In Varāha Purāṇa also yajña is linked with the preservation of the world order and the safeguarding of the earth itself. Such a framing supports the view that ecological care was understood as a part of a sacred responsibility (Made & Somawati, 2021).

Ecological responsibility is seen in integration with spiritual and social life in Bharatiya traditions. Obligation towards divine forces are reflected in rituals as reverence for rivers, mountains, forests, rainfall etc. Such reverence helps in regulating human conduct by placing restraint on extractions and by shaping habits of gratitude. Obligations towards sages and knowledge traditions support the preserving and transmission of agricultural knowledge, medicinal plant use, seasonal cycles, and ecological observations through oral traditions. In this way environmental understanding remained embedded in lived culture (Saxena, 2025; Singhal & Sharma, 2025). This continuity is found in text and practice as well. Applied studies of traditional farming communities indicate that yajña based values can still shape contemporary practices of ecological care, intergenerational knowledge transfer, mutual aid, and collective resilience (Suresh & Sreejith, 2023).

In the contemporary period, when ecological crises are often approached through policy and compliance frameworks, revisiting the continuity of Pañca Ṛṇa becomes specially meaningful. It shows that sustainability in Bharatiya traditions is embedded within a larger worldview in which ecological care and everyday living are connected.

2.3 Indigenous ecological practices and environmental ethics

"mātā bhūmi putruhāna pṛthvyā"
"Earth is my mother and I am her son."

The Sanskrit ślok stated above is from the Arthaśāstra expressing the relationship of the human being with the earth (nature), here, it has been compared with the relationship between a mother and a child stating that human being the child of mother earth should not harm the environment or its flora and fauna. For example, Kautilya wrote a comprehensive book on economics and statecraft in the 4th century BC named the ārthaśastra, containing detailed guidelines about managing forests and preserving them for coming generations. Moreover, there were some dynasties like Solanki, Chola and Asaf Jahi who used similar long-sighted methods to manage the limited natural resources. Some innovative examples were the advanced irrigation methods and water management techniques used by the indus valley civilization. The "more crop per drop" approach of the government through Pradhan Mantri Krishi Sichayi Yojna highlights Chanakya's advocacy in favour of sustainable water management (Ashika et al., 2025).

Other Hindu scripture, like Vedas also advise us to treat the environment with respect and responsibility (Naees, 1993). We get worried about the air quality these days, but the Atharvaveda has stated long back that clean air is a medicine and it emphasises on the need to preserve the air quality. The ozone hole being a problem today was stated in the Ṛgveda, which described a shield around the Earth that was similar to the ozone layer and cautioning us against any activity that could harm it. The careful usage of natural resources has been stated in many other Bharatiya scriptures, apart from Arthaśāstra and Vrikśayurveda (Pradhan, 2026). For example, Ṛgveda suggests using the resources without greed while describing the Earth as mother. The Atharvaveda has praised water and earth, hailing their purity and the need to preserve them. Similarly, Saṃhitās describe various rituals that necessarily promote the conservation of nature by requiring the use of biodegradable materials (Panda, 2024)). Similarly, other important scripture such as Manusmṛti have keenly described water conservation and protection of forest, stating some severe punishment for the people who damage nature (Jha, 1999; Mistri, 2026). In the face of intensifying ecological challenges such as climate change, deforestation, biodiversity decline, and pollution of water and air the urgency of integrating spiritual and religious perspectives has grown significantly (Arsal, 2026). Bharat's contribution to worldwide sustainability can be immense if we turn to traditional wisdom with modern science.

The Kautilya's Arthaśāstra, even though having its main focus on statecraft and economic practices, describes the importance of environmental management and preservation. It states how crucial it is to emphasise forest resources, safeguarding wildlife and constructing water reserves for management. The scripture emphasises on the King's responsibility to safeguard sustainable use of resources. An idea that is relevant to current state responsibilities with respect to environmental protection. The Vrikśayurveda, which actually means the science of plant life, provides a detailed knowledge about environmental management, agricultural practices and plant care. This text has a number of detailed topics about plants, best, medicines, planting methods, selection of seeds and soil health. This Bharatiya scripture focuses on the harmony between human activity and the ecosystem, which makes it especially profound today (Jasrotia & Manhas, 2024). This scripture already emphasises the practices which are being followed through organic farming. It promotes techniques and ideas that foster biodiversity by naturally increasing soil fertility and organic treatments for plant disease diseases. Shifting to natural compost and biopesticides can reduce expenses in connection to cultivation, improve soil fertility and add to biodiversity. Historical techniques like the Barahnaja system in Uttarakhand in which 12 or more crops like kidney beans, amarnath, finger millet are grown in a single farm. The method can resolve pest outbreaks and improve soil fertility without dependence on chemical pesticides (Bora & Kumar, 2025).

The indigenous ecological practices are the ongoing ethical examples that have traditionally modulated the relationship between human and earth. These stated ideas are not just the factual response to environmental problems, but a representation of deeper moral worldwide views in which the balance of ecosystem, societal harmony and spiritual responsibilities are inseparably knitted (Tiwari, 2016). In the culture, such practices developed within a normal framework that feuds nature as holy, life as codependent and human existence indebted to multiple beings, right from birth (Muthuswamy, 2018; Muralivallabhan, 2022). This ethical idea has a philosophical imprint in the doctrine of Pañca Ṛṇa, which envision the human life as responsible towards nature, ancestors, sages, fellow beings and society at large (Das, 2020).

Indigenous ecological practices are centralized towards understanding nature unlike the modern techniques majorly concentrating on the interest of humans beyond that of the ecosystem. On the other hand, the indigenous practices keep humans in a participatory position rather than the dominant one (Sivaramakrishnan, 2015). The Vedic, Upaniṣadic and Purāṇic thoughts have always emphasized on the co-dependence and self-discipline and discouraging the exploitation of natural resources (Das, 2019, Singh, 2001) unlike the current times where wars and attacks are being done with the mere objective of capturing the natural elements. Such thoughts incline directly with the Deva Ṛṇa, where humans their interdependence with the cosmic and natural forces, and Bhūta Ṛṇa, which extends moral thoughts towards non-human life forms.

In the Bharatiya culture people have traditional and indigenous ways of living that naturally protects the environment as a sustainable daily culture and beliefs. Sacred groves, reverence for rivers and mountains, ritualized agricultural cycles and restriction on hunting & use of forest represents concrete evidence of the ethical worldview. The sacred groves are preserved through religious practice of setting aside small patches of forests where they do not cut the trees, which helps keep rare plants and animals safe. Empirical studies show that these practices have a significant role in conserving endemic species and maintaining ecological balance over many centuries (Ashika et al., 2025). Practice of sacred groves can help in restoration of degraded terrain and aforestation in such landscapes (Anthwal et al., 2010). These moral and ethical practices do not come out of law enforcement but by a deep sense of obligation and gratitude towards nature, connecting strongly with the logic of Bhūta Ṛṇa, where harm to other life forms is considered a violation of moral duty and not just an environmental malpractice.

The famous environmentalist Rajendra Singh, known as the "Waterman of India," claims that the traditional water management systems show the ethical depth of indigenous ecological practices. Stepwells, tanks, temple ponds, and rainwater harvesting structures were designed and maintained through the collective effort of everyone which were guided by principles of moderation, equity, and long-term sustainability. The Arthaśāstra and other traditional practices focused on careful regulation of water resources to prevent their depletion and make sure that everyone has equal access to them (Ashika et al., 2025). Because water has not been devised as a private property but a joint trust, this reflects the sense on Manuṣya Ṛṇa which emphasizes on societal responsibility and obligation. It also spiritise Pitṛ Ṛṇa, which considers natural resources as an inherited legacy, that should be preserved rather than exploited for immediate gain.

The multi-generational aspects of indigenous ecological ethics are very significant. The use of resources was attached to an awareness that current actions will have consequences for future generations. Such thoughts discouraged extravagant behavior and encouraged habits that maintained soil fertility, forest conservation, and water availability over long periods. Researchers indicate that such ethical orientation made conservation not merely a strategic choice but a moral duty (Das, 2020). Within the Pañca Ṛṇa framework, this responsibility has been most clearly depicted through Pitṛ Ṛṇa by honouring our ancestors extends beyond rituals but to preserving the ecological and cultural foundations they created to future generations.

The ethical inheritance of ecological knowledge is equally important in the indigenous societies, relating to agriculture, forestry, medicine and climate, which were preserved through vocal traditions, rituals, folklore and daily practices. This kind of transmission highlighted practical learning, moral implementation and relevant examples rather than abstract facts. Experts believe that these traditions teach people to be modest, careful and caring towards nature. They argue that modern world views often ignore their values because they focus too much on speed, profit and constant growth at any cost (Muniapan, 2013; Das, 2020). Instead of just trying to use up everything as fast as possible these traditional practices remind us of humility, restraint and sensitivity conceptualising the Ṛṣi Ṛṇa, which views wisdom as a sacred trust to be preserved, practiced, and transferred responsibly rather than exploited actively.

For the indigenous groups, taking care of nature is like taking care of each other. These practices ensure (i) equity wherein everyone gets their fair share; (ii) reciprocity, if you take something from nature then give something back; (iii) collective decision making. The Bharatiya culture is deeply rooted towards helping each other, the social fragmentation makes us lose the connection with earth (Das, 2020), this tendency is countered through implanted environmental care as a social responsibility and collective moral promise. This integration reflects the moral vision of Pañca Ṛṇa, where the obligations towards nature, society, knowledge traditions and ancestral legacies are mutually binding (Dwivedi, 1993; Gupta & Maurya 2026).

Indigenous practices challenge the belief that humans are superior to all other species, and the earth exists solely for their use, rather than considering humans as masters of nature, these traditions emphasize on coexistence, balance, and ethical control. Nature is neither treated as a beautiful, dreamy landscape nor like a grocery store or a machine that only exists to give things, instead, to be seen as a moral partner. The Bharatiya philosophical traditions do not see humans and nature as two separate entities (Das, 2019) but all beings as a same big family, keeping a holistic vision and no rigid walls because if one hurts nature, they are essentially hurting themselves. Such perspectives are strongly relevant in current environmental discussion, particularly in debates surrounding sustainability, ecological justice, and alternative development tracks (Chapple and Tucker, 2000).

Indigenous ecological practices when viewed through the lens of Pañca Ṛṇa depicts that environment responsibility comes organically and not through any external regulation or law. In this view, sustainability is not just a government goal or a project, it is what happens naturally when people live according to dharma or their moral duty. Instead of just taking what they please, people make an effort to give back to nature, their community and ancestors. By respectfully returning what they use, they ensure that the world stays balanced and healthy for everyone (Dwivedi, 1993). When people act with this sense of duty, protecting the planet becomes a normal part of life rather than a special rule. This ethical continuity ensures an effortless transition to the discussion of Pañca Ṛṇa as a comprehensive moral structure for environmental ethics in Bharatiya thought.

Bharatiya culture represents that protecting nature has always been organic, instead of having environmental enforcements like there are today, these ideas were woven into religious stories, prayers and daily habits. Caring for earth did not start because of modern crises like global warming; it started simply as the right way to live based on long standing cultural values (Das, 2019; Das, 2020). Being kind to nature was just part of being a good human.

One of the most significant scriptural examples of ecological ethics is the protection and worship of trees, repeatedly stated in the Puraṇas. The matsya purana teaches that planting one tree is just as important as raising ten sons (daśahradasamaḥ putro, daśaputrasamo drumaḥ), making the protection of nature a deep family duty rather than just a chore. Varāha Purāṇa expresses planting trees helps a person do good deeds while destroying the forests is seen as a moral wrong. By linking nature to a person's soul and family, these scriptures encouraged people to save the environment out of love and respect instead of a government fine (Mukil, 2026).

Rāmāyaṇa and Mahābhārata treat forests as peaceful places for learning and growth, not as wild areas to be destroyed. These "Āśrama forests" acted as early nature parks where harming animals or taking too much wood was strictly forbidden. According to Das (2020), these rules were based on Bhūta Ṛṇa, the idea that we have a duty to care for the creatures we share the earth with. This kept the environment healthy because people followed their spiritual values towards non-human beings inhabiting the ecosystem.

The spiritual connection for rivers and water bodies offers another compelling example of indigenous ecological ethics rooted in scripture. rivers like the Gaṅgā and Yamunā are regarded as living goddesses. Because of this, polluting the water is seen as a moral wrong against a mother-figure, not just a broken rule (Haberman, 2006). Vedas and Puraṇas also talk about keeping temple ponds and tanks clean; while these were used for prayers, they also helped store water for the community and kept the local environment healthy. These traditions are based on the duty to respect nature's power and ensure that everyone in the village has enough clean water to live (Ashika et al., 2025; Sharma & Biswas, 2024). These practices reflect the ethical dimensions of Deva Ṛṇa, acknowledging dependence on natural forces, and Manuṣya Ṛṇa, ensuring equitable access to shared resources.

Scriptural injunctions promoting non-violence (Ahiṃsā) toward all living beings further reinforce environmental ethics (Jayanthi, 2001). While often discussed in moral or spiritual contexts, Ahiṃsā carries profound ecological implications. The *Viṣṇu Purāṇa* explicitly states that harming non-speaking creatures displeases the divine, thereby extending moral consideration beyond human society (Das, 2019). The *Narasiṃha Purāṇa* condemns ritual observance devoid of compassion toward animals, underscoring that ethical conduct toward nature supersedes ritual formalism. Such teachings align with Bhūta Ṛṇa and anticipate contemporary debates on intrinsic value in environmental ethics (Das, 2020).

Animal symbolism and protection also play a vital role in indigenous ecological ethics. Several animals such as the cow, peacock, snake, tiger, and elephant are revered as sacred or associated with deities, effectively ensuring their protection through religious sentiment. For instance, the cow's sanctity discouraged slaughter (Jayanthi, 2001; Smart & Thakur, 1993) and promoted sustainable agrarian practices, while reverence for snakes (*Naga worship*) protected key species within ecological food chains (Das, 2020). These practices illustrate how ecological balance was maintained through cultural symbolism rather than scientific management alone.

The ethical transmission of ecological knowledge is another scripturally supported practice. The Vedic and Upaniṣadic traditions emphasize knowledge (*Vidyā*) as sacred and relational, not extractive (Radhakrishnan, 1994; Ganguly 2024). Knowledge of seasons, agriculture, medicinal plants, and ecological rhythms was transmitted orally through the guru-shishya tradition, reinforcing Ṛṣi Ṛṇa, the debt owed to sages and knowledge traditions (Muniapan, 2013). This ensured continuity, contextual relevance, and ethical application of ecological wisdom across generations (Roy, 2025).

Collectively, these scriptural examples demonstrate that indigenous ecological practices in Bhārata were underpinned by a coherent ethical framework rooted in duty, gratitude, and reciprocity. When viewed through the lens of Pañca Ṛṇa, these practices reveal a civilizational ethic wherein environmental protection is not an external obligation but an intrinsic aspect of righteous living (*dharma*). Such an understanding offers enduring moral insights for contemporary environmental challenges, reaffirming that sustainability is most effective when grounded in ethical consciousness rather than regulatory compliance alone.

2.4 Mapping Indigenous Practices to Pañca Ṛṇa Framework

The discussions so far show that indigenous ecological practices in the Bharatiya traditions are understood as expressions of a deeper framework in which responsibility is distributed across multiple relationships which sustain human life. The Pañca Ṛṇa framework is especially useful here because it helps to interpret ecological conduct as a duty-based response to interdependence. Mapping indigenous practices with this framework makes visible how environmental ethics, social responsibility, knowledge transmission, and intergenerational continuity are connected within a unified civilizational wisdom.

This concept resonates with indigenous ecological worldviews that understand life as mutually interconnected (Kumar & Prasad, 2025). Encompassing obligations toward divine forces, humans, ancestors, sages, and living beings, reflects a comprehensive ethical system extending beyond ritual into everyday ecological conduct (Adhikary, 2023). This becomes significant in contrast to anthropocentric frameworks that position humans as masters of nature (San Román & Molinero-Gerbeau, 2023). The following subsections map indigenous ecological practices to each dimension of Pañca Ṛṇa, demonstrating how this framework offers a coherent structure for understanding sustainability.

Deva Ṛṇa refers to the obligation towards the divine or cosmic forces that make life possible (Muniapan, 2013; Sarkar, 2021). It includes reverence for natural elements such as the sun, rain, rivers, wind, earth, and faire, all of which sustain existence directly or indirectly. The ecological significance of this obligation lies in growing restraint and gratitude towards the natural systems on which life depends (Saxena, 2025) along with the acknowledgement of ritual practices (Maurya, 2025). Indigenous practices linked to rivers, forests, seasonal festivals, sacred landscapes, and agricultural cycles can be understood as expressions of Deva Ṛṇa. These practices reinforced the idea that nature is not a passive stock of resources for unrestricted extraction (Chakraborty, 2002) but is a part of a larger order that must be engaged with care. In contemporary times, this has direct relevance for ecological crises driven by extractive attitudes, particularly climate change, deforestation, pollution, and the overuse of common resources. The implications of Deva Ṛṇa lies in the shift it creates from entitlement to responsibility. Modern environmental governance often struggles because natural systems are treated only through economic valuation or regulatory control. While regulation remains necessary, it adds an internal ethical orientation by framing ecological care as a duty arising from dependence and gratitude. This can enrich present sustainability discourse by encouraging behaviour that is preventive rather than merely corrective. In this sense, Deva Ṛṇa relates directly to contemporary concerns around resource depletion and ecologically indifferent development.

Rşi Ṛṇa represents the obligation towards sages, teachers, and knowledge traditions (Saxena, 2025). Its ecological significance emerges from the recognition that environmental wisdom has also been preserved through long standing traditions of observation, practice, and transmission (Kumar & Prasad, 2025). Agricultural knowledge, medicinal plant understanding, seasonal awareness, water management techniques, and ethical teachings on restraint all form part of this inherited knowledge. Indigenous ecological traditions such as Vrikśayurveda, agrarian wisdom, and oral pedagogic practices reflect Rşi Ṛṇa because they preserve and transmit ecological knowledge (Ketecha & Pandey, 2025) for collective well-being. Applied studies of traditional farming groups also show that yajña based principles can inform ecological care, respect for living beings, intergenerational transfer of indigenous knowledge, and mutual support during vulnerability (Suresh & Sreejith, 2023) and efforts to conserve water resources through traditional local wisdom (Putra et al., 2025). This dimension is highly relevant today because many environmental problems are intensified by a break between knowledge and values. Technical knowledge improves efficiency but without ethical direction it may also intensify exploitation. Rşi Ṛṇa offers a different perspective by treating knowledge as a trust that must be preserved, interpreted wisely, and applied responsibly. This has strong implications for current sustainability movements. Climate resilience, regenerative agriculture, biodiversity restoration, and sustainable livelihoods all depend along with innovation on continuity of knowledge. Rşi Ṛṇa therefore addresses a central contemporary challenge of the loss of ecological memory in societies increasingly detached from local environments. It also offers a corrective to consumerists and short term approaches by reminding us that wisdom transmission is itself a form of ecological responsibility.

Pitr Ṛṇa is commonly understood as the obligation towards ancestors. It reflects gratitude for what has been inherited and responsibility for what must be passed on (Maurya, 2025). In ecological terms, this makes Pitr Ṛṇa one of the most powerful components of the Pañca Ṛṇa framework because it links inheritance with intergenerational duty (Sarkar 2021; Hourdequin, 2025). Traditional practices related to land stewardship (Wilson & Noyes, 2025), water conservation, preservation of seeds, continuity of agricultural systems, and respect for inherited ecological resources can be understood through Pitr Ṛṇa. These practices assume that natural resources are not the unrestricted property of the present generation. Rather, they are received from the past and must be maintained for the future. The present environmental crisis is deeply connected with the weakening of long-term thinking (Jacoba, 2025). Consumerism and growth centered models often privilege immediate gain over future stability. Pitr Ṛṇa challenges this orientation by asking whether present conduct honors what has been inherited and safeguards what must endure (Saxena, 2025). Its relevance is therefore especially strong in the context of groundwater depletion, soil degradation, loss of biodiversity, and climate burdens that are transferred across generations. By embedding ecological care within a chain of continuity, Pitr Ṛṇa provides a moral foundation for long term sustainability.

Bhūta Ṛṇa refers to the obligation towards other living beings. It reflects the recognition that human life is sustained within a larger network of non-human existence and that animals, birds, insects, plants, and other forms of life are not morally irrelevant (Saxena, 2025). This widens the field of ethical responsibility beyond human society and places biodiversity and ecological coexistence within the domain of duty (Sarkar, 2021). Practices such as sacred groves (Rai, 2025), care towards animals, preservation of species associated with local ecosystems (Damke, 2025), and avoidance of unnecessary harm all express the spirit of Bhūta Ṛṇa. Their relevance today is especially striking because modern ecological crises include not only climate change and pollution, but also mass biodiversity loss, habitat destruction, and species decline (Nobles, 2025). These developments show limitations of anthropocentric models that value non-human life only in terms of immediate utility (Andriana, 2025). Bhūta Ṛṇa offers a broader ethical basis for environmental thought by affirming that other forms of life deserve protection not merely because they are useful to humans, but because coexistence itself is part of right living. This perspective strengthened current discussions on ecological justice, conservation ethics, and biodiversity governance. It also provides an important corrective to industrial models that normalize the destruction of habitats in pursuit of economic expansion (Owolabi et al., 2026). Through Bhūta Ṛṇa indigenous ecological ethics anticipate the contemporary recognition that sustainability must include responsibility towards the more than human world.

Manuşya Ṛṇa refers to the obligation towards fellow human beings and society. In ecological terms, this obligation is crucial because environmental degradation is never only a problem of nature in abstraction. It affects communities, livelihoods, health, water access, food systems, and social justice. Sustainable living therefore requires not only care toward nature, but also responsibility towards the human consequences of ecological disruption (Chandel et al., 2024; Saxena 2025). The same applied literature also shows that mutual aid during economic crises and the collective sharing of losses in times of vulnerability can be understood as forms of socially embedded duty and collective responsibility (Suresh & Sreejith, 2023). Traditional practices of collective water management (Lema, 2025) shared maintenance of ecological commons, hospitality, local cooperation, and community centered resource use reflect Manuṣya Ṛṇa. These practices recognize that ecological wellbeing and social well-being are inseparable. This is highly relevant in the present context, where environmental crises disproportionately affect vulnerable communities (Choi et al., 2025), deepen inequality, and create new forms of ecological injustice. The value of Manuṣya Ṛṇa lies in its ability to connect sustainability with social ethics. Contemporary environmental governance often suffers when ecological policy is separated from community participation and lived responsibility. Manuṣya Ṛṇa offers an alternative by emphasizing mutuality, cooperation, and shared stewardship. This makes it relevant to present debates on climate adaptation, just transitions, local governance, and community resilience. It also challenges hyper individualistic and consumerist tendencies by reminding us that ecological responsibility must be socially distributed and collectively enacted.

Viewed together, the five Ṛṇas create a multidimensional framework for ecological ethics. Deva Ṛṇa addresses responsibility toward natural forces and helps counter extractive attitudes. Ṛṣi Ṛṇa protects ecological knowledge and resists the erosion of wisdom traditions. Pitṛ Ṛṇa embeds intergenerational responsibility and opposes short term exploitation. Bhūta Ṛṇa expands moral concern to non-human life and strengthens biodiversity ethics. Manuṣya Ṛṇa links ecological responsibility with social justice, cooperation, and collective well-being. This mapping is important because it shows that Pañca Ṛṇa is not a symbolic philosophical concept alone. It functions as a practical ethical framework through which environmental responsibilities can be distributed across nature, society, memory, knowledge, and life itself. In present sustainability discourse, where ecological problems are often addressed through fragmented sectors, the framework offers a more integrated orientation. It does not replace policy, science, or technological intervention. Rather, it complements them by supplying a moral basis for restraint, continuity, reciprocity, and shared responsibility. For this reason, the Pañca Ṛṇa framework remains analytically valuable in the contemporary period. It helps explain how indigenous ecological practices were sustained historically, and it also offers a conceptual bridge for rethinking sustainability today in ways that are ethically grounded, socially meaningful, and intergenerationally responsible.

3. REVIEW OF GOOD PRACTICES OF ORGANIZATIONS

Certain good practices of organizations have been presented as examples which are aligned with the theme which this chapter is trying to address.

3.1 Deva Ṛṇa- Tata Power's Community Solar Microgrids

Tata Power's solar microgrids are an example of how we pay back to nature in the form of Deva Ṛṇa. Diesel generators are now replaced by solar energy grids contributing their bit to the panchmaha-bhutas a combination of prithvi, jal, vayu, agni & akash in an effort to reduce carbon emissions & usage of fossil fuels. TP renewable microgrids have reached 3,00,000 residents across 200 villages leading to savings of more than 3 million litres of diesel also reducing carbon emission by 8000 metric tonnes on an annual basis. *Ghar Ghar Solar abhiyan* has made sure that the environment is in no way exploited to achieve energy efficiency. Overall ecological resilience is the centre of sustainable manufacturing.

According to Vedic Culture, these initiatives taken for clean energy can classify as yajña to the devas as it proves to be a sustainable way of maintaining the ecological balance. Therefore, The microgrids offered by Tata Power are a synergic execution consisting of ecological responsibility, rural development & business innovation.

3.2 Ṛṣi Ṛṇa- Gita Press Gorakhpur's Preservation of Traditional Texts

Ṛṣi Ṛṇa is best portrayed by Gita Press, Gorakhpur as it ensures that the lineage of knowledge is not limited to an era. Since 1923, more than 417 million books have been printed in 14 different languages. Million copies of our scriptures such as Bhagavad Gita, Ram Charit Manas, Puraṇas & Upanishads are printed & distributed widely across the globe. More than 200 million smaller volumes are created specifically for children so that the message of Vedic Ṛṣis is preserved & is accessible by everyone. This is an excellent effort that enables households to repay their long due debt to sacred sages.

The editorial keeps a defined & consistent tone with an aim to preserve manuscripts, giving clear explanations and completely ad free printing. The concept of Dharma is upheld as knowledge is prioritized keeping behind the profits. With more than 3500 manuscripts & over 100 interpretations of the Gita, this wisdom of our Ṛṣis is kept protected for our coming generations. With a daily production of 70,000 books the organization aims to keep the authenticity of texts, accessibility of its reach & eventually leads to creation of a well informed community that rests on the scriptures.

3.3 Pitṛ Ṛṇa- Banasthali Vidyapith's Heritage and Bharatiya Values

Banasthali Vidyapith is an example of Pitṛ Ṛṇa or the debt to ancestors because of its foundational heritage and deep commitment to rooted Bharatiya values. Founded by Pandit Hiralal Shastri and Smt. Ratan Shastri in 1935 in memory of their daughter Shantabai, who passed away at age 12 when she made ready 600 mud bricks for a girls' school, the institution's family legacy is a tribute to the hopes of the parents and ancestral dharma. The institution has promoted self reliance and empowerment in its own ways by enduring modernity with a balance.

Today, under the leadership of Shastri family descendants, Banasthali represents holistic education through Panchmukhi Shiksha-five-fold model: physical, practical, aesthetic, moral, intellectual-interpreting shastriya & spiritual values of simplicity, self-reliance, and khadi-wearing with Western science. Enrolling approximately 15,000 students in different courses from pre-primary to PhD within a fully residential campus spread over 850 acres, it attracts more than 90% of its students from outside Rajasthan, including international students. Certain practices, such as naming sponsor buildings as "wisdom temples", and rural ethos help to maintain the roots of Bharatiya culture, creating strong women who respect the family and tradition. This transmission would repay Pitṛ Ṛṇa by ensuring that ancestral values leave a mark to lead future generations.

3.4 Manuṣya Ṛṇa- Zoho's Rural Technology Centers

Tech hubs created by Zoho at the essence of Manuṣya Ṛṇa which is an obligations toward humans that can be discharged via donation, hospitality & public service leading to society welfare in general.

This initiative started as an AI- Robotics campus in Neduvathoor involving 15 hubs hiring more than 3000 rural youngsters in different programming languages. Zoho provides training spanning 9 months to create more graduates across technical disciplines. They also contribute by teaching children via organizing cleanliness drives & free software distribution drives.

Zoho schools are a radical effort in the education ecosystem as they provide stipend instead of charging fees from students for skill based education. This attempt to scale benefits to the marginalized sections helps in maintaining the integrity of the community. Investment in future generations of our society leads to greater returns & fulfillment of Manuṣya Ṛṇa across society at large.

3.5 Bhūta Ṛṇa- WWF India's Biodiversity and Footprint Work.

WWF India has worked extensively on India's biodiversity & its ecological footprint leading to a repayment of Bhūta Ṛṇa, which is the debt owed to beings other than human particularly animals, plants & ecosystems. This is a practice of exercising harmony with nature.

WWF India aims to reduce human footprint encouraging less overexploitation & wastage. A proper assessment of biologically productive space is conducted to ensure less disturbance in natural habitat. The initiatives are as follows:

- Godavari River Basin conservation- Denmark-based consulting firm Ramboll has collaborated since 2023 aiming the conservation of the tiger, herbivores, and tribes
- Nalsarovar Ramsar Wetland conservation- Nalsarovar Ramsar Wetland is restored, focussing over 10,000 farming families to engage in regenerative cotton farming, agrochemical-free farming methods, and 'smart' water resource management
- Conservation at Ganga dolphins- Surveys are done to concentrate on the habitat along 165 km of the important area through community engagement
- Environment reforestation of more than 9,00,000 trees to build a large-scale forest landscape is another way of fulfilling obligations towards biodiversity.

The 'Living Planet Reports' emphasize the depletion of Indian wildlife and the urgent necessity to reduce footprints as a moral responsibility—implementing the 'Bhūta Ṛṇa' principle for a healthy planet.

3.6 Community Led Initiatives & Restoration Projects

India has been a ground to various community initiatives fostering balance in nature & environment contributing to the greater good.

- Earth5R is an excellent example where modern cities are brought closer to biodiversity via different initiatives. Urban gardening initiatives are integrated with modern cities to conservation of native species & their secure existence at large. They also launch composting initiatives in Mumbai involving 10,500 families to manage waste organically leading to diversion of 1.90 tonnes of waste from landfills & reduction of methane emissions.

- JSW's Mangrove initiative has been an exemplary effort to secure our coastline. The initiative restored more than 70 hectares of mangrove ecosystems & planted over 3.5 Lakhs saplings in a coordinated effort with communities. The program shifted the focus from CSR to ecological balance leading to a greater impact.
- Self employed Women's Association (SEWA) Gujarat also does its bit in Manuṣya Ṛṇa by giving women an excellent exposure of different income sources enabling them to contribute to debt repayment & financial well being. 2.1 million women across more than 18 States have benefitted through this model creating a collective income of more than Rs 3200 crore annually.

4. DISCUSSIONS AND CONTEMPORARY IMPLICATIONS

The preceding sections show that Pañca Ṛṇa along with a being scriptural source is also a duty based ethical framework through which ecological responsibility may be understood across nature, knowledge, ancestors, non human life, and society. This makes it highly relevant in the present context, where climate change, biodiversity loss, pollution, ecological injustice, and resource depletion increasingly reveal the limits of compliance driven sustainability models. Purāṇic reflections on yajña similarly suggest that ethical transformation matters more than symbolic performance (Made & Somawati, 2021). In a related way, applied sustainability research has observed that dominant approaches often prioritize technological intervention while underemphasizing collective action, sociocultural context, and traditional ethical foundations (Suresh & Sreejith, 2023). Pañca Ṛṇa offers a broader orientation by locating sustainability within obligation, interdependence, restraint, and continuity rather than external regulation alone.

4.1 Connecting Indigenous Ecological Thought to Policy Frameworks

The relevance of Pañca Ṛṇa becomes especially visible in its convergence with several contemporary sustainability movements. Nature based solutions, regenerative agriculture, responsible consumption, and community conservation all rest on concerns that resonate strongly with this framework. Deva Ṛṇa and Bhūta Ṛṇa support ecological care toward forests, wetlands, rivers, and biodiversity. Ṛṣi Ṛṇa and Pitṛ Ṛṇa reinforce the preservation of ecological knowledge and the responsible stewardship of inherited resources. Manuṣya Ṛṇa emphasizes shared responsibility, inclusion, and community resilience. Studies of tribal life in India also point to low ecological footprint, ecological balance, and preservation of resources for future generations, suggesting that sustainable consumption in Indian traditions often rested on restraint and continuity rather than accumulation (Kumar & Prasad, 2025).

This interpretive relevance may also be brought into conversation with present policy frameworks. Pañca Ṛṇa aligns broadly with sustainability goals related to climate action, ecological protection, biodiversity, education, inclusion, and intergenerational responsibility. It also speaks meaningfully to climate governance concerns reflected in the Paris Agreement, especially mitigation, adaptation, and justice across generations. The framework can deepen them by providing ethical motivation and cultural grounding.

Table 1. Pañca Ṛṇa and its contemporary sustainability relevance and broad SDG alignment

Pañca Ṛṇa	Contemporary relevance	Broad alignment with SDG
Deva Ṛṇa	Respect for natural systems, climate responsibility, ecological restraint	SDG 6- Clean Water and Sanitation SDG 7- Affordable and Clean Energy SDG 13- Climate Action SDG 15- Life on Land
Ṛṣi Ṛṇa	Preservation of knowledge, ecological literacy, ethical transmission	SDG 4- Quality Education SDG 9- Industry, Innovation and Infrastructure SDG 12- Responsible Consumption and Production
Pitṛ Ṛṇa	Intergenerational responsibility, continuity, stewardship of inherited resources	SDG 11- Sustainable Cities and Communities SDG 12- Responsible Consumption and Production SDG 13- Climate Action SDG 16- Peace, Justice and Strong Institutions
Bhūta Ṛṇa	Protection of biodiversity, non human life, habitat care	SDG 14- Life Below Water SDG 15- Life on Land
Manuṣya Ṛṇa	Social justice, shared stewardship, community resilience and inclusion	SDG 1- No Poverty SDG 3- Good Health and Wellbeing SDG 10- Reduced Inequalities SDG 11- Sustainable Cities and Communities SDG 16- Peace, Justice and Strong Institutions

4.2 Barriers and Tensions in Adopting Indigenous Ecological Practices

Bringing Pañca Ṛṇa into present environmental discourse is not entirely straightforward. Traditional sustainability oriented ways of life have come under growing pressure from globalization, urbanization, and wider socio economic change, all of which have disturbed the balance between indigenous ecological practices and modern consumption patterns (Kumar & Prasad, 2025). Consumerist ways of relating to nature, shaped more by exploitation than reverence, stand in clear contrast to a framework grounded in restraint, reciprocity, and interdependence (Singh, 2013). Another difficulty lies in translating civilizational ethics into modern institutional language, since policy systems often depend on measurable indicators, whereas Pañca Ṛṇa works through moral orientation, cultural meaning, and lived responsibility. There is also the risk that such concepts may be used only symbolically, without leading to any real change in practice. In addition, the weakening of community structures and intergenerational continuity has reduced the social foundation through which such ethical frameworks can remain active in everyday life. Even so, the value of Pañca Ṛṇa does not lie in suggesting that the past offers a ready made answer to the present. Its importance lies in reminding us that ecological responsibility can be re grounded in duty, reciprocity, continuity, and shared care. For that reason, its relevance today depends on careful interpretation, critical adaptation, and meaningful translation into contemporary institutions and practices.

Table 2. Pancha Ṛṇa Solutions to IKS Challenges

Challenge	Ṛṇa Mapping	Historical Origin	IKS Solution
Technical Complexities	**Ṛṣi Ṛṇa**	Western scientism dismissed methodologies as "unscientific"	Hybrid validation frameworks blending śāstra-pramāṇa with empirical testing
Cultural Sensitivity	**Pitṛ Ṛṇa**	Colonial Sanskritization marginalized regional traditions	Community co-creation, elder validation in NEP curricula
Resistance to Change	**Manuṣya Ṛṇa**	Post-1947 modernization rejected "backward" traditions	Pilot programs demonstrating IKS benefits (e.g., yoga in wellness)
Disparities in Access	**Manuṣya Ṛṇa**	Urban-centric education (96% funding to 4% institutions)	Digital IKS repositories, vernacular content
Preservation of Languages	**Pitṛ Ṛṇa**	English Education Act 1835; 220 languages endangered	Mother-tongue IKS pedagogy, oral archives
Digital Obsolescence	**Ṛṣi Ṛṇa**	Tech platforms prioritize English/Western formats	Indic script AI, multilingual traditional knowledge databases
Knowledge Violence	**Ṛṣi/Pitṛ Ṛṇa**	Macaulay's 1835 Minute dismissed Indian learning as "worthless."	Relational pedagogies, indigenous research ethics
Resource Depletion	**Bhūta/Deva Ṛṇa**	Green Revolution	Traditional zero-waste systems, seasonal agriculture

5. FUTURE RESEARCH DIRECTIONS

This chapter is based on a conceptual study. However, for future research, field based studies may be conducted to develop detailed organisational case studies of Tata Power, Gita Press, Banasthali Vidyapith, Zoho, WWF India, SEWA, and similar institutions. Detailed research can also be conducted on specific Ṛṇas as well connecting them to current environmental frameworks in greater detail. The ecological dimensions mentioned across other Bharatiya scriptures can also be studied with their contemporary relevance.

6. CONCLUSION

This chapter shows that the Bharatiya concept of Pañca Ṛṇa offers a meaningful way of understanding ecological responsibility, not only as a technical or policy issue, but as a question of duty, relationship, and continuity. Rather than treating sustainability as something to be addressed only after a crisis has emerged, this framework places environmental care within a wider moral vision in which human life is sustained through its relationship with nature, knowledge, ancestors, other living beings, and society. In this way, Pañca Ṛṇa adds moral depth to sustainability discourse by bringing in gratitude, restraint, reciprocity, and responsibility.

The discussion also shows that this framework is aligned to scriptural thought and continued through lived practices, institutions, and traditions over time. It can be witnessed in indigenous ecological practices, community traditions, knowledge systems, and selected contemporary organisational examples. By examining these through Deva Ṛṇa, Ṛṣi Ṛṇa, Pitṛ Ṛṇa, Bhūta Ṛṇa, and Manuṣya Ṛṇa, the chapter has analyzed that environmental responsibility in the Bharatiya tradition is understood in an interconnected way, extending across nature, society, memory, and future generations. Its significance lies in offering an ethical orientation that remains deeply relevant to current concerns such as ecological stewardship, intergenerational justice, biodiversity protection, social responsibility, and sustainability education. Yet its application in the contemporary world is not simple. Consumerist models of development, weakening community continuity, symbolic use of indigenous concepts, and the difficulty of translating moral frameworks into policy language all remain real challenges. A more sustainable future will depend, at least in part, on recovering ways of living that recognise dependence, honour inheritance, protect life, preserve wisdom, and act responsibly toward both present and future generations.

7. REFERENCES

Adhikary, C. D. (2023). Hindu World-Views and Social Work Practices. *Indigenising and Decolonising Social Work Education*, 124.

Andriana, W. D. (2025). The Impact of Anthropocentrism on Human–Nature Relations and Its Implications for Environmental Ethics. *Journal of Language and Literature*, *25*(2), 625–637. DOI: 10.24071/joll.v25i2.10876

Anthwal, A., Gupta, N., Sharma, A., Anthwal, S., & Kim, K. H. (2010). Conserving biodiversity through traditional beliefs in sacred groves in Uttarakhand Himalaya, India. *Resources, Conservation and Recycling*, *54*(11), 962–971. DOI: 10.1016/j.resconrec.2010.02.003

Arsal, D. A. T. (2026, January). INTERFAITH ECOTHEOLOGY: UNIFYING RELIGIOUS PERSPECTIVES IN RESPONDING TO THE ENVIRONMENTAL CRISIS. In *International Student Conference* (Vol. 1, No. 1, pp. 8-21).

Ashika, J., Shirly, S., Sharma, S., & Raymer, M. (2025). *Eco-wisdom of ancient India: Environmental conservation and sustainability through religious scriptures, Vedic teaching and Arthashastra.* In *Proceedings of the 3rd International Conference on Revitalizing Indian Knowledge Systems*. https://doi.org/ DOI: 10.2991/978-2-38476-477-8_23

Bora, A., & Kumar, A. (2025). The Role of Indian Wisdom in Addressing Climate Change and Environmental Crises. *Indian Knowledge Systems for Social Development: Traditional to Modernity Pedagogy*.

Chakraborty, S. (2002). *Managerial Effectiveness and Quality of Worklife Indian Insights*. McGraw Hill.

Chandel, P. S., Agarwal, P., Indolia, U., & Parashar, A. (2024). Reviving Ancient Indian Hospitality: Integrating Guest-Worship Traditions into Modern Homestay Tourism.

Chapple, C. K., & Tucker, M. E. (2000). *Hinduism and ecology*. Harvard Center for the Study of World Religions, Harvard University Press.

Choi, S., Weng, S. S., Feliciano, G., Benton, A. D., & Kim, Y. (2025). Environmental Justice and Vulnerable Populations: Perspectives of Environmental Activists. *Social Work and Social Welfare*, *6*(1), 316–326. DOI: 10.25082/SWSW.2025.01.001

Damke, S. W. (2025). CONTRIBUTION OF INDIAN WOMEN IN TACKLING ENVIRONMENT PROTECTION AND CLIMATE CHANGE. *Journal of East-West Thought (JET) ISSN (O): 2168-2259 UGC CARE I, 15*(1), 659-665.

Das, S. M. (2019). *Scriptural importance of Hindu environmental ethics.*

Das, S. M. (2020). *Indian environmental ethics. SSRN.* https://ssrn.com/abstract=3655033

Dwivedi, O. P. (1993). Human responsibility and the environment: A Hindu perspective. *Journal of Hindu-Christian Studies*, *6*(1), 8. DOI: 10.7825/2164-6279.1077

Ganguly, A. (2024). ECOLOGICAL WISDOM IN VEDIC LITERATURE: A COMPREHENSIVE ANALYSIS. *ENVIRONMENTAL SCIENCE*, 1.

Gauthier, P. E., Chungyalpa, D., Goldman, R. I., Davidson, R. J., & Wilson-Mendenhall, C. D. (2025). Mother Earth kinship: Centering indigenous worldviews to address the Anthropocene and rethink the ethics of human-to-nature connectedness. *Current Opinion in Psychology, 64*, 102042. DOI: 10.1016/j.copsyc.2025.102042 PMID: 40288260

Gita Press. (2024). *Śrīmad Bhagavad Gītā* (Sanskrit text with Hindi translation and commentary).* Gita Press Gorakhpur.

Gita Press. (2025). *Kalyan: Paryavaran ank* (99th special annual issue).* Gita Press Gorakhpur.

Gupta, A. K., & Maurya, S. K. (2026). Revivalism in Indian educational philosophy: Reinterpreting ancient wisdom for modern times. *Cogent Education, 13*(1), 2584914. DOI: 10.1080/2331186X.2025.2584914

Haberman, D. L. (2006). *River of Love in an Age of Pollution: The Yamuna River of Northern India*. University of California Press.

Hameed, K., Sirwan, K., Omar, Z. O., Mohammed, E. F., Sourkan, S., Salih, A. M., & Yaqub, K. Q. (2025). Addressing the issue of poverty, economic growth, health matters, and environmental challenges. *International Journal of Scientific Research in Modern Science and Technology, 4*(3), 18–32. DOI: 10.59828/ijsrmst.v4i3.303

Hourdequin, M. (2025). Intergenerational ethics and climate change. *Wiley Interdisciplinary Reviews: Climate Change, 16*(1), e934. DOI: 10.1002/wcc.934

Jacoba, R. C. (2025). Sacred ties to ancestral land: Reclaiming indigenous practices in environmental stewardship in the Cordillera. *Media: Jurnal Filsafat dan Teologi, 6*(1), 1-30.

Jain, P. (2024). Sustaining Dharma, Sustainable Ecology: Dharma as Rural Environmental Ethics. In *The Routledge Companion to Indian Ethics* (pp. 223-232). Routledge India.

JASROTIA, R., & MANHAS, A. S. (2024). Ancient Wisdom for Modern Environmental Problem. *Journal of Applied Social Science, 11*(3&4), 137–141.

Jayanthi, V. (2001). Vegetarianism in India. *Peritoneal Dialysis International : Journal of the International Society for Peritoneal Dialysis, 21*(3, suppl), 322–325. DOI: 10.1177/089686080102103S57 PMID: 11887845

Jha, G. (1999). *Manusmrti 10 Volumes (Complete Series): With the'Manubhasya'of Medhatithi*. Motilal Banarsidass.

Jindal, K., Singhi, R., & Srivastava, A. (2025). Indian Ethos for Sustainable Business Development. In *Business Management and Ethics* (pp. 171–189). Emerald Publishing Limited. DOI: 10.1108/978-1-83549-254-320251009

Khatib, S. F. (2025). Social Trust, Environmental Violations, and Remedial Actions in China: Shahab, Y., Ye, Z., Liu, J., & Nadeem, M.(2024). Journal of Business Ethics, 1–18. . *Social and Environmental Accountability Journal, 45*(1), 81-83.DOI: 10.1007/s10551-024-05818-5

Khoshoo, T. N. (1999). The dharma of ecology. *Current Science, 77*(9), 1147–1153.

Kumar, A., & Prasad, A. (2025). Sustainable Consumption Through the Lens of Indian Tribal Wisdom. [IJSSR]. *International Journal of Social Science Research*, *2*(2), 186–195.

Lema, M. W. (2025). Sustaining rural livelihoods through participatory water governance: A review of community-driven water resource management models in East and Central Africa. *Frontiers in Water*, *7*, 1704600. DOI: 10.3389/frwa.2025.1704600

Liang, N., & Segalas, J. (2025). A deep and shallow sustainability intervention framework: A Taoist-inspired approach to systemic sustainability transitions. *Sustainability (Basel)*, *17*(11), 5170. DOI: 10.3390/su17115170

Made, Y. A. D. N., & Somawati, A. V. (2021). THE CONCEPT OF YAJNA IN VARAHA PURANA. *Vidyottama Sanatana: International Journal of Hindu Science and Religious Studies*, *5*(2), 159–168.

Malhotra, R., & Babaji, S. D. (2020). *Sanskrit non-translatables: The importance of Sanskritizing English*. Manjul Publishing.

Maurya, N. K. (2025). *Sacred Lineage and Divine Guardians: The Role of Kuldevi, Devatha, and Pitra in Material and Spiritual Growth*. Writer's Pocket.

Mistri, S. (2026). THE USEFULNESS OF MANUSMṚTI IN RESOLVING THE CURRENT ENVIRONMENTAL CRISIS.

Mukil, M. V., Shinde, K., Rashed, T., & Bhavani, R. R. (2026). Religious-Cultural framework for sustaining environmental balance in rural ecosystems: Stewardship of village-temples in Tamil Nadu, India. *Local Environment*, •••, 1–27. DOI: 10.1080/13549839.2026.2631508

Muniapan, B. (2013). *Ancient Indian wisdom for sustainability and management.*

Muniapan, B. (2013). The roots of Indian corporate social responsibility (CSR) practice from a Vedantic perspective. In *Corporate social responsibility in Asia: Practice and experience* (pp. 19–33). Springer International Publishing.

Muralivallabhan, T. V. (2022). Protection of environment and promotion of development: The Hindu wisdom of sustainable development. *National Security, 5*(2), 146-180.

Nobles, E. C. (2025). The ethical foundations of biodiversity metrics. *Current Opinion in Environmental Sustainability*, *72*, 101503. DOI: 10.1016/j.cosust.2024.101503

Owolabi, B. A., Sanuade, A. O., Fajinmolu, A. O., Atabo, L. O., Okeyode, O. S., Adeyemi, A. A., & Oganla, A. E. (2026). Understanding the Forest Decline and Habitat Loss. In *Environmental Health and Toxicology* (pp. 382–393). CRC Press. DOI: 10.1201/9781003500452-24

Panda, T. (2024). Sustainable living in ancient Indian teachings: Glimpses from classical Indian scriptures and traditions. *The Social Science Review: A Multidisciplinary Journal, 2*(3), 81-91.

Pradhan, M. (2026). The Cry of the Earth: Literature as a Catalyst for Sustainable Behavioral Change. In Giri, A. K., & Deshpande, R. S. (Eds.), *Quest for Planetary Well-Being*., DOI: 10.1007/978-981-95-2077-0_18

Putra, M. D. P., Hastjarjo, S., & Satyawan, I. A. (2025). Environmental Communication Based on Local Wisdom in Efforts to Preserve Water Sources in Pedawa Village. *Jurnal Ilmu Sosial dan Humaniora, 14*(1), 195-209.

Radhakrishnan, S. (1994). The Principal Upanishads,(From Taittiriya Brahmana section).

Rai, G. K. (2025). Protector of the Green: Sacred Groves as Images of Ethical and Spiritual Wholeness. *International Journal for Multidisciplinary Research, 7*(3), 01-06.

Richardson, M. (2025). Modelling nature connectedness within environmental systems: Human-nature relationships from 1800 to 2020 and beyond. *Earth (Basel, Switzerland), 6*(3), 82. DOI: 10.3390/earth6030082

Roy, B. (2025). The Teachings of the Puraṇas and Modern Science in Indian Knowledge System: A Review. [IJSSR]. *International Journal of Social Science Research, 2*(3), 333–341.

S. (2025, October 11). *Sridhar Vembu's vision: How Zoho transformed Tenkasi into rural tech hub*. The Week.https://www.theweek.in/theweek/business/2025/10/11/sridhar-vembus-vision-how-zoho-transformed-tenkasi-into-rural-tech-hub.html

San Román, Á., & Molinero-Gerbeau, Y. (2023). Anthropocene, Capitalocene or Westernocene? On the ideological foundations of the current climate crisis. *Capitalism, Nature, Socialism, 34*(4), 39–57. DOI: 10.1080/10455752.2023.2189131

Sarkar, B. K. (2021). Giving as a Road to Fulfilment. *Bulletin of the Ramakrishna Mission Institute of Culture.*

Saxena, P. (2025). *Svayambodha and Shatrubodha: Hindu View of Self and the World.* BluOne Ink Pvt. Ltd.

Sharma, S. K., & Biswas, D. (2024). Consciousness: Revisiting the Sacred. *Environmental Humanities in India, 25*, 33. Singh, K. Relocating Identities. *India International Centre Quarterly, 28*(3), 100–108.

Singhal, A. K., & Sharma, U. (2025). Ethical Discourses in Ancient Indian Culture for Sustainable Protection of the Environment.

Sivaramakrishnan, K. (2015). Ethics of nature in Indian environmental history: A review article. *Modern Asian Studies, 49*(4), 1261–1310. DOI: 10.1017/S0026749X14000092

Smart, N., & Thakur, S. C. (1993). *Ethical and political dilemmas of modern India.* No Title. DOI: 10.1007/978-1-349-23057-0

Srivastava, J. (2010). *The Relevance of Gandhian Thought for Contemporary Management Systems.* Navjeevan Publications.

Suresh, V., & Sreejith, S. (2023, September). Strengthening collective action and sustainability of traditional farming organizations from an Indian philosophical perspective. In *ECKM 2023 24th European Conference on Knowledge Management Vol 2.* Academic Conferences and publishing limited.

Tiwari, D. N. (2016). Spiritual ecology and environmental ethics. *Cultura (Iasi, Romania), 13*(1), 49–68. DOI: 10.5840/cultura20161313

Vivekananda International Foundation. (2023, April 4). *Dr. Rajendra Singh: Waterman of India*. https://www.vifindia.org/india-positive/2023/april/04/dr-rajendra-singh-waterman-of-india [page:2]

Wilson, C., & Noyes, G. (2025, January). Indigenous Stewardship of Ancestral Lands Activates Land and Culture. In *Parks Stewardship Forum* (Vol. 41, No. 1, pp. 66-72). George Wright Society. DOI: 10.5070/P5.42004

WWF-India. (n.d.). Reducing India's global footprint. https://www.wwfindia.org/about_wwf/reducing_footprint/

ADDITIONAL READING

Kishore, R. R. (2025). Environment, Hinduism and the words of wisdom. In *Multicultural and Religious Perspectives on Protecting the Environment, the Biosphere, and Biodiversity* (pp. 284–299). Routledge. DOI: 10.4324/9781003632726-33

Neethu, M. S., & Bhuvaneswari, R. (2024). Eco-Dharma and Climate Change: An Ecospective Rendition of Valmiki Ramayana and Hindu Eco-Praxis for Environmental Conservation and Sustainable Living. *The International Journal of Religion and Spirituality in Society*, *15*(1), 145–163. DOI: 10.18848/2154-8633/CGP/v15i01/145-163

Pathik, P. (2019). The historical and philosophical exegesis on Yagya in Ancient India. *Interdisciplinary Journal of Yagya Research*, *2*(1), 20–28. DOI: 10.36018/ijyr.v2i1.19

PEDERSEN. K. P. (2026). The Rights of Nature and Hindu Ethics. *The Bloomsbury Handbook of Hindu Ethics*, 251.

Singhal, A. K., & Sharma, U. (2025). Ethical Discourses in Ancient Indian Culture for Sustainable Protection of the Environment.

KEY TERMS AND DEFINITIONS

Bharatiya scriptures: A vast collection of ancient literature (Vedas, Upaniṣads, Itihāsas, Purāṇas).

Environmental Ethics: Philosophy that studies the moral relationship between humans and the natural environment.

Indigenous Ecological Thought: Place based knowledge and practices on relationships among living beings and their environment.

Pañca Ṛṇa Framework: An ancient Bharatiya concept that sees human life as bound by five obligations.

Chapter 8
Advocacy for Revitalisation of African Kingdoms Amid the Call for National Dialogue in South Africa Using Ndumi dzi luma Mahosi Perspective

K. D. Mpeiwa
http://orcid.org/0009-0007-9638-4596
University of South Africa, South Africa

ABSTRACT

African Kingdoms overstayed their welcoming as ANC partners to the loss of own political identity and spiritual integrity. The chapter aims for advocacy for revitalisation of African Kingdoms as strategic pillar of South African democracy, and custodians of African Knowledge Systems. Qualitative Exploratory design is employed using secondary data analysis. The chapter coincides with a call for National Dialogue to inform newly formed Government of National Unity. ANC bid to partner with African Kingdom leaders. Ironically, thirty years since the dawn of democracy, African Kingdoms seen begging of government led by the ANC for respect and proper recognition as public entities without success. ANC policy approach in comparison with their Inkata Freedom Party come handy in secondary data analysis. The chapter calls for revitalisation of African Kingdoms as the epitome of African Knowledge Systems for sanity. Critical question would be what is it that African Kingdoms should advocate for going into the partnerships drawing from thirty years experiences under the leadership of the ANC?

1. INTRODUCTION

The *Ndumi dzi luma Mahosi* found in Tshivenda expression portraying conspiracies led by Chief Advisor aiming to overthrow the kingdom. The paper employs it in breaking away from the debate on the relevance of African Kingdom in South African democracy. It is a matter discussed beyond chieftaincy disputes to analyse policy direction of the African National Congress (ANC). Ironically, the contribution coincides with the landsliding collapse of ANC dominance as the ruling party of the country after the

DOI: 10.4018/979-8-2600-2890-2.ch008

2024 National Government Elections. It coincides with the formation of coalition government called Government of National Unity (GNU). According to Zweni, Faku and Yan (2025) the formation of GNU triggered by failure of political parties to win majority votes during 2024 National elections in South Africa. ANC that enjoyed though leadership since 1994 became the biggest looser despite securing 40.18%, Democratic Alliance (DA) 21.81%, Mkondo we Sizwe 14.58%, and Economic Freedom Fighters (EFF) 9.52% (Electoral Commission of South Africa). The challenge facing coalition partners is on the need to reconcile parties' policy position into GNU. DA opposes ANC's Broad Base Black Economic Empowerment while EFF advocates for expropriation of land without compensation.

Uncertainties in the country builds on lack of democratically approved policy position guiding the country. This propels double calls for the National Dialogue and the renewal of the ANC casting a vote of no confidence on her current situation. According to Shomolekae (2024) the ANC has just declared interests to partner with African Kingdom Leaders going into the proposed National Dialogue. This contribution view that interest as a platform enabling African Kingdoms to negotiate on terms for mutual beneficiation, and if need be, also try different political parties. The overarching aim for this contribution is to advocate for revitalisation of African Kingdoms into formidable institutions holding public entity statuses. Critical question is on what African Kingdoms carrying into the National Dialogue's negotiation table with political parties? The chapter breaks away from the debate on the relevance of African Kingdoms in the democracy of South Africa offering them a second chance to take informed decision in picking ideal coalition partner.

The ANC (2025) commits to partner for Industrial Growth and Jobs through re-industrialisation, localisation, beneficiation, and youth enterprise. Building capable and Clean State through professionalisation of public service with zero tolerance for corruption. Building of reliable Municipal Services for improved provision of water, sanitation, roads, and waste management. Human development through roll-out of National Health Insurance program and Gender Equality. Again, it commits to Global Engagement Initiatives using pro-active diplomacy in support of BRICS, African Unity, European Union, United States of America and Global South as they conclude. This urges leaders of African Kingdoms to come up with own strategic policy direction attractive to ideal partners. It makes this chapter relevant in exploitation of propelling grounds to profile African Kingdoms in global democratic processes. The literature review expands on those debates followed by presentation of research design and methods, discussion of findings, intervention and conclusion.

2. LITERATURE REVIEW

2.1. Conceptual Framework

Faku and Yan (2025) opine that the founding members of the ANC hail from African Kingdoms that view leaders as divine shepherds. This attracted criticisms and infiltrations from various political organisations to either mobilise or destabilise them. It is worth mentioning that the aspects of National Dialogues and Government of National Unity are not new to South Africans as it was so in 1994. This contribution claims that the difference now is that most South Africans are aware of their democratic rights, and its processes. Following Delius (2020) ANC then lacked patience to educate general members of the public about their democratic rights taking comfort that people will catch up along the way. However, it had to do a lot with its own fate to mend the psychological and social networks broken already with African Kingdoms and their individual leaders. It had to deal with de-militarisation of liberation forces across all political parties (Lebaka, 2018).

African Kingdoms and their leaders suffered attacks blamed as traitors, witches, darlings of the white led apartheid systems, and all forms of propaganda to the demise of many including assassination of their leaders. Some songs sang during the National Liberation Struggle attests to this sentiment. The debate on the relevance of African Kingdoms in South African democracy shows them with limited choices but to align with the ANC to survive her wrath risking total assimilation. Those that opposed to her bid suffered total exclusion that Chief Mangosuthu Buthelezi perceives as political suicide (Inkata Freedom Party, 2018). ANC targeted Kingdoms led by Bantustan Administrative Systems and their constituencies, hence the demise of many who turned down her offer. Instead, the wave of songs, slogans, and public utterances redirected from challenging white people supremacy to those African Kingdoms viewed unpersuadable. Holistically, the white people eschewed most blame to their African Kingdom leader's counterpart placing them against their own militant ANC. The move to join the ANC paid dividends as to redirect the waves of her wrath straight into the heart of her momentous stance. Consequently, South Africa is experiencing formations of many different political parties marking disintegration of the forces of National Revolution or Alliance Partnerships. *Ndumi dzi luma Mahosi* posits a subordinate turning against those who built and defended him as subjects yet lacking the divine sceptre to lead indigenous communities.

Tshivenda idiom holds the saying *Nyavhumbwa wa dakaila wa kanda vho u vhumbaho* suggesting that on your way up on the ladder never under-rate those who made you grow or holding your fortresses. Legal battles prove that most Chiefdoms in South Africa currently led by *Ndumi* than rightful heirs due to unresolved succession disputes in royal courtyards. This shows anomalies compromising traditions that kept the communities together under leaders that royal courtyards appoint. With the National Dialogue in the making, policy positioning takes a centre stage in negotiation processes. It is important to reflect on mixed reactions on the relevance of African Kingdoms holistically.

Mawere et al. (2021), Bhuda and Maditsi (2023; 2024) contend that the institution of traditional leadership is still a trusted institution for governance in rural South Africa. This is despite that the South African post-apartheid government did not clarify the role and significance of traditional leaders in local government. Ntonzima and Bayat (2012) contesting those sentiments opining that Traditional Leadership lost influence since the start of democracy in South Africa. Carter (2025) aligns to the suggestion that Traditional Leadership play catalytic role connecting their members to government institutions. Ndlovu-Gatsheni (2014) challenge contrasting stances positing traditional leadership as irrelevant, barbaric, conservative, static, and uncivilized. Ray (2003) argues that Traditional Leaders carry the hopes of their communities as agents of security, custodians of heritage, justice, strategy, spirituality, and symbols of good governance. Klaas-Makolomakwe and Raniga (2021) encourage further exploitation on the relevance of Traditional Leadership in South African democracy. The next section gives a descriptive overview of the concept 'leadership' useful in contextualisation of this contribution.

2.2. Descriptive overview

Northouse (2010) gives descriptive overview helpful in the understanding of the concept 'leadership'. For example, leadership as a developing process builds on mentorships, hereditary, revolts, charisma, and ability to mobilize support and compassion to the goals in pursuit by the leader. According to Dickson (2023) leadership theories evolve with time owing to influential socio- economic and political environments facing the country. He opines, *Recent studies proposed a dynamic view about leadership and noted that a leader assumes and relinquishes position of authority based on the situation which is a significant departure from traditional beliefs.* Dickson (Ibid) sees existence of paradigm shifting from traditional to contemporary leadership theoretical approaches. In other words, decisions on one's ascension to lead should count on individual's ability and command of tribal knowledge and resources mobilisation for community development.

Dodo (2013) shows different attributes of leaders like the Great Man, Trait, Contingency, and Emerald amongst others. The Great Man points to a charismatic leader believed mightiest, wise, visionary and symbol of hope to their people. On Trait leadership, Fleenor (2000) opines that qualities of the leader assessed according to their personalities inclusive of physical, value and competence. Behavioural attributes partly share similarities with Trait model with slight difference on vulnerability of leaders to exploitation to fit a particular course. For example, the country that is in war consider bravery attributes than compassionate leader useful in dispute resolution cases. On contingency, Amghar (2022), and Chiwisa and Qutieshate (2024) show it is circumstantially employed in cases such as death, sudden resignations, and political recalling amongst others. In Africa tradition Tshivenda expresses such leaders as *Khosipfareli* employed to fill Chiefdom reigns while intended Heir is young of age to lead at death of their predecessor. The Path-Goal leadership come closest to it as a transitional tool, but their attributes suggest leaders leading by example (Bans-Akutey, 2021; Fleenor, 2000; & Northouse, 2010). Pathfinders also exist marking a new leadership trajectory in the organisation or country. For example, Mr Barrack Obama and Mr Nelson Mandela served as first black Presidents of the United States of America and South Africa respectively.

According to Chiwisa and Qutieshate (2024) and Dong (2023) contemporary theoretical approach holds the view that leaders compensate loyalty, dedication and delivery of strategic goals of their institutions. Those sentiments find expression on daily public political canvasses rewarding loyalty off course using monetary and promotional accolades. Leaders in Economics, Business Administration, and Sustainable development thrive utilising this model. It is development-oriented through exploitation of opportunities to grow the economy through access, control and sustainable use of means of production for community good. Dong (ibid), Amghar (2022), Bans-Akutey (2021) and Fleenor (2000) portray Laissez-fair leadership as contradictory to Transactional Leadership counterpart. According to Jarc (2015) Laissez-fair leaders allow their constituencies freedom to express themselves with little supervision. Mixed reactions to it suggests some subordinates misuse those freedoms resulting into collapse of production, and temptations to overthrow leadership, meanwhile others utilise freedom given responsibly to the enhancement of productivity and innovation. Transformational Leadership encourages innovation, creativity and portray dynamic acumen aligning to time changes for adaptability (Korejan & Shahbazi, 2016). On Servanthood leadership approach Hai and Van (2021) opine,

One of the most discussed leadership styles is the servant leadership, which implies that a person must be human before becoming a leader. Ever since Greeleaf published servant leadership style theory, emphasis shifted into portraying the leader as servant, proving its influence and effectiveness on development of society.

Under the Emerging Leadership Theoretical Framework, Nikolić, Kvasic and Grbic (2020) argue that Authentic leadership gain momentum as global governance is rising. Attempts to develop descriptive qualities of Authentic leader suggest the following. Provision of moral compass aiming to address issues of broken fibre globally. Foresight to find, assess and intervene to risks compromising social, economic and political sustainable development initiatives. Ability to mobilize human, financial, capital and entrepreneurship resources for community good. Oshame and Maureen (2023) bring another contemporary leadership trait in Strategist, Team leader, Shared value, and Ability to handle complex situations that once left unattended pose risks to the community or a particular institution. All those qualities fit in any of the already discussed theories either as complementary or distinctive qualities of a particular leader (Bhuda et al., 2021).

2.3. Global Leadership Perspective

However, a lot is known about the rest although there is a need to zoom into the relevance of Global Governance and Leadership complementing the current paper. Chiwisa and Qutieshate (2024) opine,

> The emergence of globalization has brought new realities that require a better understanding of how to lead and motivate groups to maximize the quality of multicultural group work. leadership is inherently universal, transcending borders, and epochs, the diversity of cultures and settings prompts exploration into distinct leadership traits and approaches.

African Knowledge Systems refutes this stance using Tshivenda idiomatic expression which suggests that *Vhuhosi a vhu weli mulambo* meaning that authority exists within demarcated barriers. It literary disputes existence of a universal leader giving reason why colonialism and imperialism collapsed. Globalisation exists as an ideological approach developed under the camouflage of World Orders. The orders that introduced from first to recent fourth Industrial Revolution places the United States of America and her western constituencies in thought leadership role (Gkika et al., 2022; Chakrabarty & Bass, 2013; Hyeonjung, 2021). It aims to tighten access control in the harvesting of global entrepreneurship, labour, capital and financial resources either through coercive or consensual approaches. Mapungubwe Institute for Strategic Reflection (MiSTRA, 2024) led a dialogue titled 'Thucydides Trap or Orderly Multipolar World? Deliverable extracts show rising of Asian countries like China amongst others into global thought leader like their Western Superpower counterpart. Stiglitz (1998) uses the epic of the 'Miracle of the East' as a justifiable ground about those contests. MiSTRA (Ibid) urges African Leaders across their levels of influence to explore opportunities within those global contests to address challenges facing the continent. It is important to give a Pan African Leadership perspective as discussed below.

2.4. Pan African Leadership Perspective

Mohammed (2025) appreciates the rising of African thought leaders in Global Leadership matters. This is despite of lack of shared collective view imbuing Africa's inspiration directed to the global communities in search of solidarity. Commentators range from those in academia, political parties holding less influence in their respective countries, media institutions, and emerging scholars through research and innovations, yet lacking global influence. Mohammed (ibid) sees positioning of African leaders in institutions such as World Trade Centre in Ngozi-Okonjo-lweala its Director-General, and Dr Vera Songwe as promising. Nemashakwe (2021) postulates that Pan-Africanist leadership envision a development-oriented situation directing the continent out of abject poverty. The desired goal is to achieve sustainable development by growing investment opportunities in the continent. The challenge though is that those holding public political offices criticised for lack of charisma and drive to invest on the people compared to their traditional counterpart. The latter criticised for surrendering political power through ballot electoral systems to the former, yet holding the Spiritual Sceptres leading divided members of communities shared across different political organisations. Sarum (2023) opines that African states once had robust leaders that fought against colonialism to achievement of independence and later democracy. However, a thin line exists separating current leaders from their predecessors before the eyes of their western counterparts. The use of distinctive concept 'Traditional' interchangeably used in place of 'African Leaders' sending contradicting messages to global communities. It is in public knowledge in Africa that Traditional Leadership defines individual bestowed with authority to lead a tribe. Those that ascend to leadership roles through government election processes regarded as politicians. This in turn lowers their traditional counterpart statuses to ordinary leaders under the President of the country. The call for African Leaders to adapt to the 21st century aligns to the Global Governance manifesting out of the 'World Orders' upholding Fourth Industrial Revolution also known as 4IR. Ideally, Africa needs to contribute to global sustainable development initiatives.

2.5. Global initiatives for Africa development

The Secretary-General Antonio Gutteres in the address of the United Nations Conference 2025 held in Tokyo urges African Leaders to,

> Co-create solutions to give Africa Stronger role in Global Politics, Financial Institutions, and Debt relief. He suggests that Japan Technological leadership can help narrow the gaps in Africa's development of required infrastructure in the catch up with the rest of the world. Africa must have stronger voice in shaping decisions affecting her future. That includes long-overdue call for reformation of Security Council where she lacks permanent membership.

Those views manifested into plausible bilateral talks between Greece and Egypt for solidarity to the Palestinians warfare, cease fire initiatives over Gaza Strip and delivery of humanitarian relief aid, and trade relation matters (The Arab Republic of Egypt Presidency, 2025). In another case, Portal do Governo Mozambique (2025) reports on bilateral meeting between Japan and the Republic of Malawi discussing comprehensive cooperation covering Nacala Region inclusive of Zambia, Malawi and Mozambique in the bid to apply resolutions of the recent United Nations Tokyo Conference. Still, Ministry of Foreign Affairs Japan and its Nigerian collaborate on trade relations with North Korea over Nuclear Missiles, Regional and international challenges inclusive of abductions amongst others.

Africa Energy Chamber hosts Africa Energy Week in Cape Town, South Africa in October 2025. The aim was to consult extensively as intervening to energy crisis in the continent (Africa Energy Chamber, 2025). This shows African leaders' commitment to deliver on sustainable development goals Agenda 2030. Jordan, Ng'weno, and Spencer-Oatey (2020) warn about the need to bridge skills gaps between the so-called politicians and tribal leaders for smooth transition. Those sentiments are desirable, although this contribution suggests their timing comes amid scrambling in monopoly of global resources by World's Superpowers. Positively, global leaders view their African counterpart in same lenses as those leading tribes. Its leaders stand equal opportunities to approach world leaders just as their so-called political leaders do which levels the ground for revitalization of African Kingdoms. It is important to bring in the discussion aspects of heritage conservation using African Heritage Management System threshold.

2.6. African Heritage Management Systems

This section discusses African Heritage Management Systems in the bid to highlight the role of African Kingdom Leaders in conservation. It shows challenges such as broken moral fibre facing societies in the 21st Century, collapsing and marginalisation of African Kingdoms, and anarchy sponsored politically to destabilise communities. According to Abungu (2016) Heritage Management Systems is a tool useful in safeguarding of social values, morals, customs, and spiritual sanitation for community benefit. African Knowledge Systems plays both lead and complimentary roles in building socially cohesive communities. African Kingdoms claim thought-leaders in laying down the foundations of acceptable norms and standards defining a certain community.

The *Ndumi dzi la Mahosi* perspective, and criticisms fashioned against African Kingdom Leaders often exposed. Those agitating for the collapse of African Kingdoms found lacking answers to the broken spiritual and social bonds in the community. Biblically, (Bible hub online, 2025, Proverbs 22: 28) retorts, ''Son, do not remove an ancient boundary stone which your forefathers placed''. The verse advocates for both spiritual and physical landmarks also defined as tangible and intangible heritage resources in the community. Emphasis in this instance is on the role of African Kingdoms as sacred places, while their leaders as Chief Priests. According to Mathenjwa (1991) African Philosophy holds the view that a human being interconnects the land and own biological nature. This suggests that humans are curiously aware of circumstances surrounding them including activities happening in the celestial, maritime, land, and any bio-chemical spaces sometimes as agencies frequenting them for the goodness of the community. The person appointed through politics lacks command or respect of members of the community. This is simply because such respect builds from social and cultural interconnections with both spiritual and physical landmarks defining the boundaries of those communities. ANC led Government introduced municipal systems, community-based organisations or civic systems, employs ward councillors, and municipal demarcation systems criticised for undermining ancient boundaries to the disturbances of African Kingdoms. Heritage professionals play important role in the revitalisation processes of African Kingdoms. The next section substantiates those sentiments after the discussion of research design, material and methods employed.

3. MATERIAL AND METHODS

Exploratory qualitative research conducted sharing experiences of institutions of Traditional Authorities in South Africa. Secondary data analysed with focus on discussion papers informed negotiations that led the beginning of democracy in South Africa. Those include policies, debates across various consultation platforms, songs, slogans, compromises reached, and manifestations building from those dialogues. According to Singh (2021) Exploratory research is an inclusive approach seeking various ways to explain data and reflection on its epistemological foundation. It is useful in studying the new phenomena and even in search of alternate uses of the same. The paper employs exploratory research in the quest to advocate for revitalisation of African Kingdoms for heritage conservation in South African democratic systems. It is worth mentioning that democracy as a concept is relative as countries share different experiences owing to the inherited Monarchs from colonial and imperial epochs. ANC policy plays a catalytic role guiding this paper. The paper also employs the Inkata Freedom Party (IFP) policy for comparison, balance reporting, and triangulation analysis. It is important to discuss the findings as shown below.

4. DISCUSSION OF FINDINGS

4.1. Reluctancy to sharing of power with internal constituencies

The overarching findings suggest that the ANC was never ready to share power with anyone including African Kingdoms. In its thirty years tenure as a ruling party, it managed to suppress their discerning views of becoming public entity with entitlements to lead and run their administrative duties. They expressed discontentment over the formation of local municipal government as they find it explicit marginalisation. Failure of government to meet their demands incited reluctances to cooperate with municipalities on matters of land management. Ironically, there are instances where African Kingdom Leaders escalated their concerns through channels such as ANC's leadership elective conferences, public consultation organised by the parliament, and media platforms amongst other.

4.2. African Kingdoms Lacks Organisational Structure

According to the IFP (2018) ANC made several promises to develop African Kingdoms that never materialised. This is despite of mediation initiatives led by its leaders that some became presidents of the country. Instead, the ANC defended its policy position that profiles African Kingdoms as strategic partners with some of their leaders in its full membership. The implication to this suggests that any matters requiring ANC's attention would have first discussed for adoption at the organisation's branch meetings and conferences. It therefore urges those supporting the discerning view of African Kingdom Leaders to win support of their motion that proved difficult to breakthrough. Democratic processes proved not favourable for African Kingdoms to gain automatic support they would have loved to enjoy.

4.3. Lack of Institutional Policy Framework amongst African Kingdoms

The major challenge learned in this regard shows that African Kingdom Leaders affiliate to CONTRALESA as recognised body by government. Policy review analysis conducted for this chapter suggests that CONTRALESA does not have National Policy but a constitution. It becomes difficult to imagine them to make meaningful contribution in negotiations with various political parties going to the proposed National Dialogue.

4.4. Uncertain future amid loss of power of ANC

This contribution exploited the future of African Kingdoms amid loss of political power of the ANC. South Africa experiences the growing interests in politics resulting in the formation of different parties. Some formations triggered by disgruntlements from parties such as ANC to the rising of black peoples led minority parties. It is worth mentioning that the formation of uMkhonto we Sizwe (MKP) caused major shaking with its dominance of areas such as Kwazulu-Natal Province (KZN) previously dominated by ANC and IFP. One raises this aspect to show the impact this has on IFP for her commitment to defend the course for preservation of African Kingdom's heritage. It becomes difficult for any existing party leaders to imagine winning government election, yet to surrender power to African Kingdom or any of their leaders.

Charismatic leaders or the Great Man theoretical backgrounds in the KZN plays a catalytic role guiding power exchange between parties. For example, in 1994 President Nelson Mandela of the ANC won support of the people ahead of Prince Chief Mangosuthu Buthelezi's led IFP. It kept support base with the help of other presidents but mainly Jacob Zuma. The void becomes visible at the formation of the MKP breaking away the ANC under disgruntled Jacob Zuma. The loss of majority votes to the ANC attributed to break away experienced. African Kingdom Leaders urged to make informed decision whether to stay with their long-term ally or explore opportunities at their disposal. Thirty years of partnering with the ANC offers them clear view for self-introspection to decide on the new adventure. ANC learned a hard way about the value of African Kingdoms, and their leadership influence over their constituencies. The overall assessment in post 2024 National and Provincial Government Elections shows that majority of black people abstained from voting that proved costly to the ANC. Leaders of Alliance partners seen dissatisfied with the ANC accusing her of excessive use of power, abuse of state resources, and exploitation of voters for sectarian gains.

4.5. Resistance to change effects

In view of Klaas-Makolomakwe and Raniga (2021) one reads of rift between ANC and African Kingdom Leaders that the former criticised for failure to embrace reconciliation reached. The latter criticised for opposing the democratic processes inclusive of surrendering land resource to municipalities, and petitions against municipal service delivery strategies amongst others. Those rifts retrospectively draw reference from leaders believed to have aided colonialism and apartheid government. Those include leaders of former homelands during Bantustan apartheid epoch. For example, Khosikhulu Ramaano Patrick Mphephu of Venda Administration, Kgoshi Cedrick Phathudi of Lebowa Administration, Kgoshi Lucas Mangope of Bophuthatswana, and Hosi Hudson Ntsanwisi of Gazankulu amongst many that have since collapsed. The Zulu Monarch resiliently survived those criticisms with the IFP as its political thought leader. The formation of Government of National Unity (GNU) in 1994 saw commitment to build a Rainbow Nation defining its multiracial communities. Contrary, the current lacks a slogan and direction behind which people at grassroots mobilised. ANC's recovery strategy is useful such that the next Local Government Elections voters turn-out and results will inform a new growth path. The readiness of African Kingdoms depends on their terms and successes of their strategic partners during elections. Lessons drawn from slogans such as ''Rainbow'' play significant role as to align these discussions to global development initiatives.

4.6. Lack of mass mobilisation slogans promoting GNU National Dialogue

South Africa prides herself as a Rainbow Nation since the dawn of democracy. Mpofu (2020) traces the origin of the concept 'Rainbow Nation' from the metaphoric expressions giving African Knowledge Systems and biblical symbolism substance. Symbolically, the rainbow marks the subsides of rain that often attracts raging and destructive storms compared to the damages caused by apartheid in South Africa. It envisioned a united and socially cohesive nation. It called for embracing of multiracial societies, non-sexist society existing as one human race as democracy intensifying. Mpofu (2020) raises matters such as Xenophobia, moral fibre, substance abuse, unemployment, corruption, and maladministration, and sponsored propaganda tainting images and integrity of African Kingdoms. This contribution argues that at transition from apartheid to democracy, Traditional Leaders committed themselves to the ideal they lack knowledge of. ANC as lead agent during negotiations assumed the role of a representative of black South Africans (Godihno, 2012). Lack of influence of Traditional Leaders in the outskirts of political thought led by the ANC becomes clear as the rainbow subsiding. The same goes to the church as symbolically posits a body of Jesus Christ experiences loss of power and influence in government unlike during apartheid (Lithole, 2022). Another important slogan used for mass mobilisation is the ''building of one human race'' discussed below.

4.7. The "One human race" slogan

Rabie (2021) shares five stages to lay descriptive overview of ''One human race'' phenomenon.

Firstly, it deals with the concept of human origin against biblical stance paving a way for discussion on Darwinism theory of human origin. It is important in the cosmological understanding of human origin as a counter to the notion that humans originated from evolving apes. Africa is known as home to all human beings given discovery of Australopithecus fossils in places such as Makapansvalley World Heritage Site in Limpopo Province. Scholars refutes this stance arguing it belittles Africans as the only confirmed earliest being on earth. Besides, African Knowledge Systems retorts that some clans trace their origin from the celestial, marine, fire and land (Mpeiwa, 2025).

This transcends race relations quality to contribute to nation building goal. Secondly, Interview on Race and Ethnicity captured in Brown, McIlwraith and Tubelle de Gonzales (2020) suggesting, *Discussions about race and racism are often highly emotional and encompass a wide range of emotions, including discomfort, fear, defensiveness, anger, and insecurity in society and it difficult for some individuals to discuss race dispassionately.*

Race and ethnicity seen as social constructs design to intensify class relations that later created first, second and third world divides. People mobilised themselves against racial and ethnic divides advocating for 'One Human Race' description. Reality shows that Statistics South Africa still utilises race and ethnicity in the quantification of fair sharing of resources. The conceptualisation of tribal institutions offers little contribution towards realisation of human race goal. Thirdly, Rabie (2021) mentions matters of religion in responding to fundamental issues affecting human lives. Adding on Mpofu (2020) diagnosis, Kgatle (2021) zooms into problems facing Christian Churches seeing the rise of Charismatic Prophetic Ministries amongst displayed social ills. Reasons cited include Gender-Based Violence implicating church leaders some occurring in church premises. Dube (2020) raises concern on conflicts between South African Prophets and their foreign counterparts working in the country. Another concern as Kobo (2019) opines relate to celebrity lifestyle of Senior Prophets, manipulation of members into eating grass and hazardous biochemical substances. Newcombe (2014) mentions the rise of occultism in the society as in churches inviting youth either as substance sales representatives or consumers. Family institutions experience collapse on marriages, loss of essential goods, all forms of violence from sexual assaults, rape, house breakings to mention but few. The role of Traditional Leadership as in their Chief Priesthood appears threatened as it used to be.

Fourthly, Rabie (Ibid) questions human orientation to the concept 'truth' and its implications in the society. This section moves from discussing the philosophical description of the concept 'truth' to reflect on Truth and Reconciliation at assumption of democracy in South Africa. Barry (2006) shows discontent whether it really promoted the truth or reconciliation. Instead, participants withheld lots of information as many were still in trauma, others held evidence implicating senior leaders but could not freely expressed themselves, Traditional Leaders saw assassination of their predecessors' called names, witches with some torched and killed by mobs portrayed as comrades. It stands to reason that going into negotiations to democracy Traditional Leaders were already disorganised to make proper representation in their defence. Race and uncomfortable black South Africans state of emotions indeed played elusive role going into the 1994 National Government elections.

Fifthly, Rabie (Ibid) questions the criteria humans utilise to find and fight their adversaries. In view of Robinson et al. (2018) in the article 'Modern Political Warfare: Current Practices and Possible Responses, instruments in use include Military coercive means and Consensual Subjugation through debt, aided Economic Policy Development, Uninterruptable Access Control to harvests of factors of production, Solidarity to Global and Continental Affairs amongst terms and conditions commonly imposed to the debtor. Violation to set terms often attract sanctions inclusive of increased debt repayment tariffs de-railing agreed settlement date. Newspaper headlines attest as follow. *Market summary: Impact of tariffs and South Africa's Economic challenges US Tariffs and Reaction from Global Markets.* Investments and Markets Insights (Botha, 2025). *Trump presses ahead with 30% tariff on South Africa.* British Broadcast Corporation News Johannesburg (Ngcobo, 2025a). *Tariff wars: How SA will be affected and what can it do to work around it.*

4.8. Inkata Freedom Party Criticism of ANC

> *During the ANC's land summit last month, former ANC Secretary General and former Head of State Mr Kgalema Motlanthe spoke bluntly about traditional leadership. In a scathing attack, he called traditional leaders "village tinpot dictators". Not a single leader in the ANC spoke up. No one contradicted Mr Motlanthe or called him to task. The fact is that this senior leader was voicing a long unstated position of the ANC.* Chief Mangosuthu Buthelezi (IFP, 2018). Those were laments delivered during Lekgotla of the National House of Traditional Leaders and the Executive Committees of Provincial Houses held on 27 June 2018.

The relationship between Inkata Freedom Party (IFP) and the ANC portrays two models of the National Liberation Struggle of South Africa. The first model suggests that ANC mandated its member Chief Mangosuthu Buthelezi to set up a party that became the IFP. The aim was use it to build constituencies of traditional leaders and majority Africans behind the struggle to liberate South Africa. It is the move short lived when the apartheid government countered to the setup of Bantustan and homelands working with traditional leaders. According to the Secretary General of the South African Communist Party Chris Hani (1991) IFP was a darling to the apartheid government. Its leadership enjoyed many privileges and protection of government while their ANC, SACP, AZAPO, PAC amongst other harassed, imprisoned and forced to exile. This implies that the highest point of their struggle was based on the desire to protect and intensify Zulu Monarch than liberating South Africa. The second model shows rift between the IFP and ANC as portrayed in the article "There is still a rift between Me and the ANC" (IFP 2018). It is important to show that as the rift was intensifying, the IFP Leader in Chief Mangosuthu Buthelezi subjected to hate speeches in derogative songs portraying him as *Impimpi* meaning spy against liberation forces.

The British Broadcast Corporation Newsletter (2023) titled "Chief Mangosuthu Buthelezi: the man who divided South Africa" furthers the debate on hate speeches the Zulu Kingdom Prime Minister endured. The IFP (2013) hold the view that the rift turned bloody as it claimed over 5000 lives of members of both parties between 1985 and 1992. Peace talk began at the release of Nelson Mandela from

Prison marking transition from physical combatants into negotiations (Nelson Mandela Foundation, 2025). The IFP in all negotiation platforms succeeded in defence and call for proper recognition of the Zulu Monarch. Democratic processes would suggest that collective view across institution of traditional leadership and their respective tribal authorities took stance either in support or against IFP position. Amongst recognized kingdoms during apartheid was Vhavenda of King Mphephu Ramabulana lineage. They could not voice own position as the kingdom was in disarray following assassination of Patrick Mphephu followed by the 1990 Venda coup d'état led by General Gabriel Ramushwana. In Gazankulu, as in Lebowa history shows lack of one dominant kingdom claiming authority over their traditional affairs. One infers that their leaders in Dr CN Phatudi and Professor Hudson Ntsanwisi endowed with western education as their academic qualifications attest (Makgamatha, 2024).

The IFP stood resolute against all odds in defence of Zulu Monarch as a sole voice in support of African Kingdoms and their integrity. This however does not demean contribution of the Congress of the Traditional Leaders of South Africa (CONTRALESA) shared sentiments with IFP. The shortcoming with CONTRALESA as criticism unfolds suggest its allegiances with the ANC viewed compromising its integrity. It is often bound by the collective views of constituencies that do not share their interests. Those claims are justifiable in view of ANC President Motlanthe opinion against institutions of traditional leadership depicting them as ''village Tinpot of dictators'' and failure of the party to discipline him. President Jacob Zuma held slightly different view to his predecessors in the ANC with tries to coronate Mr Tony Mphephu Ramabulana as king of Vhavenda (Tshikhudo, 2016; Government Department of Traditional Affairs, 2016). Another case is the amenability of President Ramaphosa to coronate Princess Masalanabo Modjadji as the Balobedu Rain Queen of Modjadji Dynasty (Motasa & Nortjé-Meyer, 2021; Ramothwala, 2025; Writer, 2024). Again, the decision to uphold crowning of King Misuzulu of the Zulu Monarch vindicate ANC led government (Tshikalange, 2025). Its amenability to amend Chapter 12 of the constitution to articulate the role of traditional leaders in the country backs their stance. Koenanne (2018) sees a trend of lack of consistency as leaders of the same organisations remain divided as from its formation. It is important to reflect on mixed political reactions to the position of ANC presidents using ordinary people's perspectives.

4.9. Mixed reactions to ANC presidents' positions

According to Ubink and Duda (2021) South African government spent at least fifteen years trying to centralise Senior Traditional Leadership role. Resilience is visible against total emancipation of Tribal Institutions and Leadership despite lack of public demonstration to this effect. Instead, government in turn criticised for playing ''dwindling carrot mind games'' luring traditional leaders with promises never kept (Chauke, 2015). It uses coronation program as a systematic tool to achieve its goal, it also uses recognition of tribal authorities, mobilisation of those leaders to join ANC for placements in government positions, destabilisation of kingdoms employing divide and rule strategy, and derailments on chieftaincy and land claims. Democratic processes left vulnerable to exploitations citing reasons ranging from poorly developed legal frameworks, perpetual claims sometimes from members of the same family, failure by government entities responsible for traditional affairs matters to advise properly, and political interferences.

According to IFP (2018) Chief Mangosuthu Buthelezi blames government for deceit given its unstated position on need for traditional authorities. Those critics emanated from the resolution of the 54th ANC National Conference held in December 2017. It aimed for democratisation and control of communal land that tribal authorities protected as own capital assets. It is the conference debated expropriation of land for placement under the control of government. Central to this debate was the Ingonyama Trust Land estimated at 2.8 million hectares under stewardship of the Zulu Monarch. Ngcobo (2025b) titles an article ''South African president signs controversial land seizure law'' justifying government ambivalence on matters of people empowerment. This paper argues that ANC experienced lack of resistance within majority traditional leadership constituencies.

However, the Zondo Commission on state capture (2022) exposes vulnerability of ANC to exploitation by external forces to a point of surrendering state sovereignty. This presses for a question on who stands to benefit at the total loss of land control by tribal authorities in South Africa? In other words, who is mastering and steering decisions of the ANC and her led government, and their interests of majority South Africans? It shows a sense of urgency to resolve matters politically motivated inclusive of marginalisation of tribal authorities. Strategies employed includes setups of district and local government with the latter tacitly dethroning tribal authorities in a humiliating manner. Still in public spaces traditional leaders in government pay-roll fronted as key stakeholders yet lacking self-determination virtue. They held positions determined through human resources management processes giving government rights to employ and expel, and all reservations not to fill vacancies left by immediate family members. Traditional leaders in government payroll feel trapped in the compromised socio-political networks demanding their conformity. In view of Goitsemodimo (2019), Ngcobo (2025a), Tshikalange (2025), and Tshikhudo (2016) broken trust exists between ANC and most Tribal Authorities in South Africa. The paper further explores opportunities to revitalise traditional leadership in South Africa using National Dialogue proposed by the GNU. The chapter calls for revitalisation of African Kingdoms as a strategic intervention to safeguard African Knowledge Systems. Significantly, it serves to respond to the 21st Century development challenges facing South Africa.

5. STRATEGIC INTERVENTION

5.1. Professionalisation of bodies for traditional leaders

The successes ANC enjoyed over the past thirty years of democracy was to make her alliance partners to adopt her policies regardless of areas of complete differences. It derailed them from rising as political entities to the level of contesting government elections. As a result, ANC rapid loss of influence and dominion in South African politics raises concern on the future of individual Alliance partners. It calls for revitalisation of professional bodies such as CONTRALESA. It must grow its constituencies in administrative and financial management into formidable entity driven by the needs of its members in traditional leaders. In as much as ANC is calling for self-renewal, one hopes for the organisation less excessive use of power as proven over the past thirty years into democracy. The approach was necessary in the early years of transition to apply excessive force to manage resilience to keep apartheid systems. Ironically, the approach never relaxed even though the overarching goal achieved. As it stands whoever ascends to position of power still uses the same strategy to divide and rule Alliance partners. Formidable professional bodies advocating for the interests of traditional leaders should adopt applicable policies. Bannister (2007) encourages professionalisation of administration of African Kingdoms suggesting giving them more responsibilities and resources for public good.

5.2. Policy Intervention

The policy positions of individual role players in the proposed GNU National Dialogue can decide its success or failure. The ANC kept the local government systems despite resistance by collective body of traditional leaders. They led mediations with government for rescinding of constitutional position in that regard. Policy is the merged commitment showing the position of the organisation and its direction. The ''dwindling carrot'' strategy ANC employs believed to have paralysed its alliance partners commitments to their policies. In all fairness, ANC has been equally vulnerable to lose her policy identity by assimilation to any of their alliance partner's. Resilience saw ANC succeeding to build own brand even employing own policy position tacitly binding her partners. Going into the National Dialogue, it is still policy-based contests where the top four political parties have set own tones. Professional bodies for traditional leaders urged to develop own position for proper bargaining in charting own way forward. The approach to 1994's negotiation shows attributes of decisions made on ''gentleman's agreement'' suggesting ANC partners never negotiated terms upholding their policy positions. GNU National Dialogue offers platforms for self-corrections and institutional revitalisation. General observation out of this contribution finds that professional bodies such as CONTRALESA still lacks a National Policy. Iya (2014) opines that the call for development of good policy is as important as finding the right leader to implement them. This suggests inclusive approach in the search for leaders and implementation of those policies.

5.3. Prioritisations of revitalisation of tribal authorities

Professional bodies in the affairs of traditional leaders needs to prioritise revitalisation of tribal authorities. Lessons drawn from the 1994-2024 experiences show institutions losing control over its cultural obligations to their immediate communities. The so-called democratic processes meant opening of vicious circles influencing decision-making on succession of traditional leaders. Proximities to the ANC leaders over electoral circles holds blame for perpetual chieftaincy claims and settlement derailments (IFP 2018). One employs contests over Vhavenda Kingdom to elaborate further. The working relationship between Vhavenda Kingdom led by Mr Tony Mphephu Ramabulana and ANC President Mr Jacob Zuma was well. However, it created triumphalist impression favourable to the Ramabulana at expense of his Tshivhase counterpart given the sociological orientation of Vhavenda history. Those privileges were short-lived at ascension of Mr Matamela Cyril Ramaphosa ironically of Tshivhase lineage into the president of ANC and of the country. It saw reversal of fate reducing Mr Tony Mphepphu Ramabulana into *Ndumi* than Paramount King (Tshikhudo, 2016; Ntlama, 2020). It caused serious destabilisation of chiefdoms that paid allegiances to him. Example is the Masia Tribal Authority because Chief Masia Nthumeni was the spokesperson Vhavenda Kingdom under Mr Tony Ramabulana. The instrumental figure destabilising Masia Tribal Authority Mr Livhuwani Matsila of Masia Clan a confidante to President Ramaphosa and Chief Tshivhase. Revitalisation of Tribal Authorities is essential given political interferences compromising their integrity.

5.4. Promotion of programs to build strong constituencies

In view of the Department of Traditional Affairs and Corporate Governance Medium-Term Expenditure Framework 2025/2026 per budget votes, there is no commitment to develop Tribal Authorities into a force they used to be. The rationale suggests that the views of traditional leader's delegates to ANC decision-making platforms outvoted. A counter would be to play within the system to lobby support or increase numbers of traditional leaders to enhance their chances. To achieve this goal, a roadshow promoting professional bodies for traditional leaders recommended. The call for National Dialogue used as a drawing card to mobilise traditional leaders to take part through formidable structures with amplified voice. Traditional leaders have a duty to win back support from internal constituencies against ANC's led ward councillorship, and utilisation of civic association in the affairs of communities (SALGA, 2013). This suggests development of attractive and fundable programs to reach out to their people. To achieve this goal, it is important to recognise human capital at disposal and collaborate with them for community development initiatives.

5.5. Areas for future research

It is worth mentioning that the National Dialogue is still at proposal phase, despite having the first meeting that the presidency organised at the University of South Africa from 15 to 16 August 2025. It was organised by the Office of the President of the country that have since attracted criticisms for lack of clarity on matters affecting the people of South Africa. Criticised for lack of inclusivity to allow ordinary citizens opportunity to make direct contribution into the process. Future research should follow the development of this aspect. Significantly, African Kingdom Leaders state of readiness to exist outside the ANC led processes needs exploitation. The study shows limitation in policy development that renowned organisations responsible for traditional leadership found lacking. Future research needs to explore training programs relevant for African Kingdom Leaders to exist as independent decision makers going forward. There is a need for proper audit of the remaining cultural heritage resources that many abandoned in the demise of certain kingdoms. Impact assessment plays significant role in safeguarding those resources inclusive of sacred sites, objects, and oral traditions associated with a particular kingdom.

CONCLUSION

As the paper concludes, discussions show that the ANC was never ready to share power with anyone including tribal authorities and their leaders. As it stands, ANC succeeded in decapacitation of tribal authorities through improvisation of local government, civic organisations in communities, and marginalisation of traditional leaders by reducing them into government servants. It succeeded in eradication of most political rivalries as they were during the National Liberation Struggle. Alliance partner's gradually lost influence and integrity to rise as standalones at the loss of ANC's majority votes of South Africans. The paper touched base on vulnerability of ANC to exploitation by its electoral funding constituencies that led to state capture. Safe to say in the process, it lost a sense of Africanness turning into machineries reprogrammed against all that majority South Africans in rural communities treasured as own heritage.

Those attributes would refer to destabilisation of social patterns, leadership, injection of unruly behaviour using individual leader's proxies to ANC abusively. It would include sponsored motives to political deployments to strategic institutions with sole aim to destroy them. Its loss of 2024 Government National Elections worth celebrating as blessing in disguise instigating for renewal using National Dialogue platforms. It is undisputable that the ANC lost mandate of its majority constituencies built over 100 years ago. The paper attributes this victory to institutions of traditional leadership for resilience against all odds holding the forts. History shows from the days of imperialists and monarchs, every leader holds a sceptre portraying spiritual and leadership direction of the country or kingdom. Election criteria in most succession strategies employable in the ANC leaves much to desire, hence the brewed uncontrollable corruption tainting its leadership tenure. Future studies need to focus on South African politics in anticipation of total erosion of ANC, not as a wish but possibility.

REFERENCES

Abungu, O. G. (2016). *Traditional Management Systems at Heritage Sites in Africa* (Abungu, O. G., Ndlovu, N., Thiombiano, F. E., Zulu, J., & Varissou, S., Eds.). African World Heritage Fund.

Africa Energy Chamber. (2025). Empowering Africa's Energy Sector for Growth. https://energychamber.org

African National Congress. (2025). The role of Traditional Leadership in South African Governance. Carter, E.

Amghar, A. (2022). Revisiting the contingency theories of leadership: Key features, meanings and lessons. *Revue Marocaine de Gestion et de Société*, *1*, 76–93.

Bannister, P. (2007). Regulating 'Tradition' South African Izangoma and the Traditional Health Practitioners Act 2004. *The Cambridge Journal of Anthropology*, 27(1), 25-61. www.jostor.org/stable/2380708

Bans-Akutey, A. (2021). *The Path-Goal Theory of Leadership*. Academia Letters., DOI: 10.20935/AL748

Barry, S. (2006). Reconciliation: The South African Truth and Reconciliation Commission's contribution to dealing with the past, reconciling and building the nation. *In die Skriflig*, *40*(4), 691–714. DOI: 10.4102/ids.v40i4.365

Bhuda, M. T., & Maditsi, M. E. (2023). Do Traditional Leaders still govern Indigenous communities? Insights from riots and looting in South Africa during COVID-19. *Indilinga*, *22*(2), 137–153.

Bhuda, M. T., & Maditsi, M. E. (2024). Did traditional leaders succeed in restoring law and order during the 2021 riots and looting in South Africa? Insights from Indigenous scholars. *African Journal for Peace and Conflict Studies*, *13*(2). Advance online publication. DOI: 10.31920/2634-3665/2024/v13n2a9

Bhuda, M. T., Marumo, P., & Motswaledi, T. (2021). The Involvement/Non-Involvement of Traditional Leadership in Covid-19 South Africa. *Gender & Behaviour*, *19*(1), 17480–17489.

Bible Hub Online. (2025). Proverbs, 22:28. https://biblehub.com/proverbs

Botha, F. (2025). Market Summary: Impact of Tariffs and South Africa's Economic Challenges Us Tariffs and Reaction from Global Markets. *Investments and Markets Insights*. https://oig-invest.com/wp-content/uploads

British Broadcast Corporation. (2023). Chief Mangosuthu Buthelezi: The man who divided South Africa''. *BBC Newsletter*, www.bbc.com, 08 September 2023.

Chakrabarty, S., & Bass, A. E. (2013). Encouraging Entrepreneurship: Microfinance, Knowledge support, and the costs of operating in institutional voids. *Thunderbird International Business Review*, *55*(6), 545–462. DOI: 10.1002/tie.21569

Chauke, M. T. (2015). The Role of Women in Traditional Leadership with Special Reference to the Valoyi Tribe. Kamla-Raj. *Studies of Tribes and Tribals*, *13*(1), 34–39. DOI: 10.1080/0972639X.2015.11886709

Chiwisa, C. C., & Qutieshate, A. (2024). A Comprehensive Literature Review of Global Leadership Styles: Assessing Universality and Impact on Workforce Motivation. *International Journal for Multidisciplinary Research (IJFMR)* 6(1), January-February 2024. www.ijfmr.com

Chris Hani. (1991). Chirs Hani on Socialism. ENCA News Interview. www.enca.com

Delius, P. (2020). Chiefly Succession and Democracy in South Africa: Why History Matters. *Journal of Southern African Studies*, *47*(2), 209–227. DOI: 10.1080/03057070.2021.1855042

Dickson, R. K. (2023). Analysis of The Traditional Leadership Theories: A Review of Contemporary Leadership Approaches and Management Effectiveness. *Information and Knowledge Management*, *13*(5). Advance online publication. DOI: 10.7176/IKM/13-5-02

Dodo, O. (2013). Traditional Leadership Systems and Gender Recognition: Zimbabwe. *Journal for Gender and Women*, *1*(1), 29–44.

Dong, B. (2023). A systematic review of the Transactional Leadership Literature and Future Outlook. *Academic Journal of Management and Social Sciences*, *2*(3), 21–25. DOI: 10.54097/ajmss.v2i3.7972

Dube, B. (2020). Go and prophesy in your own land: Foreign Prophets and popularism in South Africa. Evoking the need of Jonathanic Theology for Peaceful Resolution in Difference. *Religions*, *11*(42), 42. Advance online publication. DOI: 10.3390/rel11010042

Fleenor, J. W. (2000). Trait Approach to Leadership. In Rogelberg, S. (Ed.), *G* (pp. 831–833). Sage Encyclopedia of Industrial and Organizational Psychology.

Gkika, E. C., Ntanos, S., & Komisopoulos, F. (2022). The impact of global competition dimensions on economic development. www.researchgate.net

Godihno, T.A. (2012). The ANC's Land Reform Policy: An unsustainable and unsatisfactory compromise. *Pretoria Student Law Review,* 6.

Hai, T. N., & Van, Q. N. (2021). Servant Leadership Styles: A Theoretical Approach. *Emerging Science Journal*, *5*(2), 245–256. DOI: 10.28991/esj-2021-01273

Hyeonjung, C. (2021) President Biden and Climate Change: Policy and Issues. The Asan Institute for Policy Studies. https://ssrn.com/abstract=3847183, May 2021.

Inkata Freedom Party. (2018). Address by Prince Mangosuthu Buthelezi MP Inkosi of the Buthelezi Clan Traditional Prime Minister to the Zulu Monarch and Nation and President of the Inkatha Freedom Party. Lekgotla of the National House of Traditional Leaders and the Executive Committees of Provincial Houses. https://ifp.org.za

Jarc, J. (2015). Laissez-Faire Leadership. www.researchgate.net

Jordan, E., Ng'weno, B., & Spencer-Oatey, H. (2020). Developing Global Leaders: Insight from African Case Studies. Palgrave: MacMillan.

Kgatle, M.S. (2021). The balance between theory and praxis in South African Pentecostalism: Patmos Bible School as a model. *Theologica,* 31, 84-98. DOI: 31, 6.DOI: 10.18820/23099089/actat.sup

Klaas-Makolomakwe, N. G., & Raniga, T. (2021). A critical review of the role and functions of Traditional Leaders. *The Thinker: Opinion*, *86*(8), 54–60. DOI: 10.36615/thethinker.v87i2.534

Kobo, F. A. (2019). Spirituality trapped in androcentric celebrity cults in South Africa post- 1994. *Hervormde Teologiese Studies*, *75*(3), 1–7. DOI: 10.4102/hts.v75i3.5593

Korejan, M. M., & Shahbaz, H. (2016). An analysis of the Transformational Leadership Theory. *Revue des Sciences Fondamentales et Appliquées*, *8*(3), 452. DOI: 10.4314/jfas.v8i3s.192

Lebaka, M. E. K. (2018). Music of politics and Religion supporting constitutional values in South Africa. *Studia Historiae Ecclesiasticae*, *44*(1), 1–15. DOI: 10.25159/2412-4265/2343

Lithole, K. D. (2022). Afro-centred Heritage Management Approaches for Mapungubwe World Heritage Site along Vele Colliery Coal Mining in Limpopo Province, South Africa, Doctor of Philosophy in African Studies thesis, University of Venda,www.univen.ac.za

Makgamatha, M. (2024). Makgamatha-2024-analysing-the-pros-and-cons-of-party-coalitions- in-south-africa-unravelling-jacob-zuma-s-rope-a-dope. *Journal of Public Administration*, *59*(3), 531–540. DOI: 10.53973/jopa.2024.59.3.a12

Mathenjwa, L. F. (1991). Zulu Literary Artist's conception of celestial bodies and associated natural phenomena. Doctor of Literature and Philosophy Thesis, University of South Africa. https://uir.unisa.ac.za

Matshidze, P. E. (2013). The role of Makhadzi in traditional leadership among Venda. Doctor of Philosophy Thesis in Anthropology and Development Studies, University of Zululand. https://uzspace.unizulu.ac.za

Mawere, J., Matshidze, P., Kugara, S. L., & Madzivhandila, T. S. (2021). *The Role and Significance of Traditional Leadership in South African Local Governance. In book: Handbook of Research on Protecting and Managing Global Indigenous Knowledge Systems*. IGI Global Publisher., DOI: 10.4018/978-1-7998-7492-8.ch014

Ministry of Foreign Affairs Japan. (2025). Japan-Nigeria Foreign Minisers' meeting'. Tokyo: APO Group Distribution on Behalf of the Ministry of Foreign Affairs Japan. 20 August 2025.

MiSTRA. (2019). *Traditional Leaders in a Democracy: Resources, Respect and Resistance*. Johannesburg: Mapungubwe Institute for Strategic Reflection. www.mistra.org.za

MiSTRA. (2024). *Thucydides Trap or Orderly Multipolar World? Challenges and Opportunities for Africa.* Mapungubwe Institute for Strategic Reflection. www.mistra.org.za

Mohammed, J. (2025). *The rise of African thought Leaders in Global Leadership*. African Leadership Magazine. The Rise of African Thought Leaders in Global Leadership- African Leadership Magazine.

Motasa, K. O., & Nortjé-Meyer, L. (2021). Patriarchal Usurpation of the Modjadji Dynasty: A Gender-critical Reading of the History and Reign of the Modjadji Rain Queens. *Pharos Journal of Theology ISSN, 2414-3324 online 102.* https://orcid.org/0000-0002- 8350-8189 and Doi: https://doi.org/DOI: 10.46222/pharosjot.102.1.8

Mpeiwa, K. D. (2025). Develop a model leveraging from the shared experiences of the communities living in the heritage sites. UNESCO International Conference Paper on Cultural Heritage in Africa: A dialogue on the concept of authenticity 6th-9th May 2025. Nairobi: UNESCO.

Mpofu, B. (2020). South Africa: Re-imagining the rainbow nation. *Sociology International Journal, 4*(3), 67–68. DOI: 10.15406/sij.2020.04.00226

Ndlovu-Gatsheni, S. J. (2014). Global Coloniality and the challenges of creating African future. *Strategic Review for Southern Africa, 36*(2), 181–202.

Nelson Mandela Foundation. (2025). Chapter Ten: Increased Repression and Continued Attack in 1995 and 1996. www.omalley.nelsonmandela.org

Nemashakwe, P. (2021). Evaluation of the State and causes of ineffective leadership in Africa. *The Fountain: Journal of Interdisciplinary Studies*, *7*(1), 1–17.

Newcombe, S. (2014). *Cults: History, Beliefs, Practices*, In (eds) Terry Muck, Harold A. Netland, and Gerald R. McDermott: Handbook of Religion: A Christian Engagement with Traditions, Teachings, and Practices (pp.568-572). Baker Academic

Ngcobo, K. (2025a). Trump presses ahead with 30% tariff on South Africa. Johannesburg: *British Broadcasting Corporation News.* www.BBC.com

Ngcobo, K. (2025b). ''South African president signs controversial land seizure law''. www.BBC.com

Nikolić, G., Kvasic, S. G., & Grbic, L. (2020). The Development of Authentic Leadership Theory. Conference: PAR International Scientific and Professional Conference 'High Impact Leadership'. https://www.researchgate.net/publication

Northouse, P. G. (2010). *Leadership: Theory and Practice*. Sage Publications.

Ntlama, N. (2020). The Attempt to Change the Nature of Succession to the Chieftaincy in the Institution of Traditional Leadership: Mphephu v Mphephu-Ramabulana 2019 7 BCLR 862 (SCA)" PER / PELJ 2020(23) - .DOI: 10.17159/1727-3781/2020/v23i0a6804

Ntonzima, L., & Bayat, M.S. (2012). The role of Traditional Leaders in South Africa - A relic of the past, or a contemporary reality? *Arabian Journal of Business and Management Review* (OMAN Chapter) 1.

Oshame, D., & Maureen, O. E. (2023). Critical analysis of four leadership theories and principle. *World Journal of Advanced Research and Reviews, 17*(01), 1387–1395. DOI: 10.30574/wjarr.2023.17.1.0187

Portal do Governo Mozambique. (2025). *Japan-Malawi Foreign Ministers' meeting*. APO Portal do Governo Mozambique Distribution.

Rabie, M. (2021). One Human Race: Shared Destiny. https://www.researchgate.net/publication/350451969

Ramothwala, P. (2025). State no longer backs Modjadji Coronation. www.sowetanlive.co.za

Ray, D. I. (2003). Rural Local Governance and Traditional Leadership in Africa and the Afro- Caribbean: Policy and Research Implications from Africa to the Americans and Australasia. In Ray, D. I., & Reddy, P. S. (Eds.), *Grassroots Governanc? Chiefs in Africa and the Afro-Caribbean* (pp. 1–30). University of Calgary Press. DOI: 10.2307/j.ctv6gqrhd.5

Robinson, L., Helmus, T. C., Cohen, R. S., Nader, A., Radin, A., Magnuson, M., & Migacheva, K. (2018). *Modern Political Warfare: Current Practices and Possible Responses*. RAND Corporation., www.rand .org/t/RR1772 DOI: 10.7249/RR1772

Sarum, O. O. (2023). Current issues on leadership in Africa: The need for urgent reorientation in quest for a sustainable continent in the 21st Century Technological explosion. www.academia.edu/103775373/ 22 June 2022.

Shomolekae, T. (2024). ANC to work with traditional leaders in preparation for national dialogue. https:// www.polity.org.za

Singh, A. (2021). An Introduction to Experimental and Exploratory Research. SSRN Electronic Journal. DOI:DOI: 10.2139/ssrn.3789360

South African Local Government Association. (2013). Local Briefing: Challenges regarding participation of traditional leaders in Municipal Councils. www.SALGA.org.za

Stiglitz, J. E. (1998). *More Instruments and Broader Goals: Moving toward the Post- Washington Consensus*. UNU World Institute for Development Economics Research.

The Arab Republic of Egypt Presidency. (2025). President El-Sisi speaks with Greece Prime Minister Mitsotakis. Cairo: APO Distribution on behalf of Presidency Arab Republic of Egypt.

Tshikalange, S. (2025). Ramaphosa and King Misuzulu win SCA appeal in Zulu royal dispute. Timeslive Newspaper, https://www.timeslive.co.za/news/south-africa/2025-10-06- ramaphosa-and-king-misuzulu-win-sca-appeal-in-zulu-royal-dispute.

Tshikhudu, E. (2016). Coronation ceremony halted by court''. *Limpopo Mirror Newspaper*, www .limpopomirror.co.za. Accessed 17 October 2025.

Tshitangoni, M., & Francis, L. (2017). Relevance of Traditional Leadership in Rural Community Development amidst Democratic Institutions in Southern Africa: A Critical Review. *Studies of Tribes and Tribals*, •••, 70–83.

Ubink, J., & Duda, T. (2021). Traditional Authority in South Africa: Reconstruction and Resistance in the Eastern Cape. *Journal of Southern African Studies*, *47*(2), 191–208. DOI: 10.1080/03057070.2021.1893573

United Nations. (2025). Co-create solutions to give Africa Stronger role in Global Politics, Financial Institutions, and Debt relief. Secretary-General Antonio Gutteres Address. New York: The United Nations Tokyo Ninth International Conference. 20 August 2025/APO GROUP Distributions.

Writer, S. (2024). A New Dawn for the Rain Queen amid succession disputes. www.mopanitimes.co.za/ 2024/12/a-new-dawn-for-the-rain-queen-amid- succession-dispute/accessed/01 August 2025.

Zondo Commission. (2022). The Zondo Commission: A bite-sized summary. Public Affairs Research Institute. https://pari.org.za/zondo

Zweni, A. G., Faku, E. M., & Yan, B. (2025). 'The Formation of a Government of National Unity in South Africa: Issues, Trends, and Prospects'. *Journal of Nation-building &. Policy Studies*, *9*(1), 145–164. DOI: 10.31920/2516-3132/2025/v9n1a7

Chapter 9

The Role of British Governors in Reshaping Indigenous Knowledge Systems in India

Snehil Gangaprasad Jaiswal
http://orcid.org/0000-0001-8859-9799
Dr. Babasaheb Ambedkar School of Law, Rashtrasant Tukadoji Maharaj Nagpur University, Nagpur, India

Pravina Khobragade
Dr. Babasaheb Ambedkar School of Law, Rashtrasant Tukadoji Maharaj Nagpur University, Nagpur, India

ABSTRACT

Colonial governance in India extended beyond political control to the regulation and transformation of Indigenous Knowledge Systems (IKS). British Governors played a key role in reshaping indigenous health practices, education, language, heritage, environmental knowledge, and local economies through parliamentary laws, executive authority, and provincial powers. While some indigenous practices were selectively documented for administrative purposes, many were marginalized under colonial legal and revenue frameworks. Across presidencies, standardization, regulation, and knowledge extraction reflected a broader system of epistemic control embedded in law and governance. In this context, it's aimed to examine the constitutional position and statutory powers of Governors, highlighting their sector-specific interventions actions and the lasting impact of colonial administration on postcolonial legal structures and contemporary debates on revitalizing Indigenous Knowledge Systems.

INTRODUCTION

Background and Rationale Problem Statement

British officials in India carried out imperial commands through a complicated administrative framework that centralized executive power, legislative power, and discretionary control. Parliamentary statutes such as The Regulating Act of 1773, The Pitt's India Act of 1784, the Charter Acts, and subsequently

DOI: 10.4018/979-8-2600-2890-2.ch009

The Government of India Act, 1858, The Government of India Act, 1919, and The Government of India Act, 1935 delegated authority to the Governor, who was also known as the provincial governor, Lieutenant Governor, or Governor-General, and was leading provincial administration (Jain, 2014). Although sometimes shielded from local responsibility, the Governor served as more than just a ceremonial head until 1947, playing a crucial role in the constitution and exercising broad administrative, legislative, and discretionary responsibilities (Austin, 2018).

Governors played a crucial role in bridging the gap between the provincial and local realities of Indian society and the British imperial policies made in London. During times of civil upheaval, diseases, or hunger, they were able to use their emergency powers to control legislation, administrative rules, revenue, and land settlements. The governors' responsibilities went much beyond simple administrative oversight; they also played an integral part in the creation, categorization, regulation, and application of information on Indian culture. They shaped the recognition, transformation, or marginalization of indigenous activities in the colonial legal system through cultural control, surveys, educational reforms, and commissions pertaining to health, education, language, art, environment, and economic life (Dirks, 2011).

Sustainable development, pluralistic health systems, cultural preservation and postcolonial governance are among the few domains in which Indigenous Knowledge Systems has experienced academically and policy-driven surge of interest recently (Agrawal, 1995). Instead of considering Indigenous Knowledge Systems as passive or inferior replacements for Western epistemologies, IKS are now being recognized as living systems of knowledge that are embedded in cultural practices, ecological interactions and community behaviors (Dei et al., 2000). This increased interest has also involved a re-examination of colonial history in order to better understand the ways in which legal and administrative measures disrupted, appropriated or selectively protected indigenous knowledge. The understudied office of the British-era governor in this respect becomes a significant site of entanglement where indigenous knowledge and colonial power intersected in concrete and lasting ways.

Problem Statement

British Colonialism and knowledge production and imperial rule's impact among Indian society has been studies widely by the various researchers. However, the role of British governors involved with Indigenous Knowledge Systems has received less attention by the various scholars. Most of the previous work has tended to overlook the figure of the Governor in favor of examining colonial rule, missionary activities or the epistemic violence caused by colonial knowledge regimes. As a result, there has been a lack of comprehensive analysis of the Governor's legal and administrative authority in connection to IKS. This authority manifests itself through discretionary orders, legislative assent, policy execution, and institutional favoritism.

This gap is most noticeable in studies that focus on specific industries, such as those dealing with indigenous economies, health, education, heritage, and language. The operationalization of knowledge control on the ground by colonial governance structures and the way governors achieved a balance between imperial goals and practical reliance on indigenous systems are obscured by the lack of a coherent, law-focused study. In order to comprehend the current difficulties in incorporating IKS into contemporary legal and governmental frameworks, it is crucial to fill this knowledge gap (Bteille, 2017).

Research Questions

This chapter is structured around the following research questions that aim to fill the identified gap:

- How much did the British governors of colonial states acknowledge, incorporate, or repress Indigenous knowledge systems?
- How were indigenous economic structures, language traditions, educational institutions, health practices, and cultural preservation impacted by governmental policies and choices?
- How did the Governor's involvement with Indigenous Knowledge Systems manifest itself in policy acts, administrative measures, and legal instruments including statutes, regulations, executive orders, and institutional reforms?

These questions facilitate a critical analysis of the Governor's function as a link for indigenous knowledge and an instrument of colonial control, uncovering trends of marginalization, appropriation, and selective recognition.

Objectives

There are three main goals that this chapter aims to achieve. Its primary objective is to examine the constitutional and legislative framework of colonial India in order to better understand the administrative and legal responsibilities of the British governors in relation to Indigenous Knowledge Systems. The second objective is to analyze the ways in which indigenous knowledge was regulated, changed, or eliminated as a result of gubernatorial policy interventions in the areas of health, education, heritage, language, and economic development. Finally, this chapter aims to assess how these colonial interventions have shaped modern legal views on IKS, pluralism, and cultural autonomy, as well as their long-term effects on postcolonial governance.

These goals helps to provide a historically grounded explanation of why many Indigenous Knowledge Systems need to be revitalized today and how colonial legal-administrative processes still affect policy decisions in independent India.

HISTORICAL OVERVIEW OF THE OFFICE OF THE GOVERNOR IN BRITISH INDIA

Evolution of the Governor's Constitutional Position

The role of the Governor in British India was shaped by a succession of legislative acts that gradually concentrated power and established imperial dominance. The first legislative action by the British Parliament to oversee the management of the East India Company was the Regulating Act of 1773. It granted the Governor-General of Bengal the authority to oversee the administrations of Bombay and Madras (Misra, 1970). Despite its narrow focus, the Act heralded the rise of a unified executive branch that would be tasked with overseeing the finances, administration, politics, and relationships with Indian society.

This authority was further established by the Pitt's India Act of 1784, which established a dual system of governance that separated the Company's business operations from the political control exercised by the Board of Control. The expansion of the Governor-General's authority, particularly in the fields of diplomacy, the military, and civil administration, strengthened his position as an instrument of imperial strategy (Metcalf, 1994). These initial laws proved to be crucial because they formed the fundamental groundwork to the discretionary rule, thus vested the governors with a wide degree of authority to resolve local interests especially those that were related to the institutions and customs of the indigenous (Cohn, 1996).

The Charter Acts of 1813 and 1833 brought about a radical change in the colonial rule. The 1813 Act ended the monopoly of trade enjoyed by the Company and emphasized the administrative responsibility as well as missionary work and the role of the state in education. This is where the interest of the state in the indigenous societies began to grow as the object that needed regulation and reform. The Act of 1833, brought together the legislative powers to the Governor-General-in-Council and a standardized law, which further increased the influence of the Governor as a legislator (Kumar et al., 2005). The subsequent concentration of the power of the legislation allowed the Governors to influence the perception, codification, or supplantation of indigenous standards and epistemologies by means of statutory and regulatory measures.

A watershed moment in colonial administration occurred with the passage of the Charter Acts in 1813 and 1833. The Company's trade monopoly was broken by the 1813 Charter Act, which also encouraged missionary work and governmental engagement in education and placed an emphasis on administrative accountability. Native American communities were increasingly targeted by the state at this time for reform and regulation. A more prominent position for the Governor as a lawmaker was established with the Charter Act of 1833, which established a unified legal system and consolidated legislative power under the Governor-General-in-Council (Jain, 2014). According to Dirks (2001), due to the concentration of legislative authority, governors were able to use regulatory and statutory processes to either recognize or remove indigenous norms and knowledge systems.

The constitutional status of Governors was significantly changed when the Company's powers were transferred to the British Crown with The Government of India Act, 1858. While the Governors were now answerable to the Secretary of State for India for their role as Crown agents, they maintained considerable discretionary and executive power within their respective provinces (Misra, 1970). Although the Government of India Act, 1919 and The Government of India Act, 1935 brought about some changes, such as diarchy and provincial autonomy, the Governors still held important powers, especially in matters of public order, finances, and "special responsibilities" (Austin, 2018). The Governors' power of direct interference in the field of education, health, land politics and forest governance was substantially enhanced by these Acts, which brought them directly under shield from council of ministers in vital areas.

In these constitutional developments, the exercise of authority by the Governor through his utilization of executive, legislative and discretionary powers strengthen their position in colonial policymaking process. Governors were able to exert a great deal of control on Indigenous Knowledge Systems through official legislation, executive decrees, or informal administrative procedures (Cohn, 1996; Dirks, 2011).

Governor's Relationship with Native States and Communities

Indirect rule systems allowed governors to manage connections with Native States and indigenous groups in addition to directly administering British provinces. One demonstration of this policy was the Political Agent and Resident system that permitted British officials to counsel, supervise and if necessary dictate to native rulers in princely states (Metcalf, 1994). The Residents ensured that imperial interests were upheld in governance, taxation and law despite the state sovereignty.

Governors, who were responsible for authorizing treaties, monitoring Residents, ultimately oversaw these agreements and they took actions when instability or suspected misgovernance was detected. Land revenue and forest tenure, judicial administration were typical provisions of treaties, the fields in which indigenous customs and investigations are interdependent. At times, governors might use these instruments to preserve indigenous institutions in order to facilitate their control of the population who were subject to them; elsewhere they would erode their autonomy in the name of 'good governance' or reform (Guha, 2008).

Customary recognition and bureaucratic control were the cornerstones of colonial administration at the village level (Bteille, 2017). Customary authorities, village chiefs, and local consultative organizations were gradually integrated into the administrative structure, frequently serving as intermediaries for taxation and conflict settlement (Bayly, 1997). In reality, indigenous systems were exposed to colonial inspection and reinterpretation through this strategy, which gave the impression of respect. These arrangements were approved by the governors, who were the highest provincial authorities, and it was their job to decide extent to which statutory law would respect or disregard indigenous customs.

Indirect control in princely kingdoms and direct rule in provinces made it possible for governors to interact with indigenous communities in varied ways. It also made it easier to learn about the local economy, social structure, and customs, all of which were crucial for the colonial government to run smoothly.

Role in Knowledge Administration

The Governor's office had an important but underappreciated role in the dissemination and creation of information. Governors served as patrons and supervisors of various surveys, commissions, and ethnographic documentation that sought to systematize information about people, resources, and customs of India (Bayly, 1997). The close relationship between knowledge and government was evident in the governor's control over land revenue settlements, forest surveys, archeological expeditions, and census activities (Dirks, 2001).

There were several uses of administrative and ethnographic records, including the facilitation of taxes and regulation, the justification of policy interventions, and the facilitation of legal codification. It was common for governors to endorsed or initiated commissions to investigate indigenous knowledge, including their medicine, education, and customs, if they thought such information may be valuable for commercial exploitation or better governance (Arnold, 1993). But this was a selective participation. The cultural and epistemological autonomy of indigenous knowledge was often taken away and classed within colonial epistemic frameworks.

Also, governors were involved at the crossroads of many different types of knowledge producers. Colonists' medical officers recorded indigenous peoples' methods of treatment, surveyors plotted out the land and resources, and language experts created dictionaries and grammars (Kumar et al., 2005). The Governor's office controlled and legitimized these activities by determining which kind of knowledge would be institutionalized and which would be ignored.

By exercising administrative control over the creation of knowledge, the Governor demonstrates his role as both a political administrator and as a crucial epistemic authority within the colonial state. Governors were instrumental in the colonial project of knowledge co-optation, which has lasting effects on how Indigenous Knowledge Systems in India are understood today (Dirks, 2011). Governors shaped the recording, interpretation, and deployment of indigenous knowledge.

LITERATURE REVIEW

Colonial Knowledge Production

In postcolonial literature, the link between colonial authority and the creation of new knowledge has been extensively explored. In order to grasp how colonial states contributed in the construction of the "Orient" as an object of knowledge, government and control through systematic representation and epistemic stratification, Said's concept of Orientalism is crucial (Said, 1978). Said contends that colonial knowledge was inherently biased, serving to justify imperial involvement by portraying indigenous civilizations as unchanging, unreasonable, and in dire need of change (Said, 1978). The British administrators' classification, codification, and regulation of indigenous knowledge systems, traditions, and laws within the Indian context may be better understood using this framework.

(Guha, 2008) and other members of Subaltern Studies moved the focus from elite colonial narratives to those told by colonized people themselves. This corpus of work argues that colonial history was inherently hierarchical and fails to adequately address the systemic marginalization of indigenous perspectives and ways of knowing. Aside from resistance, peasant consciousness, and nationalist historiography, the institutional mechanics of colonial governance, including the particular role of Governors, have been comparatively neglected by subaltern scholars, despite their significant work in exposing the asymmetries of colonial knowledge (Chakrabarty, 2007).

The connection between power and administration is best illustrated by Bernard S. Cohn's writings on colonialism and knowledge (Cohn, 1996). According to him, the British state's desire to understand India during colonial authority was a "cultural project" that included censuses, surveys, legal codification, and ethnographic research. Administrative procedures solidified social and cultural realities into permanent categories for control, as demonstrated by his concept of governmental archive ethnography. Many of these knowledge-producing exercises were monitored and legitimized by governors, making this scholarship especially pertinent to understanding the governor's office.

A large amount of research on the codification of British traditions, languages, and behaviors supplements these theoretical viewpoints. Studies on the development of Anglo-Hindu and Anglo-Muhammadan law, the standardization of vernacular languages, and the control of customs shows how colonial law subordinated indigenous norms to Western legal rationality while selectively acknowledging them (Derrett, 1999; Washbrook, 1981). However, this literature frequently views codification as an impersonal bureaucratic procedure, failing to adequately take into consideration the governors' discretionary influence over these results at the provincial level.

Indigenous Knowledge Systems in South Asia

The South Asian Indigenous Knowledge Systems scholarly enquiry has cast some light on the richness, variety, and persistence of knowledge systems during both pre-colonial and colonial periods (Dei et al., 2000). There has been an extensive literature reporting the Ayurveda, the Siddha, and Unani systems in the health sector, which have a complicated theoretical basis, pharmacological knowledge, and community-based medical practices (Arnold, 1993; Leslie, 2022). Medical historians have shown that the colonial biomedicine did not completely replaced traditional healing systems by examining medical educational reforms and legislation.

According to studies on native education, gurukala, pathshalas and madrasas were thriving educational institutions before the invasion by the British. These institutions were deeply embedded in the local economy, social structure and common knowledge practices. They served as a part and parcel of daily living instead of being exclusive hubs of education. Contrary to colonial portrayals that depicted them as stagnant or ineffective, pioneering studies have demonstrated that these institutions were active, adaptive, and contextually relevant (Kumar et al., 2005). Colonial educational reforms often served to destroy these structures, while extending access to Western knowledge, by removing the possibilities of patronage and rearticulating legitimate knowledge through a state-mandated curriculum.

In addition to health and education, anthropological and cultural studies have explored indigenous material culture, performing arts, oral traditions, and traditional knowledge (Blackburn, 2016; Srinivas, 2003). The studies highlighted the close relationship that exists among indigenous knowledge, livelihood, and cultural identity. It is widely recognized that the evolution of cultural practices throughout imperial control was greatly influenced by colonial interventions in these areas, such as museumization and selective patronage (Dirks, 2011).

Despite the plenty of research on Indigenous Knowledge Systems, much of it discusses the structural effects of colonization without investigating into the specific roles played by the legal and administrative actors who oversaw these changes in practice (Cohn, 1996).

Governors' Role in Cultural Management

Administrative ethnography, surveys, and historical preservation campaigns had a significant influence in influencing British cultural management policies in India. Research into the Archaeological Survey of India, censuses, and anthropological compilations shed light on the efforts of colonial administrators to record and save certain elements of Indian culture. They frequently presented preservation as a responsibility to civilization (Chakrabarty, 2007). These initiatives were intricately linked to governance, which allowed for taxes, social control, and monitoring.

Regardless, the role of governors in these processes has received little theoretical attention. Previous research has ignored the Governor as a constitutional figure who authorized, oversaw, or limited these activities related cultural management. Governors are frequently depicted in the literature as passive representatives of imperial policy rather than as active decision-makers with choice over issues pertaining to culture and knowledge (Metcalf, 1994).

Governors' support for museums, educational institutions, and cultural surveys is acknowledged in few of historical studies; however, these narratives rarely contextualize these actions within a more comprehensive legal and constitutional framework. Consequently, there has been a lack of comprehensive analysis of the Governor's dual position as a regulator and curator of indigenous knowledge. This is especially true when it comes to initiatives that target specific sectors such as health, education, heritage, language, and economic life (Dirks, 2011).

Gap Identification

The aforementioned review demonstrates a substantial gap in the existing research. The field of administrative ethnography, colonial knowledge production, and Indigenous Knowledge Systems has been extensively investigated; however, there is a lack of concentrated analysis on Governor-led interventions in various sectors. More specifically, the role of governors in shaping the Indigenous Knowledge Systems in colonial India through the use of legal instruments, discretionary powers, and administrative procedures has received less attention.

This gap is most noticeable in interdisciplinary studies that aim to establish a relationship between knowledge, governance, and law. This chapter aims to bridge the gap and provide an in-depth understanding of colonial co-optation of knowledge by emphasizing the Governor's role as a prominent legal-administrative actor. This research not only enhances historical knowledge but also illuminates contemporary debates over the ability of postcolonial governance frameworks to rejuvenate and integrate Indigenous Knowledge Systems.

RECOGNITION OF INDIGENOUS KNOWLEDGE IN HEALTH SYSTEMS

Governor-led Medical Reports and Surveys

Medical reports, surveys and institutional investigations conducted under provincial authority were significant vehicles through which the office of the Governor managed the relationship between British and indigenous health systems in India. The Governors appointed medical boards and committees to record indigenous medicinal practices, pharmacopoeias and therapeutic modalities from the early nineteenth century. The motivation for such efforts was in part administrative pragmatism, particularly in colonial India, where western medicine proved inadequate to cope with the scale of disease, famine and public health disaster in colonial India (Arnold, 1993; Harrison, 1994; Mitra, 2012).

In a number of presidencies, such as Bombay, Madras and Bengal, medical surveys recorded the therapies applied by Ayurveda, siddhaand Unani. Herb use, surgical techniques and disease categorization were commonly described in these surveys (Leslie, 2022). Many of the reports presented to Governors often recognized the extensive dependence of local populations on indigenous healers and the ingrained character of these systems within community life. In the Bombay Presidency, colonial authorities periodically considered collaboration with Ayurveda practitioners to improve limited colonial medical services, especially in rural and semi-urban regions where European physicians were rare.

However, this kind of documentation was based on a system of classification that took useful information, mainly botanical and medicinal expertise, and used Western scientific standards to judge indigenous systems. Governors, as the sanctioning and supervisory authority, played a significant role in establishing which forms of indigenous medical knowledge were regarded administratively important and deserving of restricted institutional recognition.

Colonial Legal Frameworks Affecting Indigenous Health

The Governor played a pivotal role in determining the destiny of indigenous medical systems through the legal regulation of health throughout colonial administration.

During times of public health emergency, colonial executives were granted extraordinary powers to control population movement, enforce medical interventions, and override customary practices by statutes such as the Epidemic Diseases Act, 1897, provincial vaccination enactments, and quarantine regulations (Arnold, 1993; Harrison, 1994). Governors were authorized to issue executive orders and restrictions under these statutes, frequently avoiding local approval or conventional therapeutic techniques.

The late nineteenth and early twentieth centuries were marked by epidemics that triggered colonial governors to have far-reaching prerogatives to pass segregation laws, require vaccinations, and issue quarantine orders. These interventions often negatively judged indigenous health practices by labeling local health practices as unscientific or dangerous. However, colonial powers periodically employed locals as healers to announce communal medical programs or provide basic health care, which testifies to a partial dependence on indigenous epistemologies (Leslie, 2022).

The legislation was both a site of limitation and a means of discriminating intercourse. Governors used their statutory authority to decide the permissibility, control, or repression of aboriginal practice according to the ever-increasing demands of the public health and the imperial power.

Duality of Recognition and Suppression

The period of health governance during the colonial period, managed by gubernatorial office, evinced non-recognition and recognition inextricably. The indigenous medical knowledge, especially in the botany and pharmacology, was incorporated into the colonial medical practice; the herbal remedies and phytotherapeutic traditions were listed, analyzed, and, in some cases, were included in the colonial pharmacopoeias (Cohn, 1996). This method helped to develop a medical epistemology on a global scale, although it sometimes pulled medicines out of their cultural and epistemic scale.

On the other hand, the indigenous medical systems were sidelined because they were holistic. The institutionalization of scientific medicine by creating licensing conditions, professional regulatory systems and medical schools created structural barriers to the native practitioners. The western-trained, licensed doctors were given special treatment by the licensing laws of the government, making the traditional healers either an obscure issue or an illegitimate practice (Arnold, 1993). Through the proclamation of such laws and organizations, governors supported the dominance of knowledge by placing indigenous medicine as secondary or inferior.

This epistemic conflict is an example of a larger trend of colonial seizing of knowledge. The practices of indigenous people were only recognized as long as they were used to benefit the imperialists. Conformity to administrative or political goals made selective recognition; but intellectual independence and institutional independence was often destroyed along the way. Colonial interaction in the indigenous knowledge was not about real recognition but reconfiguration and control of indigenous knowledge in an imperial structure.

Critical Assessment

A critical reflection on the role of the gubernatorial involvement in the sphere of indigenous health care shows a fine balance between the control and practical dependence. Governors understood that the colonial health system was underprepared to meet the healthcare demands of a diverse population of large numbers of people. This awareness was expressed in periodical crises as either an implicit or explicit reliance on local-based indigenous healers, midwives, or community-based providers, who had credibility and authority in their community (Leslie, 2022).

This kind of reliance did not result in long-term institutional support or legal acceptance. The indigenous practitioners were used during emergent periods, but when the state was consolidating, they were dispossessed. The long-term effects of such procedural episode have had catastrophic effects. Colonialism shattered India's diverse medical landscape, privileging biomedicine and relegating indigenous modalities to positions of informality (Arnold, 1993).

In post-colonial India, attempts to revive and incorporate traditional medical systems like the formal recognition of AYUSH–have had to negotiate the historical baggage of partial recognition and systemic suppression (Anuja Priya, 2026). Understanding the impact of the Governor on colonial health policy is critical as background to contemporary debate on medical pluralism, regulatory reform and the revitalization of Indigenous Knowledge Systems within a modern legal framework.

INDIGENOUS HERITAGE, TRADITIONS, AND CULTURAL PRESERVATION

Archaeology and Heritage Surveys

The colonial encounter with the indigenous such cultural heritage in India was legally mediatized through archaeological surveys and historical conservationism through the approval of provincial Governors. The establishment and expansion of the Archaeological Survey of India (ASI) in the mid-nineteenth century illustrate this approach (Chakrabarti, 2003). It was the Governors who had issued permits for excavations, disbursed money and supported a policy of preservation, in such areas with famous historic-religious monuments.

In the presidencies of Madras and Bombay, Governors advocated the surveying and documenting of temples, monuments, inscriptions and ancient sites with a view to fulfil a duty of both civilizational as well administrative for the colonial state (Guha, 2008). This intervention included controlling access to religious sites, monitoring repair activities and overseeing festivals associated with sacred sites. While on paper these measures were proposed to protect cultural property, in reality they redefined the indigenous heritage as something that was owned by the state and subject to colonial conceptions of historical significance and aesthetic value (Chakrabarti, 2003).

The protection of temples and monuments under the provincial authorities generally involved introducing homogenizing administrative rules that replaced local stewardship. Community-based forms of maintenance, ritual use and knowledge transfer were left for archaeological concerns; the living tradition was turned into curated objects. As the ultimate authority of approval, Governors had a significant impact on reshaping the relationship between indigenous communities and their cultural landscapes.

Regulation of Cultural Practices

Related to programs of preservation, colonial government instituted control mechanisms that substantially affected indigenous cultural practices. Legislative framework included criminal law and social reform that targeted acts deemed to be inconsistent with colonial morality, and civilizational norms (Mani, 1998). This is demonstrated through its level of interception, as in the case of anti-nautch laws which intended to forbid performing ancient dance forms associated with temple

(Soneji, 2012). Curbs on access to temples, ritual acts or festival organization were often justified as precepts of public order, morality, or sanitation.

Significant power was exercised by the Governors in these areas through prescription of legislation and discretion. Provincial laws that influenced cultural and religious matters included those for which governor's consent was required, a way which enabled governors to interpret the legal definition of "tradition" (Metcalf, 1994). More generally, such policies were advertised as reformist or protective, while frequently they have eroded indigenous systems of patronage, performance and knowledge transfer.

This disciplinary approach established a larger colonial predilection to separate a culture out of its socio-economic context. Governors contributed to disfranchising the community through the expression of criminal laws and administrative laws in relation to cultural practices and undermined the legitimization of indigenous knowledge systems expressed in rituals, performances and oral traditions (Guha, 2008).

Case Studies

The administration of the Ajanta Caves by the Bombay Presidency is a good example of the dual role of guardians and deniers of heritage (Chakrabarti, 2003). To preserve and safeguard the caves, local scholars were granted the authority to compile and conserve the caves with the approval of those in government, which later gained international fame as being of great archaeological value. At the same time, conservation laws in colonial times limited the right of indigenous access and use in ritual practices, transforming the caves into cultural artifacts of historical and tourist attractions, not as a part of a living landscape. The office of the Governor therefore acted as an intermediary between international heritage rhetoric and local cultural demands, which in most cases sided with the former (Dirks, 2011).

An example of this is the colonial rule overlapping with the native knowledge of Assam in the setting up of tea plantations and the administration facilities that they had (Guha, 2008). Rules on land use, labor distribution and access to forests as issued by governors became pivotal elements that determined the connection between tribal communities and their nature. Whereas ethnographic writings retain field-based ecological knowledge of the aboriginal people, colonial economic interests changed this favor to the plantation agriculture. Indigenous practices were carefully documented and stored, but at the same time were not part of daily social and institutional life, and this is where knowledge was written and categorized and not properly supported in the context of its origin (Agrawal, 1995).

Critical Reflection

An elemental critique of the colonial past rule can identify an ironic form of reasoning that preservation is a form of control to dictate culture. Governors encouraged archaeological and cultural projects that preserved the specific monuments and customs, despite the fact that they were accompanied by reinterpreting the native heritage to conform to the colonial system of administration and knowledge. Preservation in this respect was also related to surveillance, control and re-distribution of power within the community.

This contradiction explains the reasoning of colonial epistemic extraction. The culture of indigenous people was never exterminated completely or worshipped, but rather carefully selected, re-used and manipulated with certain purposes. Modern traces of these approaches can be found in the present-day discussions around heritage management in India, where there are still conflict points in the approaches toward conservation at a state level and community involvement, or the renewal of the status of living traditions (Dei et al., 2000). The role of the Governor in this historical process is important to develop more holistic and culturally resonant strategies of protecting the heritage after colonization.

INDIGENOUS KNOWLEDGE SYSTEMS IN EDUCATION AND RESEARCH

Pre-colonial Education Models

Prior to the establishment of the colonial rule by the British, education in India was influenced by numerous local institutions which had their foundations in the social, religious and vocational practices. The Gurukuls were residential learning institutions that were concerned with knowledge spread through oral tradition, developing moral values and teaching philosophy, science, and rituals (Sharma, 2005). In eastern India, the focus was on logic, grammar and jurisprudence. Pathshalas offered fundamental training in local languages, mathematics and applied skills to meet the needs of a wide range of population (Dharampal, 2021).

These institutions were supported by the community, religious endowments and local governance, which made sure that their curriculum and courses were relatable and acceptable by the society (Sharma, 2005). The importance of indigenous epistemologies in creation and sharing of knowledge is that it focused on contextual education, morality and passing of knowledge across generations (Zaman, 2007). However, initial colonialists considered these practices as old and poor compared to the European patterns of education.

Governors' Policy Interventions

Interventions that redirected the epistemic foundations of learning in India strongly influenced colonial educational policy. English education and Western knowledge were explicitly favored by Macaulay in his Minute on Education, which argued that European literature and science were superior to local learning traditions (Macaulay, 1862). Although presented as a policy proposal, its execution was largely dependent on the power of the governors, especially at the provincial level where the governors had control over the education department, institutional approvals, and funding distribution.

The centralization of education, promotion of English as a medium of teaching, and introduction of uniform curriculum and examinations is specifically noted in Wood's Despatch of 1854 (Ghosh, 1975). In carrying out these directives, provincial governors were instrumental in establishing educational institutions including as universities, colleges, and schools for training the teachers. The Hunter Commission, which was established in 1882, strengthened the role of the state in education by placing an emphasis on elementary school and centralizing bureaucratic control (Nurullah & Naik, 1951). Frequently favoring administrative convenience and imperial preference over indigenous educational autonomy, governors determined the extent and pace of educational change through executive decrees, grants in aid schemes and budgetary jurisdiction.

Recognition vs. Replacement

The colonial education system as managed by Governors revealed a perpetual tension of minimal recognition and systematic destruction of Indigenous Knowledge Systems. Some of the traditional institutions such as Sanskrit colleges and Oriental learning institutes was amongst those which received instrumental funding support. The funding support for the institutions was always conditional. Indigenous curriculum were frequently customized to conform to Western academic standards, resulting in the erosion of their epistemological foundations and pedagogical methodologies (Dharampal, 2021).

Simultaneously, European systems of thought were endorsed as the framework for state regulated education. The teaching medium of the English language was established as the most important channel for administrative employment, career advancement and social mobility, marginalizing indigenous languages and educational practices (Macaulay, 1862). Through endorsing and consolidating a hierarchy of knowledge via policy implementation, Governors facilitated the gradual dislodging of conventional institutions. The outcome was a type of pedagogical policy, which respected Western knowledge, and devalued indigenous knowledge as something peripheral and symbolic.

Impact Analysis

The colonial educational project combined to produce grave impact on the indigenous knowledge systems. Withdrawal of patronage, bureaucratic constraints and curricula sidelining reduced the status of the prospects for gurukuls, madrasas and pathshalas. Knowledge transmission networks collapsed, academic communities lost support of the institutions they represented, and local knowledge traditions became further isolated from a formalized educational system (Sharma, 2006).

Hybrid systems of education, at the same time emerged on account of colonial domination. The curriculum having blend of Indigenous knowledge and Western education were utilized for teaching at Sanskrit colleges, Orientalist centres and Anglo-vernacular schools (Zaman, 2007). These institutions showed both strength and constraints, preserving bits of indigenous knowledge within frameworks largely influenced by colonial forces.

Table 1 illustrates how executive gubernatorial directives turned imperial educational laws into life-world changing new directions in Indigenous Knowledge Systems

Table 1. Colonial Educational Policies and Their Impact on Indigenous Knowledge Systems

Policy / Intervention	Year	Governor's Role	Impact on Indigenous Education
Macaulay's Minute	1835	Implementation and funding oversight	Marginalization of indigenous curricula
Wood's Despatch	1854	Institutional establishment	Centralization and standardization
Hunter Commission	1882	Policy execution	Expansion of primary education, reduced autonomy

Most of these impact in educational sector were driven by the governor. Governors were empowered to influence the type of information that may be preserved, modified and supplanted through their statutory authority, financial control and administrative discretion (Metcalf, 1994). The role of the Governor in shaping educational outcome was crucial. The impact of colonial history continues to influence today's conversations about educational reform, curriculum development, and the revitalization of Indigenous Knowledge Systems in postcolonial India.

REVITALIZATION OR DECLINE OF INDIGENOUS LANGUAGES

Language Policy Under Governors

During the colonial rule, the policy on language was one of the major sites for interactions between power and IKS as Governors had much sway over linguistic changes and administrative decisions (King, 1998). In the early nineteenth century, a major change was the gradual shift from Persian, the long-standing administrative language of Mughal rule, to vernacular languages in courts and local administration. The shift happened gradually, leading to a significant change in the language and administrative framework of governance. The transition was directed and superintended by the governor who implemented this change in the understanding that it would lead to better administrative effectiveness, would result in a wider spread of colonial justice and facilitate greater ease of tax management (Bayly, 1997).

By means of departmental regulations and administrative circulars, provincial governors actively encouraged vernacular language use in education, trials and tax declarations. The promotion of printing presses in Indian languages during this time resulted in the widespread distribution of textbooks, legal manuals, and religious literature (Bayly, 1997). Instead of promoting the independent growth of indigenous language traditions, these changes mainly supported the growth of public discourse and helped increase literacy rates. They stayed closely aligned with the main goals of colonial governance, especially in terms of administrative control, standardization, and oversight.

Provincial language committees were constituted to standardize scripts, terminology, and official usage to solve the problem of linguistic diversity. These committees would be the linguistic aristocrats who determined which forms of speech were "official" or "standard," and which to be neglected as informal. Despite initial impressions that vernacularization would make language use more accessible to everybody, it actually established new social stratifications among indigenous languages. This allowed more people to participate while simultaneously highlighting the differences between legitimate and authoritative figures (King, 1998).

Codification and Standardization

Colonial support and administrative interests were also centrally implicated in the process of standardization and codification of indigenous languages. Colonial scholars, missionaries and Indian middlemen compiled dictionaries, lexicons and grammars with the support of the government. Funding, publication, and institutional support for these efforts were supported by governors to the extent of their perceived utility for administration, education, and legal codification.

Literary, aristocratic, or similarly upper-class population were frequently preferred when context-dependent languages were transformed into standardized forms. (Pollock, 2009). The bureaucratic demands of the colonists resulted into codification standards that were not in consonance with oral tradition, regional dialects and performative linguistic practices. Linguistic knowledge was accessed, compiled and stored, leading to increased administrative efficiency but narrowed linguistic variety and epistemic diversity (Cohn, 1996).

As an epistemic gatekeeper, the Governor was instrumental in decisions that codified languages, officially recognized scripts, and chose standardized vocabularies. The impact of this decision reverberates today through cultural identity, educational equity and the transmission of indigenous knowledge encoded in languages.

Tensions

There was a contradiction in the colonial language policy of encouragement of English as an administrative rule while acknowledging vernaculars for limited governance at lower levels. The efficiency of English as the medium of doing business and law combined with elite communication created a caste system that propagated western thinking (Macaulay, 1862). This top-down hierarchy was strengthened by the introduction of English as the legislative and executive business language, and the encouragement of the state of education in English.

At the same time, the vernacular languages were accepted or even promoted at the lower level of administration, leading to the establishment of the dichotomous hierarchy of language. This system aided the status Quo in the colonial system but was very divisive to the society, because mastery of the English language also came with social status, higher chances of employment and mobility (Pollock, 2009). The official recognition was limited to the fields of control and therefore the native languages did not have the capability of producing high-level scholarship and intellectual activity.

Long-term Legacy

Modern language identities and literary tradition could be related to the spread and standardization of the vernacular languages under the governor support. During the late colonial era, when the print culture was spreading and scripts were standardized, the language based world of the public was formed. These occurrences not only allowed more people to discuss specific issues, but also gave way to efforts of social reforms and they also contributed a lot in bringing about the nationalist mobilization (King, 1998).

Conversely, colonial language dominances still dominate postcolonial state policies and educational systems as the dominant role of English in the law, government and tertiary education remains overwhelming. The modern efforts to identify, safeguard and recover indigenous languages in India are still affected by the policies that came into effect during the colonial era. Though these previous actions helped some languages to become visible and have institutional backing, they provided structural boundaries and subjugations that continue to play a role in influencing the manner in which languages are perceived, instructed, and maintained.

The involvement of the Governor in the language policy formation depicts a larger pattern of colonialism in knowledge. The native languages came back and were standardized, but their application was highly supervised. Their intellectual autonomy and traditions of deeper knowledge were inhibited, and became more subject to the requirements of the colonies, by the preservation and administrative use of these languages (Bayly, 1997; King, 1998).

Table 2. Language Policy Interventions and Their Impact on Indigenous Languages

Policy Instrument	Governor's Role	Outcome for Indigenous Languages
Vernacularization of administration	Executive endorsement	Expanded use, regulated standardization
Language committees	Institutional oversight	Codification, hierarchy of dialects
Promotion of English	Policy prioritization	Linguistic stratification

Table 2 brings together major interventions in language policy and highlights their long-term implications for Indigenous Knowledge Systems.

INDIGENOUS MUSIC, DANCE, AND PERFORMING ARTS

Governor-led Patronage

A politics of selective patronage determined the ways in which provincial Governors identified themselves with local performance traditions, music and dance. Most governors believed in conserving and institutionalizing art forms that they deemed to be classical in accordance with the colonial aesthetic principles and orientalist research (Chatterjee, 1993). This assistance resulted in the creation and financial support of museums, cultural shows and archives collections to help conserve the Indian traditions of music and dancing in the past. Although these attempts were supposed to safeguard the cultural heritage, they also practiced a certain methodology of selecting, framing and formalizing some traditions as opposed to others (Guha, 2008).

Governors assisted cultural organizations that collected and preserved items like books, instruments, sculptures, and pictures of performing traditions. Even though these programmes helped to preserve some forms of art and also gave them official status, they also re-packaged the indigenous performances into sanitized and politically neutral plots that separated them out of their ceremonial, social and economic contexts (Soneji, 2012). The colonial epistemic, which placed the importance of documentation more than lived experience, dignified the existence of certain traditions but recoded their meanings.

Regulatory Frameworks

Besides patronage, the legal systems that were introduced through the colonialism also had great negative effects on economic and social status of performing societies. Policies like the Criminal Tribes Act of 1871 and anti-nautch laws were determined to enroll and supervise those tribes that had a track record with regard to dancing and performance (Soneji, 2012). Although these policies were deemed as legitimate on the grounds of the public morality, social reform and colonial decency, they ended up criminalizing age-old traditional forms of practices that had played a central role in sustaining indigenous knowledge and social identity (Dirks, 2011).

The governors had a considerable influence in this area through legislative approval, rule-making powers, and executive enforcement. Determining the boundary of legally authorized cultural expression was in the hands of provincial governors, who were obliged to legalize legislation that controlled performance customs. The tendency of colonization to equate governance with the moral control weakened the freedom and legitimacy of native cultural societies by giving the criminal and police law the effects on artistic works.

Dual Impact

There arose a paradoxical situation on the issue of indigenous performance traditions. The colonial state offered patrons and encouragement to the particular type of art, but at the same time, it formulated laws which curtailed or even made most of the traditional performances illegal. The classical conventions which were based on either courts, temples, or upper-caste traditions were those which could be more easily preserved with the aid of institutions and recognized officially (Chatterjee, 1993). Over time, these forms changed to align with colonial views on propriety, professionalism, discipline, and moral respectability. They were slowly recognized as symbols of national culture and progress, while other performance traditions were pushed to the sidelines.

In contrast, indigenous community-based performance traditions and folk forms were stigmatized and subsequently declined. The economic and social bases of such performance communities were weakened by such divestment and legal stigmatization following laws such as the Criminal Tribes Act, 1871 (Cohn, 1996). The traditional knowledge transmission systems or indigenous performing arts ecosystems where learning is based on oral tradition, apprenticeship and embodied practice were destroyed.

This irony is a reflection of colonial appropriation of knowledge. Governors served as patrons of culture and had the power to choose the creative forms to preserved and to suppress. These ideas are still being discussed in modern discussions about cultural policy, heritage legislation, and how to bring Indigenous knowledge systems back into the spotlight in the performing arts.

Table 3. Gubernatorial Interventions in Indigenous Performing Arts

Mode of Intervention	Governor's Role	Impact on Performing Arts
Cultural patronage	Institutional support	Preservation of elite traditions
Criminal legislation	Legislative assent	Marginalization of communities
Museumization	Epistemic framing	Detachment from lived practice

The Table 3 summarizes the contrasting effects of gubernatorial patronage and regulation on indigenous music, dance, and performance traditions.

COLONIAL POLICY FRAMEWORKS AND INDIGENOUS KNOWLEDGE

Statutory Interventions

Statutory interventions were crucial to the colonial administration of India in reshaping the indigenous relationships to the land, forests, and communal forms of government. Legislative reforms like the Indian Forest Acts of 1865, 1878, and 1927, as well as land revenue codes and regulations pertaining to tribal administration, were crucial in this change because they recast Indigenous Knowledge Systems as things to be regulated instead of as independent systems of government (Gadgil & Guha, 2014). By using their executive and discretionary powers, provincial governors were instrumental in making these legislation work at the local level.

The Indian Forest Acts of 1865, 1878, and 1927 profoundly transformed indigenous ecological knowledge systems by establishing governmental authority and ownership over forest resources. Traditional forest activities including shifting agriculture, grazing, hunting, and harvesting were restricted by these acts, which divided forests into three types: reserved, protected, and village (Guha, 2008). The colonial forest policies interrupted the ecological knowledge of the indigenous people which was deeply rooted within the community management systems and traditional practices. These local cultures instead of being supported were commercialized and became a priority of state-controlled conservation and conservation over traditional epistemologies and forest governance (Scott, 2020). Governors established a formal system of colonial environmental control in their role as provincial heads, through administration of forest departments, signing of subsidiary legislation and allowing the coercive regulations to be implemented.

Community land management was eliminated by colonial land revenue codes that put in place individualized property and assessment of the state. These introductions upset the indigenous agricultural knowledge transfer and sharing of the resources. The regulations applied to tribal governments placed members of aboriginal societies under different regulations that were supposedly created to assist them but were in fact used to spy and control them (Gadgil & Guha, 2014). Governors had a plethora of resources to enforce these laws and determine the extent, to which statutory framework or customary governance systems are given weight in the governance structures (Misra, 1970).

Knowledge Extraction and Control

Ethnographic surveys were not just academic exercises; they were used as practical tools for taxation, social control, and administration. Collecting detailed information about communities became a central part of colonial governance, allowing the state to monitor, classify, and regulate society more effectively (Cohn, 1996). Administrative ethnographers studied traditional resource management patterns, social strata, occupational practices and customary law. Governors frequently provided funding, supported and even sanctioned those surveys as they contributed to administrative efficiency and rational management of budgets.

In order to gain knowledge, indigenous people had to make their practice-based, contextualized and flexible forms of knowledge fit into fixed categories in which the bureaucratic rule could operate (Scott, 2020). Ethnographic knowledge also assisted colonial authorities in more precisely governing indigenous communities in land settlement, forest regulation and even criminal classification. However, this shift turned indigenous knowledge into an administrative statistic, separated from lived experience and wrenched out of its cultural and epistemological moorings.

In their capacity as supervisors of survey operations, census exercises, and departmental reporting, governors were instrumental in establishing the credibility of this regime of knowledge. They structured a governance paradigm based on categorization, monitoring, and control by incorporating ethnographic results into laws, executive orders, and administrative practice (Scott, 2020). Therefore, the colonial state had epistemic and economic reasons to extract indigenous knowledge.

Governors as Policy Agents

In addition to enforcing laws as written, governors had an active role in shaping policy by deciding how to apply and interpret codifications of customary law. While colonial legal systems did incorporate indigenous practices into formal law, it was selective and mediated by colonial legal logic; this was most noticeable in areas of personal law, land tenure, and community governance.

The recognition, modification, or displacement of customs was impacted by governors through legislative assent, executive rule-making and administrative interpretation. The flexibility of the executive authority of the governors enabled them to strike a compromise between practical administrative demands and the greater imperial goals. Traditional practices in certain situations were retained as they aided governance or helped make the society stable. In other situations they were altered or eliminated whenever traditional practices became inconsistent with the economic interests or moral values of the colonial power (Misra, 1970).

The role of the governors was not limited to the emulation of the policies they were to implement, but they also had the proactive modeling of the legal terrain of Indigenous Knowledge Systems. Governors helped to institute colonial knowledge hierarchies through the legal and administrative machinery by determining the translations of custom practices into formal law. By so doing, they served as judges of the native knowledge on which tradition would undergo formalization and how it would be reformulated in a colonial legal context (Dirks, 2011; Scott, 2020).

Table 4. Colonial Statutory Frameworks and Their Impact on Indigenous Knowledge

Statutory Framework	Governor's Role	Impact on Indigenous Knowledge Systems
Forest Acts (1865, 1878, 1927)	Oversight of forest administration and enforcement	Marginalization of ecological knowledge and community resource management
Land Revenue Codes	Approval of revenue settlements and regulatory rules	Disruption of communal land practices and agricultural knowledge
Tribal Administration Laws	Supervision of protective and regulatory regimes	Reconfiguration of indigenous governance structures

Table 4 demonstrates how the legal actions that were established with the influence of gubernatorial power changed the indigenous knowledge systems into the field of control of the colonists. It shows the ways in which the law was turned into a tool of removing, rearranging, and exerting control over existing local knowledge cultures in the larger context of imperial rule.

TOURISM AND CULTURAL ECONOMY UNDER BRITISH GOVERNORS

Early Heritage Tourism

Colonial India's heritage tourism emerged as a byproduct of state-led initiatives to promote sites of historical significance (historical, natural and cultural) within and beyond the country as one of the valuable imperial assets. Governors promoted religious sites, archaeological monuments, and ancient ruins as places of discovery and preservation, often presenting them as cultural attractions for European officials, scholars, and visitors (Cohn, 1996). These efforts were framed as acts of protection and guardianship. However, they also reshaped indigenous knowledge systems through colonial ways of interpreting history and heritage. In doing so, colonial rule positioned itself as the rightful custodian of India's past and its historical records (Said, 1978).

Hill stations like Shimla, Ootty and Darjeeling clearly demonstrate the concept of heritage tourism. These places, which were built and enlarged with the support of the governor, served as the administrative headquarters, a haven for leisure, and a representation of power (Kennedy, 2023). Hill stations were developed as climate getaways, but in reality, indigenous landscapes were altered and local populations were displaced. In order to accommodate administration and tourism, governors legislated the purchase of land, the construction of infrastructure, and regulatory regimes that placed colonial planning rationalities ahead of indigenous spatial knowledge.

Heritage tourism was further institutionalized through the display of antiques, crafts, and cultural practices to imperial audiences at colonial exhibitions and fairs, which were often arranged with the support of governor (Dirks, 2011). These events had the effect of reducing indigenous culture to a commodity, while simultaneously establishing value systems in which colonial interpretations took precedence over indigenous understandings.

Use of Indigenous Cultural Sites as Revenue Sources

In addition to symbolic representation, Governors increasingly acknowledged the economic potential of indigenous cultural places as revenue sources. Provincial laws were authorized by the governor and put in place to collect an entrance charge from visitors to historical sites, museums, and archeological sites. The money collected would typically go toward administrative costs or site maintenance (Metcalf, 1989). Although these actions were vital for conservation, they changed traditional ways of accessing, caring for, and using cultural property, and also commercialized it.

Governor-approved restoration grants focused on places associated with strong tourist attraction or imperial ties, thus neglecting community-managed areas that are not recognized under colonial aegis. Cultural landscapes were transformed into regulated commodities in which indigenous custodianship was replaced by commercial management priorities. This illustrates the way in which indigenous cultural economies and local knowledge practices were downgraded through the colonial accession of heritage to fiscal and administrative structures (Cohn, 1996; Guha, 2000).

Institutionalizing Tourism Knowledge

In the colonial period, reports, gazettes and guide books were compiled and managed by governors to institutionalize tourist knowledge. Through the compilation of such sub-sets of narratives that reinforced imperial-military-administrative-commercial interests, such publications became a standardized facts about geography, history, monuments and cultural habits (Cohn, 1996; Dirks, 2011). Gazetteers were prepared to record details about populations, land, and resources mainly for administrative use by the government. In contrast, guidebooks shaped how visitors and tourists viewed India. Together, these publications helped to construct and spread a colonial understanding of India's society, culture, and heritage (Said, 1978).

Table 5. Gubernatorial Interventions in Tourism and Cultural Economy

Area of Intervention	Governor's Role	Impact on Indigenous Cultural Economy
Heritage promotion	Policy endorsement	Commodification of cultural sites
Revenue mechanisms	Regulatory approval	Shift from custodianship to commercialization
Knowledge production	Oversight of publications	Standardized colonial narratives

Table 5 synthesizes how British Governors institutionalized tourism as a cultural-economic sector, revealing the intersection of governance, knowledge production, and indigenous heritage under colonial rule.

The production, review, and distribution of these volumes were overseen by governors to ensure they were in accordance with administrative goals and provincial policy (Metcalf, 1989). Through the mechanism of managing a production and distribution system for tourism knowledge, Governors further entrenched their position as cultural arbiters of value and economic possibility. This epistemic authority contributed to the long-term shaping of tourism practices in India, influencing both colonial and post-colonial understandings of heritage, culture, and Indigenous Knowledge Systems.

IMPACT ON INDIGENOUS ECONOMIC SYSTEMS

Traditional Economic Structures

Before colonization, Indigenous Indian economies were characterized by distributed and community-based economies that led to exchange of information and goods (Ray, 2003). The rural craftspeople worked to ensure that the standards of living in their communities were maintained through generation of products like textile, metalwork, pottery and farming tools. Social and cultural norms governed the organization of labor, pricing mechanisms, and the transmission of technical knowledge from one generation to another, all of which were integral parts of these economic arrangements (Bayly, 1997; Roy, 2012)

An other critical component of this framework was indigenous medical economies. Localized markets built on community trust, apprenticeship-based learning, and region-specific botanical knowledge were the foundation for practitioners of Ayurveda, Siddha, Unani, and different folk healing traditions (Leslie, 2022; Mukharji, 2016). Similarly, craft traditions served as conduits for the transmission of embodied skill, ecological knowledge, and ritual practice from one generation to the next, in addition to providing a means of subsistence.

Governors' Economic and Legal Policies

Through a combination of legislative and administrative actions, British governors were instrumental in reorganizing indigenous economic institutions. Traditional land-use practices that had supported communal economic life were upended by land revenue reforms such as the Permanent Settlement, the Ryotwari System, and the Mahalwari System, which reshaped property relations (Stokes, 1992). These policies undermined community resource management and disrupted indigenous peoples' ability to make a living since they emphasized profit maximization and private property rights (Breman, 1996).

The Indian Forest Acts of 1865, 1878, and 1927 established limitations that limited access to communal resources vital for artisanal production, pastoralism, and indigenous medicinal practices. For example, shifting cultivation and gathering forest produce were made illegal as a result of conservation and taxation regimes imposed by governors who used their discretionary executive power to reclassify forests as state property (Saberwal & Rangarajan, 2009) . Incorporating indigenous economies into colonial fiscal frameworks through changes in monetization and taxes hastened the decline of subsistence-based systems and increased reliance on uncertain market interactions (Roy, 2012).

Promotion of Cash Crops vs. Indigenous Practices

Cash crops including indigo, tea, coffee, cotton, and opium were progressively prioritized in colonial economic strategy overseen by governors (Macarthur, 1984). Plantation economies were bolstered by governors who provided land concessions, invested in infrastructure, and provided legal protections to European commercial firms (Metcalf, 1989). Indigenous subsistence, crop diversification, and local exchange agricultural practices were systematically replaced by the initiative (Ray, 2003).

Such developments as the spread of the cash-crop agriculture among the tribal and rural people resulted in the loss of land, oppressive labor methods and the destruction of the environment. Methods based on indigenous knowledge like the regeneration of soil, seed preservation, the seasonal variations and the multi-cultivation were replaced with a monoculture system that led to the imperial trade. Over time, these changes eroded indigenous economic resilience and fostered enduring patterns of rural distress, indebtedness, and dispossession that persisted into the colonial era (Breman, 1996; Scott, 2020).

Table 6. Gubernatorial Economic Policies and Their Impact on Indigenous Knowledge Systems

Policy Domain	Governor's Action	Effect on Indigenous Economic Systems
Land revenue	Settlement implementation	Disruption of customary land relations
Forest governance	Enforcement of Forest Acts	Loss of resource-based knowledge
Agricultural policy	Promotion of cash crops	Marginalization of subsistence practices

Table 6 highlights how gubernatorial economic governance systematically reshaped indigenous economic systems, revealing the centrality of legal authority in the transformation of knowledge-based livelihoods during the colonial period.

DISCUSSION AND ANALYSIS

Comparative Analysis Across Presidencies

Bombay, Madras, Bengal, and United Provinces show geographical variance and structural continuity in gubernatorial interaction with Indigenous Knowledge Systems (IKS). Governors in the Bombay Presidency were pragmatic, engaging with local medicinal practices, archeological preservation, and language documentation where it benefited administrative efficiency or revenue.

Through the use of district collectors and ethnographic surveys overseen by the governor, the Madras Presidency showed a more organized administrative engagement with indigenous knowledge by documenting local customs, temple economies, and educational practices in great detail.

Due to its status as the first British colonial administration, Bengal served as a test bed for various codification initiatives, most notably those pertaining to education, language, and law. The governors of Bengal were active in establishing English education and legal uniformity, which in turn led to similar efforts in other regions of the globe (Cohn, 1996; Metcalf, 1994). The United Provinces, on the other hand, had longer-lasting indigenous institutions because of well-established agricultural systems and influential local elites. The colonial legal and administrative frameworks gradually replaced these institutions.

Indigenous knowledge systems were impacted by the distortion, alteration, and constraint imposed by colonial legal and bureaucratic governance in every Presidency, regardless of their location.

Table 7. Comparative Gubernatorial Approaches to Indigenous Knowledge Across Presidencies

Presidency	Dominant Approach	Key Sectors Affected	Nature of IKS Engagement
Bombay	Pragmatic–utilitarian	Medicine, archaeology	Selective incorporation
Madras	Administrative–ethnographic	Temples, education	Regulation and documentation
Bengal	Centralized–codificatory	Law, language, education	Systematic replacement
United Provinces	Hybrid	Agrarian knowledge, customs	Gradual marginalization

This comparative framework, as shown in Table 7, focuses on differences between regions while also showing how epistemic regulation was used in all Presidencies.

Patterns of Recognition in Relation to Domination

Across many fields, there has been a consistent pattern of recognizing indigenous knowledge as an integral part of domination methods. Governors typically acknowledged indigenous customs when they could be analyzed, classified, and leveraged to advance colonial administration objectives. The Indigenous systems were neither fully saved nor totally abolished instead, they were altered to be used as a form of control, thus, exemplifying the colonial knowledge appropriation.

Recognizing the codification of conventions, institutionalization of specific traditions and the process of integrating indigenous behaviors within legal systems, governors managed to standardize and systematically reduce vibrant and even complicated bodies of knowledge. All other activities that were perceived to contravene imperial morals and economic effectiveness were either not allowed or greatly malevolent.

This two-sided process reveals that the colonial status was contingent and instrumental, which strengthens the unequal relations of power between the colonial nation and the indigenous groups.

The Law–Knowledge Nexus

The Governor's office in India during the colonial period was a key place where knowledge and legal frameworks came together. The Governors controlled the scope of indigenous knowledge system, legitimacy, and their institutionalization through legislative approval, discretionary ordinances, executive rules, and administrative circulars. Educational, forest, public health, and cultural statutes changed epistemic landscapes.

Beyond simply enforcing provincial policies, governors had other responsibilities. By exercising their judgment in interpreting laws in light of local reality, they influenced whether indigenous knowledge was recognized, modified, or suppressed. When it was more convenient from an administrative standpoint to be flexible, imperial officials could use their discretion to selectively adopt traditional norms; when imperial economic or moral concerns were more pressing, they may strengthen exclusion in other circumstances.

As a result, the law-knowledge nexus reveals how the exercise of legal authority as a means of epistemic governance ingrained colonial power in intellectual and cultural life.

Postcolonial Reflections

Policy decisions made under British rule have an impact on how Indigenous Knowledge Systems in India engage with the government now. Postcolonial administrations have maintained the legal, administrative, and regulatory frameworks that were established during colonial operations. Efforts to revitalize indigenous languages, crafts, traditional medicine, and cultural heritage have been supported by legislative frameworks that have been in existence since colonial authority. These frameworks have both remained consistent and been the subject of debate.

By examining the historical role of governors as mediators of indigenous knowledge, one can gain a better understanding of the current conversations surrounding educational decolonization, cultural autonomy, and IKS integration. There needs to reimagine policy and legal frameworks that go beyond control and towards true acknowledgment, involvement, and knowledge equity because regulation-oriented paradigms that originate from colonial rule are still very much alive and well (Santos, 2016).

REFERENCES

Agrawal, A. (1995). Dismantling the Divide Between Indigenous and Scientific Knowledge. *Development and Change*, *26*(3), 413–439. DOI: 10.1111/j.1467-7660.1995.tb00560.x

Anuja Priya, S. (2026). Negotiating science, power, and identity: The institutionalization of Ayurveda in contemporary India. *Social Sciences & Humanities Open*, *13*, 102535. DOI: 10.1016/j.ssaho.2026.102535

Arnold, D. (1993). *Colonizing the body: State medicine and epidemic disease in nineteenth-century India*. University of California Press.

Austin, G. (2018). *The Indian constitution: Cornerstone of a Nation* (Thirty-first impression). Oxford University Press.

Bayly, C. A. (1997). *Empire and Information: Intelligence Gathering and Social Communication in India, 1780–1870* (1st ed.). Cambridge University Press., DOI: 10.1017/CBO9780511583285

Blackburn, S. H. (2016). *Singing of Birth and Death: Texts in Performance*. University of Pennsylvania Press., DOI: 10.9783/9781512800579

Breman, J. (1996). *Footloose Labour: Working in India's Informal Economy* (1st ed.). Cambridge University Press., DOI: 10.1017/CBO9781139171076

Bteille, A. (2017). *DEMOCRACY AND ITS INSTITUTIONS*. OXFORD University Press.

Chakrabarti, D. K. (2003). *Archaeology in the Third World: A history of Indian archaeology since 1947*. D.K. Printworld.

Chakrabarty, D. (2007). *Provincializing Europe: Postcolonial thought and historical difference*. Princeton university press.

Chatterjee, P. (1993). *The nation and its fragments: Colonial and postcolonial histories*. Princeton University Press.

Cohn, B. S. (1996). *Colonialism and its forms of knowledge: The British in India*. Princeton University Press.

Dei, G. J. S., Hall, B. L., & Rosenberg, D. G. (Eds.). (2000). *Indigenous knowledges in global contexts: Multiple readings of our world*. Published in association with University of Toronto Press.

Derrett, J. D. M. (1999). *Religion, law and the state in India* (Repr.). Oxford Univ. Press.

Dharampal. (2021). *The beautiful tree: Indigenous Indian education in the eighteenth century*. Rashtrotthana Sahitya in association with Centre for Policy Studies, Chennai.

Dirks, N. B. (2011). *Castes of Mind: Colonialism and the Making of Modern India*. Princeton University Press. DOI: 10.2307/j.ctt7rq9d

Gadgil, M., & Guha, R. (2014). *This fissured land: An ecological history of India* (3. impr). Oxford University Press.

Ghosh, S. C. (1975). Dalhousie, Charles Wood and the Education Despatch of 1854. *History of Education*, *4*(2), 37–47. DOI: 10.1080/0046760750040204

Guha, R. (2000). *The unquiet woods: Ecological change and peasant resistance in the Himalaya* (Expanded ed). University of California Press.

Guha, R. (2008). *India after Gandhi: The history of the world's largest democracy*. Harper Perennial.

Harrison, M. (1994). *Public health in British India: Anglo-Indian preventive medicine, 1859-1914*. Cambridge University Press.

Jain, M. P. (2014). *Outlines of Indian legal and constitutional history* (7th ed.). LexisNexis.

Kennedy, D. (2023). *The Magic Mountains: Hill Stations and the British Raj* (1st ed.). University of California Press., DOI: 10.2307/jj.8085339

King, R. D. (1998). *Nehru and the language politics of India*. Oxford Univ. Pr.

Kumar, D., Raychaudhuri, T., & Habib, I. (Eds.). (2005). The Cambridge economic history of India. *Volume II, C.1757-2003* (Édition augmentée). Orient Longman, in association with Cambridge University Press.

Leslie, C. M. (2022). *Asian Medical Systems: A Comparative Study* (1st ed.). University of California Press.

Macarthur, J. D. (1984). The political economy of underdevelopment A. K. Bagchi Modern Cambridge Economics, Cambridge University Press, Cambridge, 1982, 280 + viii pp. *Public Administration and Development*, *4*(1), 104–105. DOI: 10.1002/pad.4230040112

Macaulay, T. B. (1862). *Macaulay's Minutes on Education in India*. (Original work published Baptist Mission Press)

Mani, L. (1998). *Contentious traditions: The debate on Sati in colonial India*. University of California Press. DOI: 10.1525/9780520921153

Metcalf, T. R. (1989). *An imperial vision: Indian architecture and Britain's raj*. University of California Press.

Metcalf, T. R. (1994). *Ideologies of the Raj*. Cambridge University Press.

Misra, B. B. (1970). *The Administrative History of India (1834-1947)*. Oxford University Press.

Mitra, R. (2012). The Famine in British India: Quantification Rhetoric and Colonial Disaster Management. *Journal of Creative Communications*, *7*(1–2), 153–174. DOI: 10.1177/0973258613501066

Mukharji, P. B. (2016). *Doctoring traditions: Ayurveda, small technologies, and braided sciences*. The University of Chicago Press. DOI: 10.7208/chicago/9780226381824.001.0001

Nurullah, S., & Naik, J. P. (1951). *History of education in India (during the British period)*. Macmillan Bombay.

Pollock, S. I. (2009). *The language of the gods in the world of men: Sanskrit, culture, and power in premodern India* (1. paperback print). Univ. of California Press.

Ray, R. K. (2003). *The felt community: Commonalty and mentality before the emergence of Indian nationalism*. Oxford University Press.

Roy, T. (2012). *India in the world economy: From antiquity to the present*. Cambridge University Press. DOI: 10.1017/CBO9780511920516

Saberwal, V. K., & Rangarajan, M. (Eds.). (2009). *Battles over nature: Science and the politics of conservation* (2. impr). Permanent Black.

Said, E. W. (1978). *Orientalism* (1st ed.). Pantheon Books.

Santos, B. de S. (2016). *Epistemologies of the South: Justice against epistemicide*. Routledge., DOI: 10.4324/9781315634876

Scott, J. C. (2020). *Seeing like a state: How certain schemes to improve the human condition have failed* (Veritas paperback edition). Yale University Press.

Sharma, R. S. (2006). *India's Ancient Past.*

Soneji, D. (2012). *Unfinished gestures: Devadāsīs, memory, and modernity in South India*. University of Chicago Press.

Srinivas, M. N. (2003). *Religion and society among the Coorgs in South Asia*. Oxford University Press.

Stokes, E. (1992). *The English utilitarians and India* (Third impression). Oxford Univ. Press.

Washbrook, D. A. (1981). Law, State and Agrarian Society in Colonial India. *Modern Asian Studies*, *15*(3), 649–721. DOI: 10.1017/S0026749X00008714

Zaman, M. Q. (2007). *The Ulama in contemporary Islam: Custodians of change* (Paperback ed.). Princeton University Press.

ADDITIONAL READING

Chatterjee, P. (2004). *The politics of the governed: Reflections on popular politics in most of the world*. Columbia University Press.

Ostrom, E. (2012). *Governing the Commons: The Evolution of Institutions for Collective Action*. Cambridge University Press.

Pande, I. (2012). *Medicine, race and liberalism in British Bengal: Symptoms of empire*. Routledge.

Sarkar, S. (1983). *Modern India, 1885-1947*. Macmillan.

Skaria, A. (1999). *Hybrid histories: Forests, frontiers and wildness in western India. Oxford Univ. Press. Headrick, D. R. (1981). The tools of empire: Technology and European imperialism in the nineteenth century. Oxford University Press. Radhakrishnan, S., & Mohanty, J. N. (2008). Indian philosophy* (2nd ed.). Oxford University Press.

Smith, L. T. (2021). *Decolonizing methodologies: Research and indigenous peoples* (Second edition, reprinted). Zed Books. DOI: 10.5040/9781350225282

KEY TERMS AND DEFINITIONS

Administrative Ethnography: Colonial documentation of customs, social hierarchies, and local practices primarily undertaken to facilitate governance, taxation, and legal regulation.

Codification: The transformation of fluid customary practices and oral traditions into fixed legal or administrative rules aligned with Western legal rationality.

Colonial Co-optation of Knowledge: The process by which colonial authorities selectively extracted, classified, and repurposed indigenous knowledge while undermining its epistemic autonomy and community control.

Cultural Commodification: The transformation of indigenous heritage and practices into economic or touristic assets, detached from their original social and ritual contexts.

Epistemic Governance: The governance of society through the control, validation, and institutionalization of knowledge, determining whose knowledge is recognized and whose is marginalized.

Gubernatorial Discretion: The legal and administrative authority exercised by Governors to interpret, implement, or modify colonial policy frameworks at the provincial level.

Indigenous Knowledge Systems (IKS): Cumulative, practice-based systems of knowledge developed through long-term interaction between communities, culture, and environment, transmitted across generations through social institutions rather than formal state mechanisms.

Law-Knowledge Nexus: The reciprocal relationship between legal authority and knowledge production, where law shapes what counts as legitimate knowledge, and knowledge, in turn, legitimizes governance.

Chapter 10

Challenges and Opportunities for Children Under Extended Family System in Ward 14, Bikita District, Zimbabwe in the Face of Austerity Era

Witness Chikoko
http://orcid.org/0000-0002-9459-3972
University of Zimbabwe, Zimbabwe

Reason Mupanga
http://orcid.org/0009-0007-0561-0739
University of Zimbabwe, Zimbabwe

ABSTRACT

This qualitative case study interrogates the challenges and opportunities for children under extended family systems in Ward 14, Bikita district, Masvingo Province, Zimbabwe, during the austerity era. Guided by Ubuntu Perspectie, the research employed purposive and convenience sampling to recruit twelve children aged between seven and seventeen years under kinship care arrangements. Data collection utilized in-depth interviews, focus group discussions, and key informant interviews, with analysis conducted through thematic content analysis. Findings reveal that children under kinship care face multifaceted challenges encompassing child marriages, school dropouts, child labour in artisanal mining and domestic work contexts, stigma and discrimination, sexual deviance including commercial sex work and intergenerational sexual relations, substance abuse, and acute shortages of essential provisions such as food and healthcare services. The study advocates for strengthened social protection mechanisms, including expanded government-led child protection programmes

DOI: 10.4018/979-8-2600-2890-2.ch010

INTRODUCTION AND BACKGROUND

Extended family constitutes a structure referring to people related through blood, including mothers, fathers, grandparents, and children (Ringson and Chereni, 2020). The terms kinship and extended family system are synonymous, with kin sometimes used interchangeably with relative (Selwyn and Nandy, 2014). In the United States, kin encompass individuals related to the child by marriage, blood, or adoption, while some states extend this definition to friends, clan members, tribal affiliates, and neighbours (Selwyn and Nandy, 2014). Kiraly (2019: 449) and United Nations (2010:06) define kinship care as "family-based care within the child's extended family or with close friends of the family known to the child, whether formal or informal in nature." Zonta (2016: 04) characterises extended families as "living plans of groups of individuals whose relationship extends outside the nuclear family," comprising multiple generations including uncles, aunts, and grandparents (Chiimba, 2020).

Extended family systems represent crucial mechanisms for addressing the needs of orphans and vulnerable children globally. Zimbabwe's National Orphan Care Policy of 1999 identifies extended families as key mechanisms for caring for vulnerable children. However, this historically strong African institution has been severely compromised by modernity, eurocentrism, globalisation, and urbanisation (Kang'ethe et al., 2022; Mushunje, 2014), despite its foundation in Ubuntu philosophy (Mushunje, 2014; Ringson and Chereni, 2020).

Indigenous knowledge systems embedded in Ubuntu/Hunhu philosophy (Mushunje, 2014; Samkange and Samkange, 1980) have historically governed the functioning of extended family systems in Zimbabwe, encoding norms of ukama (relational obligation), zvishuwo (collective moral aspirations for children's welfare), and dare-based community accountability that ensured the social protection of orphaned and vulnerable children within kinship networks.

The "austerity era" denotes Zimbabwe's specific economic configuration from the late 2010s onwards, characterised by persistent commodity price increases, currency volatility from partial dollarization, severe wage erosion particularly affecting civil servants and rural households, rapid economic informalization replacing formal employment with precarious livelihoods, and marked contraction of international donor funding for social protection programmes (Chereni, 2020).These austerity-driven processes weaken extended family systems' capacity to care for vulnerable children by simultaneously deepening household deprivation, reducing formal welfare services, and undermining the material basis sustaining kinship obligations. This intersection creates compounding effects: deepening household poverty diminishes caregivers' provisioning capacity while donor-funded programme withdrawal eliminates institutional safety nets that once supplemented family-based care.

Despite growing recognition of kinship care challenges globally and regionally, academic literature interrogating challenges and opportunities for children under extended family systems in rural Zimbabwe during the austerity era remains scarce. While scholars including Mushunje (2014) and Ringson and Chereni (2020) explored extended family systems' role in Zimbabwe, these studies inadequately articulated lived experiences and specific challenges facing children in rural kinship care during economic hardship. Furthermore, existing research has not adequately captured how traditional family system erosion converging with contemporary economic crisis uniquely impacts rural children in kinship arrangements. Children's voices and experiences remain largely absent from academic discourse, creating significant knowledge gaps regarding their daily realities, coping mechanisms, and survival strategies.

This study comprehensively interrogates challenges and opportunities for children under extended family systems in Ward 14, Bikita district, Zimbabwe, amid the austerity era, understanding complex emotional and relational dynamics within extended family units and how individual challenges affecting children reverberate throughout entire family systems. The findings ultimately advocate for strengthened social protection mechanisms tailored to children in kinship care within rural communities, contributing to improved child welfare outcomes in Zimbabwe's most vulnerable populations.

The study is anchored in indigenous knowledge perspectives, recognising that extended family systems in Zimbabwe are not merely residual welfare arrangements but living expressions of Ubuntu/ Hunhu philosophy — an African indigenous knowledge system that positions communal care, mutual obligation, and the collective protection of children as foundational moral principles governing social life (Ringson and Chereni, 2020; Samkange and Samkange, 1980).

LITERATURE REVIEW

The extended family system has historically served as the primary social safety net for vulnerable children in Zimbabwe. However, the institution faces unprecedented challenges amid economic austerity measures characterizing Zimbabwe's economic landscape. This literature review examines scholarly perspectives on multifaceted challenges and emerging opportunities for children under extended family care, drawing from global, regional, and local research to contextualize experiences of children in Ward 14, Bikita District.

The review organizes around four analytical clusters: (i) institutional support and service access; (ii) household economic strain; (iii) psychosocial and behavioural outcomes; and (iv) relational dynamics and placement experiences. Each cluster is examined across global, regional, and local contexts. Rather than treating studies in isolation, the review synthesises common mechanisms and identifies what is established, contested, and unresolved in literature.

Consistent with an indigenous knowledge perspective, this review acknowledges that the extended family system in Zimbabwe is embedded in Ubuntu/Hunhu philosophy and governed by indigenous social norms, customary obligations, and intergenerational knowledge systems that have historically regulated child care responsibilities within kinship networks (Ringson and Chereni, 2020; Mushunje, 2014).

Global Context

Globally, extended family networks play crucial roles in child welfare social protection systems. United Nations Guidelines for Alternative Care of Children (2010) emphasize family-based care represents the most appropriate environment for child development, with permanency and stable attachments serving as critical goals. However, global research reveals extended family systems can perpetuate child rights violations, including child labour and sexual abuse, particularly affecting orphaned girls withdrawn from school to provide domestic labour (Mushunje, 2014; Taylor et al., 2020).

A consistent finding is institutional support deficits for kinship carers. Kiraly (2019) and United Nations (2010) establish that quality kinship care requires adequate formal support, yet empirical evidence consistently documents such support's absence. In the United States, Gentles-Gibbs and Zema (2020) found seven out of eight grandparents had no assigned caseworkers, resulting in limited support and reduced service access. Farmer (2009) reported similar patterns in the United Kingdom, where kin carers received no financial support from Children's Services departments. Taylor et al. (2020) documented that Scottish carers experienced support lack alongside carer stress and difficult contact with biological parents. These studies demonstrate institutional support deficits are structural features of kinship care systems across high-income countries, raising questions about such deficits' amplification in resource-constrained settings like rural Zimbabwe.

Children under kinship foster care in Victoria, Australia experienced mental health challenges including depression and anxiety resulting from stressful transitional phases including puberty and placement changes (Fergeus et al., 2019).

Household economic strain is equally well-documented globally. Farmer (2009) reported financial challenges in the United Kingdom were compounded when kin carers were responsible for children with special needs, resulting in struggles paying for school uniforms, trips, and school-related activities. Xu et al. (2022) found that in the United States, grandparent-headed households experienced material hardship further exacerbated by COVID-19 pandemic, manifesting in increased costs for shelter, food, clothing, and school fees. These economic pressures are active mechanisms directly compromising children's access to education, nutrition, and healthcare.

In the United Kingdom, kin carers faced numerous challenges including overcrowding, financial hardships, and poor health, with some becoming too old and developing health problems including disabilities (Farmer, 2009). In a United States study, care arrangements under grandparent-headed households were detrimental to children's well-being, with grandchildren becoming victims of physical and verbal abuse by grandparents who demonstrated limited parenting capabilities (Gladstone and Brown, 2007).

Kinship care in Spain has been marred with challenges including poor coordination between foster care and biological families (Inchaurrondo et al., 2015). Drug or substance abuse is another challenge, where some kinship carers' substance abuse creates marital problems subsequently affecting children's well-being (Inchaurrondo et al., 2015).

Relational dynamics reveal more contested terrain. While Gladstone and Brown (2007) document abuse cases and parenting deficits among United States grandparent carers, Ihara et al. (2012) show relational tensions in Asian and Pacific American grandparent-headed households often stem from cultural clashes rather than inherent caregiving inadequacies. Among Korean American families, grandchildren's preference for host culture over caregivers' cultural expectations created intergenerational friction (Ihara et al., 2012). This finding suggests relational challenges are shaped by specific socio-cultural and economic contexts, informing this study's attention to both structural constraints and relational dynamics in Ward 14, Bikita District, Zimbabwe.

Scholars such as Ringson and Chereni (2020) and Mugumbate and Chereni (2019) argue that mainstream global child welfare frameworks have systematically undervalued indigenous knowledge-based care systems, positioning Western institutional models as universal standards while neglecting the Ubuntu-grounded collectivist arrangements that have sustained kinship care across African communities for generations.

Regional Context

In Ghana, children under extended family systems faced financial challenges, with families receiving no support from Government Department of Social Services, making children's lives very miserable (Manful and Cudfoe, 2018). Similarly, in Buffalo City Metropolitan, South Africa, grandparents lacked access to financial resources, compromising their ability to provide adequate psychosocial support to grandchildren (Folorunsho and Tanga, 2021).

In Hill Crest location, Alice township, Eastern Cape, South Africa, grandparents faced numerous challenges including delinquent behaviours where grandchildren stole valuables (Kang'ethe et al., 2022). In Mdlankomo location, Libode, Eastern Cape, some grandparents were unhappy about grandchildren's delinquent behaviours involving substance misuse (Soganga and Kang'ethe, 2023).

In KwaZulu Natal, South Africa, grandchildren were stealing household appliances for selling to raise money for buying substances (Kasiram and Holscher, 2015). Cases existed where grandchildren connived with thieves for stealing valuables from grandparents' houses (Kasiram and Holscher, 2015). Some grandparents of Hill Crest location were overwhelmed with caring-related duties at peer socialization's expense, with no time for socializing due to grandchildren's delinquent behaviours (Kang'ethe et al., 2022).

Delinquent children made carers succumb to diseases, with some grandparents developing stroke and high blood pressure resulting from grandchildren's delinquent behaviours (Kang'ethe et al., 2022). In another Buffalo City Metropolitan study, grandparents could not provide much-needed psychosocial support because of old age-related diseases such as diabetes and high blood pressure (Folorunsho and Tanga, 2021).

Children under kinship in Botswana have behavioural challenges including withdrawn and rebellious behaviours possibly resulting from feelings of rejection, being unloved, or inadequacy (Malinga-Masumba, 2015). Some children did not relate well with caregivers, complaining about favouritism over household chores where grandparents preferred certain children and less-favoured children performed more chores (Malinga-Masumba, 2015). Kinship caregivers complained orphaned children did not want to do household chores, creating tension (Malinga-Masumba, 2015).

In Mdlankomo location, grandparents and grandchildren had relational challenges where grandchildren accused grandparents of misusing their social welfare grants, reporting them to social workers (Soganga and Kang'ethe, 2023).

Another challenge has been environmental change (Malinga-Masumba, 2015). Some orphaned children relocated from urban to rural areas upon parents' death, significantly affecting children with no prior rural experiences (Malinga-Masumba, 2015).

Regional literature reveals patterns consistent with global evidence: household economic strain and institutional support deficits are primary vulnerability drivers for children in kinship care across sub-Saharan Africa. However, regional evidence highlights dimensions less prominent globally, particularly intergenerational relational conflict, physical and psychological tolls on elderly caregivers, and environmental dislocation's impact. Importantly, behavioural challenges should be understood not as children's inherent pathologies but as responses to constrained environments characterised by deprivation, limited supervision capacity, and formal support systems' absence. This structural reading is essential for avoiding moralistic interpretations and developing policy responses addressing root causes rather than symptoms.

Across sub-Saharan Africa, indigenous knowledge systems have long provided complementary social protection mechanisms — including community granaries, age-grade mentorship systems, and clan-based dispute resolution councils — that historically supported extended family care systems; the marginalisation of these indigenous systems within formal social protection frameworks has left kinship households without culturally grounded buffers against poverty and institutional abandonment.

Local Context

Research from Zimbabwe confirms the extended family institution remains one of primary indigenous social protection systems, with the majority maintaining ties with extended kin, particularly during times of need (Dhemba et al., 2002). However, the system's capacity has been severely compromised by multiple intersecting crises. The current socio-economic environment has made financial assistance increasingly unreliable, exacerbated by structural adjustment programs, poverty, high unemployment, HIV/AIDS, and COVID-19 pandemic. Extended family system fabric has been affected as members migrate and adopt western lifestyles focusing on individualism, resulting in weakened rural-urban support networks.

In a Zvishavane study of orphans under grandmothers, children faced numerous challenges including lack of birth registration, food insecurity, and limited access to education and health (Matshalaga, 2004). Some children experienced psychosocial challenges such as depression and anxiety (Matshalaga, 2004).

Mushunje's (2014) study discussed extended family system roles in Zimbabwe, including aunts' and uncles' roles in imparting knowledge to boys and girls. Aunts and uncles are critical in training on values at family and community levels, teaching them how to behave when married (Mushunje, 2014). Extended family systems are also responsible for producing law-abiding citizens promoting accountability and responsibilities. Mushunje (2014) observed Zimbabwe's extended family system is under severe attack. Globalisation, HIV and AIDS, Western lifestyles, and macro-economic instability have significantly affected extended families' traditional child care roles (Mushunje, 2014). Extended families are struggling to care for their own children given prevailing economic uncertainty (Mushunje, 2014).

Existing Zimbabwean scholarship makes important contributions. Mushunje (2014) establishes extended family system's institutional role and historical significance in child socialisation. Ringson and Chereni (2020) contribute by framing kinship care through indigenous knowledge perspectives, demonstrating Ubuntu-based family support mechanisms' efficacy. However, these studies do not resolve several critical questions. Neither captures children's own accounts of lived experiences within kinship care in rural ward contexts during current economic crisis periods. Their analyses focus on institutional roles and cultural frameworks rather than day-to-day realities, coping strategies, and children's survival mechanisms. Moreover, neither systematically examines how austerity-era conditions' convergence, including price increases, currency instability, wage erosion, and donor contraction, specifically compromises kinship caregiving capacity in rural communities. This is the gap this study addresses, by centring children's voices in Ward 14, Bikita district, and documenting both challenges and opportunities persisting within their kinship arrangements.

Ringson and Chereni (2020), drawing on indigenous knowledge frameworks, demonstrate that the extended family system in Zimbabwe is not merely a pragmatic welfare arrangement but a culturally embedded institution grounded in Ubuntu/Hunhu philosophy, wherein caring for orphaned and vulnerable children is understood as a sacred intergenerational obligation (mwana wevhu) that sustains community cohesion and moral order.

THEORETICAL FRAMEWORK

The *Ubuntu* or *hunhu* as an African philosophy originated among the Bantu speaking people (Ramose, 2002; MBiti, 1969). The African Moral Philosophy, also known as *Ubuntu* or *Hunhu.* Samkange and Samkange (1980) are some of the earliest scholars writing about *Ubuntu* and they define it as a code of behaviour in which a human being treats others with courtesy, kindness, consideration and friendliness. *Ubuntu* is more than being a human being but is a profound concept and principle which guides behaviour. *Ubuntu* is a value system principally associated with black and Bantu speaking peoples of Sub-Saharan Africa (Samkange and Samkange, 1980). *Ubuntu* may have some ideas that are similar to other cultures but the concept of *Hunhuism* itself is unique to Bantu speaking peoples, regardless of individual tribal differences (Samkange and Samkange, 1980). It has different names across various languages and dialects and some of these include *vutho* and *hunhu. The* basic tenets of the African moral philosophy are brotherhood, collective unity, goodwill, collective identity, generosity and protection of community welfare and interests (Tutu, 1999). The underlying principle is that humanity is achieved through others hence the popular adage 'a person is a person through other persons' (Metz, 2007). Samkange and Samkange (1980) who also refer to *Ubuntu* as *Hunhuism* and emphasize on that any behaviour that is not acceptable is termed as '*kushaya unhu*', meaning that it reflects a lack of *Ubuntu or Unhu*. Conversely, any behaviour that is pro social and in line with norms is considered to be '*kuva ne hunhu'* meaning that it embodies the ideals of *Hunhuism* (Samkange and Samkange, 1980).

The *Ubuntu* or *hunhu* is a philosophy which is grounded in the African thinking as noted by Chikoko and Ruparanganda (2020), Mugumbate and Chereni (2019), Mugumbate and Nyanguru (2013), Ramose (1999) and Mbiti (1969). The *Ubuntu* or *Hunhu* philosophy is premised on key characteristics such as humanness, kindness, love, social justice, compassion, moral values, and a Xhosa proverb *umuntu ngumuntu ngabantu* (I am what I am through others) (Samkange and Samkange 1980, Shutte 1993, Mbiti 1975, Mangena, 2007, 2012 & 2016, Hanks 2008, Wagid and Smeyers 2012, Morreira 2013). The *umuntu ngumuntu ngabantu* proverb demonstrates the communitarian nature of humanity. The Ubuntu philosophy is linked to the concept of cooperation and working together as individuals, groups and communities (Mupedziswa et al 2019). The ubuntu philosophy is imbedded with the principles of individuals and families caring for one another out of a moral obligation (Mupedziswa et al 2019). The *Ubuntu* or *Hunhu* philosophy means humanity towards others (Mapaure 2011). It is known as humanness (Van Breda, 2019).

RESEARCH METHODOLOGY

A qualitative case study research design was adopted to understand the experiences of children under extended family systems in Ward 14, Bikita district, Masvingo Province, Zimbabwe. Through purposive sampling, twelve (12) children under extended family systems were identified, including those in grandparent-headed households and those staying with aunts and uncles. Some research participants had been left by parents who migrated to diaspora including the United Kingdom, Australia, and South Africa, while others had divorced parents. Research participants aged between eight and eighteen years. Data was drawn through in-depth interviews, focus group discussions, and key informant interviews.

In-depth interviews were conducted at convenient places including when research participants were herding cattle, ploughing fields, or coming from school. Interviews lasted thirty (30) to forty-five (45) minutes and focused on challenges and opportunities children under extended family systems encounter. Focus group discussions were divided based on age and gender, focusing on personal experiences related to challenges and opportunities. Through convenience sampling, five (5) key informants were identified: a social worker under the Department of Social Development, a traditional chief, a church pastor, a school headmaster, and a councillor. Key informants had expert knowledge about children's lives under kinship foster care in Ward 14, Bikita district, Masvingo Province, Zimbabwe.

Ward 14, Bikita district constituted the study's target area. Ward 14 is a rural area where the majority rely on peasant farming with very few livelihood opportunities. Although scant data exists about the exact number of children under extended family systems, reports indicate increasing vulnerability of orphans and vulnerable children.

The study received ethical clearance from the University of Zimbabwe and the Department of Social Development under the Ministry of Public Service, Labour and Social Welfare. Informed consent was obtained and granted by social workers employed by the Department of Social Development. Research participants were explained the study's purpose and were free to withdraw at any stage. To promote privacy and confidentiality, data was anonymized through use of alphabetical letters.

Data was analysed through thematic content analysis at two levels: during and after data collection. Themes and sub-themes that emerged during analysis included lack of support from child welfare agencies, school dropouts, lack of basic needs such as food and health, child marriages, child labour, delinquent sexual behaviours (commercial sex, masturbation, bestiality, intergenerational sex), misuse of substances, and stigma and discrimination.

RESEARCH FINDINGS

The findings are presented thematically, highlighting the challenges and opportunities faced by children living within the extended family system in Ward 14 of Bikita District, Zimbabwe, during the austerity era. Verbatim quotes are incorporated to strengthen and support the findings.

A. CHALLENGES FOR CHILDREN UNDER EXTENDED FAMILY SYSTEM

1. Misuse of substances

Abuse of substances is one of the behaviours common among children under extended family in ward 14 of Bikita district. During in depth interviews, one of the participants called B had this to say;

> Take chamba. I started to take chamba so as to cool down and also due to peer pressure. The majority of my colleagues, they take chamba[1] We normally take it when we are coming from Marirangwe High School.

Another participant, called F said the following;

> I take on alcohol. My first experience taking alcohol has been during the tradition rain making ceremonies. Some two years ago, my grandmother was preparing the mukwerera. I took some traditional brew during the ceremony. I got extremely drunk. Form that day onwards, I have started taking some homemade alcohol.

During focus group discussions, the participants agreed that misuse of substances was very common among adolescent children under extended family system. The participants added that, some of the commonly abused substances included alcohol, marijuana, aphrodisiac ones among others.

One of the key informants also confirmed that, children under extended family system of ward 14, Bikita district, Zimbabwe were abusing substances. He added that the adolescent children were misusing substances such as *ndari* or *ngoto(*homemade brew), marijuana among others. The traditional leaders also added that peer pressure and some of the psychological challenges, facing the children accounted for increase in misuse of substances.

Shona indigenous knowledge systems historically regulated the use of traditional brews such as hwahwa (traditionally fermented beer) through ceremonial contexts governed by elders and spiritual protocols; the unregulated misuse of substances by adolescents under extended family systems reflects the erosion of these indigenous regulatory frameworks and the loss of elder-guided moral socialisation.

2. Sexual behaviours (commercial sex work, bestiality, rape)

Sexual abuse such as commercial sex, masturbation, incest, intergenerational sex, forced sex, bestiality was some of the sexual deviances associated with children under extended family system of ward 14, of Bikita district. During in depth interviews, one of the research participants, called C said this;

I sell sex. I raise money for survival through selling sex with gold panners. With my friends, we travel to Bikita Mineral (where there is mining of lithium), and sell sex to artisanal miners.

Another participant called, E said the following;

Both of my parents died when I was a toddler. I am told they succumbed to the deadly HIV and AIDS disease. I was left in the care of my grandmother who is very poor. I did not go to school. How do you expect me to survive. I do not have any skills for survival besides selling sex to artisanal miners at Diyo Shopping Centre.

An adolescent boy said this;

I engage in sex with animals. I normally have sex with goats or cows. I enjoy having sex with the animals because I do not have a girlfriend.

Another adolescent boy said the following;

I masturbate as I peep women bathing in Mujiche river. Sometimes I engage in masturbation when I am under my blankets.

During focus group discussions, the participants agreed that, children under extended family system were involved in sexual deviance such as commercial sex work, masturbation, bestiality, forced sex or rape, intergenerational sex among others. The participants added that, the children were impoverished to the extent that raising money for survival was one of the most viable options for them.

A village head also confirmed that, sexual deviance is one of the challenges common among children in grandparent headed households of ward 14, Bikita district. He added that some of the sexual deviance include; forced sex, intergenerational sex, incest, commercial sex, bestiality among others.

Within Shona indigenous knowledge systems, adolescent sexual behaviour was regulated through culturally sanctioned institutions such as nhango (age-grade peer socialisation and moral instruction led by elders), which provided structured guidance on sexuality, relational ethics, and gender responsibilities; the collapse of these indigenous regulatory systems has left adolescents without culturally grounded frameworks for navigating sexual development.

3. School drop outs

School drop outs are some of the challenges affecting children under extended family system in ward 14, of Bikita district. During an in-depth interview, one of the participants, called G, had this to say;

As an orphan I only went to school as far as grade seven. My paternal grandmother could not raise money for secondary school. I should have enrolled for form one at Marirangwe High School but I could not because of lack of school fees.

The participants agreed that school drop outs were increasing in ward 14 of Bikita district. The children in grandparent headed households were part of the pupils pulling out of schooling. They were dropping out of schools because of lack of school fees among others. Some of the children were also dropping out of schools as they did not have school uniforms.

The village head also confirmed that, lack of school fees and unforms among children in grandparent headed households were contributing towards school drop outs. He added that the grandparents were having financial challenges in raising money towards school fees and uniforms for their grandchildren.

Traditional Shona indigenous knowledge recognised education as a communal responsibility captured in the concept of kudzidzisana (learning together), wherein elders, aunts, uncles, and community members collectively contributed to the intellectual and moral formation of children; the commodification of formal schooling has displaced these complementary indigenous educational systems, deepening exclusion for orphaned children.

4. Lack of basic needs such as food and health

Evidence from field work suggests that lack of basic needs is one of the challenges that affect grandparent headed households of ward 14 Bikita district. During in depth interviews, one of the participants, called F, said the following;

I stay with my grandmother. We normally face shortages of basic needs. Some of the basic needs such as food supplies are always in short supply. We do not have enough money to buy sugar, bread, cooking oil and maize flour for sadza[2].

Another participant, called, A had this to say;

My health is poor. I am always sick and it is extremely to have access to medical care. The other time I visited the clinic, there was no medicine. We were told to buy medicine from the private pharmacies and it is extremely difficult for us given the financial status of our grandmother.

During a focus group discussion, the participants all agreed that, lack of basic needs was a major challenge among grandparent headed households. The participants indicated that, grandparent headed households did not have enough food supplies.

One of the social workers also confirmed that, grandparent headed households of ward 14 Bikita district did not have enough food supplies. The social worker also added that the grandparent headed households also did not have access to medical care. She added that access to medical care was important given that some of the children under their care have pediatric HIV among others.

Indigenous knowledge practices such as nhimbe (communal labour and resource-sharing) and the tradition of ukama-based food reciprocity historically ensured that no household within a community faced absolute food deprivation; the weakening of these practices under austerity conditions has left grandparent-headed households without customary buffers against food insecurity.

5. Lack of support from child welfare agencies

Lack of support from child welfare agencies is one of the challenges that affect children in grandparent headed households of ward 14 Bikita district. During in depth interviews, one of the participants, called, L said the following;

There is limited support from social workers for children in grandparent headed households. The child welfare agencies both from the state and non-governmental organisations have not been supporting us. I am very much worried by this lack of support from child welfare agencies.

Another participant, called K, had this to say;

I am about to sit for public examinations at grade seven. I do not have a birth certificate. The social worker has not been supportive towards my issue of a birth certificate. I am very much disturbed.

The participants during focus group discussions, agreed that limited support by child welfare agencies was a common challenge affecting children in grandparent headed households of ward 14 Bikita district. The participants added that some of the child welfare agencies last visited the orphans and vulnerable children many years ago.

One of the social workers also confirmed that lack of support from child welfare agencies was a challenge affecting children in grandparent headed households of ward 14, Bikita district. The social worker added that some of the non-state actors in the provision of child welfare services were closing as a result of a number of factors such as donor flight among others.

Historically, indigenous knowledge systems in Zimbabwe provided complementary social protection through zunde ramambo (the chief's communal granary), a traditional institution in which community members contributed grain to a communal store specifically intended for widows, orphans, and vulnerable households; the erosion of such indigenous safety nets has amplified the impact of formal welfare deficits.

6. Child Marriages

Child marriages are common among adolescent children under grandparent headed households of ward 14, Bikita district. During in depth interviews, one of the participants, called I had this to say;

I am 17 years old and I got married last year when I was 16 years. I decided to get married as a way of running away from the challenges I was facing as an orphan under grandparent headed household.

Another participant, said the following;

Getting married is a better option for us. It provides opportunities for new life. I got married recently as a way of escaping challenges associated with orphanhood.

During focus group discussion, the participants agreed that, child marriages were common among children in grandparent headed households of ward 14, Bikita district. The participants added that child marriages provided better opportunities for them given the challenges associated with orphanhood.

One of the social workers confirmed that child marriages has been associated with a number of children in grandparent headed households of ward 14, Bikita district. The social worker explained that, in the eyes of the adolescent girls, child marriage was a viable option, although that further exposed them to abuse, vulnerabilities among others.

In Shona indigenous knowledge tradition, marriage is governed by the institution of roora (bride wealth), which is a carefully negotiated social contract between families premised on the maturity and consent of both parties; the premature entry of adolescent girls into marriage due to economic desperation represents a corruption of this institution and a failure of the community's customary protective mechanisms.

7. Child Labour

Child labour is one of the challenges affecting children in grandparent headed households of ward 14, Bikita district. One of the participants had this to say;

We engage in child labour. Some of us we engage in herding cattle. Others are in artisanal mining at Bikita Minerals. We are not in schools.

Another one, called J said the following;

Life is tough for us as children staying with grandparents. I have to supplement household income through working in other people's fields. The money I generate sustains our household. I stay with my maternal grandmother and my younger siblings.

During a focus group discussion, the participants agreed that, child labour was rampant among children in grandparent headed households of ward 14, Bikita district. The participants added that through child labour the children were able to raise income for the households.

A village head also confirmed that, some children under grandparent headed households of ward 14, Bikita district were involved in child labour practices. He added that the majority of adolescent children were involved in artisanal mining, some were also in vending, others were employed as domestic workers among others.

Indigenous knowledge traditions in the Shona context distinguish between constructive intergenerational labour socialisation — where children learn farming, herding, and craftsmanship under adult guidance as part of kurera (upbringing and moral formation) — and exploitative child labour driven by household deprivation, which represents a breakdown of the community's protective obligations.

8. Stigma and Discrimination

Stigma and discrimination is one of the challenges affecting children under grandparent headed households of ward 14, Bikita district. One of the participants had this to say;

Orphans are stigmatised or labeled in this area. We are treated differently. We are not even recognised during community meetings.

Another participant had this to say;

It is a curse to be an orphan in this ward. As orphans we do not exist in many people's minds in this area. We are treated badly. No one recognise us. It is as if we are not human beings.

Another participant said the following;

No one cares about us as orphans. We are given derogatory names. We are not in schools.

During focus group discussions, the participants agreed that, stigma and discrimination was one of the challenges affecting children under grandparent headed households. The participants added that stigma and discrimination is demonstrated by the derogatory names given to the orphans among others.

A councilor of ward 14, Bikita district also confirmed that, stigma and discrimination is one of the challenges affecting children in grandparent headed households. He added that stigma and discrimination among these children was illustrated in many ways such as derogatory names, denied access to basic services among others.

From an indigenous knowledge perspective, such treatment of orphaned children violates the foundational principle of unhu (humanness), which recognises every child, regardless of parental status, as a full member of the community deserving dignity, care, and inclusion in all communal spaces and activities.

B. OPPORTUNITIES AND PROTECTIVE FACTORS WITHIN THE EXTENDED FAMILY SYSTEM

While the preceding findings document the significant challenges facing children in kinship care in Ward 14, the data also reveal elements of resilience, opportunity, and protective functioning within the extended family system that merit recognition. These opportunities, though operating within severely constrained conditions, are important for understanding the full picture of children's experiences and for informing policy interventions that build on existing strengths rather than focusing exclusively on deficits.

1. Kinship Care Placements as Social Protection

First, the continued existence of kinship care placements itself represents a form of social protection. Despite the severe economic pressures documented above, grandparents, aunts, and uncles in Ward 14 have continued to absorb children who have lost parental care through death, migration, or divorce. This persistence of kinship obligation, rooted in the Ubuntu philosophy, demonstrates that the extended family system, while severely strained, has not entirely collapsed. Participant E's account powerfully illustrates this continued kinship absorption despite extreme hardship:

This persistent kinship absorption reflects the indigenous Shona knowledge concept of ukama (relatedness and mutual obligation), which mandates that blood relatives and clan members assume responsibility for orphaned children as a moral and spiritual duty, encapsulated in the proverb mwana wevhu ndewese (the child of the soil belongs to all).

> Both of my parents died when I was a toddler. I am told they succumbed to the deadly HIV and AIDS disease. I was left in the care of my grandmother who is very poor.

Similarly, Participant J described a household where the grandmother has taken on responsibility for multiple children:

> Life is tough for us as children staying with grandparents... I stay with my maternal grandmother and my younger siblings.

Participant F further attested to this continued kinship functioning

> I stay with my grandmother. We normally face shortages of basic needs. Some of the basic needs such as food supplies are always in short supply. We do not have enough money to buy sugar, bread, cooking oil and maize flour for sadza.

Participant A also acknowledged the grandmother's continued care within constrained circumstances:

> My health is poor. I am always sick and it is extremely difficult to have access to medical care... it is extremely difficult for us given the financial status of our grandmother".

Several participants, including those facing significant hardship, expressed awareness that their caregivers were doing the best they could under difficult circumstances. What is notable across all these accounts is that despite documenting severe deprivation, every participant remained within a kinship care arrangement, and several implicitly acknowledged the care effort of their grandparents even while describing the limitations of that care. The majority of participants (eight out of twelve) were living in grandparent-headed households, while two were in the care of aunts and two were staying with uncles, demonstrating the breadth of kinship absorption across different family branches.

2. Community Leadership Structures as Protective Platforms

Second, the involvement of community leadership structures, including traditional chiefs, village heads, church pastors, and councilors, as key informants in this study reflects the continued engagement of local institutions in monitoring and advocating for children's welfare. The village head's awareness of child labour practices was detailed and specific: he confirmed that:

Some children under grandparent headed households of ward 14, Bikita district were involved in child labour practices....the majority of adolescent children were involved in artisanal mining, some were also in vending, others were employed as domestic workers among others.

The village head also demonstrated awareness of educational challenges, confirming that:

Lack of school fees and uniforms among children in grandparent headed households is contributing towards school drop outs. The grandparents are having financial challenges in raising money towards school fees and uniforms for their grandchildren.

The councilor's knowledge of stigma and discrimination was equally detailed. He confirmed that:

Stigma and discrimination is one of the challenges affecting children in grandparent headed households.

The social worker key informant provided institutional-level insight, confirming that

Lack of support from child welfare agencies was a challenge affecting children in grandparent headed households of ward 14, Bikita district.... some of the non-state actors in the provision of child welfare services were closing as a result of a number of factors such as donor flight among others.

Collectively, these key informant accounts suggest that community-based protective structures remain active, even if under-resourced. These structures represent potential platforms for strengthened intervention.

The engagement of traditional chiefs (madzishe) and village heads (masvikiro/sabhuku) reflects the indigenous knowledge institution of dare (the traditional community council), which historically served as the primary governance structure for resolving disputes, protecting vulnerable community members, and upholding collective moral obligations toward children.

3. Children's Adaptive Coping Strategies and Self-Advocacy

Third, the coping strategies documented among children, while sometimes involving harmful behaviours, also include adaptive responses such as supplementing household income through work in neighbours' fields, maintaining social bonds with peers, and expressing agency through focus group participation. Participant J's account of working to sustain the household demonstrates a form of resilience and responsibility, even as it simultaneously reflects the burden of child labour:

Life is tough for us as children staying with grandparents. I have to supplement household income through working in other people's fields. The money I generate sustains our household. I stay with my maternal grandmother and my younger siblings.

Another participant described the collective nature of children's coping strategies:

We engage in child labour. Some of us we engage in herding cattle. Others are in artisanal mining at Bikita Minerals. We are not in schools.

The willingness of all twelve participants to share their experiences in interviews and focus group discussions itself reflects a capacity for self-advocacy that can be harnessed in future interventions. The accounts of participants who spoke with remarkable candour about their circumstances demonstrate this self-advocacy. Participant I, for example, demonstrated agency in narrating her own decision-making:

I am 17 years old and I got married last year when I was 16 years. I decided to get married as a way of running away from the challenges I was facing as an orphan under grandparent headed household.

Participant C also articulated her survival strategy with directness:

I sell sex. I raise money for survival through selling sex with gold panners. With my friends, we travel to Bikita Mineral where there is mining of lithium, and sell sex to artisanal miners.

While these coping strategies are harmful and reflect the severity of deprivation, the participants' capacity to articulate their experiences, analyse their own circumstances, and engage in collective discussion during focus groups demonstrates a form of agency and self-advocacy that constitutes a protective resource for future intervention design.

4. Emotional Bonds and Cultural Continuity

Several participants indicated that despite their challenges, the emotional bonds with their grandparent caregivers provided a degree of stability and belonging. The language used by participants throughout their accounts reveals the relational anchoring provided by kinship care. Participant F consistently referred to *"my grandmother"* when describing household circumstances, Participant G referenced *"my paternal grandmother"* with a sense of familial belonging, and Participant J spoke of *"my maternal grandmother and my younger siblings,"* indicating a functioning, if strained, family unit. Participant E's account, while documenting extreme hardship, also reveals an unbroken kinship bond:

Both of my parents died when I was a toddler. I am told they succumbed to the deadly HIV and AIDS disease. I was left in the care of my grandmother who is very poor. I did not go to school. How do you expect me to survive.

Even in expressing frustration, Participant E's narrative is situated within a continuous caregiving relationship with the grandmother, reflecting the persistence of emotional attachment and belonging within kinship arrangements.

The continued functioning of cultural practices, such as the rain-making ceremonies described by Participant F, suggests that cultural continuity within kinship arrangements offers a form of psychosocial anchoring that formal welfare systems cannot easily replicate. Participant F's account of this cultural practice is particularly illustrative:

I take on alcohol. My first experience taking alcohol has been during the traditional rain making ceremonies. Some two years ago, my grandmother was preparing the mukwerera. I took some traditional brew during the ceremony.

While Participant F's account is situated within the problematic theme of substance misuse, the underlying cultural practice of the *mukwerera* (rain-making ceremony) reflects the continued transmission of cultural traditions within kinship households. The grandmother's active role in preparing the ceremony demonstrates that kinship arrangements serve not only as material care environments but also as sites of cultural reproduction and intergenerational continuity.

Indigenous knowledge systems recognise such intergenerational cultural transmission as ndangariro (living memory), wherein ceremonies, rituals, and oral traditions performed within kinship households serve as foundational mechanisms for transmitting identity, values, and a sense of belonging to children across generations.

DISCUSSION

This study makes a distinctive empirical contribution by centering voices and lived experiences of children in kinship care in rural Zimbabwe during the austerity era, a perspective largely absents from existing literature. Findings reveal that children under extended family systems in Ward 14, Bikita district, face a constellation of interrelated challenges shaped by the intersection of household deprivation, institutional support deficits, and broader economic constraints of the austerity era. Simultaneously, data reveal elements of resilience, cultural continuity, and community engagement representing opportunities for strengthened intervention. This discussion interprets key findings through Ubuntu Philosophy and the austerity framework established in this study, while connecting them to broader literature reviewed above.

Research findings suggest children under extended family systems in Ward 14, Bikita district face myriad challenges including child marriages, school dropouts, child labour, stigma and discrimination, lack of support from child welfare agencies, substance misuse, and lack of basic needs such as food and health. Drawing from Ubuntu Philosophy, challenges documented in this study can be understood as expressions of systemic dysfunction within extended family units, rather than isolated individual failings. Withdrawal of institutional support from child welfare agencies, documented through participants L and K's accounts and confirmed by key informants, represents a change in external support systems cascading through family units, weakening caregivers' capacity to meet children's basic needs. This in turn drives children into coping strategies such as child labour, commercial sex work, and early marriage, each further destabilizing family systems by removing children from education, exposing them to health risks, and altering family relational dynamics.

These challenges are further illuminated through the indigenous knowledge lens of zvido zvemhuri (family obligations), which historically positioned the entire kinship network as duty-bound to ensure the physical, emotional, and moral well-being of every child within the lineage.Children's experiences of stigma and discrimination, as described by participants reporting derogatory names and exclusion from community meetings, reflect their peripheral positioning within both family and community systems. Through the Ubuntu Philosophy, such marginalisation represents disruption of connectedness degrees essential for healthy family functioning (Mabetha et al., 2024). Similarly, substance misuse and sexual behaviours documented among adolescent participants can be interpreted through the theory's focus on adaptation to changes within family systems: as families' protective and supervisory capacity diminishes under austerity-era constraints, children develop alternative behavioural patterns reflecting absence of adequate guidance, structure, and material provision.Within indigenous Shona knowledge frameworks, such exclusion contradicts the principle of unhu (personhood through communal belonging), as the community's dignity is inseparable from the dignity accorded to its most vulnerable members, including orphaned children.

Critically, austerity-era conditions identified in this study, including price increases, wage erosion, informalization, and donor contraction, function as structural drivers weakening family systems from outside. Non-governmental organisations' closure due to donor flight, as reported by social worker key informants, directly reduces institutional support available to kinship households. Grandparents' inability to afford school fees, medication, and food, documented across multiple participants, reflects material impact of austerity on household resources. These are not failures of family systems in isolation but consequences of broader political-economic processes systematically undermining conditions under which kinship care can function effectively.

Children in grandparent-headed households of Ward 14, Bikita district exhibited behavioural problems including sexual behaviours in various forms such as bestiality, commercial sex, and masturbation. Research findings corroborate observations by Kelley et al. (2023), noting some grandchildren under grandparents in the United States had behaviour-related problems resulting from grandparents' high levels of psychological problems or distress. In Hill Crest, Alice township, South Africa, grandchildren under grandparent households exhibited challenging behaviours including stealing and substance misuse, severely affecting grandparents' psychological and physical well-being (Kang'ethe et al., 2022). In another study of grandparent-headed households of Mdlankomo location, Libode, Eastern Cape, South Africa, grandchildren also exhibited challenging behaviours such as substance abuse (Soganga and Kang'ethe, 2023).

However, it is essential to move beyond describing these behaviours as evidence of family dysfunction and recognise structural conditions producing them. Literature from both global and regional contexts consistently demonstrates behavioural challenges among children in kinship care are associated with poverty, limited institutional support, psychosocial distress, and inadequate supervision capacity among elderly caregivers (Farmer, 2009; Fergeus et al., 2019; Kang'ethe et al., 2022). In Ward 14, these structural drivers are amplified by austerity-era conditions simultaneously increasing household deprivation and reducing formal support services availability. Family systems, as conceptualised by Bowen, cannot be held solely responsible for outcomes fundamentally shaped by forces external to family units.Indigenous knowledge systems in the Shona context recognise this interdependence through the concept of ukama (relatedness), which frames child welfare as a communal rather than individual responsibility; when ukama is weakened by external structural forces, children inevitably bear the consequences.

Children in grandparent-headed households of Ward 14, Bikita district faced challenges such as lack of access to basic needs. Their households were poverty-stricken to the extent that accessing basic needs such as food and health was very difficult. Evidence suggests due to lack of basic needs, some children resorted to various options such as child labour (domestic work, artisanal mining) to supplement household income. Similarly, grandchildren of Buffalo City Metropolitan, South Africa faced challenges induced by limited financial resources, resulting in lack of access to basic needs such as health and food (Folorunsho and Tanga, 2021).

Alongside these challenges, findings also reveal important opportunities and protective factors informing future interventions. Persistence of kinship care obligations, continued engagement of community leadership structures, and cultural practices providing psychosocial anchoring for children all represent strengths within extended family systems that can be built upon. Through Ubuntu Philosophy, these represent areas where systems' capacity for adaptation and interconnectedness remains functional, even under severe strain. Ubuntu-based collectivist orientation underpinning kinship care in Zimbabwe (Mushunje, 2014; Ringson and Chereni, 2020) provides cultural foundation for interventions strengthening family-based care rather than replacing it with institutional alternatives. Active involvement of village heads, councillors, and church leaders in monitoring children's welfare suggests community-based child protection structures, if adequately resourced and coordinated with formal welfare systems, could serve as effective platforms for supporting kinship households.

From an indigenous knowledge standpoint, these protective factors resonate with the Shona concept of roorano (solidarity through kinship networks), wherein the extended family serves not merely as a material safety net but as a living repository of cultural wisdom, moral guidance, and collective identity for children.

FUTURE RESEARCH DIRECTIONS

The book chapter focused on challenges and opportunities for children under extended family system in ward 14 of Bikita district, Zimbabwe. The study suggests the following future research directions;

Firstly, there is need for future research studies to focus on the interface of traditional and modern child sensitive social protection systems toward children in ward 14 of Bikita district, Zimbabwe,

Secondly, there is need for another study that focuses on the strategies that enhance the role of extended family systems towards care of children in ward 14 of Bikita district using mixed methodological approach,

Thirdly, there is need for a study that focuses on political economy of extended families taking care of the children in ward 14 of Bikita district, Zimbabwe.

CONCLUSION

Children under extended family systems in Ward 14, Bikita district, Zimbabwe face myriad challenges including lack of support from child welfare agencies, stigma and discrimination, substance misuse, shortage of basic needs such as food and health, school dropouts, child marriages, child labour, and sexual behaviours (commercial sex, forced sex, masturbation, bestiality). Utilising Ubuntu Philosophy, this study demonstrates these challenges are not isolated individual problems but systemic expressions of kinship care systems operating under severe strain. Austerity-era conditions including price increases, wage erosion, informalization, and donor contraction function as structural drivers weakening family systems from outside, reducing both household resources and institutional support. Within this constrained environment, children develop coping strategies, some adaptive and some harmful, reflecting family systems' diminished capacity for protection, supervision, and guidance. The study identifies important opportunities including persistence of kinship obligations, community leadership engagement, cultural continuity, and children's resilience, providing foundations for strengthened interventions. These findings contribute child-centred, qualitative evidence from rural Zimbabwe during the austerity era, addressing gaps in existing literature

Rooted in the indigenous knowledge principle of nhaka (cultural heritage and intergenerational responsibility), the recommendations emerging from this study call for the revitalisation of community-based care systems that honour ancestral obligations to protect children through collective action.

RECOMMENDATIONS

The study proffers a number of recommendations. Some of them include;

- The Department of Social Development should register and monitor all the children under extended family system in rural Zimbabwe. A caseworker or social worker should be assigned to supervise the children under kinship care in rural Zimbabwe,
- The Government of Zimbabwe and other duty bearers should provide more services to children under extended family system in rural Zimbabwe. This could be though expanding the existing

social protection mechanisms to take on board children under kinship care in rural communities of Zimbabwe,

- The community-based child protection structures such as community child care workers should provide more support to children under kinship care in rural communities of Zimbabwe,
- There is need for building capacities of grandparents headed households on child care. The training programme could include child rights (abuse and neglect) among others,
- There is need for hamornisation of indigenous and modern social protection programmes so as to ensure improved performance towards the children under grand parent headed households among others.

REFERENCES

Chiimba, C. M. (2020). *Experiences of young orphans who transition into extended families in Zimbabwe* [Unpublished doctoral thesis]. Faculty of Education, University of Pretoria.

Chikoko, W., & Ruparanganda, W. (2020). *Ubuntu* or *Hunhu* perspective in understanding substance abuse and sexual behaviours of street children of Harare Central Business District, Zimbabwe. *African Journal of Social Work, 10*(1).

Farmer, E. (2009). How do placements in kinship care compare with those in non-kin foster care: Placement patterns, progress and outcomes? *Child & Family Social Work, 14*(3), 331–342. DOI: 10.1111/j.1365-2206.2008.00600.x

Fergeus, J., Humphreys, C., Harvey, C., & Herrman, H. (2019). Supporting foster and kinship carers to promote the mental health of children. *Child & Family Social Work, 24*(1), 77–83. DOI: 10.1111/cfs.12583

Folorunsho, F., & Tanga, P. T. (2021). Grandparenting and extent of fulfilling the psychosocial well-being of adolescents from single parent homes in Buffalo City Metropolitan, South Africa. *African Journal of Social Work, 11*(2).

Gentles-Gibbs, N., & Zema, J. (2020). It's not about them without them: Kinship grandparents' perspectives on family empowerment in public child welfare. *Children and Youth Services Review, 108*, 104650. DOI: 10.1016/j.childyouth.2019.104650

Gladstone, J. W., & Brown, R. A. (2007). Grandparents' and social workers' experience with child welfare system: A case for mutual resources. *Children and Youth Services Review, 29*(11), 1439–1453. DOI: 10.1016/j.childyouth.2007.07.002

Ihara, E. S., Tompkins, C. J., & Sonethavilay, H. (2012). Culture and familism: An exploratory case study of grandparent-headed households. *Journal of Intergenerational Relationships, 10*(1), 34–47. DOI: 10.1080/15350770.2012.645737

Inchaurrondo, A. M., Bailon, A. B., Vicente, C. P., Tio, E. V., & Bolos, A. M. (2015). Risk and protective factors associated with kinship care. *Child & Adolescent Social Work Journal, 32*(5), 417–427. DOI: 10.1007/s10560-015-0382-8

Kang'ethe, S. M., Mugedya, S., & Nomngenyiya, T. (2022). Experiences of grandmothers looking after grandchildren exhibiting delinquent behaviours in Hill Crest location, Alice township, Eastern Cape: Implications for social work. *Social Work/Maatskaplike Werk, 58*(3 & 4).

Kasiram, M., & Holscher, D. (2015). Understanding the challenges and opportunities encountered by the elderly in urban KwaZulu Natal, South Africa. *South African Family Practice : Official Journal of the South African Academy of Family Practice/Primary Care, 57*(6), 380–385. DOI: 10.1080/20786190.2015.1078154

Kelley, S. T., Whitley, D. M., & Lamis, D. A. (2023). An intervention to improve the well-being of families in which African American grandmothers are raising grandchildren: A longitudinal mediation analysis. *Children and Youth Services Review, 151*, 107021. DOI: 10.1016/j.childyouth.2023.107021

Mabetha, K., DeWet-Billings, N. C., & Odimegwu, C. O. (2024). Exploring the effects of kin's caregivers' family functioning on child survival in South Africa: An application of the Family Systems Theory. *Child & Family Social Work, 29*(1), 58–77. DOI: 10.1111/cfs.13051

Malinga-Masumba, T. (2015). The nature of relationships between orphans and their kinship carers in Botswana. *Child & Family Social Work, 20*(3), 257–266. DOI: 10.1111/cfs.12121

Manful, E., & Cudfoe, E. (2018). Is kinship failing? Views on informal support by families in contact with social services in Ghana. *Child & Family Social Work, 23*(4), 617–624. DOI: 10.1111/cfs.12452

Matshalaga, N. (2004). *Grandmothers and orphan care in Zimbabwe.* SAFAIDS.

Morreira, S. (2013). *Transnational human rights and local moralities: The circulation of rights discourses in Zimbabwe and South Africa* [Unpublished doctoral thesis]. Department of Social Anthropology, University of Cape Town.

Mugumbate, J., & Chereni, A. (2019). Using African Ubuntu theory in social work with children in Zimbabwe. *African Journal of Social Work, 9*(1), 27–34.

Mugumbate, J., & Nyanguru, A. (2013). Exploring African philosophy: The value of *ubuntu* in social work. *African Journal of Social Work, 3*(2), 82–100.

Mushunje, M. T. (2014). Interrogating the relevance of the extended family as a social safety net for vulnerable children in Zimbabwe. *African Journal of Social Work, 4*(2), 1–10.

Ramose, M. B. (2002). *African philosophy through Ubuntu.* Mond Books Publishers.

Ringson, J., & Chereni, A. (2020). Efficacy of the extended family system in supporting orphans and vulnerable children in Zimbabwe: An indigenous knowledge perspective. *African Journal of Social Work, 10*(1).

Samkange, S., & Samkange, T. M. (1980). *Hunhuism or Ubuntuism: A Zimbabwe indigenous political philosophy.* Graham Publishing.

Selwyn, J., & Nandy, S. (2014). Kinship care in the UK: Using census data to estimate the extent of formal and informal care by relatives. *Child & Family Social Work, 19*(1), 44–54. DOI: 10.1111/j.1365-2206.2012.00879.x

Soganga, A., & Kang'ethe, S. M. (2023). Exploring reciprocity in grandparent-grandchildren care: The case of Mdlankomo location, Libode in the Eastern Cape, South Africa. *Social Work/Maatskaplike Werk, 59*(1).

Taylor, E., DiFolco, S., Dupin, M., Mithen, H., Wen, L., Rose, L., & Nisbert, K. (2020). Socioeconomic deprivation and social capital in kinship carers using a helpline service. *Child & Family Social Work, 25*(4), 845–855. DOI: 10.1111/cfs.12763

Xu, Y., Zhao, Q., Schuler, B. R., & Levkoff, S. E. (2022). Material hardship among custodial grandparents in COVID-19 and its association with grandchildren's physical and mental health: A latent class analysis. *Children and Youth Services Review, 132*, 106340. DOI: 10.1016/j.childyouth.2021.106340

ADDITIONAL READING

Farmer, E. (2009). How do placements in kinship care compare with those in non-kin foster care: Placement patterns, progress and outcomes. *Child & Family Social Work, 14*(3), 331–342. DOI: 10.1111/j.1365-2206.2008.00600.x

Folorunsho, F., & Tanga, P. T. (2021). Grandparenting and the psychosocial well-being of adolescents from single parent homes in Buffalo City Metropolitan, South Africa. *African Journal of Social Work, 11*(2).

Mugumbate, J., & Chereni, A. (2019). Using African Ubuntu theory in social work with children in Zimbabwe. *African Journal of Social Work, 9*(1), 27–34.

Mushunje, M. T. (2014). Interrogating the relevance of the extended family as a social safety net for vulnerable children in Zimbabwe. *African Journal of Social Work, 4*(2), 1–10.

Ringson, J., & Chereni, A. (2020). Efficacy of the extended family system in supporting orphans and vulnerable children in Zimbabwe: An indigenous knowledge perspective. *African Journal of Social Work, 10*(1).

Samkange, S., & Samkange, T. M. (1980). *Hunhuism or Ubuntuism: A Zimbabwean indigenous political philosophy*. Graham Publishing.

KEY TERMS AND DEFINITIONS

Austerity era: The austerity era in Zimbabwe denotes the period from the late 2010s onwards characterised by persistent price increases, currency instability, wage erosion, economic informalisation and contraction of donor funding, conditions that weaken extended family systems and limit their capacity to meet children's basic needs and protection requirements.

Extended family system: The extended family system refers to a kinship structure that goes beyond the nuclear family to include grandparents, aunts, uncles and other relatives who share responsibility for the care, socialisation and welfare of children, including orphans and vulnerable children, often functioning as a primary social safety net in Zimbabwe.

Kinship care: Kinship care is a form of family-based alternative care in which children who lack adequate parental care are placed with relatives or close family friends, such as grandparents, aunts or uncles, in either formal or informal arrangements that provide day-to-day support and protection.

Ubuntu/Hunhu philosophy: Ubuntu or Hunhu is an African moral philosophy originating among Bantu-speaking peoples, grounded in values such as humanness, collective unity, compassion, social justice and the idea that "a person is a person through other persons," which underpins communal child-rearing, mutual obligations and the expectation that kin care for orphans and vulnerable children.

ENDNOTES

1 Chamba is marijuana

[2] Sadza is a staple food in Zimbabwe

Chapter 11

Incorporating Indigenous Knowledge Systems as a Strategy to Effective Cross-Cultural Communication in the Military:

A Case of the South African National Defence Force (SANDF)

Mamogobo Rosinah Mamabolo
http://orcid.org/0000-0002-0574-1751
University of Limpopo, South Africa

ABSTRACT

This chapter examines the importance of Indigenous Knowledge Systems (IKS) as key strategies for enhancing cross-cultural communication within the South African National Defence Force (SANDF). The chapter employs Oetzel's (2005) Effective Intercultural Workgroup Communication Theory as the guiding theoretical framework to deepen understanding of the cultural factors that influence workgroup diversity and cross-cultural communication among service members. Employing a qualitative research method through an analysis of documents, the chapter indicates that IKS philosophies can enhance solidarity, reduce conflict and cultural misunderstanding, as well as improve intercultural and cross-cultural interactions within the SANDF. The chapter concludes that integrating IKS as a strategy to cross-cultural communication creates a representative, effective, culturally competent, and united military force.

INTRODUCTION

The world over, communication remains an integral part of human life, which plays a crucial role in the effective management of various types of organisations across numerous fields. Its significance thrives on its ability as a social interaction of creating meaning, exchanging, sharing ideas, informa-

DOI: 10.4018/979-8-2600-2890-2.ch011

tion, opinions, facts and experiences between communicating parties (Abuarqoub, 2019). According to Heinecken (2020A), globalisation and the integration and inclusion policies brought about diversity, which created the need for cross-cultural communication in various workplaces. Aspects such as labour, education and training environments have become the primary areas of diversity, and the military context is no exception. The modern military system comprises uncertainties and constant changes that challenge the "warfighting standard of the past" decades (Heinecken, 2011: 26). Armed forces have become more diverse in terms of religion, gender, race, language and culture with various forces being deployed and trained with or within other nations that feature cultural differences (Soeters & Van Der Meulen, 2007). The increasing diverse military and military recruitment processes pose a challenge to the traditional military institutional values, attitudes, beliefs and norms. In South Africa, the 1994 integration and absorption of the South African Defence Force (SADF) and revolutionary armies into the South African National Defence Force (SANDF) evolved with a multi-cultural military context which brought about convergence among people from various ideological backgrounds and ethnic groups. These diverse groups may have conflicting values and communication challenges characterised by language barriers, stereotypes, racism, ethnocentrism and prejudice, which must be managed carefully if cultural minorities are to be recruited and retained within the military (Pinch, MacIntyre, Browne & Okros, 2004; Dodd, Bester & Van Der Merwe, 2020).

Diverse military contexts require diversity management as it demands that people work together to deal with diversity issues such as culture, language and cross-cultural communication between parties. Amongst others, the SANDF has implemented a National Defence White Paper (1996), Defence Act (2002), Promotion of Equality and Prevention of Unfair Discrimination Act (2000), White Paper on Affirmative Action in the Public Service (1998) and Defence Review (1998; 2015) that addressed issues related to transformation, diversity, diversity management and equal treatment of diverse members within the SANDF. However, despite these transformative post-apartheid legislative frameworks, a critical communication practice gap persists within the force. Implemented policies aimed at addressing the past injustices have succeeded in the demographic representation, consequently, communication processes shaped by gendered power relations, language and culture remain unreconstructed. Additionally, cross-cultural communication required in diverse military settings remain problematic and constitutes as the primary gap in this chapter. Therefore, this chapter argues that incorporating IKS into the SANDF can enhance cultural competence and create effective communication and cross-cultural communication amongst service men and women. The chapter aims to examine the importance of IKS as a key strategy for enhancing cross-cultural communication within the SANDF. Through the systematic literature review, the chapter seeks to analyse the relevance of IKS to cross-cultural communication and its role in promoting cohesion and understanding amongst culturally diverse members of the SANDF.

This chapter commences by providing the historical background of the South African military. Secondly, a theoretical framework used to guide the chapter is presented. The section is followed by a comparative analysis of Western linear and dialogic (IKS) approaches to communication, a review on the cross-cultural communication barriers, the integration of IKS into the military and IKS as a complementary mechanism to enhance collaboration and mutual understanding amongst SANDF personnel. The chapter further presents the research methodology, findings and discussions as well as the recommendations for future research.

THE HISTORICAL BACKGROUND OF THE SOUTH AFRICAN MILITARY

Pre-1994, the South African military, previously known as the Union Defence force (1912) and later transformed to the SADF (1957) was characterised by a uniform racial group dominated by White males with the aim to defend South African citizens and their property as well as to defend the country against any internal or external threats (De la Rey, 1990; Scott, 2002; Lamb, 2018). The racial uniform nature of the SADF was driven by the country's discriminatory legislative frameworks that excluded previously disadvantaged racial groups such as Black Africans, Indians and Coloureds. The unequal discriminatory laws were however challenged by the emergence of the liberation amies or paramilitary wings of the African National Congress (ANC), uMKhonto we Sizwe and the Pan African Congress (PAN), Azanian People's Liberation Army (APLA), as well as guerrilla and homeland armies that promoted democracy, equality and racial inclusion of diverse members (Williams, 2001; Veldtman, 2018).

During 1990, the statutory forces which included the conventional SADF and the homeland armies known as the Transkei, Ciskei, Venda and Boputhatswana Defence Force as well as the non-statutory forces or liberation armies, the uMKhonto we Sizwe and APLA (Lamb, 2018; le Roux, 2005; Van Der Waag, 2015) negotiated and integrated to form a transformed, balanced, modern, technologically advanced defence force that promotes non-partisan as well as gender and racial equality, the SANDF (Davis, 2018; Veldman, 2018; Magadla, 2023).

The SANDF was formed on the 27 of April 1994 and promoted inclusion of diverse members within the force regardless of gender, culture, religion, racial and linguistic backgrounds. The force evolved with the transformation of defence policy, structure and adaptation of new force design (Van Der Waag, 2015 in Mamabolo, Le Roux & Muswede, 2025). Diverse members from various backgrounds and worldviews were recruited to join the force and served in various mustering (Heinecken, 2020A). Figure 1 illustrates the statutory and non-statutory forces that integrated to form the SANDF.

Figure 1. The evolution of the SANDF (Adapted from Turner, 2015)

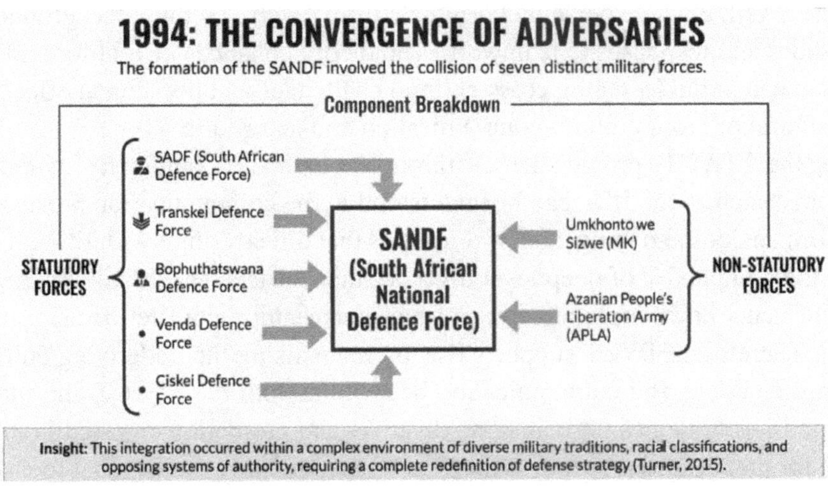

*(Kolodziej-Smith, 2016; Oetzel, 2005; 2017).

THEORETICAL FRAMEWORK

This chapter is guided by the Effective Intercultural Workgroup Communication theory (EIWCT), which provides a comprehensive framework for understanding how individual's attitudes, cultural values and situational aspects (deep-level aspects) affect communication, behaviour and decision-making process amongst diverse groups. EIWCT challenges the traditional approaches to diversity that addresses surface-level diversity including gender, race and age through social categorisation perspectives or information exchange (Oetzel, 2005; 2017). EIWCT emphasises on addressing the deep-level diversity that encompass intragroup variances such as personality, values, attitudes and culture which have significant role in influencing communication and group outcomes (Harrison, Price, Gavin & Florey, 2002; Oetzel, 2017).

EIWCT asserts that both situational factors (such as status inequalities, group history and task structure) and individual or group-level factors (including self-construal, face concerns and group diversity) interact to influence the communication climate and the efficacy of group interactions (Oetzel, 2005; Oetzel, McDermott, Torres & Sanchez, 2012). The theory posits that both heterogeneous and homogeneous cultural traits within a group can influence the quality of group performance and the interaction process (Oetzel, 2017). Amongst others, individualistic - those believing in individual appraisal - and collectivistic - those believing in collaboration and group work - cultural orientations affect communication outcomes as well as the quality of group performance in diverse working environments (Kolodziej-Smith, 2016; Oetzel, 2017; Tiatira, Herwandito & Krisnawati, 2024). Therefore, Oetzel (2005; 2017) suggest that for effective workgroup communication, diverse members must understand the distinct cultural orientations portrayed by diverse members and the influence they pose on communication.

In the context of this chapter, the EIWCT was employed to facilitate an in-depth understanding of how cultural characters shape cohesion and effective cross-cultural communication in the SANDF. Unlike traditional models or approaches to diversity, EIWCT focusses on the communication process. The theory examines how situational and cultural factors shape the communication process of diverse members. It focuses on understanding how diverse cultural traits affects the process of transferring information, building relationships and resolving problems. This is pertinent in diverse military contexts, particularly in the SANDF where personnel originate from diverse cultural backgrounds affecting their behaviours, attitudes and responses to teamwork. The theory enhances insight for understanding deep-level diversity that can aid in lessening cross-cultural challenges and implement effective strategies for enhanced intercultural or cross-cultural communication and teamwork.

Furthermore, the EIWCT, provide lenses through which the complexity of cultural embedded knowledge systems, such as the IKS, can be understood in the communication process. IKS are rooted in deep cultural dimensions and indigenous worldviews that directly align with EIWCT that focusses on comprehending the complexity of deep level diversity than surface level diversity. In diverse military institution, IKS influence how people perceive and interpret meaning, conflict, decision making, authority and cooperation. Therefore, EIWCT supports IKS by recognising the underlying cultural orientations that shapes group dynamics and communication behaviour. Both the EIWCT and the IKS contribute to culturally responsive strategies for managing diversity and promoting cross-cultural communication as they advocate for organisations to transition from Western centric models that focus on surface level diversity to a more indigenous perspective that emphasis on understanding the deep cultural variations of individuals (refer to Table 1). This approach will not only improve communication, but will also support the development of cohesive and culturally competent military force.

WESTERN LINEAR VS INDIGENOUS KNOWLEDGE SYSTEMS (IKS) DIALOGIC APPROACHES TO COMMUNICATION

This section provides a comparative table (Table 1) contrasting Western linear and IKS dialogic approaches that serve as lenses to comprehend and highlight the demerits of relying solely on Western linear models to communication in diverse organisational contexts such as the SANDF.

Table 1. Western Linear vs Indigenous Knowledge Systems (IKS) Dialogic Approaches to Communication

Dimension	Western Linear Approach	IKS Dialogic Approach
Communication Style	Linear, direct, sender–receiver model	Circular, relational, dialogic and participatory
Knowledge Orientation	Objective, universal, and often decontextualised	Contextual, experiential and rooted in local realities
View of Meaning	Meaning is transmitted and fixed	Meaning is co-created through interaction
Decision-Making	Individualistic, often hierarchical	Collective, consensus-driven
Cultural Focus	Emphasis on surface-level diversity (e.g., race, gender, age)	Emphasis on deep-level diversity (values, beliefs, worldviews)
Conflict Resolution	Problem-solving oriented, often task-focused	Relationship-centred, restorative and community-oriented
Role of Authority	Formal, rank-based authority structures	Respect-based, often influenced by elders, tradition and social roles
Approach to Knowledge Systems	Western-centric, privileging scientific rationality	Inclusive of indigenous epistemologies and lived experiences
Communication Goal	Efficiency, clarity and task completion	Understanding, harmony and relational cohesion
View of Diversity	Managed through categorisation and compliance	Embraced as dynamic and embedded in cultural identity
Relevance to Military Context (SANDF)	Focus on command, control and standardised procedures	Enhances cohesion, mutual respect and culturally responsive teamwork

*(Kolodziej-Smith, 2016; Oetzel, 2005; 2017).

Table 1 illustrates the epistemological and communicative variations between Western linear approaches and IKS dialogic approaches. The comparisons indicate that Western communication approaches are linear, they perceive communication as a transactional process where information is transferred from the sender to the receiver. This viewpoint associates with traditional communication models and theories that focus on control, clarity, task execution and quantifiable results. Western approaches are further related to hierarchical authority and bureaucratic structures with centralised decision making (Gudykunst, 2003; Oetzel, 2005). In contrast, the IKS dialogic perspective sees communication as a relational, circular and participatory process in which meaning is created together through ongoing interaction. Instead of thinking about communication as mere transmission of information, IKS is concerned with joint understandings, mutual respect and harmony in social relations. This mirrors indigenous epistemologies that understand knowledge to be socially embedded, situated in context, and transmitted through lived experience, oral traditions and collective engagement (Dei, 2000; Odora Hoppers, 2002). The effect of this is that communication in IKS is less concerned with simply quickness and efficiency than it is with maintaining relationships and facilitating inclusive participation.

Western linear models largely favour individual autonomy, managerial authority and top-down leadership structures. Such characteristics may underpin rapid responses and defined lines of command, especially in highly formalised institutions. Western linear models also suppress dissenting voices and limit the ability to collaborate. On the contrary, IKS dialogic approaches emphasise consensus-building, communal responsibility and authority grounded in social legitimacy, wisdom and communal trust (Mbigi, 2005; Nzimakwe, 2014).

Another important point from Table 1 is about diversity management. For decades, Western research has centred on surface-level diversity categories (such as race, gender and age) using compliance-based diversity frameworks. Although these measures are important, they might not consider deeper cultural factors that account for values, belief systems, communication norms and worldviews. The IKS dialogic view is related to the EIWCT, which emphasises that deep-level diversity is important in shaping interaction processes and group effectiveness (Harrison et al., 2002; Oetzel, 2017). This indicates the necessity to attend not only to the apparent demographic differences but to the deeper underlying cultural orientations driving behaviour and interpretation in successful intercultural and cross-cultural communication.

In addressing issues around conflict, the Western perspective is generally linear, result-driven, whilst the dialogical character of IKS promotes restoration, reconciliation and the promotion of social cohesion. IKS approaches are thus important in multicultural organisations where interpersonal conflict may arise, affecting teamwork, morale and the effectiveness of an organisation. Conventional ways of mediating and repairing such challenges are generally not fought in an adversarial environment but are rather repaired through dialogue, mediation and collective responsibility (Murithi, 2006). In the military, particularly in the SANDF, both approaches provide significant albeit partial perspectives if used on their own. Western linear processes might be successful in operational command, procedural standardisation or rapid execution of tasks. Yet the SANDF is also functioning in a culturally diverse environment where cohesion, trust and intercultural understanding are considered equally important. The IKS dialogic method thus has an added value in generating an environment of mutual respect, culturally responsive leadership and enhanced interpersonal cohesion among personnel from different backgrounds. Combining both approaches, therefore, can enhance communication effectiveness by leveraging operational excellence and relational attunement.

Table 1 indicates that it is critical for contemporary organisations to shift the paradigm away from a single, West-centric approach to communication and towards more pluralistic forms that are informed by the perspective of the indigenous communities. Such integration is aligned with current demands for decolonised organisational approaches and culturally grounded leadership in South Africa. For the SANDF, this means adopting communication methods that take into account both the structural requirements of military institutions and the cultural realities of a heterogeneous workforce (Msila, 2015; Oetzel, 2017).

CROSS-CULTURAL COMMUNICATION BARRIERS IN MILITARY FORCES

The existing literature reviewed shows that cross-cultural communication is a practical necessity for contemporary military organisations reflecting a growing diversity of armed forces and the multinational character of contemporary military operations. Military institutions around the globe are composed of members of diverse racial, linguistic, ethnic, cultural and gender backgrounds who must collaborate to train, provide peacekeeping, undertake humanitarian operations and fight in combat. Despite this diversity contributing to representativity and institutional capacity, communication barriers that threaten operational effectiveness, command efficiency, trust and social cohesion persist (Gudykunst, 2005; Heinecken, 2011). Coordination, communication and overall military goals are directly impacted by language, which is considered a crucial component of daily military operations and training (Skilleås & Grande, 2024). Language difficulties arise during cross-cultural interactions as military organisations comprise individuals from various linguistic backgrounds (Knouse, 2001; Bakhaya, 2024). The differences across language proficiency, accents, terminology and comprehension frequently cause misunderstanding and a lack of mutual understanding during operational briefings, intelligence sharing and command execution in a multilingual force (Knouse, 2001; Bakhaya, 2024). In forces where English functions as the dominant language, such as US forces, UK forces and Israel, differences in military terminology based on proficiency level (Erwin, 2005; Poteet et al., 2008) and distinctions between British and American military jargon can cause confusion when coordinating operations. As a result, clear language usage is necessary for communication, coordination and effectiveness of the mission (Skilleås & Grande, 2024).

Extensive literature further highlights the influence of cultural diversity in cross-cultural interactions. Military personnel from different cultures have distinctive values, norms, leadership expectations, perceptions of authority and non-verbal communication attitudes. These differences affect how commands are interpreted, conflict is addressed and teamwork is performed (Georgieva & Marinov, 2017). For instance, Logan-Terry and Damari (2015) revealed that the United States soldiers participating in military training and serving overseas missions originate from different cultures and have different language capabilities and encounter difficulties with embodied cross-cultural communication (gestures and body formations). During peace support operations in India, a United States soldier and an Indian village elder encountered nonverbal communication barriers. When a soldier tried to shake hands with the Indian senior, the Indian person put his hands together and bowed his head in return (Logan-Terry & Damari, 2015). Additionally, during Afghanistan and Iraq military operations, United Kingdom and United States soldiers experienced cultural variances - related to religion, communication norms and social hierarchies - when collaborating with the local societies and authorities (Patten & Parker, 2011). These cases (challenges) demonstrate how intercultural competence is becoming central to effective functioning of military forces.

Race also emerge as a barrier to cross-cultural communication in the military. While many military institutions have now integrated diversity and inclusion into their policy framework, racial disparities in leadership appointments, promotion prospects and institutional involvement still exists. In the UK military, diversity programmes aimed at promoting inclusion remain under-utilised, with reports of racial discrimination (Kirton & Greene, 2016) and lopsided promotion. In Israel, ethnic and social splits still affect interpersonal relationships and the chances of advancement (Ben-Eliezer, 2012). These inequalities may establish distrust, perceived exclusion and communication silos that weaken cohesion.

Cross-cultural communication barriers are also persistent in African military forces such as Kenya and Nigeria. Over 40 ethnic groups, each with its own languages and customs are represented in Kenya's defence force, which complicates operational communication and command effectiveness regardless of constitutional commitments to inclusivity (African Center for Strategic Studies, 2025). Similarly, Nigeria's military faces communication challenges tied to linguistic fragmentation, cultural diversity and perceptions of ethnic imbalance in leadership representation (Ogunleye, 2020; Sulaimon & Olayiwola, 2021). Such dynamics can intensify regional tensions, reduce trust and weaken collective identity within the armed forces.

In the South African case, the SANDF is represented as a transformed and demographically representative military institution with members from diverse racial, linguistic, ethnic, religious and gender backgrounds (Heinecken, 2020; Magadla, 2023B; Dube, 2023). However, the literature suggests that demographic transformation does not automatically lead to communicative integration. Different members of the force may face daily challenges, rooted in language differences, cultural norms, value differences and differing attitudes towards communication (Heinecken, 2020). Despite the multilingual context of South Africa, English continues to be the dominant lingua franca for command, documentation, training and operational coordination. Thus, African languages continue to be marginalised in formal military discourse and training materials, which constitutes exclusionary dynamics for members using English as a second or third language (Heinecken & Nel, 2021; Mamabolo et al., 2025). The SANDF literature also emphasises how cultural differences influence communication. Differences in nonverbal communication, collectivist and individualist orientations, and conflict management styles create variations in how messages are sent, received and responded to (Mamabolo et al., 2025). For instance, eye contact is seen as disrespectful in some African cultures when it is made to elders, however, in Western societies, it can be interpreted as a sign of confidence and credibility. Similarly, members socialised in collectivist traditions may value group harmony and collective goals, while individualists value initiative and achieving individual goals. In the absence of intercultural awareness, such variations may result in tension, stereotyping and misunderstandings amongst military personnel (Mamabolo et al., 2025).

Gender inequality is also regarded as a barrier to effective cross-cultural interactions. While women's presence in the military has increased both internationally, regionally and locally, military institutions have proven to be historically masculine organisations that are guided by male norms, values and communication practices (Van der Meulen & Soeters, 2007; Haring, 2020). Women in the United States, Israel, Canada and African militaries such as Kenya, Nigeria and Ghana have had increased exposure to combat and leadership but face discrimination, underrepresentation, lower promotion and exclusion from informal support structures (Taber, 2018; Erwin & Cseh, 2023). In most military environments, command communication tends to favour assertive and hierarchical communication strategies over those that emphasise cooperation or relational communication. The same trend can be noticed within the SANDF where there is an underrepresentation of women, especially in top leadership and command positions (Radebe & Landsberg, 2022). The enduring masculine institutional culture has placed pressure on women to conform with prevailing norms to be legitimate as soldiers (Magadla, 2023A).

The presented literature suggests that cross-cultural communication challenges in military organisations are not just individual problems, but are institutional issues that are grounded in power, institutional culture, identity politics and disparities inherited from the past.

INTERGRATION OF INDIGENOUS KNOWLEDGE SYSTEMS (IKS) INTO THE MILITARY

The introduction of IKS within contemporary militaries represent a change from traditional, Western-centric and linear communication models and to relational, dialogic and inclusive approaches. This section considers the conceptual and practical aspects of IKS in modern military situations, with reference to particular forms of communication routines including *lekgotla* (deliberate council), storytelling, proverbs and call-and-response. The section reviews how the traditional communication methods from international militaries like the New Zealand Defence Force (NZDF) and the Canadian Armed Forces (CAF), regional forces such as the Botswana Defence Force (BDF), Ghana Armed Force (GAF) and the

Rwanda Defence Force (RDF) as well as the SANDF are adapted to contribute to operational effectiveness within units, and support unit cohesion while coexisting with traditional rank-based command structures.

Lekgotla (Deliberative Council)

Lekgotla, is a traditional practice, that is inclusive and consensus-oriented deliberative council (Labour Research Service, 2021; Rasweswe, 2023). In the military environment, it acts as a formalised channel of transparent discussion, turning decision-making from a purely top-down process into more informed and collaborative solutions. It provides for honest reflection and participatory planning - necessary steps to work through the various cultural differences in a diverse force (Labour Research Service, 2021).

Deliberative councils as consensus-driven models across the world are exemplified internationally by the New Zealand Defence Force (NZDF), where the integration of *Māori tikanga* (customary practices) involves consultative processes where collective knowledges are respected and yet operational tempo is kept (Scoppio, 2018).

Amongst others, the BDF has integrated *kgotla* based indigenous consultative techniques into its military-community engagement procedures. Participatory contact between local populations and military leadership is made possible by the *kgotla*, a customary public gathering for discussion and consensus-building. This strategy lowers tensions and increases trust between the military and citizens by fostering inclusivity, openness and understanding (Molomo, 2005; Sharma, 2012). The integration of such indigenous deliberative practices thus demonstrates how relational communication can complement formal hierarchical structures without undermining command authority.

Similarly, the Ghana Armed Forces (GAF) have integrated customary chieftaincy practices into their plans for civil-military cooperation. Chiefs are responsible for maintaining local communities' culture and power in Ghanaian society. In order to improve legitimacy and enable more seamless interactions during peacekeeping missions and domestic operations, the military acknowledges these customary frameworks in communication and negotiating procedures (Aning & Edu-Afful, 2020). This strategy shows an awareness that alignment with indigenous governance and communication systems is necessary for effective communication in culturally diverse situations.

The Rwanda Defence Force (RDF) has included indigenous frameworks, such as *Gacaca*, a traditional community-based judicial system, into reconciliation and communication procedures in the context of post-conflict reconstruction. The *Gacaca* concepts of conversation, collective involvement and restorative engagement have impacted wider institutional communication practices in Rwanda's security sector, despite their primary use in transitional justice (Clark, 2010; Ingelaere, 2016). These practices emphasise accountability, communal healing and shared meaning-making which are critical components of effective cross-cultural communication in post-conflict environments.

Similar to the above identified international and regional forces, the SANDF has adopted the concept of lekgotla in strategic contexts in South Africa. For example, with Defence Industry *Lekgotla* (2022/2023), which hosted Department of Defence (DOD) leadership, industry participants and academics in response to complex operational and policy challenges (Defence Web, 2022).

Narrative Briefing and Storytelling

Storytelling serves as a powerful vehicle for conveying knowledge, gaining empathy and releasing trauma (Dickson & Maylahn, 2024; Brown, 2025). It is a narrative-based reflection process incorporated in military debriefing whereby personnel can process experiences of their own which are often excluded from formal after-action reviews. The Canadian Armed Forces (CAF) has established 'Bold Eagle', as a storytelling tool that builds self-confidence and respect amongst recruits (Langmuir, 2025). Similarly, veterans and personnel in the SANDF are becoming increasingly interested in utilising storytelling as a mechanism for narrative release from trauma and coping with the intricacies of post-apartheid integration (Magadla, 2023A; Dickson & Maylahn, 2024).

Proverbs as Rhetorical and Framing Devices

Proverbs are brief, traditional sayings that contain meaning and suggest ways to behave (Adepoju, Emmanuel & Audu, 2024). In African military leadership, they function as rhetorical devices to connect abstract military goals with cultural wisdom (Mieder, 2004). For instance, "*An army of sheep led by a lion defeats an army of lions led by a sheep*", is a traditional African proverb used often in military settings, to indicate that leadership is about the quality of the leaders and not the numerical strength (Mieder, 2004). Such statements are employed by commanders to establish a spirit of cooperation and to instil the importance of ethical behaviour in subordinates (Khoza, 2011). Proverbs, especially in the SANDF, contributes considerably in explaining rules of engagement and disciplinary rules to personnel from cultural backgrounds that have a high regard for metaphorical language (Finnegan, 2012; Heinecken, 2017). Embedding military values within culturally appropriate proverbs, leaders can develop a belief in ethical practice which goes beyond compliance.

Call and Response

Call and response is an interactive communicative rhythm that engages interaction, helps communicate information, reinforces a message and develops a shared rhythm among individuals (Floyd, 1995). In military practice such a dynamic forms part of drills and briefings, to foster active participant involvement and message retention. The most well-known global example is used for instance by the NZDF, who maintains the Haka. The Haka is a traditional Māori posture dance or a collective expression of unit identity that constitutes an intense call and response, that energies the force and reaffirms the shared purpose (King, 2003). Similar synchronised affirmations are used in African militaries, such as the SANDF, as part of collective exercises and training operations to boost morale, recognise and to internalise orders from the same army (Nzewi, 2007; Heinecken, 2017).

The presented literature review argues that IKS can create more resilient and cohesive military units. Nonetheless, the criticism is that these benefits are not given off as being directly proportional to improved cross-cultural communication. The identified IKS practices are generally instruments for cultural reinforcement and in-group bonding. In this role these tools would have to take on a conscious adaptation and facilitate the deliberate intention of bridging differences rather than celebrating one tradition. This means developing new, cross-cultural processes utilising ideas from several cultures or utilising the identified IKS tools to initiate explicit conversations about the cultural differences in communication styles, value orientation and decision-making processes. Although the current review establishes a strong foundation, it would be stronger if it delineated more distinctly the role of building a cohesive unit culture from that of building the distinct skills necessary for effective communication across the cultural diversity within a unit.

INDIGENOUS KNOWLEDGE SYSTEMS (IKS) AS A COMPLEMENTARY MECHANISM TO CROSS-CULTURAL COMMUNICATION

Given the diverse nature of the contemporary organisations, cross-cultural communication has become the primary concern as members within diverse organisations are required to communicate in order to achieve the desired aim and objectives of the organisation. The contemporary diverse institutions challenge the appropriateness of the Western-centric communication frameworks in addressing communication in diverse work contexts. The effectiveness of the Western-centric communication frameworks in addressing deep-level cultural variations, distinct worldview and cross-cultural communication remains a question of interest for scholars (Oetzel, 2005; Sun & Shi, 2024). Moreover, persistent cross-cultural communication barriers that arise with Western-centric communication frameworks raise critical concerns on diversity management and effective cross-cultural communication within diverse workplaces. Military institutions are not excluded from Western-centric communication frameworks. As indicated above, legislative frameworks and traditional communication methods are implemented to promote inclusivity; however, cross-cultural communication barriers persist. Implemented initiatives focus on addressing the past imbalances (surface-level variances), reinforcing culture and in-group building and transforming the military into a more representative institution, however deep-level cultural differences that affect cross-cultural communication and cohesion within the military forces remain unaddressed (Oetzel, 2005; Mamabolo et al., 2025). These limitations have prompted growing scholarly interest in encouraging inclusive institutional cultures that considers diverse gender, languages, cultures, races and leadership styles.

Buthelezi, Ocholla and Dlamini (2024) advocate for the utilisation of IKS as a strategy or complementary framework to enhance effective cross-cultural communication across cultures. IKS refer to the grassroots indigenous knowledge systems that comprise indigenous value systems, faith systems, social norms, indigenous communication practices and indigenous philosophies of relationships developed by indigenous people across the years. While conventional non-indigenous or Western knowledge systems such as the ones described above are often based upon written and individual or hierarchical patterns of communication, indigenous knowledge systems are primarily based upon oral patterns of communication, shared meaning and relationships - factors which position indigenous knowledge systems as having great potential for improving cross-cultural communication.

IKS focusses on Afrocentric perspectives, meaning making and active participation that can enhance mutual understanding. It emphasises on respect for diverse cultural contexts and sense making which are vital in encouraging and building mutual understanding, trust and cohesion in diverse organisations (Sobane, Lunga & Setlhabane, 2024). IKS challenges the linear models of communication and promotes reciprocity in diverse contemporary institutions. Therefore, it is viewed as a critical strategy for promoting effective two-way engagements.

The SANDF is a structured, hierarchical, masculine and culturally diverse workforce that is dominated by a historical communication power imbalances that affect effective communication required for operational efficiency and cohesion (Hienecken, 2020). Given the communication challenge, a growing body of literature highlights the need for integrating IKS in diverse military institutions. Sun and Shi (2024) argue that IKS provides distinct communication practices that go beyond individualist and hierarchical approaches to communication. IKS emphasise cultural contextualisation, dialogue and shared meaning making which are required for enhanced intercultural comprehension, trust and to disrupt the dominant masculine (assertive) communication norms within the military. Inclusive, listening processes and consensus-building practices such as collective deliberation, storytelling and mutual respect foster spaces conducive to dialogue, which enable diverse perspectives among people (including women and lower-ranking personnel) to be heard (Sobane et al., 2024). This communication movement, as is stressed by IKS, transitions the discourse from being one sided, command-and-control to a multi-directional and interpersonal dialogue. Such dialogue focuses on mutual understanding of one another, reflection and taking a shared responsibility for the construction or re-construction of meanings which is quite different to how many military commands are based upon a communication style that hinders feedback and discourages differences of opinion(s). IKS works against entrenched power relations and provides pathways to inclusive participation in decision making. The principle of shared ownership of responsibility integrated into IKS rejects rank-related and individualised authority through creating opportunities to collectively take responsibility of decisions and results. This does not eliminate hierarchy, but rather locates hierarchy in a relational aspect of power in a context of a relationship where authority operates under supervision, together with responsibility and reciprocity. Such an approach has a significant impact on changing gendered communication norms, as it weakens normative and exclusionary masculine voices while legitimating alternative lines of communication traditionally associated with collaboration and inclusiveness.

In various disciplines, IKS has also been used in public engagement and knowledge dissemination strategies that emphasise local epistemologies, thereby closing the communication divide between Western scientific institutions and indigenous societies. For instance, studies on science communication have shown that traditional deficit models of science communication, in which Western science is simply disseminated to wider publics, are ineffective in engaging indigenous knowledge and perspectives. Rather, socially inclusive models of science communication must place Indigenous epistemological contributions at the forefront of creating shared understandings and collaborative solutions (Sobane et al., 2024). These models illustrate how IKS can shift cross-cultural communicative engagement from one-way knowledge transfer to two-way cultural exchange.

IKS have gained importance and is recognised as a valuable asset for social cohesion and resilience support for military organisations. The literature underlines their importance for the formation of collective identity, keeping the cultural heritage and member solidarity. However, these advantages do not guarantee enhanced cross-cultural communication. The current scholarship concentrates on the growth of cohesive unit cultures, serves as coping mechanism of past segregation and explain the rules of military engagement to diverse personnel (Finnegan, 2012; Heinecken, 2017; Magadla, 2023B). The efficacy of IKS practices as vehicles of cultural consolidation and participation mechanisms is not questioned; their effectiveness as translators between cultures, however, depends on judicious and context dependent adaptation. In this context, IKS need to be intentionally designed to facilitate inclusive communication across cultural boundaries. These could include creating hybrid communicative conceptual frameworks with cross-cultural perspectives that bridge the gap between different cultural perspectives, as well as the deliberate use of IKS-based modalities to engender dialogue around barriers in communication styles, value systems, decision making models and processes. As such, IKS could significantly contribute to the strengthening of cross-cultural communication capabilities in military situations when used with the intent that its use reaches beyond cohesion-building and extends to the intentional development of intercultural communicative competency. This chapter contends that the realisation of such benefits would not be fully realised without a clear conceptual differentiation between the purposes of unit cohesion and cross-cultural communication.

RESEARCH METHODOLOGY

This chapter employed a qualitative desktop research design to examine the importance of IKS as a key strategy for enhancing cross-cultural communication within the SANDF. Qualitative desktop research design collects non-standardised data and analyses subjective meanings by analysing texts rather than numbers and statistics (Flick, 2011). A desktop approach was utilised to gather, analyse and synthesise systematic literature review and helped to create an integrated conceptual understanding of cross-cultural communication and IKS. Peer-reviewed journal articles, books, South African Government legislation and Department of Defence doctrines, official SANDF publications and thesis and dissertations related to cross-cultural communication and IKS were reviewed to attain the objectives of this chapter. A systematic literature search using relevant keywords and phrases (for instance, cross-cultural communication, military and culture, IKS and cross-cultural communication, diversity and so forth) were used to collect data. Elicit, Scopus, Google scholar and Research Gate were employed as the academic databases. Additionally, thematic content analysis - a systematic rigorous approach to analysing documents obtained or generated during research (Zawacki-Richter, Kerres, Bedenlier, Bond & Buntins, 2020) - was used to analyse and interpret the collected data.

FINDINGS AND DISCUSSIONS

Historical Transition and Communicative Integration within the SANDF

The SANDF post 1994 evolution marked a pivotal change from exclusivity to constitutional representativity. This provided access for diverse groups to serve within the SANDF. However, implemented initiatives (policies) to inclusivity emerged with cross-cultural tensions that persist as barriers to effective intercultural and cross-cultural interactions within the force. The integration of diverse members consequently affected communicative cohesion within the SANDF. The amalgamation of statutory and non-statutory force brought together different military formations characterised by different communication practices, ideologies and cultural traditions. The former SADF functioned with Western oriented, hierarchal command structure rooted in centralised and standardised communication style while the non-statutory (liberation armies) focussed more on relational forms of engagement and community-based mobilisation strategies. With this presented, the integration of the forces into the SANDF did not only require administrative restructuring, it also required the reconciliation of distinct communication practices. Therefore, the SANDF is described as an inclusive military force that is communicatively disjointed and experiences tensions emerging from unresolved linguistic and cultural differences.

Cross-Cultural Communication Barriers within the SANDF

According to the literature review, race, language and gender emerge as the key factors shaping cross-cultural communication barriers within the SANDF. Language emerges as a key factor affecting cross-cultural communication. South Africa recognises twelve official languages; however, English remains the dominant language of instruction, documentation, command and operational communication within the SANDF. English remains a standardised language for clarity in both local and multinational operations, however, it marginalises African languages and indigenous modes of expression. The utilisation of English language may create asymmetrical power relations by privileging native speakers and those who are proficient while limiting participation, confidence and expression among non-native speakers. Language inequality therefore contributes to misunderstanding and limited cross-cultural communication, thereby affecting operational effectiveness and cohesion within the SANDF. The use of English within the SANDF further reflects the Western and Colonial military traditions that restricts vertical communication and supports hierarchical structure.

Additionally, cultural variation emerges as a barrier to effective cross-cultural communication within the SANDF. SANDF members emerge from various cultural backgrounds and have different approaches to authority, attach different meanings to non-verbal cues and have different perspectives on collectivism and individualism traditions. Non-verbal cues are complex and interpreted differently by various cultures. Physical proximity, eye contact, gestures and tone carry cultural rooted meanings. Therefore, in the absence of intercultural or cross-cultural communication competency and awareness, such variances may create misinterpretation regarding discipline, respect and competence. For instance, culturally specific norms regarding eye contact may be interpreted differently across cultures. It can either be seen as a sign of disrespect or honesty depending on one's culture. In the majority of Black African cultures, maintaining eye contact, especially with an elderly person can be seen as a sign of disrespect, however in White cultures, eye contact is viewed as a sign of honesty or trustworthiness. These cultural dynamics indicate that cross-cultural communication barriers not only emerge from interpersonal misunderstanding, but are thus rooted in deep cultural epistemologies that guides communication interpretations and meaning making. The SANDF is further characterised by a collectivistic tradition that values group harmony over individualistic traditions that emphasises on personal autonomy creating tensions and affecting collaboration amongst the distinct parties.

Gender inclusivity contributes as another key barrier to cross-cultural communication within the SANDF. Regardless of gender mainstreaming, the SANDF remains a male dominated institution embedded with masculine institutional culture. The force persists as a traditional male institution that heavily rely on physical strength and command driven communication that may affect collaborative and assertive communication style that is usually utilised by women. Additionally, women remain underrepresented, especially in leadership positions. As a hierarchal communication institution, the majority of women's voice remain marginalised while male voices and perspectives remain legitimised. This may unconsciously favour masculine communication patterns characterised by assertiveness, consequently creating tension between service men and women.

The existence of cross-cultural communication barriers suggests that Western-based communication models traditionally employed in military settings may have some shortcomings. These communication models are linear in design and rely on hierarchy and uniformity. While these principles are useful for establishing operational order and consistency, they may not be entirely invested in trust-building, understanding and dialogue in a culturally diverse setting. Transformational policies have been primarily driven by representativeness and legislative compliance. The epistemic dimensions of communication, including meaning and recognition, have not been entirely integrated into transformational policies. This suggests a need for culturally responsive frameworks that can productively engage with deep-level diversity.

IKS as an Effective Cross-Cultural Communication Tool in the SANDF

Persistent cross-cultural communication barriers faced within the SANDF represent a communicative integration gap that is currently insufficiently addressed. Despite post-1994 transformation processes achieving considerable improvements in demographic representativity, structural integration did not necessarily extend to effective communicative cohesion between members of different racial, linguistic, gendered and historical backgrounds. Consequently, misunderstandings, mistrust and divided relationships in the SANDF remain as barriers to the forces' cross-cultural communication effectiveness and internal cohesion.

Existing literature argues that IKS may lessen such barriers within the force by offering culturally sensitive and relational forms of communication (Ngulube, 2018). IKS are not just Western models of communication that are based on formalised and hierarchical written forms of communication, instead, they include indigenous forms of community communication, social norms and relational philosophies that emerged in particular cultural contexts. These approaches draw upon active involvement, shared meaning-making, oral traditions and collective efforts, which are important in intercultural and cross-cultural contexts. IKS further acknowledge indigenous epistemologies which could promote the mutual understanding, respect and recognition between cultural and linguistic groups. IKS moves communication away from a sender-receiver linear model towards an integrated transactional and dialogical model, where communication is perceived as a reciprocal process, rooted in mutual understanding and contextual meaning-making (Gudykunst, 2003; Meyer, 2017).

Further critical examination of the literature discovers that many IKS practices have been historically employed to engender internal harmony and cultural reinforcement, as opposed to being used for explicit cross-cultural dialogue. Their effectiveness often stems from deepening identity, solidarity and belonging in particular communities. It is in this sense that military institutions should tailor these practices with a deliberate focus to closing the cross-cultural communication gap. This involves generating hybrid communicative frameworks that are cross-pollinated from the inter-cultural communication models developed based on an inter-specific blend of traditions and relying on IKS for purposeful discussion process for discussing communication style, value system, leadership expectations and decision-making styles explicitly. Military institutions can move from a symbolic recognition for individual traditions to understanding and skill building of complex cultural communication capabilities among a wider range of staff. This could involve facilitator training for leaders, co-created cross-cultural communication guidelines (that makes sense, but must be done collaboratively), an inclusive dialogue, multilingual dialogue platforms and the use of IKS instruments to interpret and analyse communication and interaction practices at every level of cultural dynamics within the operations domain. From this perspective then, IKS cannot be dismissed purely as heritage practice but as a tangible tool for effectively managing diversity and building organisation unity.

The integration of IKS into the SANDF would not only signify a cultural upliftment but as a strategic obligation. Policy initiatives have also contributed to an increasingly fast-paced demographic change, however, entrenched cultural, linguistic and gender differences persist in the difficulties communicative cohesion experienced. IKS provide a relational, dialogical and contextual matrix that is conducive to trust, reduces misunderstanding and increases unity among different SANDF members. This will need to be done with careful institutionalisation alongside formal command systems as IKS can significantly contribute to a more cohesive, adaptive and operationally effective defence force.

RECOMMENDATIONS

Given the cross-cultural communication barriers and transformation challenges within the SANDF, this chapter proposes various practical and policy-oriented recommendations.

Considering the critical analysis of the literature that finds the use of IKS to promote military unit cohesion, while recognising their role as primarily a tool for cultural reinforcement, rather than for explicit cross-cultural communication, it is suggested that military institutions use IKS tools for strategically aligning and enabling an intentional intention to bridge cultural disparities. This includes building fresh hybrid, cross-cultural practices that leverage a range of traditions and employing IKS to drive explicit conversations about communication styles, value orientations and decision making. Military forces can then move beyond celebrating specific traditions and establish unique skills to support effective communication from their diverse staff by focusing on facilitator training, jointly implementing inclusive communication protocols and using the IKS tools to analyse the nuances of cultural expression and interaction in operations.

The chapter further recommends for the incorporation of IKS into the SANDF's training curricula. The SANDF should integrate IKS into military functional and promotional courses such basic military training, officer's formative courses as well as strategic and leadership courses or leadership development programmes. Cultural awareness and cross-cultural communication competency, indigenous conflict management strategies, participatory or transactional communication models and communal and indigenous leadership viewpoints methods should be developed as the core modules to address cross-cultural communication barriers. This will aid in addressing and understanding deep-level diversity. The integration of IKS into military curricula should therefore complements existing doctrines by instilling Afrocentric approaches to communication philosophies such as Ubuntu, interpersonal dialogue and shared responsibility.

Additionally, the SANDF should move from a communication model based on rigidity, hierarchy and masculinity (Western communication approach) to one that promotes relational dialogue and shared responsibility (IKS communication approach) through military training, leadership development and communication practices. This change will not only to improve cross-cultural communication but will also build a more united, agile and representative military institution.

To alleviate linguistic marginalisation of African languages and encourage cross-cultural communication, the SANDF should develop a cross-cultural communication regulatory framework that clearly incorporates IKS philosophies. The regulatory framework should promote indigenous communication practices and institutionalise cross-cultural dialogue platforms.

The SANDF should further establish cross-cultural forums and campaigns that address cross-cultural misunderstandings, enhance shared meaning making and encourage cultural exchange sessions and storytelling across diverse groups. These sessions should reflect IKS philosophies of oral traditions to enhance cohesion amongst diverse service men and women.

Moreover, the SANDF should develop monitoring and evaluation tools to assess the efficiency of IKS incorporation initiatives and levels of cross-cultural trust. Consistent monitoring and evaluation within the SANDF will ensure that transformation is beyond inclusivity and strive for meaningful cultural integration.

FUTURE RESEARCH DIRECTIONS

Despite that this chapter contributes to the body of knowledge on cross-cultural communication and IKS within the SANDF, several limitations require further research.

Empirical Studies on Cross-Cultural Communication and IKS in the SANDF

This chapter employed a qualitative desktop research design. The approach provided a theoretical and conceptual understating, nevertheless, empirical research employing in-depth interviews, focus group discussions and surveys amongst others should be conducted to validate the incorporation of IKS as a strategy to cross-cultural communication. Additionally, future research should be conducted to examine the impact the SANDF's masculine culture on cross-cultural communication and the correlation between gender and rank and how they shape the communication process within the force.

Comparative Military Studies

The SANDF and other African or multinational military forces that integrate indigenous concepts into institutional procedures should be compared in future research. Best practices for military communication frameworks with a cultural foundation would be revealed through comparative analysis.

CONCLUSION

This chapter examined the importance of IKS as a key strategy for enhancing cross-cultural communication within the SANDF. The chapter indicates that the integration of the statutory and non-statutory forces emerged with an inclusive military force that comprises diverse members originating from various gender, cultural, linguistic and racial backgrounds. However, this integration created cross-cultural communication barriers that affects collaboration and cohesion within the force. Legislative frameworks have been implemented to promote inclusion within the SANDF; however, cultural and structural barriers to effective communication persist. Therefore, IKS is perceived as an effective strategy that should be incorporated into the SANDF and aid in reducing cross-cultural tensions. The integration of IKS into the SANDF may support mutual understanding, improve cultural competence and create a unified communication climate amongst diverse service men and women. This chapter further highlights that transformation within the SANDF must not focus only on demographic inclusivity, it must also focus on communicative transformation. Addressing cross-cultural communication does not only require demographic representation, it requires organisational strategies that acknowledge linguistic and cultural diversity as well as indigenous philosophies.

REFERENCES

Abuarqoub, I. A. S. (2019). Language barriers to effective communication. *Utopía y Praxis Latinoamericana, 24*(6), 64–77.

Adepoju, A., Emmanuel, A., & Audu, J. (2024). Silencing the Guns: African Traditional Proverbs and Aphorisms for Peace and Security in 21st Century Africa. *The Journal of Society and Media, 8*(2), 537–551. DOI: 10.26740/jsm.v8n2.p537-551

Africa Center for Strategic Studies. (2025). *Reflections on 60 years of strengthening military professionalism in Kenya*. Available at: https://africacenter.org/spotlight/60-years-military-professionalism-kenya/ [Accessed: 4 April 2026].

Aning, K., & Edu-Afful, F. (2020). African militaries and the challenge of gender inclusion and equality. In Aning, K., & Edu-Afful, F. (Eds.), *Governing Africa's Security: The Challenges of Gender and Military Transformation*. Routledge.

Bakhaya, I. (2024). *The Impact of Language Barriers in Multinational Military Operations: A Case Study of NATO Forces*. AFASES Conference Proceedings. [Accessed 20 January 2026]. https://biblioteca.afahc.ro/ro/afases/2024/Volume-AFASES2024/10-IrinaBAKHAYA.pdf

Ben-Eliezer, U. (2012). Operation Defensive Shield. In *Old Conflict, New War: Israel's Politics toward the Palestinians* (pp. 123–140). Palgrave Macmillan US. DOI: 10.1057/9781137027573_8

Buthelezi, S. C., Ocholla, D., & Dlamini, P. (2024). Strategies for documenting and disseminating indigenous knowledge at a South African university. *South African Journal of Information Management, 26*(1), 1648. DOI: 10.4102/SAJIM.v26i1.1648

Clark, P. (2010). *The Gacaca courts, post-genocide justice and reconciliation in Rwanda: Justice without lawyers*. Cambridge University Press. DOI: 10.1017/CBO9780511761584

Davis, S. R. (2018). *The ANC's war against apartheid: Umkhonto We Sizwe and the liberation of South Africa*. Indiana University Press. DOI: 10.2307/j.ctt2272733

De la Rey, A. (1990). *Suid-Afrikaanse Weermag - South African Defence Force Review*. Walker-Ramus Trading Co. SA. (PTY) LTD.

Defence Web. (2022). Defence Industry Lekgotla tackles challenges facing South Africa's defence sector. DefenceWeb. Available at: https://www.defenceweb.co.za (Accessed: 28 April 2026). Dei, G. J. S. (2000). *Rethinking the role of indigenous knowledges in the academy*. International Journal of Inclusive Education, 4(2), 111–132.

Department of Defence. (1998). *South African Defence Review*. Department of Defence.

Department of Defence. (1998). *Department of Defence Instruction: Pol and Plan No 00001/ (1998 Policy on Equal Opportunity and Affirmative Action No 00001*. Pretoria: Department of Defence.

Department of Defence. (2015). *South African Defence Review*. Department of Defence.

Department of Defence. (2020). *Department of Defence Strategic Plan for 2020-2025, 12 March 2020*. Department of Defence.

Department of Defence. (2020). *Bulletin Promulgation: Appointments and/or Promotions of Cols/ Capt (SAN) for the Financial Year 20/21. 22 October 2020*. Department of Defence.

Dickson, N., & Meylahn, J. A. (2024). The role of storytelling as a possible trauma release for war veterans: A narrative approach. *Hervormde Teologiese Studies, 80*(1), 9408. DOI: 10.4102/HTS.v80i1.9408

Dodd, N., Bester, P. C., & Van Der Merwe, J. (Eds.). (2020). *Contemporary Issues in South African Military Psychology*. African Sun Media. DOI: 10.18820/9781928480631

Dube, B. (2023). *Language, power and communication in South African security institutions*. Africa Institute of South Africa.

Erwin, S.I. (2005). Language barriers hinder multinational operations. *National defence, 90*(620), pp.16-17.

Erwin, S. K., & Cseh, M. (2023). Invisible and yet hypervisible: Gendered misconduct and the US military. *Industrial and Commercial Training, 55*(3), 402–413. DOI: 10.1108/ICT-10-2022-0075

Finnegan, R. (2012). *Oral literature in Africa*. Open Book Publishers. DOI: 10.11647/OBP.0025

Flick, U. (2011). Methoden-Triangulation in der qualitativen Forschung. In *Triangulation: Eine Einführung* (pp. 27–49). VS Verlag für Sozialwissenschaften. DOI: 10.1007/978-3-531-92864-7_3

Floyd, S. A. (1995). *The power of Black music: Interpreting its history from Africa to the United States*. Oxford University Press.

Georgieva, V., & Marinov, P. (2017). Intercultural interactions in a military context. *Land Forces Academy Review, 22*(3), 153–162. DOI: 10.1515/raft-2017-0021

Gudykunst, W. B. (2003). Cross-cultural and intercultural communication. *Sage (Atlanta, Ga.)*.

Gudykunst, W. B. (2005). *Theorising about Intercultural Communication*. Sage.

Haring, E. (2020). *Gender and military organisations. In the Gender and Security Agenda*. Routledge.

Harrison, D. A., Price, K. H., Gavin, J. H., & Florey, A. T. (2002). Time, teams, and task performance: Changing effects of surface-and deep-level diversity on group functioning. *Academy of Management Journal, 45*(5), 1029–1045. DOI: 10.2307/3069328

Heinecken, L. (2011). A diverse society, a representative military? The complexity of managing diversity in the South African Armed Forces. *Scientia Militaria: South African Journal of Military Studies, 37*(1), 25–49. DOI: 10.5787/37-1-58

Heinecken, L. (2017). *South Africa's post-apartheid military: Lost in transition and transformation*. Wits University Press.

Heinecken, L. (2020A). *South Africa's post-apartheid military: Lost in transition and transformation*. Springer. DOI: 10.1007/978-3-030-33734-6

Heinecken, L. (2020B). Political Reform: Integration, Representivity and Managing Diversity. South Africa's Post-Apartheid Military: Lost in Transition and Transformation, pp.71-86.

Heinecken, L., & Nel, M. (2021). Leadership, organisational culture and transformation in the SANDF. *African Security Review, 30*(3), 215–230.

Ingelaere, B. (2016). *Inside Rwanda's Gacaca courts: Seeking justice after genocide.* University of Wisconsin Press. DOI: 10.2307/jj.36032571

Khoza, R. J. (2011). *Attuned leadership: African humanism as compass.* Penguin Books.

King, M. (2003). *The Penguin history of New Zealand.* Penguin Books.

Kirton, G., & Greene, A. M. (2016). Understanding diversity management in the UK. In *Corporate Social Responsibility and Diversity Management* (pp. 59–73). Theoretical Approaches and Best Practices.

Knouse, S. B. (2001). Diversity and Military Effectiveness: How Diversity Enhances Performance and Cohesion in Military Units. *Armed Forces and Society, 27*(3), 473–494.

Kolodziej-Smith, R. (2016). *A preliminary study of workgroup dynamics: A critique and reconceptualisation of Oetzel's effective intercultural workgroup communication theory.* Wayne State University.

Labour Research Service. (2021). *Worker participation and engagement models in South Africa.* Labour Research Service.

Lamb, G. (2018). *New Model Army: The transformation of the South African National Defence Force 1994-2004.* Stellenbosch, South Africa.

Le Roux, L. (2005). *The post-apartheid South African military: Transforming with the nation. Evolutions and Revolutions: A Contemporary History of Militaries in Southern Africa.* Institute for Security Studies.

Logan-Terry, A., & Damari, R. R. (2015). Key culture-general interactional skills for military personnel. *Procedia Manufacturing, 3*, 3990–3997. DOI: 10.1016/j.promfg.2015.07.951

Magadla, S. (2023A). *Guerrillas and Combative Mothers: Women and the Armed Struggle in South Africa.* University of Kwa-Zulu Natal Press. DOI: 10.4324/9781003455981

Magadla, S. (2023B). Military identity and post-apartheid integration in South Africa. *African Security Review, 32*(2), 120–135.

Mamabolo, M., Le Roux, J., & Muswede, T. (2025). *Cross-Cultural Communication Challenges in Basic Military Training: The Case of Recruits in the South African Air Force.* Southern African Journal for Folklore Studies.

Mbigi, L. (2005). *The spirit of African leadership.* Knowres Publishing.

Meyer, E. (2017). *The culture map: Breaking through the invisible boundaries of global business.* Public Affairs.

Mieder, W. (2004). *Proverbs: A handbook.* Greenwood Press.

Molomo, M. G. (2005). Democracy and the kgotla system in Botswana. *African Studies Quarterly*, *8*(2), 1–20.

Msila, V. (2015). *Ubuntu and school leadership. Journal of Social Sciences, 43(1), 67–76. Ngulube, P. (2018). Handbook of research on theoretical perspectives on indigenous knowledge systems in developing countries*. IGI Global.

Murithi, T. (2006). Practical peacemaking wisdom from Africa. *The Journal of Pan African Studies*, *1*(4), 25–34.

Nzewi, M. (2007). *A contemporary study of musical arts: Informed by African indigenous knowledge systems*. Centre for Indigenous Instrumental Music and Dance Practices.

Nzimakwe, T. I. (2014). Practising Ubuntu and leadership for good governance. *African Journal of Public Affairs*, *7*(4), 30–41.

Odora Hoppers, C. A. (2002). *Indigenous knowledge and the integration of knowledge systems*. New Africa Books.

Oetzel, J. G. (2005). Effective intercultural workgroup communication theory. *Theorising about Intercultural Communication*, pp.351-371.

Oetzel, J. G. (2017). *Effective intercultural workgroup communication theory*. The International Encyclopaedia of Intercultural Communication. DOI: 10.1002/9781118783665.ieicc0025

Oetzel, J. G., McDermott, V. M., Torres, A., & Sanchez, C. (2012). The impact of individual differences and group diversity on group interaction climate and satisfaction: A test of the effective workgroup communication theory. *Journal of International and Intercultural Communication*, *5*(2), 144–167. DOI: 10.1080/17513057.2011.640754

Ogunleye, T. (2020). Colonial Legacy and Ethnic Imbalance in the Nigerian Armed Forces. *African Military History Review*, *4*(3), 58–75.

Patten, E., & Parker, K. (2011). *Women in the US military: Growing share, distinctive profile*. Pew Research Center.

Pinch, F. C., MacIntyre, A. T., Browne, P., & Okros, A. C. (2004). *Challenge and change in the military: Gender and diversity issues*. Wing Publishing Office.

Poteet, S., Xue, P., Patel, J., Kao, A., Giammanco, C., & Whiteley, I. (2008). Linguistic sources of coalition miscommunication. In *NATO RTO-MP-HFM-142 Adaptability in Coalition Teamwork*, p. 10.

Radebe, S. and Landsberg, C. (2022). Transformation, representation and leadership in South Africa's security sector, *Journal of Contemporary African Studies*, 40(4), pp. 523-538. Rasweswe, M. (2023). Indigenous governance systems and participatory democracy in South Africa. *Journal of African Governance*.

Republic of South Africa. (1996). *Defence in a Democracy: White Paper on National Defence for the Republic of South Africa, May 1996*. https://www.gov.za/documents/national-defence-white-paper [Accessed 10 January 2026].

Republic of South Africa. (2000). *Promotion of Equality and Prevention of Unfair Discrimination Act, No. 4 of 2000*. Government Printer.

Republic of South Africa. (2002). *Defence Act, No. 42 of 2002*. Government Printer.

Scoppio, G. (2018). Diversity management in the military: The case of indigenous integration in the New Zealand Defence Force. *Armed Forces and Society*, *44*(2), 337–357.

Scott, N. (2002). From the SADF to the SANDF: Safeguarding South Africa for a better life for all. Violence and transition series, 7, pp.1-88.

Sharma, K. C. (2012). *Public administration in Botswana*. Botswana Institute for Development Policy Analysis.

Skilleås, H. K., & Grande, B. (2024). Language Lessons Learned? NATO Requirements and Norwegian Officer Cadets. *Scandinavian Journal of Military Studies*, *7*(1). Advance online publication. DOI: 10.31374/sjms.221

Sobane, K., Lunga, W., & Setlhabane, L. (2024). Harnessing Indigenous knowledge systems for socially inclusive science communication: Working towards a 'science for us, with us' approach to science communication in the Global South. In Rasekoala, E. (Ed.), *Race and sociocultural inclusion in science communication*. Policy Press.

Soeters, J. L., & Van Der Meulen, J. (Eds.). (2007). *Cultural diversity in the armed forces: An international comparison*. Routledge. DOI: 10.4324/9780203967409

Sulaimon, M., & Olayiwola, I. (2021). Ethnic and Religious Bias in the Nigerian Armed Forces: A Critical Analysis. *African Security Studies*, *18*(4), 101–119.

Sun, Y., & Shi, Y. (2024). Knowledge strategies for Indigenous studies in intercultural communication: Decentring Western epistemic hierarchies. *Social Sciences* [Special Issue]. *Role of Intercultural Communication in Multicultural or Culturally Diverse Societies*, *13*(10), 1060–1072.

Taber, N. (2018). After Deschamps: Men, masculinities, and the Canadian Armed Forces. *Journal of Military, Veteran and Family Health*, *4*(1), 100–107. DOI: 10.3138/jmvfh.2017-0005

Tiatira, Y. C., Herwandito, S., & Krisnawati, E. (2024). Harmonising Religious Diversity: Exploring Effective Intercultural Communication in Dukuh Sampetan Village. *International Journal of Multicultural and Multireligious Understanding*, *11*(3), 108–119. DOI: 10.18415/ijmmu.v11i3.5563

United Nations. (2023). *Women, peace and security: Report of the Secretary-General*. United Nations.

Van Der Meulen, J., & Soeters, J. L. (Eds.). (2007). *Cultural diversity in the armed forces: An international comparison*. Routledge.

Van Der Waag, I. (2015). South African defence in the age of total war, 1900-1940. *Historia (Wiesbaden, Germany)*, *60*(1), 129–155.

Veldtman, S. L. (2018). Strategic Change Management in the South African National Defence Force (Doctoral dissertation, University of Cape Town).

White Paper on Affirmative Action in the Public Service. (1998). *Notice 564 of 1998*. Government Printers.

Williams, H. L., Jr. (2022). Examining the Cross-Cultural Adaptation of Department of Defence Leaders in Joint Multinational Military Training Environments (Doctoral dissertation, Trident University International).

Williams, R. (2001). *Completing the defence transformation process: The transformation of the South African Reserve Force System*. Institute for Security Studies Papers.

Winslow, D., Browne, P., & Febbraro, A. (2007). Diversity in the Canadian Forces. In Sceters, J., & Van Der Meulen, J. (Eds.), *Cultural diversity in the armed forces: An international comparison*. Routledge.

Zawacki-Richter, O., Kerres, M., Bedenlier, S., Bond, M., & Buntins, K. (2020). *Systematic reviews in educational research: Methodology, perspectives and application*. Springer Nature. DOI: 10.1007/978-3-658-27602-7

ADDITIONAL READING

Dei, G. J. S. (2011). *Indigenous philosophies and critical education: A reader*. New York: Peter Lang.

Heinecken, L. (2017). Transitions and transformation in gender relations in the South African military: From support in warfare to valued peacekeepers. *The Palgrave International Handbook of Gender and the Military*, pp.355-368. DOI: 10.3726/978-1-4539-0131-1

Hofstede, G., Hofstede, G. J., & Minkov, M. (2010). *Cultures and organizations: Software of the mind* (3rd ed.). McGraw-Hill.

Mbigi, L., & Maree, J. (1995). *Ubuntu: The spirit of African transformation management*. Knowledge Resources.

Yankah, K. (1989). *The proverb in the context of Akan rhetoric*. Peter Lang.

KEY TERMS AND DEFINITIONS

Cross-Cultural Communication: Cross-cultural communication involves the exchange of meanings between individuals from diverse cultural backgrounds, shaped by variations in language, norms and values.

Diversity: Diversity signifies the existence and coexistence of differences among individuals within a group, organisation or society.

Indigenous Knowledge Systems (IKS): IKS identify culturally rooted and developed forms of knowledge, practice, value or belief system that are specific to certain communities and are often passed down through generations.

Intercultural: communication Competence: The ability to effectively engage with diverse groups across cultural lines by demonstrating adaptability, empathy and an understanding of cultural nuances.

Military Cohesion: Refers to the trust-based relationships and unity among unit members that contribute to enhanced operational effectiveness.

Chapter 12
African Indigenous Knowledge Systems for Climate Adaptation:
Lessons on Extreme Weather Events in Lesotho

Nthateng Mokhochane
http://orcid.org/0009-0009-2896-901X
National University of Lesotho, Lesotho

ABSTRACT

This chapter examines the role of African Indigenous Knowledge Systems (IKS), with specific reference to Basotho IKS, in response to extreme weather occurrences in Lesotho. Climate change has been reported to affect rain-fed agriculture systems that lack adaptive infrastructure. The research was conducted using the qualitative research approach. Data was collected from 16 purposively selected participants. The data was then analysed using thematic data analysis. From the research findings, it was concluded that Basotho IKS, which relies on ecological observations, rituals, ceremonies, and traditional norms in the communities, plays a crucial role in the response to extreme weather occurrences in Lesotho. The research was theoretically underpinned by the Afrocentric theory, which emphasises the use of African perspectives in knowledge systems. The research findings provide insights that can be used in the response to extreme weather occurrences in Lesotho and beyond.

INTRODUCTION

Africa is known for its richness in Indigenous Knowledge Systems (IKS), which have been deeply embedded in the culture, traditions, and experiences of people (Ossai, 2010). Indigenous Knowledge Systems refer to a collection of knowledge developed through generations of interaction with a particular environment. Indigenous Knowledge Systems are not just traditions; they have been dynamic and adaptive approaches to dealing with issues in society. These knowledge systems are not merely traditions but represent dynamic and adaptive ways of understanding and responding to environmental and social challenges. However, issues such as globalisation, urbanisation, education systems, and the adoption of Western epistemology have led to the marginalisation of Indigenous Knowledge in African society

DOI: 10.4018/979-8-2600-2890-2.ch012

(Mekoa, 2018). Therefore, the role of Indigenous Knowledge in the management of the environment has been undervalued in African society.

Generally, Indigenous Knowledge Systems refer to knowledge developed and nurtured in a particular society (Sen, 2005). This knowledge is transmitted across generations through oral tradition, observation, and lived experience, forming part of a broader cultural and ecological system.It is a collection of knowledge that has been passed down through generations through word of mouth and practical experience. It is a collection of knowledge that has been nurtured in a larger tradition or culture. It is through Indigenous Knowledge that people have been able to understand environmental patterns and have been able to anticipate seasonal patterns. In many African countries, Indigenous Knowledge has been a major approach to survival. Within the growing field of climate adaptation, Indigenous Knowledge Systems are increasingly recognised as valuable contributors to understanding environmental variability and supporting community resilience. Lesotho is a country in Africa that has been known for the role played by Indigenous Knowledge in the management of the environment.

Lesotho is a Southern African country characterised by mountainous terrain and mostly rain-fed agriculture. The country has poor infrastructure to deal with environmental challenges (Moshesha, 2023). The country experiences extreme weather conditions such as droughts, hail storms, strong winds, lightning strikes, floods, frost, and excessive rainfall. In addition, over 90% of disasters experienced in the country are climate-related disasters characterised by localized floods, early frosts, and crop failures resulting in damage to infrastructure and livelihood (World Bank, 2021). The climatic conditions are posing a great challenge to food security and water resources.

In his address on October 4th, 2024, commemorating the bicentenary of the country and the 58th Anniversary of independence from the British Protectorate, His Majesty King Letsie III emphasised the country's struggle with climate change-related crises. He stated:

> "The current crisis of hunger, poverty, disease, and climate change threatens to reverse the progress we have made over the last century. In Lesotho, climate change exacerbates environmental degradation, impacting agriculture, water resources, and energy supply, which has led to food shortages and compelled the government to declare a state of emergency in food security."

This statement highlights the severity of climate vulnerability in Lesotho and underscores the urgency of strengthening adaptive strategies that are context-specific and culturally grounded. Soil erosion, falling farm yields, droughts, and unpredictable rainfall are all too familiar, like the fingerprints of a world wracked by climate extremes. Yet, they are not random occurrences. Rather, they are all part of the pattern of climate variability affecting livelihoods and ecosystems. In this context, there is increasing recognition that adaptation strategies should not rely solely on contemporary or externally derived approaches, but should also engage Indigenous Knowledge Systems that are embedded in local experience and practice.

Climate change refers to long-term alterations in temperature, precipitation, and other weather patterns (Alexander, 2016). Africa is among the continents most affected by climate variability due to limited infrastructure and high dependence on natural resources (Gemeda & Sima, 2015). In Lesotho, historical records show repeated droughts, crop failures, hailstorms, floods, strong winds, frost, and lightning strikes. These events reduce community resilience and threaten livelihoods. Historically, Basotho communities developed Indigenous Knowledge Systems to manage environmental challenges. African communities relied on ecological observations and seasonal patterns to anticipate and prepare for environmental uncertainties. The Basotho were no exception. Their Indigenous Knowledge Systems provided guidance on when and how to respond to climatic hazards.

Traditionally, the Basotho communities designed their Indigenous Knowledge Systems to foresee and manage environmental challenges. They did this by using ecological watching, seasonal indications, rituals, taboos, and collective decision-making to develop methods of preventing and managing extreme weather conditions. These were not merely spiritual activities but rather organized methods of managing the environment that relied on collective responsibility and knowledge gathered over the years.

Different studies have been conducted on the management of IKS in the Basotho culture. Mashini & Mokhothu (2009) studied traditional methods of sorghum cultivation in the culture. They discussed the management of seeds, fertility of the soil, and the management of water in the culture. Johnson (2018) studied rainmaking rituals in the culture. The study highlighted the spiritual aspect of the culture in managing drought. Possa & Khotso (2015) studied the use of medicinal plants in the culture. However, existing studies have tended to focus on specific aspects of Indigenous Knowledge Systems, such as agriculture, rituals, and medicinal practices, with limited attention given to their use as a holistic framework for managing extreme weather conditions in Basotho communities.

It is within this context that this study examines how Basotho Indigenous Knowledge Systems guide community responses to extreme weather events. The study also considers how past knowledge can support contemporary and future climate adaptation strategies. It further situates Indigenous Knowledge Systems within broader climate adaptation debates that emphasise resilience-building and the integration of indigenous and scientific knowledge systems rather than treating them as separate or opposing frameworks. The chapter contributes to climate adaptation literature and practice by highlighting culturally rooted solutions rather than relying solely on Western-centric models. It aims to provide information that can inform policy and offer practical strategies that communities can adopt to strengthen resilience to extreme weather events.

The significance of this study is threefold. First, it contributes to the growing body of scholarship on climate adaptation by foregrounding African Indigenous Knowledge Systems. Second, it advances the decolonisation of environmental discourse by challenging the marginalisation of African epistemologies. Third, it provides policy-relevant insights that may inform culturally grounded adaptation strategies in Lesotho and similar contexts.

RESEARCH QUESTION

- What types of Indigenous Knowledge Systems do the Basotho use to cope with extreme weather events?

AIM AND OBJECTIVE

The aim of this chapter is to examine African Indigenous Knowledge Systems (IKS), particularly the Basotho IKS, on how they guide community responses to extreme weather events in Lesotho. Specifically, it seeks to:

- Identify the types of IKS used by the Basotho people and their practical application in coping with extreme weather.

LITERATURE REVIEW

This section uses a thematic approach to the review of related literature. Dawadi (2020) defines the thematic approach to a literature review as a method, which organises the previous or existing research based on recurring themes, patterns or concepts. The literature for the present study is grouped and examined under the following themes: Indigenous Knowledge Systems and other fields and Indigenous Knowledge Systems on Weather and Medical Plants.

Indigenous Knowledge Systems and Other Fields

Indigenous Knowledge Systems in Lesotho are not confined to environmental management alone. They permeate education, peace-building, health, agriculture, and higher learning institutions. Recent scholarship situates Indigenous Knowledge Systems within broader debates on community-based development and climate resilience, where they are increasingly recognised as complementary to scientific knowledge systems rather than separate from them.

Morojele (2017) presents a thorough analysis of the indigenous knowledge of Basotho male herders, presenting this knowledge as a science and a lifelong learning process. The research applied a qualitative interpretive approach, in which interviews and observations were conducted among purposefully selected herders in rural Lesotho. The research findings refuted the notion that formal literacy is a singular channel of knowledge acquisition; instead, they show that herders are able to accumulate complex practical knowledge by interacting with their environment. The research shows that, through observation, imitation, and peer guidance, herders are able to accumulate knowledge in areas such as cattle management, reading weather, food preservation, numeracy, and spatial skills. This shows that Indigenous Knowledge is an organised system of education that is premised on practical experience. The research also shows that the integration of the herders' Indigenous Knowledge into non-formal education is a way of preserving and validating this knowledge, and this is a welcome approach. The research shows that Indigenous Knowledge systems are premised on a survival mechanism and an educational model. However, the study largely treats IKS within an educational framework and does not fully explore its role in broader environmental adaptation processes such as climate variability and disaster preparedness.

Mokotso (2022) has moved the focus from survival to peace-building. Mokotso studied indigenous education practices among the Basotho people, with a focus on *lebollo* as a potential site for peace education. Mokotso employed a transdisciplinary case study approach to engage with traditional leaders and the culture. Mokotso's work situates *lebollo* in UNESCO's peace-building framework while grounding it in African Indigenous systems through literature and interviews. The research found that indigenous initiation practices are more than just rites of passage; they can be considered spaces for the formation of morals, discipline, civic duty, and conflict resolution. Mokotso's work suggests that indigenous education can be a potential site for the extension of peace education beyond the formal school system by engaging with political and social violence through culturally meaningful practices. The suggestion to work with traditional bodies to develop new models of health and peace education is a welcome shift in approach. Rather than replacing indigenous education systems with the formal education system, it is suggested that they be complemented. While this contribution is significant, it remains primarily focused on social cohesion and does not engage with environmental knowledge or ecological adaptation embedded within indigenous institutions.

At the higher education level, Kaya & Seleti (2013) investigated the importance of African Indigenous Knowledge Systems (AIKS) in the higher education sector, including in contexts that could be applicable in Lesotho. They conducted a mixed-methodology case study in the North-West University in South Africa, engaging with students and graduates to explore the implications of the inclusion of AIKS in the academic curriculum. The research findings indicate that the inclusion of Indigenous Knowledge in the academic curriculum improves the performance of students. Students who participated in the research reported that the inclusion of Indigenous Knowledge in the academic curriculum made the subject matter more meaningful as it was related to their realities rather than the Western realities presented in the curriculum. The research was conducted in the context of policy frameworks that include NEPAD. This reflects a broader continental shift toward decolonising education and recognising epistemological diversity within formal institutions.

In the domain of health as well as medicine, Motjotji et al. (2023) undertook an investigation to review ethno-medicinal, phytochemical, as well as pharmacological investigations pertaining to approximately 355 medicinal plants used in the context of Lesotho. The study's synthesis of the literature finds that the Basotho have an in-depth understanding of healing as well as the extensive use of plants in the context of healing. The study finds that plants such as Alepidea amatymbica are not only used for healing but also have spiritual implications. The study finds that the scientific community has the opportunity to tap into the wealth of knowledge embedded in the Indigenous Knowledge system pertaining to the use of plants for healing. The study is important as it finds that the Indigenous Knowledge is not antithetical to the contemporary academy; in fact, the study finds that the inclusion of the Indigenous Knowledge enriches the contemporary academy. The study finds that the intellectual property rights need to be protected in order to safeguard the Indigenous Knowledge. However, the study primarily adopts a biomedical lens, with limited engagement with the broader socio-ecological and environmental meanings of plant use within Indigenous Knowledge Systems.

The Indigenous Knowledge System, therefore, is not fragmented; it is a system of meaning, survival, and governance. It is also a science for practical application, which guides the observation of nature, the conduct of individuals, and the identity of cultures. New studies on the subject are revealing that the IK System provides the foundation for the survival of individuals, the solidarity of societies, and the development of institutions. From the review of literature, it is evident that the IK System has applicability in the fields of education, peace-building, higher education, ethno-medicine, among others. However, the literature on the subject does not recognize the IK System as a holistic approach to adapting to extreme weather. It recognizes the IK System in the survival skills of herders (Morojele & Pitikoe, 2017), the term *lebollo* in the field of peace-building (Mokotso, 2022), the transformation of the curriculum in higher education (Kaya & Seleti, 2013), and the medicinal plants in ethno-medicine (Motjotji et al., 2023), among others. Although the literature recognizes the applicability of the IK System in the fields of education, peace-building, higher education, ethno-medicine, among others, it does not recognize the IK System as a holistic approach to adapting to extreme weather. The theoretical approaches differ, including the interpretive theory in education, the trans-disciplinary theory in higher education, the policy theory in ethno-medicine, and the Afrocentric theory in environmental management. There is, however, a knowledge gap on the anticipation, prevention, and response of the Basotho IK System to extreme weather.

This study fills the identified knowledge gap with the focus now set intentionally on extreme weather conditions and climate resilience in Lesotho. The study's intention is not only to explore the Indigenous knowledge systems but also to critically examine the cumulative effect of the knowledge systems in addressing droughts, hailstorms, floods, and lightning in the region. The study's methodology is qualitative in nature due to the focus set on the community's experiences. Theoretically, the study is underpinned by Afrocentric theory that seeks to address the marginalisation of Indigenous knowledge systems in climate change discourses in the Western world. The study may contribute to the development of new knowledge to the field with the Basotho Indigenous Knowledge Systems identified as dynamic systems that are context-specific.

Indigenous Knowledge Systems on Weather and Medical Plants

Mashini & Mokhothu (2009) investigated the use of IKS in the sustainable production of sorghum by Basotho farmers. The research was conducted using participatory research methods where the farmers participated as research assistants. The research indicates that the practices and tools used in IKS are derived from the use of local plants, trees, animals, and other natural resources. The research shows that IKS in the production of sorghum is embedded in the ecological knowledge of the farmers as well as the values of the communities. The research indicates that IKS in the production of sorghum changes over time but is vulnerable owing to the prevalence of Western knowledge systems. Another research was conducted by Possa & Khotso (2015) on the use of Basotho medicinal plants. The research was conducted using questionnaires and interviews to understand the linguistic and philosophical implications of the use of the medicinal plants. The research indicates that the use of the medicinal plants by the Basotho is not merely based on the name of the plant. The name of the plant indicates the illness it cures as well as the economic benefits of the use of the medicinal plants. The use of the medicinal plants by the Basotho also indicates the cultural philosophy of the Basotho as well as economic benefits. However, the study focuses mainly on agricultural productivity and does not explicitly link these practices to broader climate variability or extreme weather adaptation.

Differently, the study by Kose et al. (2015) contributed to this documentation through conducting an ethno-botanical survey in Maseru District. The study documented medicinal plants for treating common ailments in both urban and rural areas of Lesotho. Through structured interviews, the study identified 80 plant species from 44 plant families for treating 38 ailments, mostly reproductive and respiratory problems. The study documented some of the plants for the first time, showing their continued usage among the Basotho community. The study also showed some of the challenges facing medicinal plants, such as over-exploitation, which threatened sustainability. Although this study documented medicinal plants, its focus is still narrow, looking at medicinal plants from a biomedical perspective. This highlights that medicinal plant knowledge is not purely biological but also cultural and epistemological in nature.

Additionally, Balogun et al. (2016) studied medicinal plants used by the Sotho community in managing diabetes mellitus. The review documented evidence of 23 plant species used in managing the disease in the Sotho community. The study showed the dominance of Asteraceae family plants in managing the disease, the affordability of herbal medicine, and its accessibility compared to conventional medicine. The study also showed the lack of scientific validation of medicinal plants in managing the disease, which needs further investigation. Like previous studies, this study focused on medicinal plants from a biomedical perspective, ignoring the socio-ecological perspective. Like other studies, this work focuses narrowly on biomedical applications and does not fully explore the ecological or climate-related dimensions of Indigenous Knowledge Systems.

Despite the significant contributions of past studies on IKS, gaps persist. Most of the studies focused on specific themes, such as sustainable agriculture (Mashini & Mokhothu, 2009), the naming of medicinal plants (Possa & Khotso, 2015), ethnobotany (Kose et al., 2015), and the use of herbs in the management of diabetes (Balogun et al., 2016), among others. However, no single one of these studies was designed to explore the Basotho IKS in the face of extreme weather conditions or climate change. Theoretically, the past studies on IKS in Lesotho lacked explicit references to the Afrocentric perspective. Most of the studies were ethno-botany, ethno-medicine, or sustainability-based, with references to the broader scientific paradigms. There was, therefore, a theoretical gap, as the IKS was explored from a functional rather than an epistemological perspective. With regard to the knowledge contribution, the past studies on IKS in Lesotho were significant in the documentation of agricultural practices, the use of herbs, the naming of plants, and community practices. However, the gaps that persist are the exploration of the role of such ecological observations, rituals, ceremonies, and community practices as adaptive mechanisms in the face of contemporary extreme weather conditions. There was inadequate exploration of the integration of IKS with the contemporary environmental crises or the exploration of the role of indigenous knowledge in future climate change strategies.

This chapter aims at bridging the gaps through an exploration of the Basotho IKS in the face of extreme weather conditions. This chapter has employed the qualitative explanatory descriptive method to explore the IKS based on the Afrocentric approach. Therefore, it aims at bridging the gaps through an exploration of the IKS based on a wider perspective than the conventional sectorial approach. This chapter aims at demonstrating the significance of the indigenous knowledge through an exploration of the IKS in the face of contemporary climate change.

THEORETICAL FRAMEWORK

This chapter uses Afrocentric Theory. Afrocentricity is a theory that developed in the late 20th century to address the marginalisation of African knowledge in the academy (Mgbeadichie, 2015). For decades, knowledge about Africans has been viewed through the lens of the European worldview, where Africans have been perceived as having no theory, science, or structured way of thinking. Afrocentricity seeks to correct this view by giving Africans the central place in the discussion of any issue to do with them. The theory's major contributor is Molefi Kete Asante, who posits that the study of African society must be grounded in the culture of Africans (Daniels, 2022). Knowledge about Africans must be derived from the historical and cultural experiences of Africans themselves, not from the viewpoint of outsiders. Afrocentric Theory is based on the view that the experiences, values, and knowledge systems of Africans must be the central focus in the study of society, politics, culture, and the environment (Pellerin, 2012).

Afrocentricity is a term used to describe a group of views that ask who is centered in knowledge, who defines truth, and who has knowledge that matters. Chawane (2016) describes Afrocentricity as a focus on Afrocentrism, which is described by Bangura (2019) as the centering of African culture and successes, and by Mamaleka (2020) as a guide to a transformation in attitudes and beliefs. Afrocentricity views Africans as agents in history. It encourages Africans to see themselves and be seen by themselves, rather than be seen by others (Khunwane, 2019), to be repositioned in historical, social, economic, and political spaces. Importantly, in the context of this study, Afrocentric Theory is used not only as a lens of cultural affirmation but as an analytical framework for understanding how Indigenous Knowledge Systems function as structured systems of environmental knowledge and climate response. Afrocentricity further challenges the assumption that Western scientific epistemology is the only valid framework for interpreting environmental phenomena. In this study, this is particularly relevant because Basotho interpretations of droughts, floods, hailstorms, and lightning are not treated as symbolic beliefs, but as knowledge systems developed through long-term environmental observation and cultural transmission.

There are a number of strengths associated with Afrocentricity. It re-establishes African agency in academic inquiry. It validates oral traditions as a legitimate means of knowledge. It promotes the incorporation of Indigenous epistemology into academic research. Afrocentrism, according to Churu et al. (2022), promotes cultural identities among African people. It promotes unity among African people through shared heritage and memory. It provides a means of examining Indigenous epistemology without dismissing it as mere superstition. However, there are also a number of weaknesses associated with Afrocentrism. It has been criticized for promoting a monolithic view of Africa despite its diversity. In this study, these limitations are acknowledged, and Afrocentricity is applied in a balanced manner that recognises both African epistemic validity and the contextual diversity of Basotho knowledge systems. Afrocentrism minimises contact between African people and people of other cultures. This is a weakness that needs to be considered with analytical rigor. Afrocentricity must avoid romanticism; it must be grounded in evidence (Mekoa, 2018).

Despite the criticisms levelled against Afrocentric theory, it is helpful for the study. According to Rogwers (2023), Afrocentrism values African cultures as part of the history of humanity. Dagnini (2023) states that Afrocentrism encourages unity through the use of cultural heritage. Kaulemu (2022) adds that Afrocentrism seeks to correct historical injustices through the promotion of fairness in the production of knowledge. In this chapter, Afrocentric Theory is applied specifically to interpret Basotho Indigenous Knowledge Systems as adaptive environmental mechanisms. Rather than viewing rituals, taboos, ecological indicators, and oral traditions as symbolic or informal practices, the theory positions them as structured systems of environmental observation, forecasting, and response. More specifically, the theory guides the interpretation of how Basotho communities anticipate, prevent, and respond to extreme weather events such as droughts, hailstorms, floods, and lightning. It therefore supports the central argument of this chapter that Indigenous Knowledge Systems function as adaptive and context-specific climate resilience mechanisms rather than outdated cultural practices.

RESEARCH METHODOLOGY

This section outlines the research approach, research design, research site, population, sampling, sampling techniques, methods of data collection, validity and reliability in qualitative research, data organisation and presentation, and ethical consideration.

Research Approach

The research approach that this study uses is a qualitative approach in understanding African Indigenous Knowledge Systems (IKS) and its role in climate adaptation. The qualitative approach of research is defined as a research approach that involves collecting and analysing non-numerical data in order to understand the social reality of individuals (Teherani et al., 2015).

In this study, the qualitative approach is appropriate because Indigenous Knowledge Systems are embedded in lived experience, oral tradition, and cultural practice, which cannot be adequately captured through quantitative measurement. This study therefore focuses specifically on how Basotho communities interpret, apply, and transmit Indigenous Knowledge Systems in response to extreme weather events such as droughts, floods, hailstorms, and lightning.

Sample and Sampling Techniques

A sample is a group of individuals selected from the research population to participate in the research. The selected group represents the research population. It assists the researcher in drawing conclusions about the research (Lohr, 2021). Bekele & Ago (2022) note that the sample size is determined by various factors including the research topic, nature of the questions, resources available, time constraints, as well as the requirements of the research institution. Therefore, in this research, the sample size was determined to be sixteen individuals. The individuals involved in the research were farmers, elders, as well as traditional healers aged 40+ years from Mount Moorosi in the district of Quthing. These individuals were selected for the research since they had the knowledge needed to handle extreme weather conditions.

The process of selecting the participants involved the following non-probability sampling methods: purposive sampling. Non-probability sampling is a sampling method in which the individuals selected to participate in the research are selected based on non-probability criteria (Stratton, 2023). Therefore, it means that not all individuals in the research population have an equal probability of being selected to participate in the research. Stratton (2023) adds that non-probability sampling methods include purposive sampling, snowball sampling, as well as quota sampling.

Using purposive sampling, the participants can be selected based on specific criteria that match the research topic (Etikan et al., 2016). Ratsie (2021) adds that purposive sampling provides the researcher with the opportunity to use judgment in the process of sampling. Therefore, purposive sampling provides the researcher with the opportunity to select the participants who have the knowledge of the phenomenon under study. Therefore, the researcher was able to select the individuals from different social groups with the aim of collecting a variety of perspectives related to the research topic. These groups included the elderly, farmers, as well as traditional healers.

In this study, purposive sampling ensured that participants selected were directly involved in Indigenous Knowledge practices related to environmental observation, agricultural decision-making, and traditional weather interpretation.

Method of Data Collection

This section entails the methods of data collection used in the study. The study used the interviews as the main method of data collection. Ajayi (2017) indicates that the primary methods of data collection refer to the methods that involve the collection of original information for the research purpose. These methods include the use of interviews, questionnaires, and observations. These methods of data collection focus on the collection of new information as opposed to the use of secondary sources of information or the use of the information that already exists. Interviews can be defined as the data collection method where the interviewer asks the respondents questions in person, over the phone, or online. In the study, the researcher used both face-to-face and telephone interviews. Telephone interviews were used to reach the participants who were distant from the interviewer. Interviews can be structured, semi-structured, or unstructured, as noted by Broache (2022).

This study used the semi-structured interviews as the data collection method. Osborne & Grant-Smith (2021) indicate that the semi-structured interviews involve the use of the data collection method that combines the use of both structured and unstructured elements. Although the data collection method has a predetermined set of questions, the interviewer can use the flexibility of the method to explore the research topic based on the responses received from the participants. The use of the data collection method was appropriate since it allowed the participants to express their views as they shared their lived experiences in coping with the challenges that arise as a result of the impacts of climate change. The data collection method was appropriate since it allowed the collection of detailed information about the Basotho Indigenous Knowledge Systems.

This method was appropriate because it allowed participants to express their lived experiences in their own words while enabling the researcher to probe deeper into Indigenous Knowledge practices related to extreme weather events.

Validity and Reliability in Qualitative Research

To ensure trustworthiness and credibility, the study employed data triangulation. Data triangulation involves the use of multiple data sources to investigate the same phenomenon (Donkoh & Mensah, 2023). In this study, information was collected from different categories of participants, including farmers, elders, and traditional healers, who possess indigenous environmental knowledge relevant to extreme weather management in Basotho communities. The interviews were also conducted on different days, which helped to capture varied perspectives and experiences. This approach enhanced the credibility of the findings by allowing the researcher to compare and confirm recurring patterns, views, and practices across different participants. In addition, the interviews were recorded, transcribed, and analysed systematically to maintain consistency and transparency throughout the research process.

Data Organisation and Presentation

The data collected through interviews were organised, transcribed, and analysed using thematic analysis. Thematic analysis involves identifying and interpreting recurring themes and patterns within qualitative data (Castleberry & Nolen, 2018). In this study, the interview responses were carefully examined to identify common ideas related to Basotho Indigenous Knowledge Systems and climate adaptation. The findings were then organised into major thematic categories and presented according to the patterns that emerged from participants' responses. Similar views expressed by multiple participants were summarised to avoid repetition, while unique perspectives were presented through direct quotations. This approach enabled the researcher to present the findings in a clear, systematic, and coherent manner while preserving the meanings and experiences shared by the participants.

Ethical Considerations

Ethical issues play a vital role in the conduct of credible research, especially when human participants are involved (Arifin, 2018). This study adhered to ethical principles of informed consent, confidentiality, and voluntary participation. Participants were fully informed about the purpose of the study, the nature of their involvement, and their right to withdraw at any stage without any consequences. All participants voluntarily agreed to take part in the study prior to data collection. Confidentiality was ensured by protecting participant identities and removing any personally identifying information from the data presentation. Data collected during interviews were securely stored and used strictly for academic purposes. This study was conducted as independent academic research and did not fall under a formal institutional ethics clearance process. However, ethical practice was maintained through participant-informed consent and adherence to standard qualitative research ethical principles.

FINDINGS AND DISCUSSIONS

This section presents findings from interviews conducted with sixteen participants, including farmers, elders, and indigenous healers. The data were analysed using thematic analysis to identify recurring patterns in participants' experiences and interpretations of Indigenous Knowledge Systems (IKS) in relation to extreme weather events. The analysis is structured around Basotho Indigenous Knowledge Systems as both preventive and control mechanisms of climate response.

The Basotho Indigenous Knowledge Systems are understood in this study as culturally embedded environmental knowledge that operates through two interconnected functions: prevention and control. Prevention refers to anticipatory practices used to reduce the likelihood or impact of extreme weather events, while control refers to reactive practices used after such events have occurred or become imminent.

Prevention Systems (Proactive Measures)

Prevention in this study can be defined as the practices that are put in place to mitigate the occurrence of extreme weather events. In this section, the Indigenous Knowledge Systems that the Basotho use in the anticipation of extreme weather conditions will be discussed. The IKS discussed in this section include Traditional Ecological Knowledge, Norms and Taboos, and the Use of Protective Charms. The study is guided by the Afrocentric Theory, as proposed by Asante (2017). The theory posits that the agency of the African should be central in the discourse on the topic in question in order to decolonize the discourse. The theory challenges the Eurocentric notion that IKS is ''superstition''

Traditional Ecological Knowledge (TEK)

Traditional Ecological Knowledge in this study refers to accumulated environmental knowledge developed through long-term interaction with nature, including the observation of animal behaviour, weather patterns, and astronomical signs (Mokonen, 2017). Participants consistently described TEK as a practical and reliable system for anticipating weather changes. One participant explained:

"*Ha linonyana li fofela tlaase, rea tseba hore lifefo li atametse*", (Farmer)

When birds fly low, we know storms are near.

Similarly, another participant highlighted the role of frogs:

"Ha liqaqana li lla haholo mantsiboea, re itokisetsa pula", (elder)

When frogs croak loudly in the evening, we prepare for rain.

These observations show that environmental interpretation is not random but structured through repeated lived experience. In addition, participants emphasised the role of lunar cycles in climate prediction:

"Ha khoeli e thoasa eba joale e shebile tlaase, re lebella lipula tse ntle. Ha e shebile holimo rea tseba hore selemo seo e tla bas a komello", (elder)

When the moon is new/waxing and appears to be facing downward, we expect good rains. When it is facing upward, we know that that year will be a year of drought.

This knowledge is transmitted through generations, with elders playing a key role in guiding agricultural preparation. Rather than viewing these practices as superstition, participants positioned them as practical early warning systems, especially in rural contexts where formal meteorological services are limited or generalised. A key analytical point emerging from the data is that TEK functions as a locally grounded adaptation system that enables timely decision-making in agriculture and livestock management. It also demonstrates that environmental knowledge is embedded in everyday observation rather than formal scientific structures.

The participants observed that, just like in many other African countries, scientific weather forecasts are not accessible in the rural areas. Furthermore, the forecasts provided on the radio or national media are often too generalized. In various parts of sub-Saharan Africa, Indigenous and Local Knowledge are often the early warning systems that help the population prepare for extreme weather conditions. In some pastoralist communities in Ethiopia, for example, 97% of the population rely on Indigenous knowledge for weather prediction compared to scientific forecasts (Iticha and Husen, 2019).

The participants are not talking about random beliefs, but they are talking about observations that they have learned from their ancestors. They observed that the knowledge they possess enables them to prepare early for drought, floods, frost, and hail. This, in essence, makes Traditional Ecological Knowledge a form of climate change adaptation. It is the ability to observe nature. The participants observed that the knowledge they possess is passed down from their grandparents, and it enables them to prepare their farms, prepare their livestock, and prepare water in good time. They observed that the contemporary population can also use the knowledge they possess, coupled with scientific knowledge, to improve the prediction of extreme weather. From an Afrocentric perspective, this reflects the positioning of African communities as active knowledge producers rather than passive recipients of external scientific systems. However, in this study, the emphasis is not on repeating theoretical claims, but on showing how TEK operates as a lived environmental practice embedded in experience and memory.

Norms and Taboos

Norms and taboos in this study refer to socially enforced behavioural rules that regulate human interaction with the environment and function as preventative climate governance mechanisms (Swanta, 2026). Participants identified several norms linked to seasonal awareness and environmental protection. One of the most frequently mentioned was *ho ilela Sefako*, a prohibition of certain activities during hail-prone periods.

A participant explained:

"Ka Sesotho ha re hlatoe le ho anehe liaparo motšeare, ha re hlaole masimong, ebile ha re ee mabitleng har'a mpa ea motšeare", (elder)

In Sesotho, we do not wash or hang clothes during the daytime, we do not weed in the fields, and we do not go to the graveyard in the middle of the day.

The reasoning behind these restrictions was consistently linked to environmental timing and safety rather than fear alone. Another participant explained burial practices:

"Re pata bafu hoseng hobane har'a mpa ea letsatsi ho tlisa liphetoho tse kotsi tsa leholimo tse kang sefako", (traditional healer)

We bury the dead early in the morning because midday brings dangerous weather changes like hail.

What emerges here is a system of behavioural regulation that aligns human activity with perceived atmospheric stability. Participants identified several norms, among them the norm concerning the change of season and the ability to predict the weather. One such norm is *ho ilela Sefako* norm, which refers to the prohibition of some activities when there is an expected occurrence of hailstorms. The participants said that when the season is expected to be hail-prone, people do not visit graveyards in the midday period, do not wash clothes, do not weed, and do not bring firewood into the community. According to the participants, the reason why they do not do these activities in the midday period is because it is the time when the atmosphere is not stable, which is why they are prohibited from washing clothes, weeding, bringing firewood, and performing burial activities at the time. Therefore, they change their behaviour to avoid harm. The participants further said that they bury the dead in the early morning. The participants said that when they bury the dead in the midday period, they may be visited by hail or some kind of weather. Therefore, they believe in the norm concerning the change of season and the ability to predict the weather. The participants said that these practices are part of the community's knowledge that some human activities affect the atmosphere. Therefore, they change their behaviour to avoid harm. Another norm was the one concerning the cutting and burning of trees.

Participants also discussed restrictions on tree cutting:

"Morena ha a lumelle ho rema lifate nakong ea meea e matla. Lifate li sireletsa mobu ebile li thibela tšenyo", (farmer)

The chief does not allow cutting trees during windy seasons. Trees protect the land and stop soil from being destroyed.

In most African countries, deforestation is said to be the cause of wind damage and soil erosion (Mekoya et al., 2025). The participants said that the chiefs prohibit the cutting of trees. The trees can be cut when there is a need to perform some ceremony. The participants said that these practices help in the governance of the environment. Therefore, the participants believe in the norm concerning the cutting and burning of trees.

Field burning is also restricted. The participants stated that burning is allowed only when it is necessary, such as for the preparation of new crop fields or green manuring. They further stated that burning during dry seasons is harmful to the soil, causes desiccation of the land, and increases wind disturbances. This increases the risk of erosion and crop failure. Swidden farming is controlled in the same way in West Africa. Taboos have been used to prevent burning during certain seasons to reduce soil erosion and fire risks (Brandt et al., 2018). Overall, this shows that norms are not just a product of culture; they have a role to play in the management of the environment. These findings suggest that norms and taboos function as informal environmental governance systems that regulate land use and reduce ecological risk. A key insight here is that environmental protection is not enforced through formal law alone but through cultural regulation embedded in social authority structures such as chiefs and elders. The analysis shows that these norms function as ecological management tools grounded in community authority and environmental observation.

Protective Charms

Protective charms refer to objects, substances, or symbols believed to protect individuals and households from environmental harm (Ibrahim, 2023). Participants identified two key protective practices: tšita-tlali (lightning stopper plant) and mofifi (forked roofing stick).

One participant explained:

"Re sebelisa tšita-tlali holim'a matlo hore lehalima le se ke la a otla", (healer)

We use tšita-tlali on houses so that lightning does not strike them.

Another added:

"Mofifi o kengoa marulelong kapa o nkoa ha ho tsamauoaa. O sireletsa motho khahlanong le lehalima", (elder)

Mofifi is kept in the roof or even carried when travelling. It protects a person from lightning.

Participants did not separate these practices from physical or spiritual causality. Instead, they viewed environmental danger as a combination of both. Another participant explained this integrated worldview:

"Lehalima ha se taba feela e bonahalang. Le na le lisosa tsa semoea, kahoo re itšireletsa ka litsela tsohle", (healer)

Lightning is not just a visible physical phenomenon. It also has spiritual causes, so we protect ourselves in all ways

The participants have identified two traditional protective practices employed by the Basotho people that have been associated with protection against lightning and storms. One such practice is the use of *tšita-tlali*, literally translated to mean "lightning stopper," a plant with medicinal value that is used to protect homes against lightning strikes. The other practice is the use of *mofifi*, a forked stick used in the roofing structure of a house, which is employed for the same purpose. Therefore, the two practices have been associated with two important aspects of Basotho environmental knowledge. The practices associated with the use of *tšita-tlali* and *mofifi* have been associated with a belief that environmental phenomena such as storms and lightning result from a mix of physical and spiritual factors. This is in line with the argument advanced by scholars such as Hoppers (2002), who have argued that cosmologies in Africa do not make a distinction between the physical and spiritual; in other words, spiritual forces are always at work in the environment.

Moreover, the participants pointed out that *mofifi*, which is a forked plant, can be carried in a person's pocket during travel. They think that *mofifi* provides protection from lightning strikes. These charms provide people with a sense of control over their lives when formal science and infrastructure do not fully protect them from environmental forces. This concept resonates with other studies conducted in Africa, where people have resorted to spiritual activities to ensure local control over their environment, which is full of uncertainties (Gumo et al., 2012; Ombati, 2021). Although the dominant Western science views adaptation in terms of material or technological adaptations, indigenous knowledge often associates interacting with the environment with cultural or spiritual activities to enhance the relationship between people and their environment (Dolan, 2016; Diaz-Reyes, 2019).

Control Systems

Control systems refer to reactive strategies implemented after extreme weather events occur or when drought becomes imminent. These include rainmaking rituals, hailstorm interventions and lightning cleansing.

Rainmaking Rituals

Rainmaking rituals are collective interventions performed during drought periods to request rainfall (Ombati, 2017). For the men in Basotho culture, the rainmaking rituals are called *Molutsoane*, meaning the "communal hunt for rain. Male participants described *Molutsoane* as a communal hunting ritual linked to rainmaking. One participant explained:

"Rea tsoa re eo tsoma ebe re lahlela likarolo tsa phoofolo ka nokeng hore pula e tle", (elder)

We go out to hunt and offer parts of the animal to the river so that rain can come.

During the rituals, the men in the Basotho community would wake up early in the morning and go for hunting expeditions in the different landscapes. During the expeditions, the men would meet some wild animals in the course of the journey and kill them. After reaching the river, the internal organs would be removed from the bodies of the animals and thrown into the river. Some parts of animals would also be burned as an offering to the spirits. During the rituals, no animals were allowed to go back to the community with the intestines.

There is also a ritual performed by women. Women described *Lesokoana* as a ritual involving collective movement and symbolic exchange:

"Re utsoa lesokoana motseng o mong re balehe ka lona ho le isa moreneng oa rona re ntse re bina re kopa pula", (elder)

We steal the stirring stick from another village and run away with it to our village while singing and asking for rain.

In this ritual, girls from one village stole the porridge stirring stick from the homestead of another village. On their way back to their village, they would pass the stick to each of the girls while singing songs which requested rains to fall. The girls of the village which had been stolen from would quickly gather and chase them to retrieve their stolen stick. This would go on for most of the day. After the girls returned the stolen stick to their chief, they would start singing the rains-requesting songs.

According to the participants, these two rituals are applicable to the contemporary community for planning to prepare for natural calamities such as floods, droughts, and storms. On the above two rituals discussed, Johnson (2018) explains that rain praying activities are not isolated rituals but rather common social activities undertaken by men, women, girls, and boys during specific times of the year when rains are most needed. The festivities and prayers are not just spiritual requests for rains but are also considered to bring the community together and help to identify common values and roles for specific gender and age groups during natural calamities.

These rituals demonstrate that drought is seen as a disruption that requires collective and coordinated efforts. The act of casting animal intestines into the river is a symbolic act of restoring life into the water source. When considering these rituals in relation to extreme weather conditions, they are performed to mentally and emotionally mobilise communities, enhance social unity, and highlight a collective dependency on the balance of the environment. When considering these practices from an Afrocentric perspective, they demonstrate African agency and social unity of the community.

Hailstorm Interventions

Interventions related to hailstorms can be understood as the immediate rituals that traditional healers perform in response to the threat that the hailstorm poses to the livelihood of the people. The participants claimed that traditional healers can change the hailstorm to rain (*ho se hlapolla*), disperse it (*ho se qhala*), or change its direction so that it does not reach the village. The traditional healers did not reveal the nature of the charms that they use in intervening in the hailstorm. However, they claimed that when the hailstorm comes unexpectedly, they can change its direction so that it reaches the ground as rain rather than as hail. He said, "*Ha sefako se e-tla, lingaka li tseba ho se isa nqa e 'ngoe hore se seke sa na sebakeng seo*". (When hail is approaching, traditional healers are able to redirect it to another direction so that it does not reach the area where it is heading).

Others reported that healers can disperse or redirect the hail to prevent it from striking their villages. They claimed that it is the responsibility of the traditional healers to disperse the hailstorm so that it does not reach the village. The participants claimed that the traditional healers can control the hailstorm in that it provides immediate safety to the vulnerable crops.

Lightning Cleansing

The lightning cleansing is the ritual purification that is performed after lightning strikes the homestead. The participants explained that the traditional healer cleanses the village, individuals are restricted in the village, and communal bathing and sprinkling are performed. When lightning entered the homestead (struck the village), the traditional healer is called to cleanse the village and the individuals. After the traditional healer blew the medicine for protection, the villagers gathered for communal sprinkling. The villagers then go to the fields with the cattle, men, women, and children to bathe and remove the medicinal mixture used the previous day. When they returned to the village, lightning was cast away.

Participants described communal cleansing rituals:

"*Ngaka ea setso e hloekisa motse 'me batho bohle ba itlhatsoa hammoho ho tlosa lehalima*", (elder)

The traditional healer cleanses the village and everyone bathes together to remove danger.

Another participant explained:

"*Ka mor'a ho itlhoekisa, re lumela hore lehalima le keke la khutla motseng*", (healer)

After cleansing, we believe lightning will not return to the village.

The participants also talked about the use of *Setima-mollo*, meaning fire extinguisher. *Setima-mollo* is a medicinal plant that is placed around the house.

Participants described;

"*Semela sena se sirelletsa matlo khahlanong le mollo o bakoang ke lehalima*", (farmer)

This plant protects houses from fire caused by lightning.

It is believed that the plants protect the house from fire damage caused by lightning strikes. The participants explained that the plants protect the house from fire. The participants explained that lightning strikes the house, causing fire that damages the house. The plants are believed to prevent the fire from occurring. The participants also explained that the plants are used in protection rituals performed by the traditional healer. The plants are believed to show that the house is protected. The participants pay attention to the plants that are placed around the house. This is supported by Possa and Khotso (2015). They stated that *Setima mollo* is said to keep a house safe from fire, especially lightning fire, because its name and use communicate protection according to the beliefs and aims

The analysis indicates that cleansing rituals function as symbolic restoration of environmental and social balance after perceived disruption. The process of lightning cleansing shows how Basotho communities rely on indigenous knowledge to deal with environmental events. The respondents claimed that these rituals are based on Indigenous Knowledge Systems passed down through the ages. The rituals are based on the premise of interconnectedness between human life, nature, and spirituality. The cleansing ritual aims to bring back balance to nature after the hazardous occurrence. In Afrocentricity theory, the process of lightning cleansing shows how Basotho communities place their indigenous knowledge first when dealing with environmental events. The communities do not rely on external or Western knowledge to deal with environmental hazards but rather on Indigenous Knowledge Systems developed over time.

This approach is Afrocentric and shows how Afrocentricity views place the experiences and values of Africa at the heart of understanding and dealing with environmental issues.

LIMITATIONS OF THE STUDY

This study is not without limitations. Firstly, the research was conducted with a relatively small number of participants, which limits the extent to which the findings can be generalised beyond the study context. The sample was also drawn from a single district, meaning that the perspectives captured may reflect localised experiences and may not fully represent views from other parts of Lesotho.

Secondly, the study focused mainly on participants aged 40 years and above. While this group was deliberately selected because of their role as custodians of indigenous environmental knowledge, it means that the perspectives of younger generations were not extensively captured. This limits the ability to fully explore how indigenous knowledge systems are understood and transmitted across all age groups. Thirdly, the study relied on participants' self-reported knowledge and experiences. As with most qualitative studies, this may introduce subjective interpretations shaped by memory, personal beliefs, and social positioning.

CONCLUSIONS AND RECOMMENDATIONS

The purpose of this research was to explore the existence of African Indigenous Knowledge Systems (IKS), with specific reference to the Basotho Indigenous Knowledge Systems, and their significance to the management of extreme weather events in Lesotho. This research was underpinned by Afrocentric Theory, which privileges and prioritizes the central place of African knowledge systems and cultural values in the construction of knowledge and experiences. This contrasts with Western-centric theories and knowledge systems.

The research found that the Basotho Indigenous Knowledge Systems are underpinned by preventive and control mechanisms. The preventive mechanisms of the Basotho Indigenous Knowledge Systems are exemplified by the existence of Traditional Ecological Knowledge (TEK), taboos, and the use of protective charms. The reactive mechanisms of the Basotho Indigenous Knowledge Systems are exemplified by the existence of rainmaking rituals, hailstorm intervention rituals, and the use of lightning cleansing rituals. The research found that these mechanisms of the Basotho Indigenous Knowledge Systems are important to the management of extreme weather events and are important for promoting social cohesion and cultural identity.

Based on the findings, it is concluded that the Basotho IKS provides effective and practical approaches to climate change adaptation. The knowledge systems reveal the significance of intergenerational knowledge transfer, communalism, and the unification of spiritual and ecological knowledge in responding to climate change. The study affirms the significance of African Indigenous Knowledge, which is not merely folklore, but a powerful and effective strategy in coping with the challenges of climate change.

Based on the findings, it is recommended that policy and development actors integrate African Indigenous Knowledge Systems with scientific knowledge in the quest for climate change adaptation. Moreover, it is important to ensure the protection and documentation of African Indigenous Knowledge to ensure its appropriateness in the future. Finally, it is recommended that community-based initiatives be supported, which promote cultural values and practices, thereby enhancing cultural pride and epistemological legitimacy in Africa.

REFERENCES

Ajayi, V. O. (2017). Primary Sources of Data and Secondary Sources of Data. *Benue State University*, 1(1), 1-6.

Alexander, L. V. (2016). Global Observed Long-Term Changes in Temperature and Precipitation Extremes: A Review of Progress and Limitations in IPCC Assessments and Beyond. *Weather and Climate Extremes*, *11*, 4–16. DOI: 10.1016/j.wace.2015.10.007

Asante, M. K. (2017). The philosophy of Afrocentricity. In *The Palgrave handbook of African philosophy* (pp. 231–244). Palgrave Macmillan US. DOI: 10.1057/978-1-137-59291-0_16

Balogun, F. O., Tshabalala, N. T., & Ashafa, A. O. T. (2016). Antidiabetic medicinal plants used by the Basotho tribe of Eastern Free State: A review. *Journal of Diabetes Research*, *2016*(1), 4602820. DOI: 10.1155/2016/4602820 PMID: 27437404

Bekele, W. B., & Ago, F. Y. (2022). Sample Size for Interview in Qualitative Research in Social Sciences: A Guide to Novice Researchers. *Research in Educational Policy and Management*, *4*(1), 42–50. DOI: 10.46303/repam.2022.3

Brandt, M., Rasmussen, K., Hiernaux, P., Herrmann, S., Tucker, C. J., Tong, X., Tian, F., Mertz, O., Kergoat, L., Mbow, C., David, J. L., Melocik, K. A., Dendoncker, M., Vincke, C., & Fensholt, R. (2018). Reduction of tree cover in West African woodlands and promotion in semi-arid farmlands. *Nature Geoscience*, *11*(5), 328–333. DOI: 10.1038/s41561-018-0092-x PMID: 32944066

Broache, M. P. (2022). *Planning, Conducting and Analysing Interviews*. Edward Elgar Publishing.

Castleberry, A., & Nolen, A. (2018). Thematic Analysis of Qualitative Research Data: Is it as easy as it Sounds? *Currents in Pharmacy Teaching & Learning*, *10*(6), 807–815. DOI: 10.1016/j.cptl.2018.03.019 PMID: 30025784

Chawane, M. (2016). The development of Afrocentricity: A historical survey. *Yesterday and Today*, (16), 78–99. DOI: 10.17159/2223-0386/2016/n16a5

Churu, W. B., & Nancy, W. (2022). Afrocentricity as a basis for restoration of African identity and agency: A call to authentic education in Africa. *AKU: An African Journal of Contemporary Research*, *3*(2), 27–43.

Dagnini, J. K. (2023). Afrocentrism through Afro-American Music: from the 1960's until the Early 2000's. www. revue-sociologique. org/sites/default/files/2. pdf *(Accessed 20 Jul. 2019)*.

Daniels, R. (2022). Racing toward an African Inclusion: De-essentializing the Global South. *Decolonizing communication studies*, 42-56.

Dawadi, S. (2020). Thematic Analysis Approach: A Step-by-Step Guide for ELT Research Practitioners. *Journal of NELTA*, *25*(1-2), 62–71. DOI: 10.3126/nelta.v25i1-2.49731

Diaz-Reyes, T. M. (2019). *Ecological Relationality and Models of Sustainability: Towards a Social, Intergenerational, Multi-Species Environmental Ethic through the Evaluation of Cultural-Spiritual Ecosystem Services*. Wake Forest University.

Dolan, J. M. (2016). *The restorative ecology of peace: Haudenosaunee environmental knowledge and philosophies of stewardship*. McGill University.

Donkoh, S., & Mensah, J. (2023). Application of Triangulation in Qualitative Research. *Journal of Applied Biotechnology and Bioengineering*, *10*(1), 6–9. DOI: 10.15406/jabb.2023.10.00319

Etikan, I., Musa, S. A., & Alkassim, R. S. (2016). Comparison of Convenience Sampling and Purposive Sampling. *American Journal of Theoretical and Applied Statistics*, *5*(1), 1–4. DOI: 10.11648/j.ajtas.20160501.11

Gemeda, D. O., & Sima, A. D. (2015). The Impacts of Climate Change on African Continent and the Way Forward. *Journal of Ecology and the Natural Environment*, *7*(10), 256–262. DOI: 10.5897/JENE2015.0533

Gumo, S., Gisege, S. O., Raballah, E., & Ouma, C. (2012). Communicating African spirituality through ecology: Challenges and prospects for the 21st century. *Religions*, *3*(2), 523–543. DOI: 10.3390/rel3020523

Hoppers, C. A. O. (Ed.). (2002). *Indigenous Knowledge and the Integration of Knowledge Systems: Towards a philosophy of articulation*. New Africa Books.

Ibrahim, M. (2023). The people with iron skin: Protective charms, traditional religion, and vigilante authority in Lagos, Nigeria. *Culture and Religion*, *23*(3), 240–261. DOI: 10.1080/14755610.2024.2344563 PMID: 39108276

Iticha, B., & Husen, A. (2019). Adaptation to climate change using indigenous weather forecasting systems in Borana pastoralists of southern Ethiopia. *Climate and Development*, *11*(7), 564–573. DOI: 10.1080/17565529.2018.1507896

Johnson, L. R. (2018). Basotho culture and the prayers for rain: Where climate change converges. *Southern African Journal for Folklore Studies*, *28*(2), 1–19. DOI: 10.25159/1016-8427/3799

Kaulemu, D. (2022). *The Ethics and Politics of Conceptualizing Values in Africa*. Springer International Publishing. DOI: 10.1007/978-3-031-12938-4_2

Kaya, H. O., & Seleti, Y. N. (2013). African indigenous knowledge systems and relevance of higher education in South Africa. *International Education Journal: Comparative Perspectives, 12*(1).

Khunwane, M. R. (2019). *A Comparative Analysis of the influence of Folklore on the works of the following African writers: Chinua Achebe, Eskia Mphahlele, Ngungi wa Thiongo'and Andrew Nkadimeng: An Afrocentric approach* (Doctoral dissertation).

King Letsie, I. I. I. (2024). *Speech on Lesotho's Bicentenary and the 58th Anniversary of Independence* [Audio Recording]. Government of Lesotho.

Kose, L. S., Moteetee, A., & Van Vuuren, S. (2015). Ethnobotanical survey of medicinal plants used in the Maseru district of Lesotho. *Journal of Ethnopharmacology*, *170*, 184–200. DOI: 10.1016/j.jep.2015.04.047 PMID: 25957810

Lohr, S. L. (2021). *Sampling: Design and Analysis*. Chapman and Hall/CRC. DOI: 10.1201/9780429298899

Mamaleka, M. (2020). *Developing guidelines for indigenous practices: A case study of Makhuduthamaga municipality at Sekhukhune district, Limpopo province*. Limpopo.

Mekoa, I. (2018a). Essentialising African Indigenous Knowledge Systems in the Midst of Globalization and Modernity. *African Renaissance*, 11-28.

Mekoa, I. (2018b). Advocating for Afrocentric studies and African scholarship in the midst of the hostile Western scholarship. *African Renaissance, 15*(3), 87–110. DOI: 10.31920/2516-5305/2018/v15n3a5

Mekoya, A., Tesfaye, A., & Bekele, D. A. (2025). Rainfall, air temperature, and reference evapotranspiration variability and trend and their implication for agroforestry development in the Lake Tana sub-basin of Ethiopia. *Discover Atmosphere, 3*(1), 21. DOI: 10.1007/s44292-025-00045-7

Mgbeadichie, C. F. (2015). *The critical concept of Afrocentrism in Nigerian literature*. University of Exeter.

Mokotso, R. I. (2022). Considering the Basotho indigenous education and school system as resources for peace-building education in Lesotho. *The Journal for Transdisciplinary Research in Southern Africa, 18*(1), 1213. DOI: 10.4102/td.v18i1.1213

Morojele, P. (2017). Indigenous knowledge/s of survival: Implications for lifelong learning among the Basotho herding fraternity. *Educational Research for Social Change, 6*(1), 38–55. DOI: 10.17159/2221-4070/2017/v6i1a4

Moshesha, P. 2023. *Climate change adaptation strategies and their challenges in Lesotho*. Master's thesis. National University of Lesotho: Roma.

Motjotji, L., Moteetee, A., & Seleteng-Kose, L. (2023). Medicinal plants of Lesotho: A review of ethnomedicinal, pharmacological and conservation studies: Lesotho medicinal plants. *Ethnobotany Research and Applications, 25*, 1–19. DOI: 10.32859/era.25.29.1-19

Ombati, M., 2017. Rainmaking rituals: Song and dance for climate change in the making of livelihoods in Africa. *International journal of modern anthropology, 1*(10), 74-96.

Ombati, M. (2021). Climate rituals: Cultural response for climate change adaptations in Africa. In *Decolonising Conflicts, Security, Peace, Gender, Environment and Development in the Anthropocene* (pp. 385–410). Springer International Publishing. DOI: 10.1007/978-3-030-62316-6_11

Osborne, N., & Grant-Smith, D. (2021). In-depth Interviewing. In *Methods in Urban Analysis* (pp. 105–125). Springer Singapore. DOI: 10.1007/978-981-16-1677-8_7

Ossai, N. B. (2010). African Indigenous Knowledge Systems (AIKS). *Simbiosis, 7*(2).

Pellerin, M. (2012). Benefits of Afrocentricity in exploring social phenomena: Understanding Afrocentricity as a social science methodology. *The Journal of Pan African Studies, 5*(4).

Possa, R., & Khotso, P. (2015). Naming of Basotho Medicinal Plants: Semantic Connection to their Remedies. *Southern African Journal for Folklore Studies, 25*(1), 34–47. DOI: 10.25159/1016-8427/541

Ratsie, S. 2021. Perceptions of Initiation Instructors and Parents on the Role of Initiation (Bogwera Le Bojale) in Character Building of Adolescents: A Case of Mochudi Village, Kgatleng District in Botswana. Doctoral Dissertation. The University of Zambia, Zambia.

Sen, B. (2005). Indigenous knowledge for development: Bringing research and practice together. *The International Information & Library Review, 37*(4), 375–382. DOI: 10.1080/10572317.2005.10762695

Stratton, S. J. (2023). Population Sampling: Probability and Non-Probability Techniques. *Prehospital and Disaster Medicine*, *38*(2), 147–148. DOI: 10.1017/S1049023X23000304 PMID: 36939054

Swanta, H.P. (2026). Taboos And Their Ethical Implications In The Formation Of Identity Among The Oegworok (Kagoro) Community Of Southern Kaduna, Nigeria. *Journal of Human, Social and Political Science Research.*

Teherani, A., Martimianakis, T., Stenfors-Hayes, T., Wadhwa, A., & Varpio, L. (2015). Choosing a qualitative research approach. *Journal of Graduate Medical Education*, *7*(4), 669–670. DOI: 10.4300/JGME-D-15-00414.1 PMID: 26692985

World Bank. (2021). *Climate risk country profile: Lesotho*. Climate Change Knowledge Portal. https://climateknowledgeportal.worldbank.org/sites/default/files/2021-08/15930-WB_Lesotho%20Country%20Profile-WEB.pdf

Chapter 13
Exploring Food Security Through the Lens of Vhavenda Indigenous Knowledge

Divhani Muvhango
http://orcid.org/0009-0001-9381-522X
University of Venda, South Africa

ABSTRACT

Food insecurity remains a challenge in many rural African communities despite long histories of self-sustaining food systems (FAO, 2022). Among the Vhavenda of Limpopo Province, indigenous food preservation practices historically supported year-round food availability, nutritional balance, and cultural continuity (Dei, 2018; Chilisa, 2020). This chapter explores food security in Khalavha village, focusing on indigenous science embedded in traditional preservation methods. Drawing on qualitative community-based research, it examines how colonial disruption, modernisation, and socio-economic change have marginalised Indigenous Knowledge Systems, increasing vulnerability to food insecurity. It argues that revitalising indigenous preservation practices can strengthen affordability, resilience, and culturally appropriate food security, particularly in contexts marked by rising food prices, climate variability, and unreliable energy supply in South Africa.

INTRODUCTION

Food security is widely defined as a condition in which all people, at all times, have physical, social, and economic access to sufficient, safe, and nutritious food that meets their dietary needs for an active and healthy life (Food and Agriculture Organization [FAO], 2001). Among the Vhavenda, food security extends beyond physical access to food and relates to *u kona u disa zwiliwa hayani* (the ability to provide food within the household consistently), a process deeply connected to *mvelele* (culture or way of life), environmental knowledge, and communal responsibility. While this definition remains influential, it has increasingly been critiqued for its narrow emphasis on availability and access, often neglecting issues of cultural relevance, local control, and the knowledge systems through which food is produced, preserved, and sustained (Maxwell, 1996; Patel, 2009). Recent scholarship further argues that such frameworks inadequately capture the lived realities of rural communities, particularly in the Global South, where

DOI: 10.4018/979-8-2600-2890-2.ch013

food systems are deeply embedded in cultural and ecological relationships (FAO, 2022; High Level Panel of Experts [HLPE], 2020).

Dominant food security models continue to prioritise industrial agriculture, technological innovation, and large-scale distribution systems as primary solutions to food insecurity. However, a growing body of literature challenges this techno-centric orientation, arguing that it often reinforces structural dependency, marginalises Indigenous Knowledge Systems, and fails to address local socio-economic constraints (HLPE, 2020; Patel, 2009). Contemporary debates in food sovereignty and sustainability transitions emphasise the importance of decentralised, community-driven food systems that are culturally appropriate and ecologically adaptive (Anderson et al., 2021; Clapp et al., 2022). This tension between centralised technological solutions and localised knowledge-based systems forms a critical analytical foundation for this chapter.

From an Indigenous perspective, food security extends beyond access to food and encompasses the processes through which food is produced, preserved, shared, and valued within a community. Indigenous Knowledge Systems conceptualise food security as relational and collective, linking human well-being to ecological balance, cultural continuity, and intergenerational knowledge transmission (Chilisa, 2020; Odora Hoppers, 2017). Within this framework, food preservation is not a peripheral activity but a central mechanism for ensuring food stability and resilience across seasons.

Recent studies on climate resilience and sustainable food systems increasingly highlight the relevance of Indigenous Knowledge Systems in addressing contemporary challenges. Indigenous practices are recognised for their adaptability, low energy requirements, and alignment with local environmental conditions, making them particularly effective in contexts characterised by climate variability and infrastructural limitations (Berkes, 2018; FAO, 2022; Intergovernmental Panel on Climate Change [IPCC], 2022). These findings challenge dominant assumptions that modern, energy-intensive technologies are inherently superior, instead demonstrating that Indigenous systems often provide more sustainable and contextually appropriate alternatives.

Despite this growing recognition, Indigenous Knowledge Systems remain marginalised within mainstream policy and academic discourse. They are frequently acknowledged symbolically but rarely integrated into formal food security strategies or development frameworks (Dei, 2018; HLPE, 2020; Ndlovu-Gatsheni, 2013). This marginalisation reflects broader epistemic hierarchies that privilege Western scientific models while positioning Indigenous knowledge as informal or secondary (Bhuda, 2024). Such hierarchies not only undermine the legitimacy of Indigenous practices but also restrict the diversity of solutions available to address food insecurity.

PROBLEM STATEMENT

Despite significant advances in modern food preservation technologies, food insecurity remains a persistent challenge in many rural communities in South Africa. Contemporary food preservation methods such as refrigeration, freezing, and the use of chemical preservatives are widely promoted as efficient solutions. However, these methods are increasingly proving unreliable and inaccessible in rural contexts due to high costs, dependence on electricity, frequent power outages, equipment failure, and limited household income. As a result, many households experience increased food spoilage, reduced nutritional quality, and greater reliance on purchased foods. At the same time, Indigenous food preservation practices that historically sustained communities such as the Vhavenda are rapidly declining. Colonial and postcolonial development policies systematically marginalised Indigenous Knowledge Systems by positioning Western scientific methods as superior and dismissing indigenous practices as outdated or unscientific (Huda & Koitsiwe, 2022; Bhuda et al., 2025) . This has contributed to the erosion of locally grounded knowledge related to sun drying, fermentation, smoking, and natural cooling, despite their long-standing effectiveness in ensuring food availability and stability across seasons.

As per Molokwa and Setshego (2025), the decline of indigenous preservation methods has coincided with rising food prices, climate variability, and energy insecurity in South Africa, exposing the limitations of modern, energy-intensive preservation systems . Unlike modern technologies, Vhavenda indigenous food preservation methods are perceived to be low-cost, environmentally sustainable, and based on locally available resources. This is because they do not rely on electricity or expensive infrastructure and are embedded within cultural practices that promote careful resource management and intergenerational knowledge transmission.

The core problem addressed in this chapter is the growing disconnect between contemporary food preservation approaches and the lived realities of rural communities. While modern methods continue to dominate policy and practice, they fail to provide resilient and affordable solutions for food preservation in resource-constrained settings. This study, therefore, argues that revitalising indigenous food preservation practices is not merely a cultural exercise but a practical response to current food security challenges. By repositioning Vhavenda indigenous food preservation as a form of indigenous science, this chapter highlights its potential to complement modern approaches, reduce food waste, and strengthen household and community food security.

CONCEPTUAL AND THEORETICAL FRAMING

Food Security and Indigenous Knowledge Systems

This chapter conceptualises indigenous food preservation as a form of indigenous science. Within the Vhavenda worldview, this relationship reflects "*ndivho ya vhathu na mupo*" (the interconnected relationship between people and the environment), where food preservation practices emerge through continuous interaction with ecological systems. Indigenous science is developed through long-term observation, experimentation, and adaptation to specific environmental conditions, often outside formal scientific institutions (Dei, 2018). Among the Vhavenda people, food preservation practices such as sun drying, fermentation, smoking, and natural cooling reflect systematic knowledge of seasonality, moisture control, microbial activity, and storage conditions. These practices demonstrate that indigenous knowledge is not static or primitive, but dynamic, empirical, and responsive to environmental and social change.

The study further draws on the community food security framework, which emphasises local participation, sustainability, equity, and community control over food systems (Hamm & Bellows, 2003). This framework aligns closely with Vhavenda food systems, where food production and preservation were historically organised at household and community levels, guided by customary norms and shared responsibility. Within this context, food security was achieved not through dependence on external markets or technologies, but through collective knowledge and locally available resources.

By integrating Indigenous Knowledge Systems with the community food security framework, this chapter challenges dominant food security narratives that privilege industrial agriculture and modern preservation technologies. It argues that excluding indigenous knowledge from food security strategies reinforces dependency, vulnerability, and cultural loss. In contrast, recognising indigenous food preservation as legitimate knowledge opens possibilities for culturally appropriate, cost-effective, and resilient responses to food insecurity, particularly in rural settings such as Khalavha village.

Analytical Contribution of the Study

This chapter contributes to existing scholarship by moving beyond descriptive accounts of Indigenous food preservation practices to position them as a form of indigenous science with direct relevance to contemporary food security challenges. It demonstrates that Vhavenda preservation practices are not only culturally significant but also function as adaptive, low-cost, and decentralised systems that address the limitations of energy-dependent food preservation technologies. In doing so, the chapter challenges dominant techno-centric food security frameworks that prioritise industrial and infrastructure-heavy solutions, often at the expense of locally grounded knowledge systems. By foregrounding Indigenous preservation practices within debates on food sovereignty, resilience, and sustainability, the study provides a conceptual shift from viewing Indigenous knowledge as supplementary to recognising it as central to building contextually relevant and sustainable food systems.

This chapter draws on the community food security framework, which emphasises local control, sustainability, social justice, and community participation (Hamm and Bellows, 2003). This approach aligns closely with Vhavenda indigenous practices, where food production and preservation were governed by customary norms, traditional leadership structures, and collective responsibility. Positioning indigenous food preservation as indigenous science challenges Western epistemological hierarchies and affirms the legitimacy of indigenous ways of knowing. This conceptual framing positions indigenous food preservation not as an alternative to modern science, but as a complementary system that can strengthen food security outcomes when acknowledged, revitalised, and integrated into contemporary food policies and practices.

The Vhavenda People and Food Security in Khalavha Village

Khalavha village is situated in the Vhembe District of Limpopo Province, an area where rural livelihoods remain closely linked to land, seasonal cycles, and indigenous cultural practices. The Vhavenda people of Khalavha have historically relied on a diversified food system that combined crop cultivation, livestock rearing, and the use of indigenous plants and wild foods. This diversity played a central role in sustaining household food security and reducing vulnerability to seasonal shortages (Netshiluvhi, 2014; Shackleton et al., 2019).

Food security within the Vhavenda context has traditionally been understood as the ability of households and communities to produce, preserve, and access food throughout the year without dependence on external markets. Indigenous food systems were organised around seasonal knowledge, environmental observation, and collective responsibility. Elders and women held specialised knowledge related to planting cycles, harvesting periods, and food preservation practices, ensuring that surplus production during periods of abundance was stored for times of scarcity (FAO, 2013; Kuhnlein et al., 2013).

In Khalavha village, food preservation is integral to everyday life rather than a response to crisis. Crops such as maize, sorghum, millet, pumpkins, and leafy vegetables are preserved through sun drying and fermentation, while meat is smoked or dried to extend its shelf life. These practices reduce post-harvest losses, maintain nutritional value, and ensure that households have access to food during dry seasons, droughts, or periods of economic hardship. Such practices function as informal yet they are highly effective, food security systems rooted in Indigenous Knowledge Systems.

In recent decades, however, the food security landscape in Khalavha village has changed significantly. Increased reliance on purchased food, declining participation in subsistence agriculture, youth migration to urban centres, and climate variability have weakened traditional food systems. Rising food prices and high levels of unemployment have further constrained household access to nutritious food, increasing dependence on processed and energy-intensive food products (Crush & Frayne, 2011; Stats SA, 2021).

These changes have been accompanied by the gradual erosion of indigenous food preservation knowledge. As modern preservation methods such as refrigeration and freezing become dominant, traditional practices are increasingly perceived as outdated, despite their continued relevance in contexts of unreliable electricity supply and limited financial resources. This study contends that the decline of indigenous preservation practices has contributed to increased food insecurity by reducing household resilience and increasing dependency on costly food systems.

Understanding food security in Khalavha village, therefore, requires attention to both material conditions and cultural knowledge systems. The Vhavenda experience demonstrates that food insecurity is not only a result of insufficient food production but also of the loss of indigenous practices that once ensured stability, affordability, and sustainability. Re-engaging with these practices offers valuable insights into locally grounded strategies for strengthening food security in rural South Africa.

RESEARCH METHODOLOGY

This study used a qualitative research design to examine indigenous food preservation practices among the Vhavenda people of Khalavha village and their role in household food security. A qualitative approach was suitable because the research aimed to understand lived experiences, cultural meanings, and local knowledge systems instead of measuring food security with statistical indicators. The study focused on Indigenous Knowledge Systems and explored how preservation practices are woven into social relations, ecological awareness, and the transmission of knowledge across generations.

The research was conducted in Khalavha village, situated in the Vhembe District of Limpopo Province, South Africa. Khalavha is mainly rural, with people's livelihoods mainly based on subsistence farming, seasonal harvests, livestock keeping, and traditional ecological knowledge. The Vhavenda people living here have long relied on local systems of food production and preservation to ensure food is available year-round. The choice of Khalavha village was intentional and based on theoretical considerations. The village is a community where Indigenous Knowledge Systems still exist, although they face increasing pressure from modernization, urban migration, and market-driven food systems. Khalavha faces socio-economic issues common in many rural communities in South Africa, such as unemployment, rising food costs, and reliance on social grants. These factors make it a suitable place to look beyond national food security statistics and explore people's actual lived experiences. This then provides relevant context for examining the limitations of modern preservation systems. The researcher's familiarity with the community facilitated relational accountability, trust-building, and ethical engagement. This was particularly important given the study's focus on Indigenous knowledge, which requires culturally respectful research processes and sensitivity to community intellectual heritage.

The researcher's positionality as someone familiar with the Khalavha community played an important role in facilitating trust, access, and deeper engagement with participants. This relational proximity enabled richer insights into culturally embedded practices; however, it also required ongoing reflexivity to ensure that interpretation remained analytically grounded rather than shaped by prior assumptions. While the study provides in-depth insights into Vhavenda food preservation practices, its findings are context-specific and rooted in the socio-cultural and ecological conditions of Khalavha village. As such, they are not intended to be universally generalisable. However, the study offers analytical transferability to similar rural contexts where communities face comparable challenges related to food insecurity, climate variability, and infrastructural limitations. The principles underlying Indigenous preservation practices, such as decentralisation, low-cost adaptation, and ecological alignment, may inform broader applications in other settings, while still requiring contextual adaptation.

The study targeted individuals with experiential knowledge of traditional food preservation practices. Participants included elders, women responsible for household food preparation, small-scale farmers, and selected youth familiar with indigenous preservation methods. A purposive and snowball sampling strategy was employed to identify participants with relevant knowledge and direct experience. Data was analysed using thematic analysis through multiple qualitative methods to allow for depth and triangulation. These methods included: semi-structured interviews, focus group discussions, and participant observation. Field notes were maintained throughout the research process to capture contextual insights and reflective observations. The researcher paid attention to the following ethical considerations obtained from the relevant institutional review structures before data collection. Informed consent was secured from all participants, and the purpose of the study was clearly explained. Participation was voluntary, and confidentiality was maintained. All these are considered to show respect to community knowledge and ensure that the research strengthens, rather than harming, cultural heritage.

DISCUSSION

Indigenous Food Preservation Practices among the Vhavenda

Food security among the Vhavenda people has long been maintained through food preservation practices that guarantee food availability, accessibility, and stability throughout different seasons. This study examines how Vhavenda preservation methods serve not only as cultural traditions but also as practical food security strategies developed in response to environmental uncertainty, economic challenges, and seasonal changes.

Among Vhavenda people, food preservation is traditionally understood not just as a technical process but as a holistic practice that ensures survival, health, and cultural continuity. It is practical for survival, ensuring the availability of food during periods of scarcity, drought, and off seasons. This allows households to store grains, vegetables, meat, and sorghum without spoilage. Vhavenda methods ensure cultural continuity because these methods are tied to cultural and communal traditional practices. It is natural and low-cost, using locally available resources such as fire, sunlight, wood, smoke, and salt for sustainability, leading to good health, as these preserved food substances retain their nutritional value and taste.

Stages of Indigenous Food Preservation among the Vhavenda People

Among the Vhavenda people of Limpopo Province, Indigenous food preservation practices are deeply embedded in local ecological knowledge, cultural values, and social organisation. This study presents that Vhavenda food preservation follows clearly defined stages that reflect long-standing adaptation to seasonal variability, climatic uncertainty, and household food security needs. Contextualising these stages within the Vhavenda setting demonstrates the specificity and sophistication of Indigenous Knowledge Systems at the community level.

Food Selection and Seasonal Harvesting

In Vhavenda communities, food preservation begins with deliberate selection of crops, livestock products, and wild foods based on seasonal cycles and environmental indicators. During late Summer to early Autumn (Harvest Season: January–April), this is the primary harvesting period in the Vhembe region following seasonal rains. During this time, crops such as maize, sorghum, Millet, pumpkins, and leafy vegetables are harvested at optimal maturity to ensure durability during storage. Leafy vegetables are washed, sometimes lightly boiled, and spread under direct sunlight to reduce moisture content. Sun drying inhibits microbial growth and allows vegetables to be stored for months (Jain & Sharma, 2013; Nisha, 2017). Grains are also thoroughly sun-dried before storage to prevent fungal contamination. Maize, sorghum, and millet are threshed and winnowed before drying and storage. Removing chaff reduces moisture retention and pest infestation risks. This season is critical because surplus production must be stabilised for the coming dry months. Research on Indigenous food systems confirms that drying grains and vegetables during peak harvest reduces post-harvest losses and strengthens household food stability (FAO, 2013; Shackleton et al., 2019).

Winter and Dry Season (May–August). Winter in Limpopo is characterised by low rainfall and cooler temperatures. Fresh vegetable production declines during this period. Foods consumed and preserved during this season, previously dried leafy vegetables, stored grains, dried legumes, and dried or smoked meat. Winter is a period of reliance on preserved foods, demonstrating how earlier seasonal preparation ensures food availability during scarcity.

Spring (September–October)

Spring marks a transitional period before the onset of heavy rains. Food reserves may be limited at this stage.

Foods preserved and prepared:

- Remaining dried vegetables
- Fermented grain products
- Fermented milk

Preservation techniques used

Fermentation

Fermentation of sorghum and maize is common for preparing porridges and traditional beverages. Fermentation not only extends shelf life but also enhances nutritional value and digestibility (Steinkraus, 1996; Marco et al., 2017). This technique reflects an understanding of controlled microbial activity, where beneficial microorganisms inhibit pathogenic bacteria. Fermentation is a cost-effective and sustainable method of preservation, as it does not require the use of electricity or refrigeration. The Venda people have traditionally used fermentation to preserve their crops during times of abundance, allowing them to have a supply of food during periods of scarcity (Mabinya & Mabinya, 2017). This traditional method of preservation continues to be an important aspect of the Venda culture and is still widely practised today.

Fermentation as Cultural and Microbial Regulation

Fermentation among the Vhavenda functions as both preparation and preservation. It involves the controlled breakdown of carbohydrates by naturally occurring microorganisms, producing acids that inhibit pathogenic bacteria (Steinkraus, 1996; Sadiq & Yan, 2018). In Khalavha village, fermented sorghum and maize products are commonly prepared during periods when fresh food is limited, particularly in late winter and early spring.

From my analysis, fermentation is understood locally as a "living process" that requires attentiveness. Participants rely on sensory indicators such as smell, bubbling patterns, texture, and taste to determine readiness. Timeframes are not fixed mechanically but assessed relationally. Scientific literature supports the nutritional benefits of fermentation, including improved digestibility, enhanced micronutrient availability, and increased probiotic potential (Marco et al., 2017).

This study argues that Vhavenda fermentation practices demonstrate an Indigenous understanding of microbial regulation that predates the formalization of microbiology. Food safety is maintained not through laboratory testing but through accumulated experiential knowledge passed down through generations. As emphasised in Indigenous research paradigms (Chilisa, 2020), knowledge is validated through practice, relational accountability, and community verification.

Fermentation in this context extends beyond food preparation to function as a culturally embedded system of nutritional and microbial regulation. Its continued use demonstrates that Indigenous practices provide both health and preservation benefits without reliance on industrial processing. The erosion of such practices contributes to declining dietary diversity and increased reliance on processed foods (Maepa et al., 2025).

Seasonal Logic and Food Security

The seasonal organisation of preservation practices demonstrates that Vhavenda food systems are structured around environmental predictability and uncertainty. Drying is concentrated in hot, sunny months to maximise dehydration efficiency. Smoking is more effective during cooler, less humid periods. Fermentation provides flexibility across seasons, especially when fresh food availability declines. These practices collectively ensure the three core dimensions of food security: availability, stability, and access

(FAO, 2001). More importantly, they illustrate that Indigenous preservation techniques are adaptive, low-cost, and resilient in contexts of climate variability and unreliable electricity supply.

This study highlights that elders play a key role in determining harvesting periods, drawing on Indigenous indicators such as rainfall patterns and plant behaviour. Such practices reflect detailed ecological knowledge developed over generations (Berkes, 2018; Shackleton et al., 2019).

Preparation and Cultural Pre-treatment Practices

After harvesting, food preparation among the Vhavenda involves culturally specific pre-treatment methods such as cleaning, sorting, light boiling, curing, and controlled fermentation aimed at reducing spoilage and enhancing preservation outcomes. Grains are threshed and winnowed, leafy vegetables are washed and sliced, and meat is cut into strips suitable for drying or smoking. This suggests that preparation practices are often gendered, with women holding specialised knowledge related to vegetable and grain preservation. This knowledge is acquired through intergenerational transmission within households and extended family networks. From early childhood, girls observe mothers, grandmothers, and aunts during harvesting, preparation, drying, fermentation, and storage processes. Learning occurs through participation rather than formal instruction. Over time, young girls move from observation to active involvement, assisting in sorting grains, slicing vegetables, monitoring drying stages, and inspecting stored food for spoilage. These processes also remove impurities and excess moisture, improving food safety and longevity (Kuhnlein et al., 2013; Chilisa, 2020).

Sun Drying (*u omisa nga ḓuvha*) as a Preparatory and Preservation Practice

Sun drying is one of the most widely practised Indigenous preservation techniques among the Vhavenda and begins at the preparation stage. Leafy vegetables, grains such as maize and sorghum, and seasonal fruits are cleaned, sliced where necessary, and spread on flat surfaces under direct sunlight to reduce moisture content. Although often described in simple terms, my fieldwork in Khalavha village revealed that sun drying is a carefully monitored and environmentally informed practice.

Participants explained that drying is deliberately timed during periods of intense sunlight and low humidity, typically late summer and early autumn following harvest. Drying is avoided during overcast or humid conditions because incomplete dehydration promotes mould formation. This reflects an applied understanding of moisture control and microbial inhibition. Scientific research confirms that solar drying significantly reduces water activity in food, thereby limiting microbial growth while retaining much of its nutritional value (Jain & Sharma, 2013; Nisha, 2017).

What is particularly striking in the Vhavenda context is the selective adaptation of drying methods. Certain leafy vegetables are lightly blanched before drying to reduce enzymatic spoilage, while others are dried raw to preserve flavour and texture. This suggests a nuanced understanding of food-specific preservation needs. Such practices align with broader research on Indigenous ecological knowledge, which emphasises long-term observation and adaptive experimentation (Fikret Berkes, 2018). Sun drying, therefore, represents not merely exposure to sunlight, but a culturally embedded dehydration system rooted in seasonal literacy and empirical experience.

Beyond its technical function, sun drying illustrates how Vhavenda households achieve food stability through decentralised and energy-independent systems. In contexts of unreliable electricity supply, this method represents not only preservation but a form of infrastructural resilience. Its decline therefore signals more than a loss of technique; it reflects a shift toward dependency on systems that are less accessible and less reliable in rural contexts.

Smoking as Moisture Reduction and Chemical Protection

Smoking is traditionally used for preserving meat, particularly after ceremonial slaughter or communal harvesting. Meat is cut into elongated strips and suspended above slow-burning hardwood fires. From a scientific perspective, smoking reduces the moisture content and deposits antimicrobial compounds, such as phenols and organic acids, onto the surface of the meat, thereby inhibiting the growth of spoilage organisms (Xavier & Viegas, 2014; Nychas & Skandamis, 2007). Empirical studies confirm that smoking significantly extends the shelf life of meat products (Bremner & Southgate, 2007).

However, field observations indicate that wood selection is highly deliberate. Participants differentiate between hardwoods that produce dense, slow-burning smoke and softer woods that burn quickly and offer limited preservation benefit. Certain woods are preferred for their pest-repellent properties and flavour enhancement. This suggests experiential chemical knowledge embedded in practice rather than the random use of materials.

Smoking, therefore, functions as both a dehydration and a surface-level antimicrobial treatment. In the Vhavenda context, it is also a communal activity where skills are shared and evaluated collectively. Preservation here is technical, social, and ecological simultaneously. Smoking, as practised among the Vhavenda, illustrates a layered preservation system that integrates environmental knowledge, material selection, and communal participation. Its significance lies not only in extending shelf life but in sustaining knowledge systems that reinforce collective resilience and food security.

Environmental Cooling as Indigenous Thermoregulation

In colder seasons or shaded environments, Vhavenda households utilise underground pits, clay containers, and elevated granaries to regulate temperature and airflow. These storage systems take advantage of earth insulation and natural ventilation to slow microbial metabolism and insect infestation.

Research on traditional storage systems demonstrates that controlled airflow and reduced temperature significantly enhance grain stability (FAO, 2013; Shackleton et al., 2019). My analysis suggests that these methods reflect Indigenous thermoregulation strategies. Rather than generating cold artificially, preservation is achieved through spatial positioning within micro-environments that moderate temperature fluctuations.

In contexts such as rural Limpopo, where electricity supply is inconsistent and refrigeration costs are high, environmental cooling remains both practical and economically viable.

A central finding of this study is that Vhavenda preservation practices rarely rely on a single technique. Drying may be followed by storage in ventilated granaries; meat may be salted before smoking; fermented grains may be dried further to enhance stability. This layered approach mirrors what food science refers to as "multi-hurdle preservation," where multiple barriers are applied to inhibit microbial growth (Bala et al., 2015; Gounga et al., 2019).

From an analytical standpoint, this integration reflects systems thinking within Indigenous Knowledge Systems. Preservation addresses moisture, temperature, microbial activity, pest control, and environmental variability in combination rather than isolation. Such adaptive strategies reinforce household food stability and reduce vulnerability to climatic and infrastructural shocks.

Taken together, these preservation practices demonstrate that Vhavenda food systems operate as integrated and adaptive systems rather than isolated techniques. Each method addresses multiple dimensions of food security, including availability, stability, and accessibility, while remaining responsive to environmental and socio-economic conditions. Their gradual decline, therefore, has implications beyond household practices, contributing to increased vulnerability, reduced resilience, and greater dependence on external and energy-intensive food systems.

Application of Preservation Techniques

The Vhavenda employs a range of preservation techniques adapted to local conditions. Sun-drying is commonly used for leafy vegetables and grains, while smoking is applied to meat and fish using selected wood types known to enhance flavour and deter pests. Fermentation is practised in the preparation of traditional foods and beverages derived from sorghum and maize. This study contends that these methods demonstrate an Indigenous understanding of microbial control and moisture regulation, achieved without reliance on industrial technologies (Campbell-Platt, 1994; Steinkraus, 1997).

Storage Systems and Environmental Control

Preserved foods are stored in traditional containers such as clay pots, calabashes, woven baskets, and elevated storage structures within homesteads. In some cases, underground storage pits are used to regulate temperature and protect food from pests. This paper argues that Vhavenda storage systems are integral to the preservation process, as they actively manage airflow, humidity, and exposure to sunlight. These practices reflect an intimate knowledge of material properties and environmental dynamics (FAO, 2019; Shackleton et al., 2019).

Monitoring and Adaptive Maintenance

Monitoring preserved food is a continuous process within Vhavenda households. Stored foods are regularly inspected for mould, insect infestation, or spoilage using sensory cues such as smell, texture, and visual appearance. This study highlights that such monitoring allows for timely intervention, including re-drying or re-smoking food when necessary. These adaptive practices illustrate the dynamic and responsive nature of Vhavenda Indigenous knowledge (Berkes, 2018).

Consumption, Sharing, and Knowledge Transmission

Consumption of preserved foods occurs during periods of food scarcity, ceremonial events, and daily household use. Importantly, this stage enables intergenerational knowledge transmission, as children and youth learn preservation techniques through participation and observation. This analysis emphasises that food preservation among the Vhavenda is inseparable from social values such as sharing, respect for elders, and communal responsibility. The decline of these practices, therefore, threatens both food security and cultural continuity (Kuhnlein et al., 2013).

CHALLENGES

The disruption of Indigenous food systems among the Vhavenda people cannot be understood as a simple shift in dietary preference or technological advancement. It must be situated within a broader historical and structural context shaped by colonialism, epistemic marginalisation, land dispossession, labour migration, and contemporary economic inequality.

Reframing Food Insecurity as an Interconnected System

While the chapter identifies key drivers of food insecurity, colonialism, technological dependency, knowledge erosion, and socio-economic change, these processes are not isolated. Rather, they function as an interconnected system that reinforces vulnerability within rural communities. Coloniality did not only dispossess communities materially; it also restructured systems of knowledge by privileging Western scientific paradigms over Indigenous Knowledge Systems. This epistemic marginalisation has been reproduced through formal education systems, where Indigenous knowledge is often excluded or undervalued. As a result, younger generations increasingly associate legitimacy with modern, technology-driven food systems, leading to shifting perceptions that frame Indigenous preservation practices as outdated. These changing perceptions directly influence practice, contributing to the decline of indigenous food preservation techniques. The erosion of these practices, in turn, increases dependence on energy-intensive technologies such as refrigeration, which remain unreliable and inaccessible in many rural contexts.

In this sense, food insecurity in Khalavha cannot be understood solely as a problem of food availability or access. It reflects a deeper epistemic rupture, where the disruption of Indigenous Knowledge Systems undermines the very mechanisms that historically ensured food stability and resilience. This interconnected perspective highlights that addressing food insecurity requires not only material interventions but also the restoration and recognition of local knowledge systems as central to sustainable food security.

Colonialism and Epistemic Marginalisation

Colonial administrations did not merely introduce new crops and technologies; they actively delegitimised Indigenous Knowledge Systems (IKS). Western agricultural models were institutionalised as scientific and progressive, while Indigenous production and preservation practices were framed as primitive or inefficient (Sabelo J. Ndlovu-Gatsheni, 2013; Dei, 2018; Bhuda & Marumo, 2022). This epistemic hierarchy eroded confidence in locally developed food systems that had sustained Vhavenda communities across generations. Colonial land policies further restricted Indigenous agricultural autonomy by altering land access and production patterns. As communal systems were disrupted and subsistence agriculture weakened, households became increasingly dependent on wage labour and market-based food systems. The consequence was not only economic transformation but a gradual weakening of Indigenous food sovereignty. In my analysis, this historical delegitimisation created what may be described as "internalised inferiority" toward Indigenous preservation techniques. Participants noted that traditional drying or smoking methods are sometimes perceived as signs of poverty, while refrigeration is associated with modernity and status. This shift reflects coloniality of knowledge rather than objective technological superiority.

Formal Education and Knowledge Displacement

Postcolonial education systems have continued to privilege Eurocentric knowledge frameworks. Indigenous food preservation techniques are rarely included in formal curricula. As a result, younger generations increasingly associate scientific legitimacy with industrial food processing and mechanical refrigeration. As argued by Bagele Chilisa (2020), education systems that exclude Indigenous epistemologies reproduce epistemic injustice. In Khalavha village, this exclusion has contributed to a decline in practical preservation skills among youth. Knowledge once transmitted through observation and participation is interrupted when schooling and migration remove young people from daily agricultural routines. This study found that many younger participants could describe refrigeration methods but were unfamiliar with the detailed steps of sun drying or fermentation beyond superficial awareness. This generational discontinuity represents not merely knowledge loss but a structural shift in how food security is conceptualised.

Urbanisation, Labour Migration, and Generational Gaps

Urban migration has further disrupted Indigenous food systems. As young adults leave rural areas for employment in urban centres, the intergenerational chain of knowledge transmission weakens. Preservation techniques are learned through repetition, correction, and embodied practice. When youth are absent during harvest seasons, these opportunities disappear. Research on rural livelihoods in Southern Africa confirms that labour migration alters household production systems and reduces subsistence agricultural engagement (Crush & Frayne, 2011; Shackleton et al., 2019). In Khalavha, this shift has increased reliance on purchased food, reducing incentives to maintain labour-intensive preservation practices. From my field observations, elders expressed concern that "children no longer watch how food is prepared." This statement reflects more than nostalgia. It signals a rupture in the social infrastructure that sustains food resilience.

Dependence on Energy-Intensive Preservation

Modern preservation systems, particularly refrigeration and freezing, are heavily dependent on electricity. Yet rural South Africa continues to experience power instability, rising tariffs, and infrastructure challenges. When the electricity supply fails, stored food spoils rapidly, resulting in financial loss and reduced dietary stability. Unlike Indigenous preservation methods, refrigeration is vulnerable to infrastructural failure. In Khalavha village, participants reported frequent power interruptions that compromise meat and dairy storage. This exposes the fragility of relying exclusively on energy-intensive technologies in low-income contexts. This study argues that technological adoption without contextual sustainability has increased vulnerability rather than reduced it. Indigenous methods, by contrast, are decentralised, low-cost, and climate adapted.

Nutritional Transition and Health Consequences

The erosion of Indigenous food preservation practices has also reshaped dietary patterns. Increased reliance on commercially processed foods has contributed to declining dietary diversity. Research on the global nutrition transition demonstrates that shifts toward processed, energy-dense foods are associated with rising rates of obesity and diet-related non-communicable diseases (Popkin et al., 2012; Molokwa & Khazamula, 2026). Among the Vhavenda, preserved leafy vegetables and fermented grains historically supported micronutrient intake and seasonal dietary balance. Their decline reduces access to affordable nutrient-rich foods, especially during dry seasons. The disruption of preservation systems, therefore, has direct implications for health outcomes.

Food Insecurity as Structural Vulnerability

This study contends that the disruption of Indigenous food systems has weakened household resilience and intensified food insecurity. The marginalisation of low-cost, environmentally sustainable preservation techniques has increased dependency on external markets that are economically unstable and environmentally unsustainable (Bhuda et al, 2025; Kone et al., 2025). Food insecurity in Khalavha is not solely the result of insufficient production. It is also the result of disrupted knowledge systems, eroded confidence in Indigenous science, and infrastructural dependence. In this sense, food insecurity is both material and epistemic.

REVITALISING INDIGENOUS KNOWLEDGE FOR SUSTAINABLE FOOD SECURITY

Revitalising indigenous food preservation practices among the Vhavenda people presents both a cultural imperative and a practical strategy to enhance food security. Recognising Indigenous Knowledge Systems (IKS) as legitimate and scientifically informed practices is a critical first step. These systems are not mere traditions; they embody careful observation, experimentation, and adaptation to local environmental conditions over generations (Chilisa, 2020; Dei, 2018). Community-based initiatives that document and teach these practices can bridge the knowledge gap created by urban migration and modernisation. For example, intergenerational learning programs allow youth to participate in sun drying, smoking, fermentation, and natural cooling methods under the guidance of elders. Such engagement strengthens cultural identity while simultaneously equipping households with low-cost, reliable food preservation skills (Molokwa & Setshego, 2025).

While this chapter emphasises the significance of Indigenous food preservation practices, it is important to avoid romanticising Indigenous Knowledge Systems as universally applicable or unchanging. Mpatlanyane et al. (2022) argued that certain preservation methods may be labour-intensive, environmentally dependent, and vulnerable to changing climatic conditions. In addition, younger generations may associate modern technologies with social mobility and economic progress, creating tensions between cultural preservation and contemporary aspirations. Recognising these complexities is necessary for developing balanced and contextually relevant revitalisation strategies.

Rebuilding Intergenerational Transmission

First, revitalisation must prioritise intergenerational transmission. Structured seasonal workshops, household-based mentorship, and community preservation gatherings can re-establish learning through participation. Because preservation knowledge is embodied and sensory, it cannot be sustained through documentation alone. It requires practice, repetition, and communal validation.One key strategy is strengthening intergenerational learning. Community-based workshops, seasonal preservation gatherings, and school-community partnerships can create structured spaces for youth to learn drying, fermentation, smoking, and storage techniques under elder guidance. Such initiatives restore both skill transmission and cultural continuity. Revitalisation is not about romanticising the past. It is about restoring adaptive capacity within contemporary constraints. Integrating Vhavenda preservation practices into local food security planning can enhance resilience. Traditional techniques are inherently energy-efficient and climate-responsive. Unlike refrigeration, they do not depend on continuous electricity supply or costly infrastructure (FAO, 2019). Policy recognition could include supporting community drying facilities, documenting Indigenous preservation methods and incorporating IKS into agricultural extension services, including preservation of knowledge in school curricula, such as integration moves Indigenous knowledge from the margins to formal recognition, and food sovereignty and local control.

Indigenous science centres around the interconnectedness between individuals, the natural world, and the spiritual realm, and often includes a profound knowledge of the land, animals, and plants. This type of science can offer valuable perspectives and solutions for present-day environmental challenges, as well as serve to protect and honour indigenous communities and customs. It is a holistic approach to understanding the natural world and is deeply rooted in the culture and traditions of indigenous communities. This form of knowledge often includes observations, practices, and beliefs that have been passed down through generations and are closely tied to the land and environment. Indigenous science provides a unique perspective and understanding of the world and is an important part of preserving and honouring indigenous cultures.

Revitalisation also strengthens food sovereignty, understood as community control over food production, preparation, and distribution (Patel, 2009). By reclaiming preservation practices, households reduce reliance on volatile markets and processed imports. Furthermore, revitalisation initiatives can promote food sovereignty. By reclaiming control over food production, preparation, and storage, households reduce dependency on external markets and industrial food systems. Preserved indigenous foods contribute to dietary diversity, nutritional quality, and cultural continuity. For instance, fermented grains and sun-dried vegetables retain essential nutrients while providing affordable and stable food supplies during periods of scarcity (Marco et al., 2017; Bala et al., 2015).

Institutional recognition

Second, institutional recognition is critical. Integrating Indigenous Knowledge Systems into agricultural extension services and local development planning can reposition preservation practices as complementary rather than inferior to modern technologies. School-community partnerships can incorporate indigenous preservation into life sciences, agriculture, and heritage education, restoring epistemic legitimacy while equipping youth with practical resilience skills.

At the educational level, school-community partnerships can play a transformative role. Including indigenous food preservation in life sciences, agricultural studies, and heritage education restores epistemic legitimacy to local knowledge systems while equipping learners with practical skills. This inclusion challenges the hierarchy that privileges Western scientific models and marginalizes indigenous science. It affirms that systematic observation, experimentation, and environmental adaptation within Vhavenda practices constitute valid, rigorous knowledge production. Institutional recognition must also extend to policy frameworks. Local government food security policies and climate adaptation strategies should explicitly acknowledge indigenous preservation as part of resilience planning. Formal documentation, community archives, and participatory research collaborations can safeguard knowledge without extracting it from its cultural context. In this way, revitalization becomes embedded within governance structures rather than remaining symbolic or tokenistic.

Economic potential

Third, revitalisation has economic potential. Indigenous preservation methods can support small-scale income generation through the production of dried vegetables, fermented grains, and smoked products for local markets. Because these techniques rely on locally available materials, they lower production costs and increase affordability. This creates opportunities for community-based food enterprises rooted in cultural knowledge.

Beyond household consumption, preserved indigenous foods can be integrated into local markets, school feeding programmes, and community trading networks. This creates opportunities for small-scale entrepreneurship rooted in cultural heritage. Women, who traditionally hold preservation knowledge, are particularly positioned to benefit from such initiatives if supported through training, access to markets, and cooperative structures. Strengthening local value chains around indigenous preserved foods also reduces dependence on imported and processed food products. When communities generate value from their own food systems, economic circulation remains within the locality. This enhances resilience against external market shocks and rising food prices. It also promotes food sovereignty by ensuring that communities retain control over production, processing, and distribution.

Importantly, economic revitalisation should not commercialise knowledge in ways that remove it from community ownership. Any market-based expansion must respect collective intellectual rights and ensure that benefits return to knowledge holders. When managed responsibly, indigenous preservation can function as both a cultural practice and a sustainable livelihood strategy.

Strengthening food sovereignty

Revitalisation strengthens food sovereignty. By reclaiming control over food storage and preparation, households reduce dependence on volatile markets and processed imports. Preservation practices enhance dietary diversity and stabilise food access during seasonal shortages. In contexts of climate variability and infrastructural instability, decentralised preservation systems offer adaptive flexibility that industrial models often lack.

When communities maintain control over storage systems and preservation knowledge, they reduce vulnerability to market fluctuations, rising food costs, and infrastructural failures. Indigenous preservation methods decentralise food storage, meaning food remains within households and community spaces rather than relying on distant commercial facilities. This localisation of control enhances resilience during economic shocks or climate-related disruptions. Food sovereignty in the Vhavenda context also involves protecting intellectual property rights in preserving knowledge. Documentation and revitalisation efforts must be rooted in community consent and collective benefit rather than extraction. Recognising elders and knowledge holders as custodians of food heritage reinforces respect for cultural authority and ensures ethical engagement with Indigenous Knowledge Systems.

Integration

The findings of this study have direct implications for food security policy in South Africa and beyond. Current policy frameworks, including national food security strategies and global commitments such as the Sustainable Development Goals (particularly SDG 2: Zero Hunger), emphasize increased production and technological innovation. However, these approaches often overlook the role of Indigenous Knowledge Systems in strengthening local food resilience. Integrating Vhavenda Indigenous food preservation practices into policy frameworks could enhance sustainability and accessibility, particularly in rural areas affected by energy insecurity and economic constraints. This could include incorporating Indigenous preservation techniques into agricultural extension programs, rural development strategies, and community-based food security initiatives. Recognising Indigenous knowledge within formal policy structures would not only improve food security outcomes but also contribute to epistemic justice by validating locally grounded knowledge systems.

Furthermore, integrating indigenous preservation into local policy frameworks supports participatory governance. Communities should be actively involved in shaping food security programmes that affect them. Such inclusion ensures that development initiatives reflect lived realities rather than external assumptions. By positioning indigenous preservation as central to local food sovereignty, the study affirms that sustainable food systems emerge from community agency and cultural continuity.

Integrating Vhavenda indigenous practices into local food security policies can also enhance resilience. Traditional preservation techniques are inherently energy-efficient, environmentally sustainable, and adapted to local climatic conditions. Unlike modern refrigeration or industrial preservation, these methods are not dependent on electricity or costly infrastructure, making them particularly suitable for resource-constrained rural communities (FAO, 2019). Importantly, revitalisation is not about rejecting modern science. It is about building complementary systems. Hybrid approaches that combine indigenous preservation with selective technological support can enhance both safety and sustainability. Recognising Vhavenda preservation practices as indigenous science shifts food security discourse from dependency toward local capability and resilience.

FUTURE RESEARCH DIRECTIONS

This study focused on the contribution of Vhavenda Indigenous food knowledge to food security within Khalavha village and the broader Vhembe region. However, the findings also opened several important areas that require further scholarly attention. Future research should move beyond documenting Indigenous practices and begin to critically examine how such knowledge systems can be protected, revitalised, and meaningfully integrated into contemporary food security strategies in South Africa.

One important area for future investigation is the intergenerational transmission of Indigenous food knowledge. During the course of this study, it became evident that many traditional preservation practices are gradually declining as younger generations become increasingly disconnected from Indigenous ways of living due to urbanisation, migration, western education systems, and changing consumption patterns. Future studies could therefore explore the perceptions of Vhavenda youth towards Indigenous food systems and investigate the factors influencing the decline or continuation of traditional preservation practices within rural and urban contexts. Such research may contribute to the development of educational and cultural preservation programmes aimed at safeguarding Indigenous knowledge for future generations.

Further research is also needed on the relationship between Indigenous food preservation methods and climate resilience. Rural communities continue to experience environmental uncertainties such as drought, irregular rainfall, and food shortages, yet Indigenous communities have historically developed adaptive mechanisms for survival. Future studies could examine how traditional Vhavenda food preservation practices contribute to household resilience during periods of environmental and economic instability. Comparative studies between Indigenous and modern preservation systems may also provide valuable insights into sustainable approaches to food storage and food accessibility in resource-constrained communities. Although this study primarily focused on food security, it became clear that traditional food practices also hold economic potential for rural livelihoods. Future research could investigate the commercialisation of Indigenous foods and preservation techniques within local and regional markets. Such studies may explore how Indigenous food products can contribute to poverty alleviation, women's economic empowerment, and community-based entrepreneurship while ensuring that cultural integrity and ownership of Indigenous knowledge are protected from exploitation.

There is also a need for interdisciplinary research that bridges Indigenous Knowledge Systems with scientific and technological innovation. Future scholars may explore how scientific methods can support, validate, or strengthen Indigenous preservation techniques without undermining their cultural foundations. This could encourage more balanced conversations between Indigenous epistemologies and western scientific approaches to food security and sustainability. Future studies should examine the role of government institutions, heritage organisations, and educational institutions in promoting and preserving Indigenous food knowledge. There remains limited policy implementation regarding the inclusion of Indigenous food systems in mainstream agricultural and food security frameworks. Research focusing on policy gaps, institutional support, and community participation could provide important recommendations for culturally responsive development strategies.

Future research should prioritise the documentation of oral Indigenous knowledge before it disappears. Much of the knowledge surrounding traditional Vhavenda food preservation exists within oral narratives and everyday practice rather than written archives. The continued loss of elders and cultural custodians threatens the survival of this knowledge. Detailed ethnographic and community-based studies are therefore necessary to preserve Indigenous food knowledge as both a cultural heritage resource and a practical response to ongoing food insecurity challenges.

CONCLUSION

This chapter has explored food security through the lens of the Vhavenda people of Khalavha village, highlighting the central role of indigenous food preservation practices in sustaining rural livelihoods. The discussion demonstrates that Indigenous Knowledge Systems provide valuable insights into affordable, sustainable, and culturally appropriate solutions to food insecurity. Repositioning indigenous food preservation as indigenous science challenges dominant food security narratives and supports more inclusive and resilient approaches. In the face of climate change, economic instability, and infrastructural constraints, indigenous knowledge remains a critical resource for strengthening food security.

This chapter has highlighted the central role of indigenous food preservation practices in sustaining food security among the Vhavenda people of Khalavha village. By exploring sun drying, fermentation, smoking, and natural cooling within their cultural and ecological context, the study shows that these practices are far more than technical processes, they represent a form of indigenous science that integrates social organisation, environmental knowledge, and intergenerational learning (Chilisa, 2020; Berkes, 2018). The discussion further illustrates how colonial disruption, modernisation, and socio-economic changes have contributed to the decline of these practices, leaving communities more dependent on costly, energy-intensive, and sometimes unreliable modern preservation methods. This shift has heightened food insecurity, reduced dietary diversity, and threatened cultural continuity.

Revitalising Vhavenda indigenous food preservation techniques offers practical and culturally grounded solutions. These low-cost, environmentally sustainable methods improve food availability, maintain nutritional value, and strengthen community resilience. These systems are further sustained through "*vhushaka*" (relational interconnectedness) and "*vhu ifhinduleli*" (collective responsibility), which shape how food is shared, preserved, and protected within households and communities. Integrating these practices into household routines, educational initiatives, and local policy frameworks not only safeguards cultural heritage but also enhances sustainable food security strategies. Ultimately, recognising indigenous knowledge as a valid and effective form of science challenges dominant food security narratives and provides a roadmap for inclusive, locally driven solutions. For the Vhavenda community and similar rural contexts, revitalisation of indigenous food preservation is not only a reclamation of cultural identity but a tangible pathway toward resilient, nutritious, and affordable food systems.

REFERENCES

Anderson, C. R., Bruil, J., Chappell, M. J., Kiss, C., & Pimbert, M. P. (2021). From transition to domains of transformation: Getting to sustainable and just food systems through agroecology. *Sustainability*, *13*(3), 1–24. PMID: 34676112

Bala, B. K., Mondol, M. R. A., Biswas, B. K., Chowdhury, B. L., & Janjai, S. (2015). Solar drying of fruits, vegetables, and spices: A review. *Renewable & Sustainable Energy Reviews*, *47*, 1–14.

Berkes, F. (2018). *Sacred ecology* (4th ed.). Routledge.

Bhuda, M. T. (2024). Decolonising and indigenising African Indigenous research: Why Indigenous ways of knowing, being and living matter in research. In *Higher Education Transformation in Africa* (pp. 50–69). Routledge. DOI: 10.4324/9781003528319-4

Bhuda, M. T., Maepa, N., Matsotso, A., & Makgamatha, M. G. (2025). Can Indigenous Knowledge Foster Socio-Economic and Sustainable Development in South Africa? *African Journal of Innovation and Entrepreneurship*, *4*(4), 357–382. DOI: 10.31920/2753-314X/2025/v4n4a15

Bhuda, M. T., & Marumo, P. (2022). Ubuntu Philosophy and African Indigenous Knowledge Systems: Insights from decolonization and indigenization of research. *Gender & Behaviour*, *20*(1), 19133–19151.

Bhuda, M. T., Setshego, M. V., & Koitsiwe, M. (2025). The Role of Indigenous Participants in PAR: An Application of a Decolonial Approach. In *Enabling Indigenous Knowledge Systems in Action Research and Action Learning* (pp. 291–308). IGI Global Scientific Publishing. DOI: 10.4018/979-8-3693-3542-0.ch015

Bremner, H. A., & Southgate, D. A. T. (2007). *Food processing handbook*. Wiley-VCH.

Campbell-Platt, G. (1994). Fermented foods: A world perspective. *Food Research International*, *27*(3), 253–257. DOI: 10.1016/0963-9969(94)90093-0

Chilisa, B. (2020). *Indigenous research methodologies* (2nd ed.). Sage.

Clapp, J., Moseley, W. G., Burlingame, B., & Termine, P. (2022). The case for a six-dimensional food security framework. *Food Policy*, *106*, 102164. DOI: 10.1016/j.foodpol.2021.102164

Crush, J., & Frayne, B. (2011). Supermarket expansion and the informal food economy in Southern African cities. *Journal of Southern African Studies*, *28*(5), 781–797. DOI: 10.1080/03057070.2011.617532

Dei, G. J. S. (2018). *Reframing Blackness and Black solidarities through anti-colonial and decolonial prisms*. Springer.

Food and Agriculture Organization of the United Nations. (2001). *The state of food insecurity in the world 2001*. FAO.

Food and Agriculture Organization of the United Nations. (2013). *Indigenous peoples' food systems and well-being: Interventions and policies for healthy communities*. FAO.

Food and Agriculture Organization of the United Nations. (2019). *The state of food security and nutrition in the world 2019*. FAO.

Food and Agriculture Organization of the United Nations. (2022). *The state of food security and nutrition in the world 2022*. FAO.

Gounga, M. E., Xu, S. Y., & Wang, Z. (2019). Nutritional and functional properties of fermented cereal-based products: A review. *Food Reviews International*, *35*(7), 640–661.

Hamm, M. W., & Bellows, A. C. (2003). Community food security and nutrition educators. *Journal of Nutrition Education and Behavior*, *35*(1), 37–43. DOI: 10.1016/S1499-4046(06)60325-4 PMID: 12588679

High Level Panel of Experts on Food Security and Nutrition. (2020). *Food security and nutrition: Building a global narrative towards 2030*. Committee on World Food Security.

Intergovernmental Panel on Climate Change. (2022). [*Impacts, adaptation and vulnerability*. IPCC.]. *Climatic Change*, •••, 2022.

Jain, D., & Sharma, G. P. (2013). Solar drying of vegetables: A review. *Journal of Food Science and Technology*, *50*(2), 211–219. DOI: 10.1007/s13197-010-0161-7 PMID: 23572736

Kone, M. T., Budeli, A. E., & Ramaano, R. (2025). Toward sustainable food system: The role of livestock farming in Matiyani village in the Vhembe Biosphere Reserve, South Africa. *Frontiers in Sustainable Food Systems*, *9*, 1637083. DOI: 10.3389/fsufs.2025.1637083

Kuhnlein, H. V., Erasmus, B., & Spigelski, D. (Eds.). (2013). *Indigenous peoples' food systems: The many dimensions of culture, diversity and environment for nutrition and health*. FAO & Centre for Indigenous Peoples' Nutrition and Environment.

Maepa, N. M., Matsoso, A. N., & Muedi, H. T. (2025). Documenting and Preserving the Indigenous Foods to Ensure Food Security by the Moshana People in the Northwest Province, South Africa. In *Revaluation and Preservation of Indigenous Knowledge Systems in Modern Society* (pp. 197–214). IGI Global Scientific Publishing. DOI: 10.4018/979-8-3693-7964-6.ch011

Marco, M. L., Heeney, D., Binda, S., Cifelli, C. J., Cotter, P. D., Foligné, B., Gänzle, M., Kort, R., Pasin, G., Pihlanto, A., Smid, E. J., & Hutkins, R. (2017). Health benefits of fermented foods: Microbiota and beyond. *Current Opinion in Biotechnology*, *44*, 94–102. DOI: 10.1016/j.copbio.2016.11.010 PMID: 27998788

Maxwell, S. (1996). Food security: A post-modern perspective. *Food Policy*, *21*(2), 155–170. DOI: 10.1016/0306-9192(95)00074-7

Molokwa, D. T., & Khazamula, X. E. (2026). Indigenous Food and Health Sovereignty Systems in South Africa: Challenges and Prospects. *The Palgrave Handbook of Religion, Health and Development in Africa*, 1-21.

Molokwa, D. T., & Setshego, M. V. (2025). Traditional food systems as a tool for achieving food sovereignty in South Africa. In *Resurgence and revalorization of Indigenous knowledge systems in the contemporary society* (pp. 69–90). IGI Global Scientific Publishing. DOI: 10.4018/979-8-3373-4124-8.ch004

Mpatlanyane, V. L., Tshikukuvhe, L. D., & Budeli, A. E. (2022). Exploring Vhavenda Indigenous Food Practices and Knowledge Contributions to Food (in) security. In *Handbook of Research on Protecting and Managing Global Indigenous Knowledge Systems* (pp. 61-88). IGI Global Scientific Publishing.

Ndlovu-Gatsheni, S. J. (2013). *Coloniality of power in postcolonial Africa: Myths of decolonization*. CODESRIA.

Netshiluvhi, T. R. (2014). Indigenous food systems and sustainable livelihoods in rural Limpopo. *Journal of Human Ecology (Delhi, India)*, *48*(3), 279–287.

Nisha, P. (2017). Effects of drying on nutritional quality of fruits and vegetables. *Critical Reviews in Food Science and Nutrition*, *57*(3), 566–578. PMID: 25942011

Nychas, G. J. E., & Skandamis, P. N. (2007). Meat spoilage during distribution. In Bremner, H. A. (Ed.), *Food processing handbook*. Wiley-VCH.

Odora Hoppers, C. A. (2017). *Knowledge, power and transformation in higher education*. UNISA Press.

Patel, R. (2009). Food sovereignty. *The Journal of Peasant Studies*, *36*(3), 663–706. DOI: 10.1080/03066150903143079

Popkin, B. M., Adair, L. S., & Ng, S. W. (2012). Global nutrition transition and the pandemic of obesity. *Nutrition Reviews*, *70*(1), 3–21. DOI: 10.1111/j.1753-4887.2011.00456.x PMID: 22221213

Sadiq, F. A., & Yan, B. (2018). Fermented foods and human health: Microbial contributions. *Critical Reviews in Food Science and Nutrition*, *58*(7), 1151–1166.

Shackleton, C. M.. (2019). Livelihoods and vulnerability in rural South Africa. *Land Use Policy*, *84*, 87–95.

Statistics South Africa. (2021). *General Household Survey 2020*. Statistics South Africa.

Steinkraus, K. H. (1996). *Handbook of indigenous fermented foods* (2nd ed.). Marcel Dekker.

Steinkraus, K. H. (1997). Classification of fermented foods: Worldwide review. *Food Control*, *8*(5–6), 311–317. DOI: 10.1016/S0956-7135(97)00050-9

Xavier, C. V., & Viegas, R. P. (2014). Smoking as a meat preservation technique. *Food Science & Technology International*, *20*(3), 223–232.

ADDITIONAL READING

Berkes, F. (2012). *Sacred ecology* (3rd ed.). Routledge. DOI: 10.4324/9780203123843

Food and Agriculture Organization. (2009). *Indigenous peoples' food systems: The many dimensions of culture, diversity and environment for nutrition and health*. FAO.

Linda Tuhiwai Smith. (2021). *Decolonizing methodologies: Research and Indigenous peoples* (3rd ed.). Zed Books.

Mignolo, W. D. (2011). *The darker side of western modernity: Global futures, decolonial options*. Duke University Press.

Ngũgĩ wa Thiong'o. (1986). Decolonising the mind: The politics of language in African literature (pp. 1–33). London, England: James Currey.

Prah, K. (1996). *African languages for the mass education of Africans*. CASAS.

Sillitoe, P. (1998). The development of Indigenous knowledge: A new applied anthropology. *Current Anthropology*, *39*(2), 223–252. DOI: 10.1086/204722

UNESCO. (2017). *Local and Indigenous knowledge systems (LINKS)*. UNESCO.

Vandana Shiva. (2000). *Stolen harvest: The hijacking of the global food supply*. South End Press.

Weiner, M. A. (2010). *Food plants of the world: Identification, culinary uses and nutritional value*. CRC Press.

KEY TERMS AND DEFINITIONS

Climate Adaptation: Processes and strategies used by communities to adjust to environmental and climatic changes in ways that reduce vulnerability and enhance resilience.

Cultural Heritage: The traditions, practices, knowledge, expressions, and material culture inherited from previous generations and preserved within a community.

Food Preservation: Traditional or modern methods used to prevent food spoilage and extend shelf life for future consumption.

Food Security: A condition in which all people have reliable physical, social, and economic access to sufficient, safe, and nutritious food that meets their dietary needs and preferences for an active and healthy life.

Indigenous Food Preservation: Traditional methods of storing and protecting food using locally developed knowledge and natural techniques such as drying, fermentation, smoking, and ash preservation.

Indigenous Knowledge Systems (IKS): Locally developed knowledge, practices, skills, and beliefs that are passed down through generations within a specific cultural community and used to guide daily life, survival, and decision-making.

Oral Tradition: The transmission of knowledge, customs, history, and cultural practices through spoken communication from one generation to another.

Rural Livelihoods: The means through which rural households secure income, food, and survival through activities such as farming, trade, Indigenous crafts, and community-based practices.

Sustainability: The ability to maintain practices, resources, or systems over time without depleting cultural, environmental, or economic resources for future generations.

Vhavenda: An Indigenous ethnic group primarily located in Limpopo Province, South Africa, known for rich cultural traditions, Indigenous knowledge practices, and traditional food systems.

Chapter 14
Exploring the Impacts of Colonial Policies on Indigenous Food Systems:
A Systematic Review

Mamathakgalo Penelope Seakamela
http://orcid.org/0009-0008-1233-2930
DSTI-NRF Centre of Excellence in Indigenous Knowledge Systems, Northwest University, South Africa

Mothusiotsile Edwin Maditsi
http://orcid.org/0000-0001-5796-0174
Northwest University, South Africa

ABSTRACT

A solid and functional food system that indigenous communities have resonated with for centuries at end has been the one that centers within their ways of life. Globalization and westernization have however made its way into the food system and destabilized indigenous food systems that communities have always followed. This resulted in disruption of the way of life of many indigenous communities and their food systems. This can be witnessed through the sidelining of indigenous agricultural practices by many farmers thereby embracing conventional agricultural methods and practices. There has therefore been a disconnection between indigenous peoples and their environment. This has further been exacerbated by the loss of ancestral lands by indigenous communities which played a key role in their production and consumption. A systematic literature review was utilized to source out relevant literature thereby analysing the selected papers to make inferences which are then reported herein as findings of the study.

INTRODUCTION

Indigenous food systems are essential to the long-term survival of indigenous societies because they are inherently multifunctional. These solutions enable the communities to maintain their way of life by utilizing locally available resources. Indigenous food systems are complex, involving not only the production of food but also a range of methods for its cultivation, processing, preservation, and consumption,

DOI: 10.4018/979-8-2600-2890-2.ch014

as well as the knowledge derived from interactions with the environment, customs, and practices that support food production (FAO, Alliance of Biodiversity International, and CIAT, 2021). Indigenous food systems can be defined as a set of linked methods used in cultivating, harvesting, processing, storing, and consumption of food; however, these methods differ from one community to another depending on the availability of natural resources and the cultural acceptance of different foods (Keleman Saxena et al., 2016).

The academic researcher is continuously showing that indigenous food systems are multifunctional, and these food systems are not only catering for nutritional purpose, but they have the ability to sustain the communities' social cohesion, cultural identities and ecological balance (Argumedo et al., 2021; Rahman et al., 2021). These systems have great importance in ensuring cultural preservation and environmental sustainability because they represent a variety of cultures which are interlined to both biological and diverse cultures, as stressed by Swiderska and Ryan (2021). However, there are different perceptions among scholars about the extent to which colonial upheaval impacted these food systems. While some scholars stress on cultural continuity and adaptation despite the colonial impacts (Settee & Shukla, 2020), others (Carter, 2019; Daschuk, 2013), prioritize total institutional elimination. To emphasize the current Indigenous-led responses, the ways in which colonial policies functioned will be carefully examined. This review will contribute to these debates. Indigenous food systems have experienced substantial changes despite their tenacity and cultural importance. Due to the marketing of food, displacement, and land dispossession, which resulted from colonial expansion. The basic activities like farming, hunting, and gathering have been abandoned.

Indigenous food systems became destabilized as a result of colonial interventions, which disrupted the food sovereignty of indigenous people; then they were forced to rely significantly on commercial crops. Indigenous dietary habits have changed because of the transition from subsistence cultivation and culturally based food habits to commercial agriculture, leading to the dietary habits that are culturally inappropriate and harmful to health (Bussidor & Bilgen-Reinart, 2000; Carter, 2019; Daschuk, 2013; Settee & Shukla, 2020)

RESEARCH QUESTIONS

1. Which colonial policies were used to disrupt the indigenous food systems?
2. How has the policies affected the food systems of indigenous communities?

THE THEORETICAL POSITION/FRAMEWORK OF THE PAPER

The theories used in this paper are threefold. This is meant to ground the paper's debates within the relevant theoretical contexts using three theories to drive the narrative and argument towards a decolonial stance. The diagram below depicts the interrelationship between these theories.

Figure 1. Theoretical framework (source: Berry, J. W. (2017); Orsini, A. (2024))

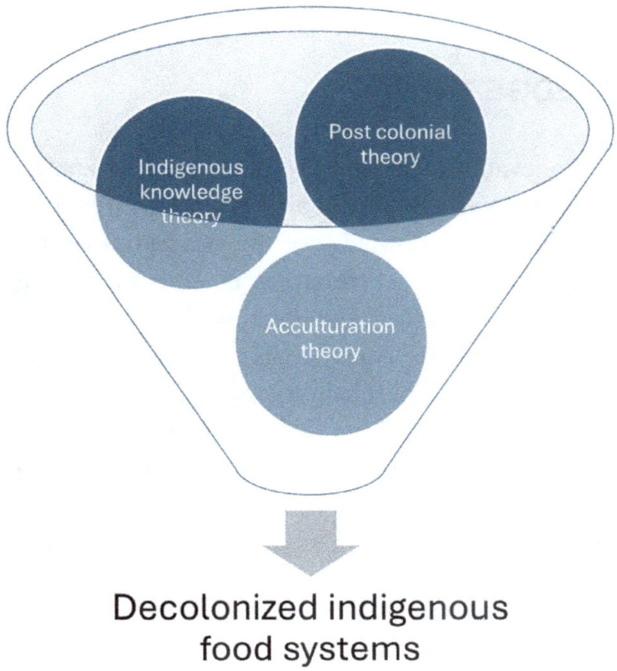

POSTCOLONIAL THEORY

Post colonial theory is the paramount theory identified for this paper. This theory assists in examining the long-term impacts of colonial policies and practices that were introduced to destabilize the political, cultural, economic, and linguistic spheres. It also looks at how colonialism influenced societal structures and how effective are those impacts are on today's social reality. Fundamentally, postcolonial theory challenges prevailing narratives created by the colonial authorities by examining power dynamics that existed before, during, and after colonization. The theory seeks to foster a more inclusive perspective on social identity and history while destroying hierarchical institutions by challenging the idea that colonial societal norms are superior and highlights the need of acknowledging other viewpoints and views irrespective of a person's ethnicity, background, or level of education (Rana, 2021). Postcolonial theory is associated with various scholars such as Frantz Fanon, Michel Foucault, Jaques Derrida, Edward Said, Gayatri, Chakravorty Spivak and Homi K. Bhabha, which each of them contributed to the theory in different years (Mueller & Allan, 2025). This theory therefore advocates for indigenous communities to start narrating their own realities regarding indigenous food systems that they are well vested in. This means rewriting and challenging past narratives about indigenous communities' food systems, which were not coherently provided by the colonial masters. This further allows indigenous peoples to challenge current policies and overall legislation speaking to food security, food systems, and food sovereignty. The postcolonial theory will be used to provide further insight into the power dynamics at play; this section will examine how colonial policies implemented distinct types of colonization through particular methods.

INDIGENOUS KNOWLEDGE THEORY

A second theory deemed relevant to this paper is the indigenous knowledge theory. This theory is deeply grounded in Indigenous knowledge. The theory outlines the importance of the customs, cultural values, societal norms, and indigenous knowledge of the indigenous people, which is associated with agriculture, food gathering, preparation, and management. This theory argues that the adaptation of external agricultural practices, monocultures, and commodification that were introduced by colonial policies damaged or suppressed indigenous food systems and compromised the biodiversity, sustainability, and cultural identity of indigenous people (Jessen et al, 2022). The theory emphasizes how resilient and adaptable indigenous communities have been in maintaining their own food sovereignty in the face of colonial interruptions. Therefore, this theory is utilized to pinpoint how indigenous communities want the promotion and revitalization of indigenous knowledge systems because of the past injustices, advancing food security and assisting in the restoration of indigenous food practices in modern contexts by emphasizing indigenous views.

ACCULTURATION THEORY

The last theory relevant to this is acculturation theory. This theory aims to examine the reasons for adaptation. This theory examines the impacts that colonial policies had on the indigenous food systems, which led to the multicultural communities and adoption of new ways of life. It also outlines the long-term impacts that are caused by the introduction of foreign economic, cultural, and political forces which indigenous communities ended up engaging in, which resulted in the loss of their food system. The theory argues that the adoption or adaptation of indigenous communities towards the colonial rules is a way of life is all because of the land dispossession, forced labour, and introduction of infrastructure. The theory is utilized to pinpoint what fostered the adoption of colonial agricultural practices, eating habits, and social structures (Barreau et al, 2019).

METHODOLOGICAL APPROACH TO THE PAPER

For this paper, to ensure a thorough and in-depth investigation of the topic, academic databases such as Google Scholar, IMDP, ScienceDirect, Sabinet and others were used to access the relevant sources. The key documents through the use of key concepts such as colonial policies on food systems, indigenous food systems, and colonization; Indigenous food sovereignty were used to acquire relevant streamlined papers to be utilized for this review. The use of the key concepts was to ensure a comprehensive and nuanced exploration of the topic and to identify key scholarly articles that critically examine the historical and contemporary impacts of colonial policies on indigenous food systems, as well as studies that explore resistance and revitalization efforts within indigenous communities. The papers focused ranged from the year 1981 to 2006. The flowchart below presented as figure 2 depicts how this was achieved.

Figure 2. Flowchart of methodological approach for review (Source: van Staden, 2025; Tak et al., 2025; Castro-Gómez, S. 2021)

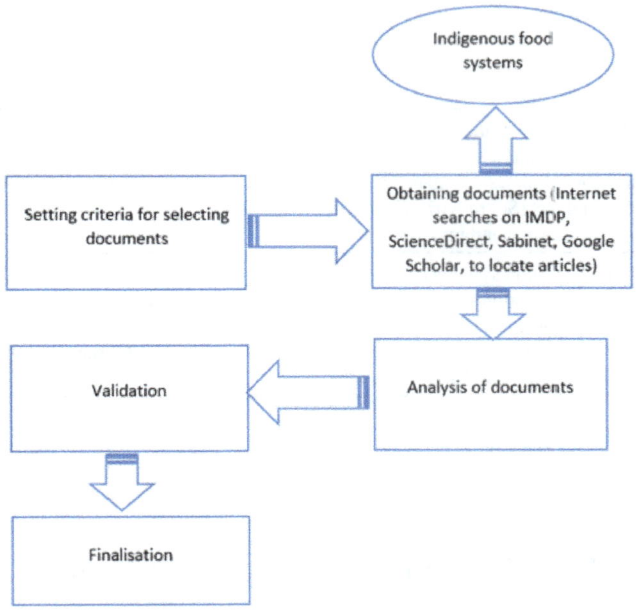

FINDINGS AND DISCUSSIONS OF THE STUDY

Table 1 below illustrates how the transition form pre-colonial era to post-colonial era shifted the food systems of indigenous communities. Within the pre-colonial era, indigenous food systems were highly governed by traditional systems and customary tenure was paramount to this system. This meant that ownership was not for one person but was circular in a sense that all would benefit from this indigenous food system. The knowledge of indigenous food systems was generational, meaning that it was passed down from one generation to the next and this kept food sovereignty intact. The introduction of the colonial era then disrupted the food systems adopted by indigenous communities in the sense that new legislative frameworks were introduced with which indigenous communities had to abide by. Some of these legislative frameworks were so severe that it led to land being dispossessed from rightful owners. Landlessness thus led to the negative impact into the indigenous food systems of the communities. Customary tenure was nullified, and this also had an impact into the food culture of indigenous communities, thus resulting and diet shifts among communities. A new form of administration was then introduced which saw what is deemed as Morden markets being introduce in the late colonial era where individuals would need to go and purchase their foods. Relocation and forced labour plagued indigenous communities and this was then followed by development of mines and railways on communal land which was primarily used for agriculture. This position by the authors is further articulated in sub-sections that follow in this paper.

Table 1. Conceptual table

Era	Key event	Impact on indigenous food systems
Precolonial	• Indigenous governance by traditional leaders • Customary land tenure/ collective ownership • Biodiversity-based farming, hunting, gathering • Food is considered spiritual and cultural	• Sustainable and multifunctional food systems • Intergenerational knowledge transfer • Food sovereignty intact
Early colonial	• Imposition of indirect rule (e.g., Nigeria, India) • Assimilation policies (e.g., Senegal) • Natives Land Act (South Africa, 1913) • Land dispossession and privatization	• Disruption of customary land tenure • Forced shift to cash crops (groundnuts, palm oil, tea, rubber) • • Beginning of food insecurity
Late colonial	• Native Administration systems entrenched • Missionary education and language suppression • Forced labor and relocation • Infrastructure development (railways, mines)	• Cultural erosion and loss of food knowledge • Dietary delocalization (shift from traditional to processed foods) • • Market dependency was established

Mamdani, 2018

INDIGENOUS FOOD SYSTEMS PRIOR TO COLONIZATION

The synthesis of the literature reveals that there are three interlinked characteristics of indigenous food systems which were classified prior to colonization, which were; to have a connection with their environment, integration of culture, and the ability to sustain themselves which was done through diversification (Rajan, 2024; Madduma Bandara, 1985; Drissi, 2022). These characteristics separated the indigenous food systems from colonial agricultural practices that eventually overtook them, because their environment possessed abundant resources for cultivation, hunting, and harvesting uncultivated wild foods. Indigenous communities instilled a strong emphasis on living in harmony with their surroundings; through that interaction, they gained a great deal of knowledge about dietary habits, health advantages, and factors that are influencing human life by closely observing farming cycles, seasonal variations, and weather trends. Food provided more than just physical sustenance for indigenous peoples, as it was ingrained in their religious, spiritual, and their cultural belief systems. From another point of view, food created a big social effect by bringing people together, strengthening links within the community, and creating shared identities. Researchers argue that as opposed to single-crop agriculture, indigenous agricultural production relied on biodiversity and environmental expertise (Snook et al., 2020; Swiderska & Ryan, 2021).

Nonetheless, there is a disagreement regarding how a "sustainable" indigenous food system operates in the present-day view. While some academics highlight the fundamental resilience of pre-colonial systems, others urge against idealizing them, pointing out that indigenous food systems, like any agricultural systems, underwent constant evolution due to internal interactions and modifications to the environment (Baena et al., 2023; FAO, 2018). While recognizing that indigenous systems were neither fixed nor flawless, the review adopts the viewpoint that suggests that the resilience of the food system over centuries indicated incredible environmental adaptation and cultural competence. Indigenous food systems include indigenous methods for cultivating, preparing, storing, and processing indigenous foods, which are deeply rooted in the practices of food cultivation and consumption.

These food systems have sustained the indigenous communities for decades through harmonious interactions with their environment. The literature shows how indigenous peoples used their food systems to preserve a strong bond with their land, health, spirituality, and cultural traditions. By promoting environmental resiliency and maintaining cultural identity, this broad strategy safeguarded the well-being of indigenous people and their environment (Swiderska & Ryan, 2021; Baena et al., 2023; FAO, 2018). Indigenous livestock and crops are generally resilient, nourishing, and sustainable because they need fewer external inputs; this is a fundamental aspect of the scholars' agreement (Argumedo et al., 2021; Rahman et al., 2021).

THE IMPACT OF COLONIAL POLICIES ON INDIGENOUS FOOD SYSTEMS

Extraction, settlement, and exploitation are the three primary forms of colonization identified in the literature, each having unique effects on indigenous food systems. In order to advance colonial economic objectives, extraction colonialism entailed the removal of natural resources. The movement and placement of colonists into native territories was known as "settler colonization," which led to the enforcement of foreign political, social, and cultural frameworks that destabilized native civilizations. Forced labor, resource depletion, dismissal of indigenous rights and sovereignty were different forms of societal inferiority, which were all included in the exploitation colonization (Machiridza et al., 2025). The postcolonial theory will be used to provide further insight into the power dynamics at play; this section will examine how colonial policies implemented distinct types of colonization through particular methods.

THE COLONIAL SYSTEMS OF ADMINISTRATION: ASSIMILATION AND INDIRECT RULE

The use of postcolonial theory, we may comprehend assimilation policies and indirect rule as complementary strategies for upholding colonial authority while controlling the expenses and liabilities of direct administration. These forms of policies were intentional tools for reorganizing indigenous society, including their food systems, rather than just being administrative comforts. Indigenous communities were structured along tribal lines before colonization, with each tribe constituting a big community governed by respectable traditional leaders with political, economic, and legal authority. However, colonial rulers employed a number of tactics that fundamentally disrupted indigenous social, political, economic, and cultural institutions in order to establish dominance and assimilation among the tribes. The colonizer used the traditional leaders to administer local government through the use of indirect rule by positioning the traditional leaders as intermediaries for colonists and indigenous people. This tactic ensured that colonial directives would be carried out. Indirect rule allowed supremacy to operate through pre-existing processes while maintaining the illusion of indigenous government. The subjugating of customary authority to colonial purposes led to the transforming of indigenous leadership into servitude representatives of colonial power; this scheme ultimately undermined the credibility and authority of indigenous structures of governance (Khan et al., 2021).

In the literature, it indicates a lot of disagreements about the legacy and efficacy of indirect rule. According to Acemoglu, Reed, and Robinson (2014), the indirect rule produced long-lasting structures of powerful hold that still influence post-colonial governance and economic growth. Others argue that despite the colonial control, the traditional leaders occasionally defended the interests of indigenous people by exercising agency within these limited frameworks (Ali et al., 2019). This review finds that the systemic impact of indirect control constantly harmed indigenous food sovereignty by changing farming operations toward colonial interests, despite the fact that the traditional leaders' replies were against the change. Assimilation strategy, which aimed to incorporate indigenous peoples into the societal frameworks of colonizers, supplemented the indirect rule. Indigenous practices, traditions, and identities were marginalized, degraded, and transformed as a result of the adoption of colonial religion, language, education, and legal frameworks (Assefa, 2011). To facilitate the exploitation of labor, land, and natural resources, indirect rule and assimilation strategies operated together to exert influence over the political, economic, and social spheres of indigenous peoples. As a result, substantial sectors and resources were dominated, which promoted economic dependence and the privatization of ancestral lands (Chanda et al., 2024).

TRADITIONAL LEADERSHIP AND NATIVE ADMINISTRATION

Native administration emerged under indirect rule and established connections with indigenous populations by the use of middlemen, in this case the traditional leaders. In order to accomplish their objectives of exploitation, extraction, and forced labor, colonizers used this strategy to gain the confidence and access of indigenous people. Indigenous communities were ruled through the customary systems by the traditional leaders serving as rulers prior to the establishment of native administration. Indigenous dissatisfaction would have probably occurred if colonists had not adhered to customary protocols, but this pre-existing system made their approach simpler to adopt and follow (Ali, 2019).

The Indian example shows how indigenous food systems were deliberately altered by indirect control in all of its phases through a variety of strategies. Firstly, the colonial rulers protected themselves from opposition by guaranteeing that land and labor were steered toward colonial aims through establishing or promoting middleman leaders, whether they were kings in lands that were subjugated in areas that they were actively controlled. Secondly, the transition from producing food for sustenance to cash crops for market shipment was forced by the income requirement incorporated into these agreements. Thirdly, farmers with customary rights became landless, which led to the reliant on unstable markets and susceptible to starvation as a result of the privatization of land and the payments it created (Ashton, 2023). According to Acemoglu, D., Reed, T., & Robinson, J. A. (2014). Produced long-lasting patterns of elite control that still influence post-colonial governance and economic growth. This analysis is expanded upon by Mukherjee (2021), who shows that the various forms of indirect rule in India, such as princely territories and territorial authority, produced unique causal processes of societal gaps and low governmental ability that remained through trajectory-dependent systems into the postcolonial period.

Although indigenous livelihoods were not instantly altered by this technique, traditional governance was seriously questioned. Indigenous people received commands from colonial officials rather than their traditional leaders, despite the fact that the customary system was ostensibly observed. While acknowledging the customary system's ability to rule people for centuries without outside intervention, Native administration also weakened it. Farming methods, soil health and customary land tenure had all been undermined when traditional leaders were compelled to discuss decisions regarding ownership of land with colonial authorities over monocultural crop farming.

The hierarchical structure has a fatal irony that is revealed by the literature. Although they had been involved in decision-making for generations before colonialism, traditional leaders were granted safety and control over decisions. The perceived threat posed by native people was fabricated to support colonial control. The use of commercialized agriculture resulted in the land that had supported communal farming for generations to be transformed to service international markets. Indigenous food systems, indigenous habitats, customary land tenure, property ownership rights, and traditional leadership itself were all gradually undermined by this approach, because their leadership had been founded on customary practices, social expectations, and cultural values instead of the political frameworks which are now enforced upon them. Acemoglu et al. (2014) argues that traditional leaders were appropriated through administrative methods they weren't entirely familiar with.

THE NATIVES LAND ACT OF 1913

Taking away of land is arguably the most significant way that colonialism disrupted indigenous food systems. Postcolonial theory allows us to view land reform as an essential redesign of indigenous connections to their territory, rather than as an indifferent administration. This transformation has an ongoing impact on food sovereignty, cultural identity, and the exchange of knowledge between generations. The land rights of indigenous tribes were significantly impacted by the Natives' Land Act of 1913. Indigenous communities used the customary land tenure, which was supported by traditional authorities, to regulate land ownership.

According to this structure, collective land ownership is associated with environmentally friendly strategies, and they have a strong spiritual bond with the environment. This system was in place till the Land Act of 1913 was established, which altered the rules related to land ownership to conform to Western perspectives. This allowed the land to shift from community to private ownership and to be viewed as a property instead of a means of subsistence. There are different elements of the Act's effects which are revealed by scholarly research. Hall (2014) shows how the Act disrupted indigenous people's employment of sustainable methods for handling land in order to prevent environmental deterioration, declining biodiversity, and erosion of cultural legacy.

The goal of the act it was to minimize the ownership of indigenous land and destabilize the communities that have occupied and governed these areas for centuries; this is highlighted by Moguerane (2016). Keegan (1985) illustrates how the Act limited the migration of the indigenous people, confined them to specific regions that reduced their access to natural resources that were historically utilized in the times of drought or hunger. Numerous detrimental repercussions resulted from the loss of land that was available to indigenous tribes. Land is necessary for the functionality of indigenous food systems; indigenous people were unable to provide for themselves because of the restricted land access which was brought about by land reduction. Indigenous people were unable to practice sustainable agriculture or cultivate wild foods as they had previously done, which led to food insecurity. Additionally, the Act limited indigenous peoples' freedom of movement, making it harder for them to acquire natural resources during times of drought or famine.

Beyond its tangible effects, the Act caused societal systems to break apart. In addition to being restricted, Indigenous tribes lost their social networks as they migrated to cities in search of work and housing on the colonial farmland. The younger generations were not exposed to traditional agricultural techniques such as harvesting, land titling, crop production, and food storage, which had weakened their understanding of environmentally friendly land use, methods used for farming, and environmental consciousness, as documented by Moguerane (2016). The erosion of indigenous tribes' identity it eroded their societal and spiritual links. The Act helped colonial rulers by promoting monoculture and commercial agriculture. In order to satisfy the requests of the global market, larger ownership of land produced higher yields. But as monoculture replaced varied farming practices, soil deterioration resulted because the cultivation of one crop species on the same piece of land degraded the soil and depleted vital nutrients. Section 27 of the Act eliminated the customary land tenure arrangements of Indigenous people by prohibiting them from renting or reselling their property when there was no more space available for subsistence cropping. They were compelled to migrate to the cities in search of jobs or to work for colonial farmers; many indigenous farmers abandoned their farms (Keegan, 1985).

REFLECTION ON SOME AFRICAN CASE STUDIES

The Effects Of Indirect Rule In Nigeria

Nigeria was one of the places that was affected by the implementation of indirect rule, which was established by Lord Lugard. This policy was predominant in the Northern Emirates like the Sokoto Caliphate. Lord Lugard ruled via these structures instead of overthrowing them. All significant decisions were under the supervision and control of a limited group of British "District Officers" who were based at the local level. Traditional leaders and their councils remained in power and were referred to as "Native Authorities" (Usman *et al*, 2016).

This reorganization had significant effects on Nigeria's indigenous food systems. Chiefs were now accounting to colonial rulers even when their customary duties were to handle the land efficiently and to ensure that there is enough food security for the community. Rather than defending traditional land ownership and food systems, their major role changed to serve colonial interests by coordinating labor for the development of cash crops. The British could essentially shift agricultural production away from the local food demands towards international markets by using these enslaved leaders to give orders that encouraged or mandated the growing of cash commodities for export, such as groundnuts and palm oil. As the traditional leader of this reconstruction of the economy in his capacity as a "Native Authority," he undermined the very food systems upon which their communities had long relied on (Goji *et al*, 2025).

The Effects Of Assimilation In Senegal

Assimilation in Senegal was achieved through thorough reorganization of political, legal, and cultural life, especially in the areas such as the Four Communes of Dakar, Saint-Louis, Gorée, and Rufisque. These communal homes were set up like French municipal governments, following the exact structure of French government entities. In contrast to previous colonial regimes, indigenous people who embraced French civil laws earned their French citizenship, which gave them the ability to choose a representative to the French National Assembly in Paris (Diouf, 1998).

The cultural shift was just as significant. Local languages were prohibited in schools and French was designated as the only native language used for administration and education. Senegalese schools followed the same syllabus as French schools, which totally disregarding or disparaging indigenous knowledge and history. For people who obtained citizenship, the French law superseded customary laws by radically changing the structures of families, ownership rights, and social relationships. The system was based on the idea that French culture was the way to "civilization" and that Indigenous customs were less valuable (Crowder, 2023).

This policy caused significant disruption to Senegal's indigenous food systems. If a Senegalese person wanted to become a citizen in the society, they had to give up their indigenous eating customs which were considered "outdated". People would portray French eating customs, crops, and cooking techniques as sophisticated and contemporary as a result of the establishment of a cultural dominance that diminished indigenous food systems within the society. This cultural imposition attacked the exact behaviors and ideas that supported indigenous food systems, so directly undermining food sovereignty. French crops, cuisine, and eating habits were marketed as better, whereas local farming knowledge, culinary customs, and food-related societal structures were deliberately discounted (Nigro, 2014).

Impacts Of Colonial Policies On Indigenous Communities

In this section there will be an examination of the various effects made by the colonial mechanisms on indigenous communities which is made easier by the previous analysis of these mechanisms. Four interrelated effects domains are land expropriation, removal, degradation of the environment, commercialization, and cultural erosion are examined using the threshold theoretical framework. Instead of addressing these separate events, this study will show how these events merged to threaten the indigenous food systems. Acculturation theory will assist in clarifying the mechanisms that occurred within indigenous communities. Indigenous Knowledge theory will give an explanation on how these consequences were so significantly harmful, and postcolonial theory will shed light on the political structures that made them possible.

Land Dispossession As A Tool Used To Rearrange Indigenous Food Systems

Land dispossession is frequently cited in the literature as the primary mechanism that colonial rulers implemented to destabilize native food systems. By utilizing the postcolonial theory, we can comprehend land dispossession as the structural basis for reorganizing indigenous livelihoods, social relationships, and educational systems rather than just as a loss of a territory. According to Ninomiya et al. (2023), land dispossession is the seizure of indigenous shared land that has been used for decades for agricultural reasons despite the imposition of private ownership privileges and legal procedures. The indigenous methods used for their survival include farming, hunting, gathering, and ceremonial rites which were marginalized because of the dispossession of land. Academic analyses show that land expropriation and agricultural degradation were used as tactical tools to interfere with standard diets, to rearrange indigenous food systems, and promote market dependency both during and after colonization (van Staden, 2025; Tak et al., 2025). The production of indigenous food was abandoned, the environment was damaged, and social instability grew, which was the result of these efforts.

According to van Staden (2025), the demise of indigenous food security and dietary modifications had a major effect on community endurance and longevity of culture. The shifts from indigenous farming practices led to the loss of indigenous farming systems; use of less environmentally friendly methods increased societal and environmental destabilization for indigenous people while causing declines in biodiversity, erosion of the soil, and decreased productivity. The transition from community sustenance and traditional customs to commercial farming presented a significant challenge for indigenous farmers.

This involved the cultivation of crops that were against their eating culture and means of their subsistence like tea, rubber, and maize. This change illustrates how colonists often overlooked indigenous well-being, customs, and self-reliance mechanisms in favor of their foreign crops that did not conform to or promote indigenous customary methods of living and food security (Peiris, 1981).

Displacement Of Indigenous Communities

Relocation disrupted both the spiritual and physical ties that bounded the indigenous peoples to their ancestry; this expanded on the land dispossession issue. Indigenous knowledge theory is especially helpful in this context because it explains why displacement was more than just a relocation; rather, it was a major disruption of the processes of knowledge transfer that were ingrained in locations that the indigenous people were staying in. Relocation destroyed the deep cultural and geographical links that they had been present in indigenous communities for centuries. One of the main reasons for their displacement was the disruption of their customs, social networks, and economic systems. Social norms become unstable as a result of the undermining of their traditional lifestyles. Significant abnormalities occurred in both the social and ecological systems which were brought about by the displacement (Sustainability Directory, 2025).

In addition to being physically removed from their homelands, indigenous peoples also lost their spiritual, cultural, and ancestral links which were severed when they were compelled to migrate from their ancestral homelands. Shiva (2015) highlights that the main reason for this relocation it was made for ensuring that the aspirations of the colonial rulers for dominance and control over indigenous lands and natural resources are made possible. This process was frequently seen by colonial rulers as a planned attempt to subjugate indigenous people and to steal their resources. This degraded the cultural systems of indigenous people; there was disintegration of their social cohesion, which resulted in the volatility of traditional relations with their ancestral land, which were among the major socio-ecological consequences of colonialism.

The primary impact of relocation was on indigenous knowledge systems, which have long been used by indigenous peoples to manage their environments and preserve their cultures. The introduction of contemporary techniques by colonial rulers, combined with relocation, resulted in the loss or deterioration of indigenous knowledge practices. Indigenous communities found it more challenging to manage their social and environmental problems in their new surroundings as a result of this shift. Colonial developments such as the introduction of equipment and extensive infrastructure which altered natural ecosystems and further diminished the cultural systems of knowledge and significantly disturbed their capacity to fully understand and evaluate their surroundings (Berkes, 2017).

ENVIRONMENTAL DEGRADATION AND ITS CONTRIBUTION TO THE DISINTEGRATION OF INDIGENOUS SOCIETY

Another aspect of the effects created by colonial policies is environmental degradation, which the Indigenous Knowledge theory sheds light on the effects it had on the environment. Indigenous communities are founded on having harmonious interactions with their ecology and biodiversity which were supplanted by the use of extractive methods that were put immediate for financial gain ahead of resilience over the long run. The local environment and ancestral homelands suffered greatly as a result of the expulsion of indigenous people because after their expulsion, the native lands were cleared for the development of infrastructure and the extraction of natural resources. This technique significantly reduced the biodiversity and degraded the environment. The destruction of forests and other natural habitats altered the ecosystems and endangered the continuing existence of species. Furthermore, extra land damage, pollution of the environment, and disturbance of the original balance of nature were often the outcomes of the employment of machinery and chemicals in construction activities. These operations led to the disintegration of indigenous society and prolonged environmental degradation (Sha, 2024; Sharma & Sharma, 2025).

Indigenous populations who rely on forest for subsistence faced a severe and complicated challenge from deforestation. According to Klein (2014), deforestation happened at the cost of indigenous communities who did not equitably profit from these operations. This deforestation was mostly caused by logging forests to create financial securities and furniture products. This indiscriminate exploitation threatens the biodiversity and accelerates climate change, which are the two problems that were not present in indigenous communities prior to the industrial growth.

The chemicals and other pollutants have been introduced through industrialization and infrastructure growth, worsening the environmental degradation. Moreover, water supplies were contaminated by machine-driven excavation methods that were employed in mineral extraction, resulting in endangering human health. Sharma and Sharma (2025) contend that these activities damaged the environment and jeopardize the social, cultural, and spiritual frameworks that are vital to indigenous cultures and well-being. The combination of these activities posed a major threat to indigenous communities in societal structure and environmental survival means. There is a compelling argument for the immediate consideration and adoption of ecologically friendly land management techniques because infrastructure and industrial activities took place on indigenous land without sufficient consideration for indigenous perspectives on the matter. The exclusion of Indigenous people from making decisions regarding their land it has negatively affected their ancestral land and cultural history, further undermining their fundamental entitlements to self-determination and sustainability.

DISRUPTION OF INDIGENOUS ECONOMIC SYSTEMS THROUGH MARKET INTEGRATION

Indigenous economic systems underwent a profound reorganization with the arrival of colonial markets. It is possible to see market integration through the lens of postcolonial philosophy as a way to integrate indigenous peoples into colonial economic institutions under essentially unfair conditions. It is believed that this economic turmoil led to the nutritional and societal shifts throughout history; this is explained by acculturation theory. Dependency on imported goods and market systems expanded the commercial agriculture and replaced the cultural-based farming, which was based on indigenous methods. Through the promotion of cash crops, the removal of commodities used for agriculture, and the relaxation of the practices related to agriculture. This shift altered smallholder farming. The deregulation and eventual replacement of indigenous production modes by commercial agriculture undermined the smallholder farming systems. In addition to modernize the economy, agricultural commercialization sought to restructure and privatize the land-use patterns to facilitate their incorporation into colonial and capitalist markets.

According to scholarly study, colonial rulers used land confiscation and the introduction of markets as two main instruments of economic subjection to transform indigenous economies (Tak et al., 2025).

According to Sansour and Tartir (2014), indigenous farmers were forced to switch to export farming in order to satisfy the needs of the global market. They further claim that in their efforts to expand agriculture, foreign firms and agricultural industries benefited from land confiscation and it led to the suppression of food sovereignty. The economic goals of colonialism and capitalism were given precedence over indigenous livelihoods, undermining and marginalizing local food sovereignty. Markets are regarded to have complex functioning, which this is drawn from the literature. Tak (2023) shows how colonial rulers utilized markets as a mechanism of dominance and control, even though he recognizes that markets can ensure food security for indigenous communities. Indigenous people there were depended on the land to produce their food, so when it was seized and it was used for other purposes, they became quite vulnerable. By integrating the global value chain, some academics are contending that commercialization could have helped the indigenous peoples in enhancing their standard of living and providing for their basic requirements. But the data constantly demonstrates that colonial rulers benefited the most since they were able to take control of resources and lands that had previously belonged to indigenous communities.

NUTRITIONAL SHIFT AND MARGINALIZATION OF INDIGENOUS DIET

The most personal aspect of the colonial policies is their nutritional effects, which have an ongoing effect on indigenous bodies and health. Here, acculturation theory is specifically helpful since it explains how the dietary changes were caused, which slowly led to the adoption of colonial norms that undervalued native foods rather than just having access to food. The market systems were created to serve international markets propelled by commercialization and land reuse, which significantly altered indigenous diets. Due to their inability to sustain themselves by using traditional farming practices that are essential to their spiritual and cultural beliefs, indigenous groups have seen a decline in their standard of living. This has resulted in a loss of their ties to their traditional food resources and customs, forcing many indigenous peoples to depend on non-local foods (Tak et al., 2025).

An excellent example of this phenomenon can be traced to the Sioux Lookout community in Ontario. Where local foods contained the dietary value that is needed to meet the body's needs, enhancing physical health and lowering the occurrence of infectious diseases, according to Parker et al. (2019). Nonetheless, the transition from native food to processed diets presents significant issues, ultimately impacting the overall health by jeopardizing food sovereignty, protecting of the ecosystem, and escalating food insecurity. The term "delocalization" was coined by Kuhnlein and Receveur (1996) to characterize the displacement of indigenous activities due to the value of indigenous food and the techniques used for its cultivation and distribution.

The social and cultural significance of local food production and distribution systems was disrupted by the delocalization process that decreased the awareness and utilization of indigenous food systems. Indigenous societies they are the only ones who recognize indigenous foods as vital sources of nutrition that improve the overall health. However, as local food systems were displaced by the global food system, the crops cultivated and collected underwent substantial changes. As a result, eating habits have changed, shifting away from familiar local foods and altering nutritional value. There are numerous factors for the nutritional change that have been identified in the literature.

Popkin (1998) and Goodland (2001) point out that dietary globalization, contemporary eating patterns, economic-driven dietary diversity, and an increase in non-communicable diseases are all complicated problems resulting from global agricultural systems. The dietary shifts have been stimulated by the erosion of indigenous food prohibitions, which were culturally safeguarding and maintaining the eating patterns and the nutritional balance by recommending particular foods for particular occasions. According to Bvenura and Afolayan (2015) and Uusiku *et al* (2010), processed meals have occasionally taken the place of native foods since they are time consuming to prepare and contain bitter flavors, which are frequently indicative of therapeutic qualities. These traits drove people to stop eating native foods, which destroyed traditional cooking techniques that were eventually supplanted by equipment. Even though they had nutritional and therapeutic benefits.

Indigenous agriculture was steadily destroyed by colonization, which propagated the notion that local food constituted outdated. According to Abukutsa-Onyango (2014), this deceit promoted exotic foods as more nutritious, depreciating indigenous food systems and making modern foods seem like appropriate substitutes. Their indigenous food culture was undermined as a result of the adoption of these alien dishes, and they were eventual accepted as integral to their identities. This process led to the continuous decrease of indigenous consumption habits and a reduction of traditional food, which in turn it contributed to the overall decline of the availability of food for indigenous populations and the integrity of their culture.

CULTURAL EROSION AND HOW IT IMPACTS INDIGENOUS FOOD SYSTEMS

The combined effect of the previously examined mechanisms is represented by cultural erosion. The theory of indigenous knowledge is crucial in this context because it clarifies how the systematic devaluing of indigenous approaches to knowing and being led to cruelty toward indigenous peoples still has an impact on communities till today. Urbanization, colonization, and globalization have had an impact on the shifts in language and the development of multicultural societies through cultural integration. The social interactions for the indigenous people have been impacted by these trends, which have made it easier for predominant languages to prevail over indigenous languages (Yılmaz, 2025).

The new social norms were produced by cultural absorption. One of the factors that was influencing young people's preference for speaking colonial or predominant languages is the linguistic variety among indigenous groups. Although linguistic diversity has benefits such as enabling young people to learn various languages, which makes it easier for them to integrate into new societies, however it has also led to language degradation and the extinction of indigenous languages, which they are regarded as the detrimental effects. Maseko and Siziba (2024) argue that the cognitive well-being and cultural continuity of indigenous people are at risk because of the extinction of indigenous languages. In addition, the endangering of cultural identities and mental health resulted in the loss of indigenous languages and eroded the social ties and interactions among indigenous people.

Language and cultural identity are crucial for socializing; it promotes children's growth and development (Machiridza *et al,* 2025. Social relations facilitate the development of distinctive identities and communal resilience by fostering the sharing of culture and language. Indigenous food systems were significantly damaged by the cultural loss brought by the colonial policies that supported Western infrastructure, industry, and agriculture. These food systems integrated indigenous knowledge of local surroundings with environmentally friendly methods of harvesting. When slavery and forced labor were introduced, indigenous tribes were dehumanized and discriminated against within the broader socio-economic framework as they gave up their traditional food habits, which advocated for morals, values, and long-standing, long-lasting connections with the land. Additionally, the notion that the Western methods of food production and consumption were more advanced was fostered by colonial authorities and organizations that promoted Eurocentric cultural norms as more sophisticated and globally relevant.

Furthermore, the colonial rulers and educational institutions are the ones that promoted that ethnocentric cultural norms as sophisticated and the global world portrayed the idea that the western approaches to food production and consumption were superior. Indigenous youth's views on land, food, and resources were shaped by colonial norms. The marginalization of indigenous beliefs, which were greatly emphasized on the use of sustainable resources and interacting harmoniously with their surroundings, was the outcome of a broader cultural invasion. Indigenous knowledge systems were continuously disregarded when it came to farming, food production, and ecological harmony. The misuse and forced labor of colonialism damaged indigenous food systems and weakened their ties to the land and resources (Issa, 2023; Fontaine, 2017).

Additionally, globalization led to the exploitation of indigenous knowledge integrated into agricultural methods, culinary customs, and associated cultural and spiritual pursuits for tourism and entertainment. According to Shin et al. (2020), indigenous culinary knowledge was lost, and the significance of indigenous food systems diminished as a result of the commercialization of local foods without their authorization. The commercial interests diluted and degraded indigenous food systems and their preserving techniques. The industrialization of indigenous food systems and practices has resulted in a misconception and deformation of true cultural manifestations.

CONTEMPORARY DRIVERS OF INDIGENOUS FOOD SYSTEM REVITALIZATION

New forces that not necessary nor resulted from colonialism have evolved in the post-colonial period. According to He et al. (2023), globalization has made it possible for Indigenous peoples to build cross-border alliances, share information, and use sites for promotion. Without interference from colonial institutions, social media enables Indigenous groups to record and disseminate traditional recipes, seed-saving methods, and eating habits. When directed by Indigenous people, cross-cultural communication initiatives have helped restore food knowledge that has been lost due to displacement. These modern instruments and networks function on Indigenous terms in ways that colonial authorities never anticipated, and they reflect post-colonial chances rather than colonial advantages.

FUTURE RESEARCH DIRECTIONS

Within the context of pre-existing cosmologies and subsistence practices, research should examine how societies either accepted or rejected foreign crops, livestock, and farming methods, such as the acceptance of wheat or maize. Additionally, research should look at how these unconventional methods have changed over time and influenced the creation of modern manifestations of food sovereignty. By giving Indigenous research paradigms priority, encouraging cooperative collaborative design procedures with Indigenous communities, and making sure that the results of studies are converted into practical policy recommendations, future investigation projects should collaborate proactively with decolonial approaches. Land restitution, seed bank restoration, and a formal acknowledgment of traditional food systems should be the main focus of such initiatives. To advance societal justice, environment durability, and cultural sovereignty in Indigenous contexts, these measures are essential to improving climate adaption techniques and reconciliation efforts.

CONCLUSION

The goal of this review was to investigate how indigenous food systems comprehend the mechanisms that underpinned these disruptions. The analysis that progressed from recording harm to elucidating how why particular colonial mechanisms produced long-lasting and interconnected effects drawing on postcolonial theory, IK theory and acculturation theory. The analysis leads back to the main point of the chapter, that colonial disruptions of indigenous food systems was a coordinated, multifunctional attack on the material, cultural and epistemological underpinnings of indigenous food sovereignty rather than a sequence of isolated shocks.

The research frequently addresses marketization, cultural erosion, nutritional change, land expropriation, and displacement as distinct consequences. This review shows that these were not ordered phases but rather interconnecting systems. Relocation facilitated commercialization by pushing Indigenous populations into wage based and cash crop markets; displacement permitted migration by removing the land foundation of subsistence; marketization caused dietary changes by substituting processed, imported, or monocrop foods for various local dietary habits; and all the four processes came together in cultural degradation. The systematic diminishing value of Indigenous knowledge, language, and identity.

It was determined that this is significant because it calls into question policy solutions that focus solely on one aspect of colonial legacy. A food sovereignty initiative that reinstates land access while neglecting cultural degradation will not succeed. Reintroducing traditional foods without resolving market dependency will jeopardize a nutritious initiative. The suggestions put forward bring attention to the interconnectedness of colonial interruptions, which necessitates holistic, multifunctional remedies. The assessment expressly disapproved of characterizing any colonial outcome as having a "positive contribution." Infrastructure, new crops, and written records sometimes referred to as colonial advantages occurred after colonial power dissolved were implemented without permission or were subsequent to exploitation. The tenacity of Indigenous peoples despite colonialism, rather than because of it, is what looks good.

Analytically, this differentiation is crucial. Acknowledging resilience avoids two mistakes. The first mistake is victimization stories that depict Indigenous peoples only as helpless victims of colonial oppression; the evidence of revitalization initiatives in this review challenges this framing. The second mistake is colonial arguments, which attribute a favorable result to colonial practices. This review criticizes this perspective as unethical and empirically incorrect. Land restoration, dietary sovereignty programs, knowledge preservation, curricular incorporation, and multicultural collaborations are examples of revival activities that are not proof that colonization was successful. They serve as proof that Indigenous agency endured both during and after colonization. They show that even if Indigenous food systems have been badly destroyed, they are still capable of regenerating when Indigenous peoples set the conditions for that regeneration.

RECOMMENDATIONS

1. Invest in Research Initiatives Led by Indigenous Peoples

Encouragement of research projects that will gather the important data from communities while making sure that any shortcomings are found, and amendments are created for the long-term survival and well-being of the community. Indigenous leadership should be prioritized when it comes to research investment so that their communities may effectively voice their issues and get the representation they need. The use of suitable methodological techniques will preserve cultural heritage and promote food sovereignty. Research ought to guarantee the revitalization, maintenance, and regeneration of indigenous food systems.

2. Incorporate Indigenous Agricultural Studies into Higher Learning Programs

To maintain and safeguard the cultural and lingual heritage. The incorporation of indigenous language and knowledge into the curriculum at schools. Despite the token of recognition, this combination will need to radically alter how the agricultural sector views the skills, expertise, and ecological sustainability.

3. Put into Practice Eco-Friendly Farming Methods

Encouragement in merging the indigenous farming practices with commercial agriculture by implementing regulations that will support the usage of sustainable resources, soil health, and biodiversity. In addition to traditional scientific methods, extension services in agriculture must be modified to emphasize indigenous knowledge.

4. Change Policies from the Colonial Era

Modifying the colonial-era policies to guarantee that Indigenous knowledge is fairly acknowledged in legal documents by using legislative structures that uphold indigenous food systems and safeguard indigenous food security. They also strive towards incorporating indigenous knowledge alongside the modern agricultural expertise.

5. Encourage Movements for Indigenous Food Sovereignty

While honoring indigenous autonomy, providing financial, technical, and political aid to the indigenous-led food sovereignty endeavors. Indigenous leadership should be given top priority by national governments, international development agencies, and civil society organizations when it comes to food system changes that impact indigenous people.

6. since colonial interruption was multipurpose, policy must also be multifaceted.

Land redevelopment is not enough on its own; it needs to be combined with educational change (incorporating Indigenous knowledge into curricula), economic restructure (altering market reliance with indigenous food systems), revitalizing culture (language reclaiming, culture reconstruction), and governance change (transferring power responsibilities to traditional leaders). These are aspects of just one initiative—reestablishing Indigenous food sovereignty—rather than distinct policy sectors.

7. Indigenous leadership and methodologies must be given top priority in research.

Many of the studies included in this evaluation were authored by non-Indigenous academics. Although useful, they cannot replace Native-led research that establishes questions, gathers information, analyzes results, and manages distribution in accordance with Indigenous customs. Indigenous-led studies ought to be funded and published as the norm rather than the exception in future research.

8. The objective is to encourage Indigenous-determined futures rather than a transition to a prehistoric past.

Prior to colonialism, indigenous food systems were neither perfect nor stagnant. They changed because of intercommunity interaction, structural dynamics, and environmental shifts. Regeneration in the modern era does not entail preserving Indigenous culinary customs from a bygone era. It entails helping Indigenous groups modify their knowledge systems to suit modern circumstances on their own terms, such as urbanization, climate change, and new technologies.

REFERENCES

Abukutsa-Onyango, M. O. (2014). Strategic repositioning African indigenous vegetables and fruits with nutrition, economic, and climate change resilience potential. Novel plant bioresources: applications in food, medicine, and cosmetics, 361-369.

Acemoglu, D., Reed, T., & Robinson, J. A. (2014). Chiefs: Economic development and elite control of civil society in Sierra Leone. *Journal of Political Economy*, *122*(2), 319–368. DOI: 10.1086/674988

Ali, M., Fjeldstad, O. H., Jiang, B., & Shifa, A. B. (2019). Colonial legacy, state-building and the salience of ethnicity in sub-Saharan Africa. *Economic Journal (London)*, *129*(619), 1048–1081. DOI: 10.1111/ecoj.12595

Argumedo, A., Song, Y., Khoury, C. K., Hunter, D., Dempewolf, H., Guarino, L., & De Haan, S. (2021). Biocultural diversity for food system transformation under global environmental change. *Frontiers in Sustainable Food Systems*, *5*, 685299. DOI: 10.3389/fsufs.2021.685299

Ashton, S. R. (2023). *British Policy Towards the Indian States 1905–1939*. Routledge. DOI: 10.4324/9781003368243

Baena, P. A., Brunei, A., Fernández-de-Larrinoa, Y., Martinez-Cruz, T. E., Milbank, C., Way, M., & Rights, I. I. P. (2023). In brief: the white/wiphala paper on indigenous peoples' food systems. Science and innovations for food systems transformation, 229.

Barreau, A., Ibarra, J. T., Wyndham, F. S., & Kozak, R. A. (2019). Shifts in Mapuche food systems in southern Andean Forest landscapes: Historical processes and current trends of biocultural homogenization. *Mountain Research and Development*, *39*(1), R12–R23. DOI: 10.1659/MRD-JOURNAL-D-18-00015.1

Berkes, F. (2017). *Sacred ecology*. Routledge. DOI: 10.4324/9781315114644

Berry, J. W. (2017). Theories and models of acculturation. The Oxford handbook of acculturation and health, 10, 15-28.

Bvenura, C., & Afolayan, A. J. (2015). The role of wild vegetables in household food security in South Africa: A review. *Food Research International*, *76*, 1001–1011. DOI: 10.1016/j.foodres.2015.06.013

Carter, S. (2019). *Lost harvests: Prairie Indian reserve farmers and government policy* (Vol. 94). McGill-Queen's Press-MQUP. DOI: 10.1515/9780773557697

Castro-Gómez, S. (2021). *Zero-point hubris: Science, race, and enlightenment in eighteenth-century Latin America*. Bloomsbury Publishing USA. DOI: 10.5040/9798881816308

Chanda, T., Akpan, W. M., Mulenga, D. M., Sain, Z. H., & Gilbert, M. M. (2024). Analyzing the Interplay Between Politics and Economic Development: An African Perspective.

Crowder, M. (2023). Senegal*: a study of French assimilation policy*. Routledge.

Daschuk, J. W. (2013). *Clearing the plains: Disease, politics of starvation, and the loss of Aboriginal life* (Vol. 65). University of Regina Press. DOI: 10.1515/9780889772977

Diouf, M. (1998). The French colonial policy of assimilation and the civility of the originaires of the four communes (Senegal): A nineteenth century globalization project. *Development and Change*, *29*(4), 671–696. DOI: 10.1111/1467-7660.00095

Drissi, S. (2022). Indigenous peoples and the nature they protect. UN Environ Programme and https://www.unep.org/new-and-stories/story/indigenous-peoples-and-nature-they- protect. Accessed 7 January 2026

FAO. (2018). *High Level Expert Seminar on Indigenous Food Systems*. Building on Traditional Knowledge to Achieve Zero Hunger.

FAO, Alliance of Biodiversity International and CIAT. (2021). Indigenous Peoples' Food Systems: Insights on Sustainability and Resilience in the Front Line of Climate Change.

Fontaine, L. S. (2017). Redress for linguistics: Residential schools and assimilation in Canada. *British Journal of Canadian Studies*, *30*(2), 183–204. DOI: 10.3828/bjcs.2017.11

Goji, D. S., Maikasuwa, W., & Ombogutsa, K. Y. (2025). Operation of Colonial Indirect Rule Policy and Political-Social Vices in Nigeria. *International Journal of Law, Politics and Humanities Research.*

Goodland, R. (2001). *The Westernization of Diets: The Assessment of Impacts in Developing Countries*. World Bank.

Hall, R. (2014). The legacies of the Native Land Act of 1913. Scriptura. *Journal for Contextual Hermeneutics in Southern Africa*, *113*(1), 1–13.

He, Y., Zeng, L., Chen, W., Gong, C., & Shen, Z. (2023). Bilateral weighted relative total variation for low-dose CT reconstruction. *Journal of Digital Imaging*, *36*(2), 458–467. DOI: 10.1007/s10278-022-00720-w PMID: 36443529

Issa, M., Khan, H. K., Hussain, M. S., & Ali, M. A. (2023). Language erosion: An overview of the declining status of indigenous languages in Gilgit-Baltistan, Pakistan. *Language*, *25*(2).

Jessen, T. D., Ban, N. C., Claxton, N. X., & Darimont, C. T. (2022). Contributions of Indigenous Knowledge to ecological and evolutionary understanding. *Frontiers in Ecology and the Environment*, *20*(2), 93–101. DOI: 10.1002/fee.2435

Keegan, T. (1985). Crisis and catharsis in the development of capitalism in South African agriculture. *African Affairs*, *84*(336), 371–398. DOI: 10.1093/oxfordjournals.afraf.a097699

Keleman Saxena, A., Cadima Fuentes, X., Gonzales Herbas, R., & Humphries, D. L. (2016). Indigenous food systems and climate change: Impacts of climatic shifts on the production and processing of native and traditional crops in the Bolivian Andes. *Frontiers in Public Health*, *4*, 20. DOI: 10.3389/fpubh.2016.00020 PMID: 26973824

Khan, R., Salim, A., Baluch, Z., Mehreen, K., & Kaleem, M. (2021). British Colonialism and the Indirect Rule: A Hierarchical Administrative Structure to Control the Unruly Tribes. *Humanities & Social Sciences Reviews*, *9*(3), 1366–1372. DOI: 10.18510/hssr.2021.93136

Klein, N. (2014). *This changes everything: Capitalism vs. The climate*. Simon and Schuster.

Kuhnlein, H. V., & Receveur, O. (1996). Dietary change and traditional food systems of indigenous peoples. *Annual Review of Nutrition, 16*(1), 417–442. DOI: 10.1146/annurev.nu.16.070196.002221 PMID: 8839933

Machiridza, E. R., Loveness, B., & Ottiliah, M. (2025). The Effects of Cultural Erosion on Indigenous Language Development at the Early Childhood Level in Multicultural Societies in Zimbabwe: A Case Study of Chimanimani Area. *International Journal of Research and Innovation in Social Science*, *9*(3), 1431–1446. DOI: 10.47772/IJRISS.2025.90300112

Madduma Bandara, C. M. (1985). Catchment ecosystems and village Tank Cascades in the dry zone of Sri Lanka are a time-tested system of land and water resource management. In *Strategies for river basin management: Environmental integration of land and water in a river basin* (pp. 99–113). Springer Netherlands. DOI: 10.1007/978-94-009-5458-8_11

Mamdani, M. (2018). Citizen and subject: Contemporary Africa and the legacy of late colonialism.

Maseko, B., & Siziba, L. (2024). Environmental conservation and the Bulawayo CBD as a linguistic landscape construction: An ecolinguistics perspective. *Journal of Asian and African Studies*, **•••**, 00219096241291054.

Moguerane, K. (2016). Black landlords, their tenants, and the Natives Land Act of 1913. *Journal of Southern African Studies*, *42*(2), 243–266. DOI: 10.1080/03057070.2016.1148954

Mueller, R., & Allan, A. (2025). *Decolonise your mind: Postcolonial theories*. IMartin. Htpps:// www .thelaosexperience.com/introduction-postcolonial-theories/ (Accessed: 20 January 2026)

Mukherjee, S. (2021). *Colonial institutions and civil war: indirect rule and Maoist insurgency in India*. Cambridge University Press. DOI: 10.1017/9781108954266

Nigro, J. (2014). Colonial Logics: *Agricultural, Commercial, & Moral Experiments in the Making of French Senega*l, 1763-1870 (Doctoral dissertation, University of Illinois at Chicago).

Ninomiya, M. E. M., Burns, N., Pollock, N. J., Green, N. T., Martin, J., Linton, J., Rand, J. R., Brubacher, L. J., Keeling, A., & Latta, A. (2023). Indigenous communities and the mental health impacts of land dispossession related to industrial resource development: A systematic review. *The Lancet. Planetary Health*, *7*(6), e501–e517. DOI: 10.1016/S2542-5196(23)00079-7 PMID: 37286247

Orsini, A. (2024). Postcolonial Theory. In *Sociological Theory: From Comte to Postcolonialism* (pp. 641–660). Springer Nature Switzerland. DOI: 10.1007/978-3-031-52539-1_16

Parker, B., Burnett, K., Hay, T., & Skinner, K. (2019). The community food environment and food insecurity in Sioux Lookout, Ontario: Understanding the relationships between food, health, and place. *Journal of Hunger & Environmental Nutrition*, *14*(6), 762–779. DOI: 10.1080/19320248.2018.1537867

Peiris, G. H. (1981). Agrarian transformations in British Sri Lanka. *Sri Lanka Journal of Agrarian Studies*, 2.

Popkin, B. M. (1998). The nutrition transition and its health implications in lower-income countries. *Public Health Nutrition*, *1*(1), 5–21. DOI: 10.1079/PHN19980004 PMID: 10555527

Rahman, D., Missouri, T., & Alexopoulos, G. (2021). The social ecology of food: Where agroecology and heritage meet. *Sustainability (Basel), 13*(24), 13981. DOI: 10.3390/su132413981

Rajan, C. (2024). *Food & Decoloniality. The Decolonial Centre.* Available at: https://decolodecolonialcentre.org/2024/08/14/food-decoloniality/ (Accessed date: 13 December 2025)

Rana, K. S. (2021). Exploring the Elements of Postcolonialism and its Exponents. *The Creative launcher,* 6(5), 44-52.

Sansour, V., & Tartir, A. (2014). *Palestinian farmers: A last stronghold of resistance.* Al-Shabaka: The Palestinian Policy Network.

Settee, P., & Shukla, S. (Eds.). (2020). *Indigenous food systems: Concepts, cases, and conversations.* Canadian Scholars.

Sha, S. (2024). Environmental degradation and migration: An analysis of tribal livelihoods. *International Journal of Political Science and Governance, 6*(2), 115–118. DOI: 10.33545/26646021.2024.v6.i2b.374

Sharma, S. R., & Sharma, M. (2025). From Environmental Degradation to Social Transformation: Exploring the Role of Eco-Justice in the Struggles of Indigenous Communities. *Open Journal of Social Sciences, 13*(5), 325–357. DOI: 10.4236/jss.2025.135019

Shin, E., Lee, S., Kang, H., Kim, J., Kim, K., Youn, H., Jin, Y. W., Seo, S., & Youn, B. (2020). Organ-specific effects of low dose radiation exposure: A comprehensive review. *Frontiers in Genetics, 11,* 566244. DOI: 10.3389/fgene.2020.566244 PMID: 33133150

Shiva, V. (2015). *Earth democracy: Justice, sustainability, and peace.* North Atlantic Books.

Snook, J., Cunsolo, A., Borish, D., Furgal, C., Ford, J. D., Shiwak, I., Flowers, C. T. R., & Harper, S. L. (2020). "We're made criminals just to eat off the land": Colonial wildlife management and repercussions on Inuit well-being. *Sustainability (Basel), 12*(19), 8177. DOI: 10.3390/su12198177

Sustainability Directory. (2025). Displacement of Indigenous Peoples. Sustainability Directory. https://pollution.sustainability-directory.com/term/displacement-of-indigenous-people/ (Accessed: 19 January 2026)

Swiderska, K., & Ryan, P. (2021). *Indigenous food systems, biocultural heritage, and the SDGs: challenges, interdisciplinary research gaps, and empowering methodologies.*

Tak, M. (2023). The Economic De-Development of Jammu and Kashmir: Uncovering the Myth of Lagging Behind. In *The Palgrave Handbook of New Directions in Kashmir Studies* (pp. 441–457). Springer International Publishing. DOI: 10.1007/978-3-031-28520-2_27

Tak, M., Hussain, S. B., Zargar, H., & Blake, L. J. (2025). Settler colonialism de-develops food systems in protracted crises. *Nature Food, 6*(8), 751–756. DOI: 10.1038/s43016-025-01211-1 PMID: 40825868

Usman, A. E., Umar, I. A., & Abubakar, H. I. 2016). *Understanding Nigerian Government and Politics.* RAFA Printing and Publishing Ltd, No 8 Keffi Street, off Makurdi Road, Lafia, Nasarawa State.

Uusiku, N. P., Oelofse, A., Duodu, K. G., Bester, M. J., & Faber, M. (2010). Nutritional value of leafy vegetables of sub-Saharan Africa and their potential contribution to human health: A review. *Journal of Food Composition and Analysis*, *23*(6), 499–509. DOI: 10.1016/j.jfca.2010.05.002

Van Staden, M. (2025). Land loss and labour control: The intersection of land dispossession and labour law in South African history. *Industrial Law Journal*, *54*(2), 324–356. DOI: 10.1093/indlaw/dwae034

Yılmaz, N. (2025). Cultural Identity and Globalization: The Impact on Indigenous Traditions. *Arts and Social Sciences Journal,* Volume 16:01

ADDITIONAL READING

Emmanuel, M. (2024). Traditional Food Systems: Preserving Culture, Health, and Sustainability. *American Journal of Food Science and Technology*, *15*(10), 1–2.

kystyna, S., & Ryan, P. (2023). Indigenous Peoples' Food Systems Hold the Key to Feeding Humanity. International Institute for Environment and Development,https://www.lied.org/indigenous-people-food-systems-hold-key-feeding-Humanity, viewed on, 11.

Malli, A., Monteith, H., Hiscock, E. C., Smith, E. V., Fairman, K., Galloway, T., & Mashford-Pringle, A. (2023). Impacts of colonization on Indigenous food systems in Canada and the United States: A scoping review. *BMC Public Health*, *23*(1), 2105. DOI: 10.1186/s12889-023-16997-7 PMID: 37885000

Matla, H. M. D., Raahalya, S., & Raj, S. (2026). Revitalizing indigenous people's food systems: Attributes, challenges, and policy directions. *Trends in Food Science & Technology*, *172*, 105694. DOI: 10.1016/j.tifs.2026.105694

Mkhize, X., & Cele, T. (2025). Market shocks, climate vulnerability, and income loss in informal indigenous food systems: Evidence from street vendors in Durban, South Africa. *Frontiers in Nutrition*, *12*, 1621204. DOI: 10.3389/fnut.2025.1621204 PMID: 40969604

Morepje, M. T., Mendi, G. T., & Zantsi, S. (2026). Indigenous Foods in South Africa: Household Attitudes, Consumption Patterns, and Market Implications. *Sustainability (Basel)*, *18*(9), 4188. DOI: 10.3390/su18094188

KEY TERMS AND DEFINITIONS

Agricultural practices: refers to the methods and activities that farmers use to produce crops and livestock which includes soil preparation and planting to harvesting.

Colonial policies: strategies implemented by colonial rulers to govern indigenous populations and their territories.

Cultural Erosion: the loss of a community's distinct cultural identity, traditions, languages, and values due to the dominance of Western culture.

Food sovereignty: the right of indigenous communities to define their own agricultural, labour and land policies that are culturally and socially appropriate.

Indigenous food systems: is the systems that are deeply grounded in culture, indigenous knowledge and surroundings.

Indigenous people: refers to people who are originating from a particular population that existed prior to colonization.

Chapter 15
Reflections on the Concept of Centrality and Kwena Livestock Management in Botswana

Nonofho Mathibidi Ndobochani
http://orcid.org/0000-0003-4997-5370
Independent Researcher, Botswana

George Okello Abungu
Okelo Abungu Heritage Consultants, Kenya

ABSTRACT

Embracing indigenous knowledge in studying past farming ideologies contextualizes the past in the present. In the same vein, present farming traditions can decode the past as there is continuity even in change. Reference is made to other parts of the world regarding the relevance of contemporary traditional practices in generating future socio-economic strategies, although the main focus is the Kwena of Botswana. This chapter reflects on the aspect of centrality, as a socio-economic and political concept. It manifests differently through time and space and is impacted by evolving farming strategies. An ethnoarchaeological study of Kwena patterns of settlement illustrates subtle but critical issues in traditional livestock management practices. Decentralization of livestock management directly impacts livestock ownership, geographical distribution of kraals, and the conceptual value of livestock. Decentralization maximizes resource availability, sustainability of livestock and the environment, a process that eventually implicates the overall Kwena patterns of settlement.

INTRODUCTION

The physical world, as denoted by plants, animals, and other significant aspects of the natural landscape, provide a context within which past societies can be studied and understood. The disunion of nature and culture is often a scholarly manifestation that disregards the complex interchange between societies and their environment, and how this evolves over time (Ndoro, 2018; Musonda-Hamusonde & Abungu, 2022). Contemporary developments result in extended occupations and a re-use of historical and natural landscapes, creating intergenerational cultural sequences that justify contextualizing the past

DOI: 10.4018/979-8-2600-2890-2.ch015

in the present. This is especially necessary because, for Africa, the past human-environment relationship was symbiotic, favorably and unfavorably (Abungu, 2016), and was very much progressive, as witnessed by various human exploitation and adaptation strategies such as farming.

It is the quest to understand the development of cultures that nurtured the growth of disciplines such as archaeology, which largely studies the imprint of past human activities - material culture. However, not all of these activities fall within the radar of scientific methodologies – what is found is far less a reflection of what actually happened. This creates limitations in the inductive and fact-based or data-based approaches to understanding the past. Moreover, by the time scientific methods are applied, a lot of valuable information would have been lost through post-depositional processes.

It is therefore not surprising that the concept of analogy became relevant and was embraced in developing archaeological theories and associated methodological approaches (Ascher, 1961; Lane, 1994; Wobst, 2004; Johnson, 2019). This was necessitated by the shifting paradigms in the discipline – that the time had come to make sense out of material culture, that there were patterns, meanings, and stories to be learnt from it (Tilley, 1994; Stiebing, 1994; Hodder, 1995; Hodder & Hutson, 2003). The interpretation of material culture further necessitated comparisons to establish resemblances, and correlations to draw similarities and parallels in the development of cultures (Ascher, 1961). In Africa, there was also a dire need to understand past and present cultures in pre-colonial and postcolonial contexts, an aspect that requires embracing alternative and indigenous sources of knowledge (Marek-Martinez, 2021).

The concept of farming is not a new phenomenon, as it has been part of humanity since time immemorial, and has helped in understanding that past societies had ideologies that advanced and sharpened over time. These ideologies were manifested tangibly and intangibly, in a gradual and evolutionary or stage by stage manner. This enabled cultural progression that was characterized by accumulation and diffusion (Hagerstrand, 1968; Rogers & Shoemaker, 1971; Eerkens & Lipo, 2007; Rogers, 2010), of intergenerational knowledge we see today.

It is therefore not surprising that a wealth of research has gone into farming and its origins, upholding it as a common ancestry to reckon with, primarily because farming is still a way of life for the larger part of the world's population. The diverse knowledge on farming is especially critical as the world is grapples with food security, a globalizing phenomenon that may only be addressed through intergenerational, collaborative, and beneficial developmental agendas. Food security and its associated knowledge of production dominated the investigations of past farming cultures. It continues to be a priority for contemporary socio-political agendas, as can be witnessed by the global concern depicted through the United Nations' Sustainable Development Goals (SDGs).

Farming is one aspect that can ensure that all sectors of societies are continually engaged in reviewing past practices and technologies, exchanging past and contemporary ideas, as well as seeking a profitable and beneficial way forward. It therefore, makes sense that, as early as 1994, Christopher Tilley was already of the view that "writing the past is drawing it into the present, re-inscribing it into the face of the present" (Tilley, 1994:74). Also, as argued by Webber Ndoro, "heritage is about today" (Ndoro, 2024), so a lot can be benefited from past and present traditional farming practices.

Embracing other sources of knowledge such as indigenous knowledge can facilitate a comprehensive approach to the study and assimilation of past and present farming practices. Mapesa (2016), has observed that indigenous or traditional knowledge system (TMS) is critical knowledge as it is time tested, spanning thousands of years. Further, traditional ecological knowledge (TEK) in Africa, which is the cornerstone of TMS, is believed to date as far back as 4,000 BC (Berkes, 1989). Christie (1991) and Berkes (1999) have described it as knowledge ingrained in the relationship of living organisms living with one another in their environment that represents experience in some cases acquired over thousands of years of direct human contact with the environment, refined and passed on through several generations. This knowledge is widespread in the entire continent and has been the cornerstone of human-nature relationship. For southern Africa in particular, the study of farming communities dominated the later parts of the 21st century. The 1970s and 1980s saw a growing interest in the study of archaeological settlements, especially discernible patterning which could be attributed to cultural development and evidence for the evolution of societal cosmologies (Schapera, 1935; Parsons, 1972; Ehret, 1973; Hays 1975; Wilson, 1982; Huffman, 1986; Clist, 1987; Pistorius, 1994; Boeyens 2003). It is apparent that livestock ownership and management, based on traditional systems, were integral in the distribution and patterning of settlements across landscapes, and the eminent mobility of past and present herder-farmer societies.

Research on traditional agro-pastoral farming in East Africa (Marshall, 1990; Lane 2013; Robertshaw, 2021) and the Sahel region (Glickman, 1972; Niamir-Fuller, 1998; Gifford-Gonzalez, 1998) shows that they are characterized by environments suitable for pastoral and agricultural activities, show socio-economic values, settlement patterning, and population mobility that are centered around food security. Although disrupted during the colonial rule, for example in eastern and southern Africa where there was large scale land alienation from communities and traditional practices at times outlawed, it became clear that environmental understanding and conservation for example cannot succeed in the absence of community knowledge. The work of Mapesa (2016) on the pastoral Maasai community of eastern Africa where there are also now large conservation areas, confirm that TMS is critical to environmental conservation and to understanding past land use systems. The north, east, and southern Africa are therefore critical knowledge hubs for agropastoral activities that date as far back as 7000 years ago (Deacon et al., 1978; Robbins et al., 2005; Sadr, 2008; Holl, 2010; Wright, 2014).

Traditional livestock management practices are noted by Paddayya (1998) in Karnataka, where there is evidence for deliberate cattle-penning practices that were integral to communal and household socio-economic activities. Further research on this part of India by Fuller (2001) highlights that the Karnataka region is not only key for understanding early pastoral activities (see also Murphy and Fuller 2016), but also critical for appreciating early agricultural civilizations and societal worldviews (Johansen, 2004; Boivin, 2004). A series of towns anchored on agricultural activities networked harmoniously with seasonal pastoral occupations that were favoured by the open savannah and shrubby landscapes, leaving behind remains of cattle pens in the form of ashmounds (Fuller, 2001).

The Uttarakhand State in India is good for comparisons on the origin and evolution of traditional farming practices as these are still in use to date. Subrahmanyeswari and Chander (2013) specifically studied these traditional farming activities to identify factors ensuring their continuity and observed that cultural elements such as farm type and diversity, breeding techniques, and animal health management were key factors facilitating the intergenerational transfer and exchange of farming knowledge. As a result, up to 80% of the Uttarakhand rural population was sustained by a heavy reliance on traditional agriculture and pastoral farming activities because they are central societal values.

In some parts of Tanzania, unlike in Uttarakhand, northern Ghana, and other places not discussed here, there are no sustainable cultural practices that facilitate integenerational transfer of indigenous agricultural knowledge (Lwoga et al., 2011). This results in resistance and or doubt regarding the traditional practices, which have been proven worthwhile elsewhere, with Bronson (1977:28) noting as early as then that "we can be quite sure that activities resembling cultivation go as far back into the Palaeolithic".

Besides sustaining contemporary livelihoods, the integration of traditional farming practices and associated cultural systems is necessary for ensuring the sustainability of modern farming strategies, an approach that is welcome to counter the current global warming impact. It is therefore not surprising that Descheemeker (et al., 2016) are calling for contemporary sub-Saharan Africa to integrate historical traditional farming practices in their diverse agricultural initiatives. In Zimbabwe, Mapfumo, Mtambanegwe and Chikowo (2016) showcase the value of integrating traditional knowledge in combating the impact of climate change in eastern Zimbabwe. Basic societal socio-economic strategies such as farming are hardest hit by climatic variations. As such, their research demonstrates that the wealth of knowledge communities have on past and current climatic variation indicators, can be beneficial to contemporary farming initiatives. Although Logwa et al., (2011) had concerns over the continuity of traditional knowledge in some parts of Tanzania, Mlay (1986) demonstrates that the adaptation strategies of shifting cultivation and pastoral activities across landscapes in Tanzania are an effective sysyem that ensures sustainability of landscapes. Consideration of traditional and indigenous knowledge is especially necessary considering that farming practices are often socio-economic constructs hinged on the sustainability of complex societal values and ethnic identities (Lane, 2013).

Reflecting on the above discussions, contemporary farmers can benefit immensely from past practices. Equally, present practices can enlighten studies on past human-environment relations and the associated reciprocal impact. A typical example is the work of research teams in east Africa, who examined the resilience of farming landscapes, and the contribution of contemporary farming practices on archaeological theory and method (Shahack-Gross et al., 2003; Augustine et al., 2003; Muchiru et al., 2008; Muchiru et al., 2009; Porensky et al., 2015). As an example, Shahack-Gross et al (2003) studied the Kisongo Maasai of Kajiado District in Kenya, observing the settlement patterning that was heavily dependent on socio-economic and political structures. The mobility of these herders results in beneficially symbiotic impact on the landscapes they occupy, even where there is no evidence of structural or material remains indicative of human occupation. They therefore studied contemporary kraals spanning a 20–30 year period and concluded that micromorphology, mineralogy, and soil chemistry are key in developing prototypes for discerning the spread of pastoralism across African savanna landscapes.

Mapesa (2016), making equally important observation on the protection of conservation areas including world heritage sites concludes that ‘as it became increasingly clear that protected areas and for that matter natural world heritage sites cannot be managed without cultural considerations of the people who live near or even lay claim over them, the concept of community conservation or collaborative management with communities was introduced in the 1990s. The concept sought to ‘involve’ communities in the protection of natural sites and ‘win’ their support with limited access to a few resources and some handouts in the name of ‘benefit sharing’’. Infield (2002:51) has also stated that “understanding the park as a cultural entity, rather than only a scientific or economic entity, would allow it to be managed so as to reflect a wider range of values”. It is therefore clear that these landscapes, whether pastoral, agricultural or conservation areas are beneficiaries of traditional knowledge systems that have shaped them and that we can and are able to interogate in the present, creating a map of landscape knowldge.

In this chapter, a background to farming is offered to highlight its relevance as an African heritage. The need for consideration of other parallels regarding the exploiation of past, present, and future knowledge on farming practices is discussed. A case study on the Kwena of Botswana traditional herding practices and their impact on the continual evolution of their settlement pattern is used to demonstrate the relevance of indigenous knowledge in understanding the beneficial history and impact of African farming practices. What impact does the decentralization of communal socio-economic values, such as traditional livestock management practices, have on the macro and micro-scale settlement patterns of past and present Kwena herder-farmers?

The Kwena patterns of settlement are discussed at the town or village level, and these are categorized as macro patterns. At this level, the management of resources such as livestock is communal and centralized. An example of macro scale is where, before the 20th century, the Kwena arranged their population into sections that comprised settlement units known as wards (Schapera, 1935; Huffman, 1986; Pistorius, 1994; Boeyens 2003) and livestock was kept in the centers of these wards. Increased stock numbers and the need to manage associated risk, the growing need for pastures and draught power at agricultural areas, as well as environmental considerations, led to the management of livestock at outer areas - cattle posts (Hitchcock, 2002; Denbow et al., 2008).

It is demonstrated here that while the central value of livestock at the macro level has been widely researched among the Kwena, embracing indigenous knowledge was key in understanding how the concept of centrality evolves and manifests itself at macro and micro scale levels. The cattle post system facilitated the mushrooming of satellite settlements and the decentralization of livestock management from the communal to the household level, resulting in micro settlement patterns that are divergent from those of main settlements. Understanding the evolution of livestock management practices among the Kwena highlights several key issues, the primary being that the absence of the physical representation of livestock in centers of main settlements does not translate to the declining value of livestock as a significant and central phenomenon in the Kwena society.

FARMING AS A HERITAGE

Farming is one of the few phenomena that globalize understanding of past and present human ideologies and practices. The general notion is that, following a reliance on naturally available resources, past societies ventured into farming as a way of ensuring the adequacy of food by domesticating plants and animals (Holl, 1998; 2022; Wright, 2014; Bellwood, 2023). Research has shown that the exercise of food production was a rather subtle and continually evolving process that required a lot of experimentation, the exchange of ideas across populations, and is still taking place to date (Marshall & Hildebrand, 2002).

As in other parts of the world, farming in Africa did not only ensure food security, it also contributed immensely to population growth and the movement of food-producing populations who sought good pastures for their livestock and cultivable soils for their crops (Robertshaw & Collette, 1983; Bellwood, 2023). The development of farming has ensured that, besides being one of the cradles of humankind, Africa is known for her contribution to the development of early farming practices that started in the Middle East and other parts of the world (Deacon et al. 1978; Hall, 1983; Marshall, 1990; Holl, 1998; Dunne et al., 2012; Russell et al., 2014; Zeder, 2014; Lander & Russell, 2018).

This momentous achievement should be continually underscored, lest the significant knowledge that Africa has is eminent but left untold - something that Chirikure et al. (2017) would pretty much discourage. Understanding the concept of farming has therefore, attracted the interests of various sectors such as archaeology, anthropology, health, technological advancement, business, politics, and socio-cultural development. Although in southern Africa the initial archaeological investigations focused largely on the acquisition and distribution of animal domesticates such as cattle (Deacon et al., 1978; Denbow 1983; Robbins et al., 2005), subsequent research has also looked at livestock management practices as a system that generated and protected this wealth.

The greatest challenge, however, lies in the fact that not all questions about the past can be understood from a laboratory. As an example, one of the issues that scholars grappled with for a while was why cultures changed over time, with initial thoughts largely attributing this to external elements. Internally influenced change or variation was only considered at later stages. Over and above that, the initial line of thinking created a scenario in which scholarly research on the past considered varying cultural patterns across landscapes as irreconcilable and often not contemporaneous. This approach therefore, often created unnecessary cultural dichotomies (Ndoro, 2018, 2024). It also prevented the celebration of the unique African culture, as witnessed by the multilayered and highly symbolic physical footprint (Ndoro, 2024; Zulu, 2024). This was largely because theoretical and methodological approaches at that time were oblivious of the fact that not all underlying issues can be understood from the laboratory, especially internally triggered cultural deviations. Additionally, the greatest concern for most African scholars has been that Africa has meaningful stories about the past. Unfortunately, these stories are in most cases compromised as they lack broad-minded integration of other knowledges and approaches, as well as emphasis or debates by African scholars (Chirikure et al., 2017; Ndoro, 2018).

A permissive and progressive attitude towards understanding the past is therefore, inevitable, especially with the case of farming. This is because although cultures evolve over time, and that some knowledge about the past has fallen through the cracks, some cultural practices are intergenerational and still prevalent to date. As a result, the greatest concern for African scholarly debates is to have knowledge about the past generated, presented, and consumed by all. The quest to interpret material culture must be anchored in the need for establishing comparisons, resemblances, and correlations to draw similarities and parallels (Ascher, 1961; Ndoro. 2018). Most importantly, the parallel interpretation and presentation of the past need to include locally generated knowledge, an approach that is deemed effective in decolonizing stories on the origins, continuity, as well as evolution of past African cultural practices (Ndoro, 2004).

THE NEED FOR PARALLELS

Cognizance of the interplay between nature and humanity and how this has continually shaped the cultural phenomena over time is critical in African heritage debates. Past societies were quite aware of this synergy, and they were deliberate in ensuring that the relationship was symbiotic, conservative, and managerial in approach. There were processes of abandonment and re-use of landscapes. The exploitation of resources tended to be selective and rotational, thereby allowing for regeneration and re-occupation of landscapes, a scenario that allows modern research to model past farming practices, especially livestock management.

In sub Saharan Africa, the Great Zimbabwe cultural landscape has seen decades of research, all of which were geared toward demonstrating its historical value, albeit the divergent conclusions. Recent comprehensive research (Pikirayi, 2019; Chirikure et al., 2017; Chirikure et al., 2021; Chirikure, 2020a, 2020b) has shown that inclusive methodologies can create new knowledge about the African past, especially that the occupation and abandonment of landscapes was a progressive exercise. Occupations were expansive, with some key socio-economic aspects of society occurring at peripheral yet significant locales that often lack structurally significant material culture (Shahack-Gross et al., 2003; Chirikure, 2020a). The resultant systems and patterns of occupation were intermittent, continuous, and extensive in character, often exploiting resources well beyond main centers (Smith et al., 2007; Huffman et al., 2013; Chirikure, 2020a), and these can be pretty much understood through contemporary farming practices.

The ability to trace where and when the domestication process started (Holl, 1998; 2022; Marshall & Hildebrand, 2002), and how it spread to other parts of the world, makes farming a common ancestry.

Realization of this scenario can facilitate acknowledgement of the fact that some societies in certain parts of the world had direct experience, ideas, and information on farming, by virtue of being physically present and performing farming activities, or having gained and acquired the information by being members of such societies. Most importantly, the unremitting nature of the domestication process and farming in general means that information sharing has been intergenerational, and this regenerative knowledge can be exploited and adapted to benefit the present (Subrahmanyeswari & Chander, 2013).

Present societies have knowledge of farming that is gained through various sources and processes. These include indigenous or local knowledge usually accumulated through direct experience (skill and capability). This knowledge would have been passed from one generation to the other - notwithstanding modification of original ideas to suit arising challenges. Indigenous knowledge can be acquired or observed (Jones, 2024), and this is gained as people observe and become aware of past and present traditional practices in their society. Members of society may take an interest in keeping or valuing the knowledge, including that of farming, and even transferring it through various forms such as writing, poetry, storytelling, and songs.

Besides the practical experience and acquired knowledge that contemporary societies have, scientific research has come in to provide empirical and scholarly knowledge and data on the history of this global phenomenon - farming. Past farming activities are studied, theoretical parameters and appropriate methodologies are established, and predictive models or prototypes are developed to contextualize and forecast the viability and longevity of contemporary farming activities/methods. This process generates valuable intergenerational knowledge, an approach that is key when considering Tilley's (2007) argument that materiality must embrace varying contexts of material culture and how it was experienced then and perceived in the present.

The largest African population is anchored in subsistence farming, as clearly shown by the fact that the continent falls into the category of 'undeveloped countries' with GDPs that are barely supporting local economies. At the same time, there is mounting pressure to ensure food security, an aspect that necessitates the engagement of all sectors of society to realize. Now, traditional farming practices are almost the only means through which food security can be achieved, especially since most of the African population is in rural areas and is already sustained by traditional agricultural activities.

In fact, Parson (1981) and Wikan (2004) demonstrate that in the 1980s, Botswana's rural population was largely sustained by cultivation and pastoral activities. The case study of Letlhakeng in the Kweneng District shows that by then about 47% of the population was actively cultivating, and livestock contributed almost 50% of the household economy. It is therefore not surprising that in terms of traditional livestock practices, Botswana's beef economy was in the 1970s heavily reliant on the cattle posts system, which contributed about 92% of the total country herd (Rennie et al., 1977). Because indigenous knowledge is integral to cultural systems and their interchange with the environment, and can be easily experienced and learnt (Jones, 2024), this knowledge can be used to sustain and improve modern agricultural development.

THE CENTRALITY OF LIVESTOCK MANAGEMENT AMONGST THE KWENA OF BOTSWANA

The Kwena patterns of settlement at town or village level have been categorized as macro, and this is where the management of resources such as livestock is communal and centralized. An example of macro scale is where, before the 20th century, the Kwena used to arrange their population into sections that comprised settlement units known as wards (Schapera, 1935). At this stage, livestock management was a shared responsibility at the ward level. The socio-cultural and political explanation of historic Tswana settlement patterning is that resources and authority were central to society (Kuper, 1975). This is believed to have resulted in the symbolic horseshoe or semi-circular shapes characterizing the settlement patterning of most historic Tswana towns (Boeyens 2003; Pistorius, 1994). Research among the Kwena facilitates understanding of how the concept of centrality manifests itself at macro scale at main the settlement, as well as at a micro scale such as at cattle posts where there is decentralization, and the management of livestock is at household level (Ndobochani, 2022).

The concepts of 'central' and 'centrality' generally entail or denote the physical center and aspect of the middle. In the heritage discourse, the physical aspect of centrality can be better understood through the tangible manifestation of culture, for example, the arrangement of structures such as houses and other buildings. Research among the Kwena has shown that the physical location or positioning of key activity areas such as kraals and socio-political spaces like meeting places in the centers is indicative that they are central to societal values. As such, in instances where centrality denotes shared resources and responsibility, as well as communal ownership and management of resources, these resources are placed in the middle or central spaces of settlement units. This way, the society and their property – houses and other structures, form a boundary around these significant and shared resources (See Figure 1).

Figure 1.

As shown on Figure 1, people are a critical component of the physical center – they are constructors and occupants of houses/dwellings that make up settlement units/wards. Together with their leadership, the occupants of wards make settlement decisions that are cognisant of population growth (members of the ward and migrants) and risk issues (external factors such as theft and wars).

Besides the tangible aspect of centrality, the arrangements of houses and structures among Tswana wards signify power and authority – as the Setswana powerful concept goes, '*lentswe la Kgosi le agelwa mosako*' implying that chiefly decisions are respected, supported, and protected. This socio-political manifestation of centrality is reflected in the physical center as households form a circle around the Chief's residence, meeting spaces, and livestock area (See Figure 1). The

Chief's key role is to ensure that there is order and formality as reflected by the patterns of settlement – occupations are not random, and in the case of the Kwena, they are blood-based (Schapera 1935). Also, Chiefs ensure food adequacy and security, as often shown by the placement of headmen in extended areas for the exploitation of resources beyond main settlements as far as 5-15km (Ndobochani 2022).

Bigger settlements are characterized by a network of wards (See Figure 2) which had communal livestock management spaces in their centres. In the case of the Kwena, mapping of these spaces maintained an approximate 100meter buffer between these communal livestock management spaces that were in the centres of wards (See Figure 3). The buffer was to cater for population growth and other socio-cultural activities. In a nutshell, while this aspect of centrality has symbolic connotations, it is also largely functional in nature.

Figure 2.

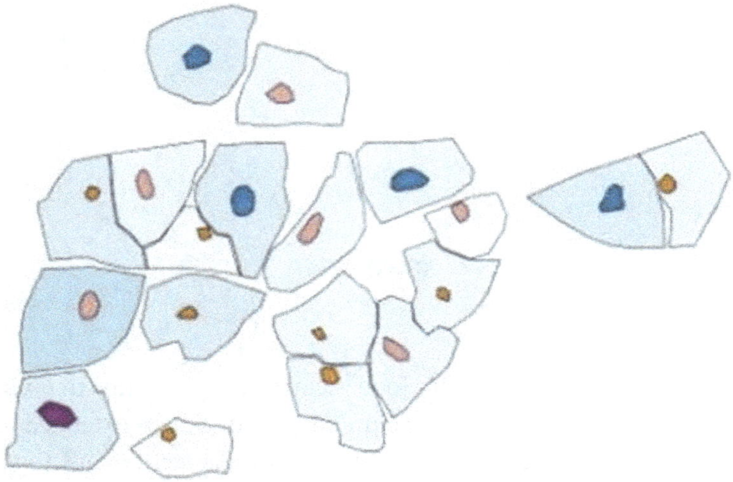

Figure 3.

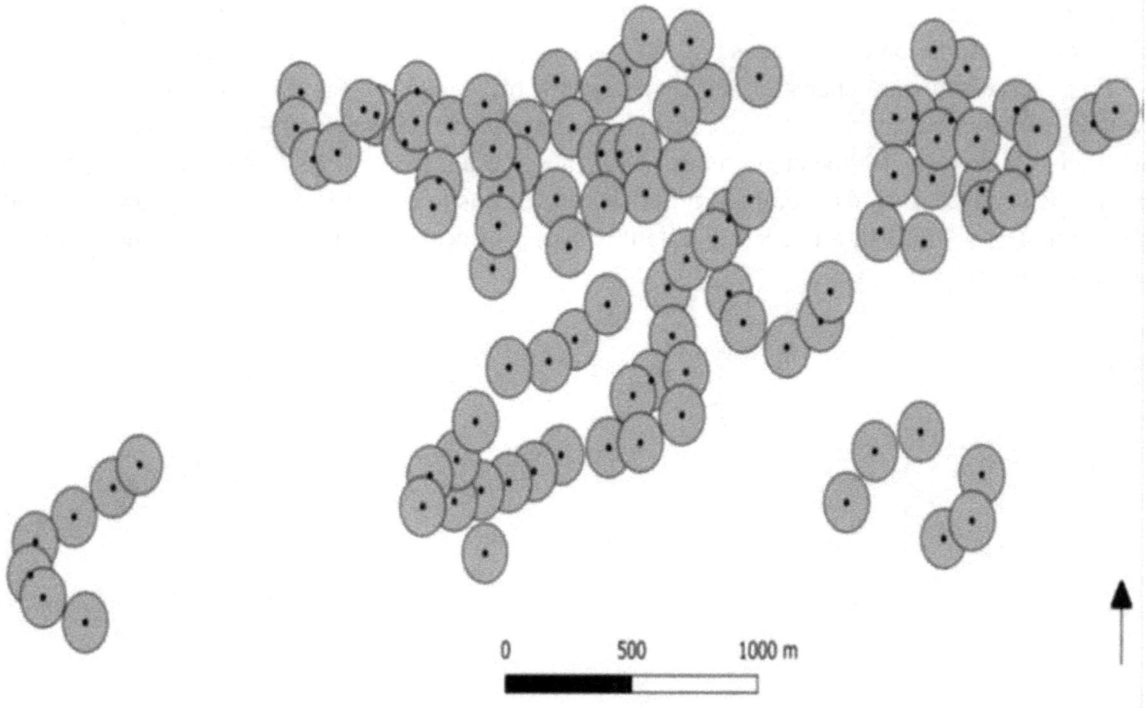

As populations expand and there is a need for food security and protection of political territories, some key socio-economic activities take place outside of main settlements for extended resource exploitation. Cultivation and livestock could now be managed outside of main settlements to secure grasslands and cultivable soils. This resulted in concepts of *moraka* (cattle post), *masimo* (cultivation areas), *masimo-moraka* (cultivation areas that later attract livestock rearing) and *moraka-masimo* (cattle posts that later attract cultivation). In all these scenarios, there is population (human and animal) movement which results in patterns of settlement that vary from the main settlement (See Figure 4). Livestock management areas are not enclosed but are instead appended to households or yards.

It should be noted here that the lack of evidence for the physical aspect of centrality (see Figure 6 showing that households or yards no londer enclose kraal areas) at main settlements does not equate to a decline in the significance of livestock among the Tswana. It largely means that livestock numbers are dwindling, the risk of theft and raids has declined, more households are acquiring livestock, and that the management of this fundamental resource could now be decentralized and is being managed elsewhere.

In this case, the new section of the main settlement lacks evidence of a typical Tswana settlement unit or ward, and this does not mean that chiefs and livestock are not central to society. Other factors that led to variation in the patterns of settlement include the centralization of political activities, where wards are no longer making socio-political decisions – a typical aspect of colonialism. Other factors were purely economic developments, such as the provision of amenities like water and electricity, leading to more linear patterns.

The Kwena of Botswana are one of the few contemporary societies that take pride in their heritage. They have an expansive tribal territory characterized by the main settlement at Molepolole, and several smaller settlements known as villages. They have cultivation and livestock management areas outside of their major settlements. In a number of times, population growth at these areas results in their conversion to settlements – necessitating the relocation of *masimo* and *meraka* to new areas. These historical cultivation areas and *meraka* are then reused to manage the growing population. This creates a scenario where these herder-farmers are continually migrating, either due to internal factors outlined here or because of external factors such as droughts, wars, and theft. The critical thing to note is that for farmers, exploitation of the environment is incremental and episodic - resources near main settlements first, then extended utilization that ranges from 5-15km and beyond in case of risk management.

As shown in this research, although cultures evolve, there is continuity in certain aspects. The only challenge is if heritage research and management do not consider variation in this continuity. Cultural evolution and the impact of modern development is inevitable, and as such, heritage research management approaches must not only manage the history but also the changes over time. This way, critical cultural structures and systems such as *masimo* and cattle posts, which were instrumental in the management of human and animal populations and continue to do so, can be considered a worthwhile heritage.

Researches such as that of livestock management among the Kwena can help comprehend the concept of 'development,' especially in relation to understanding that past societies had ideologies that advanced and sharpened over time. These ideologies were manifested tangibly and intangibly in a gradual and evolutionary, or stage by stage, manner that enabled sustainable and progressive cultural development initiatives. The key issue is to adopt approaches that, while ensuring the protection and longevity of heritage, consider the intricate values contemporary societies place on their heritage. Their need for modern development does not always withstand appreciation of heritage and its associated values. In most instances, present societies do not entirely disregard heritage but rather vary it to suit contemporary needs.

As such, an essential resource such as livestock is managed outside of main settlements to create room for human population growth within main settlements and animal population growth outside of main settlements. Permissive heritage research and management approaches can ensure that heritage landscapes are part of contemporary developmental planning and decision-making processes. Not only that, this will promote the accessibility of heritage landscapes, ensuring that they are of sustainable benefit to present societies. One of the greatest philosophies of Weber Ndoro is that 'heritage is about today'. This resonates with the Kwena case study that heritage evolution must trigger consideration of cultural variation over time so that the past remains relevant and is integrated in present developmental processes.

SUMMARY AND DISCUSSION

This chapter explores the integration of indigenous knowledge in farming practices and provides insights into past and present agricultural ideologies and their socio-economic implications. The importance of indigenous knowledge in understanding historical and contemporary farming practices is outlined with examples from Africa and other parts of the world where this knowledge is yielding results. Promoting the continuity of farming traditions and their relevance in formulating future socio-economic strategies is critical for ensuring food security in Africa.

There are clearly more than one source of knowledge regarding the understanding of humanity and its actions, especially through a common past or heritage such as farming. However, it is concerning to realize that the exploitation of knowledge about the past is often skewed towards empirical sources and processes. The challenges with an overreliance on scholarly knowledge are that cultures are dynamic and adaptive, and often responsive to external elements such as environmental factors and risks from other societies. Most importantly, in a number of times, variation in cultures over time and space are intricate, internally triggered, and have underlying causative protocols that can be learned from contemporary societies. This means that not all questions about the past, especially cultural systems that facilitated sustainable food production strategies, can necessarily be understood from a laboratory, prompting the need for alternative sources of knowledge.

The historical and contemporary significance of farming, its global impact, and its role in food security cannot go unnoticed. Farming has been a crucial part of human history, evolving through experimentation and the exchange of ideas across populations. In Africa, farming not only ensured food security but also facilitated the movement of populations seeking better pastures and soils. Africa's contribution to early farming practices, which began in the Middle East, shows that farming is Africa's heritage, and that the consumption of this heritage is inevitable. Since traditional farming practices have proven to be the backbone of most African economies, this indigenous knowledge must be used to enhance modern agricultural protocols.

Recent research in Botswana (Ndobochani 2020, 2022) has shown that livestock management practices among the Kwena of Botswana have varied temporally and spatially in response to socio-political and environmental prerequisites. This research shows that centralized management practices, as necessitated by internal and external factors, resulted in communal care of livestock at main settlements and rural areas. This resulted in expansive occupations due to the exploitation of resources at 5-15km from main settlements, and this tended to influence the Kwena patterns of settlement and population movement over time. This further supports the notion of decolonizing and opening up the heritage investigation circle to more players and the generation of new knowledge (Ndoro 2018; Chirikure 2020a, 2020b; Butler 2024; Lazzari et al. 2024).

The cattle post system is critical in developing protocols for studying past herding societies, especially in understanding the synergy between their cultures in the context of environmental considerations. Daily exploitation of resources, namely water, grasslands, and vegetation, was incremental and episodic, often leading to temporary and semi-permanent occupations of *meraka-masimo* and *masimo-meraka* settlements (Ndobochani 2022). This research provides a window to understanding past herding practices and how they contributed to the movement of livestock and human populations across the Kwena territory. It mirrors the concept of centrality in relation to historical livestock management practices. Livestock ownership and management have been a central and important aspect of farming societies, to the extent that it even impacted patterns of settlement in such societies.

It highlights that while there is continuity in some cultural aspects, there is change or variation in the physical presentation of such aspects, which is largely necessitated by the evolution of socio-cultural values and modern developmental processes. It shows that the aspect of centrality in a historical livestock management setup has evolved over time. As such, the concept of centrality in livestock is manifested differently across present-day landscapes, without necessarily implying the loss of societal values attributed to livestock as a central socio-economic element.

CONCLUSION

It is evident that livestock management is central to the socio-economic fabric of Kwena society, influencing settlement patterns and community organization. Livestock not only serves as a key resource for food security and economic stability; its historical significance is reflected in their settlement designs, with kraals and communal spaces at the center. Over time, this changed, with the decentralization of livestock management leading to a shift in ownership and settlement dynamics that impacted community structures. The shift to decentralized management resulted in new micro-settlement patterns that differ from traditional wards. However, the absence of livestock in central settlements does not indicate a decline in their value; rather, it reflects changing management practices. With the adoption of the cattle post system and decentralization of livestock management practices, livestock wealth is now more individualized, with households taking on greater responsibility for their herds.

The concept of centrality in livestock management among the Kwena reflects socio-economic and political structures that have evolved over time. The physical representation of the concept of centrality in livestock ownership and management amongst the Kwena is illustrated by the communal and centralized management of livestock in historical Kwena settlements. Before the 20th century, livestock was managed at the ward level, with shared responsibilities. The arrangement of settlements reflected power dynamics and socio-political structures, with key resources located centrally.

Livestock management remains central to the cultural and economic development of African farming societies. Livestock ownership significantly influences settlement patterns and societal structures. While some traditional cultural aspects show continuity, their practice in the modern setup necessitates variations in their physical presentation. The concept of centrality in livestock management has evolved, reflecting contemporary developmental needs. Traditional societal values attributed to livestock management continue to be relevant today, despite changes in practice.

Farming is a critical aspect of African heritage, contributing to food security and population growth while showcasing the continent's historical significance in agricultural development. Farming practices in Africa have evolved through experimentation and knowledge exchange over millennia. The continent is recognized for its early contributions to global farming practices, with evidence dating back thousands of years. Traditional farming remains vital for food security, especially in rural areas where subsistence farming is prevalent. The integration of traditional practices is essential for addressing contemporary agricultural production challenges such as food adequacy and the impact of climate change.

REFERENCES

Abungu, G. H. O. (2016). Practising Archaeology in Eastern and Southern Africa: Coming of Age or the Indigenisation of a Foreign Subject? In *A future for archaeology* (pp. 143-155). Routledge.

Ascher, R. (1961). Analogy in archaeological interpretation. *Southwestern Journal of Anthropology, 17*(4), 317–325. DOI: 10.1086/soutjanth.17.4.3628943

Augustine, D. J., McNaughton, S. J., & Frank, D. A. (2003). Feedbacks between soil nutrients and large herbivores in a managed savanna ecosystem. *Ecological Applications*, *13*(5), 1325–1337. DOI: 10.1890/02-5283

Berkes, F. (Ed.). (1989). *Common Property Resources, Ecology and Community-Based Sustainable Development*. Belhaven Press.

Berkes, F. (1999). *Sacred ecology: Traditional ecological knowledge and resource management.* Taylor and Francis.

Boeyens, J. C. A. (2003). The Late Iron Age Sequence in the Marico and Early Tswana History. *The South African Archaeological Bulletin, 58*(178), 63–78. JSTOR. https://doi.org/DOI: 10.2307/3889303

Boivin, N. (2004). Landscape and Cosmology in the South Indian Neolithic: New Perspectives on the Deccan Ashmounds. *Cambridge Archaeological Journal, 14*(2), 235–257. DOI: 10.1017/S0959774304000150

Bronson, B. (1977). The earliest farming: demography as cause and consequence. In *Origins of agriculture* (pp. 23–48). Mouton Publishers. DOI: 10.1515/9783110813487.23

Butler, B. (2024). *Reimagining Heritage Conservation'after Modernity'as Constellations and Cosmologies of Care and Protection*. Instituto Nacional de Antropología e Historia.

Chirikure, S. (2020a). New Perspectives on the Political Economy of Great Zimbabwe. *Journal of Archaeological Research, 28*(2), 139–186. DOI: 10.1007/s10814-019-09133-w

Chirikure, S. (2020b). *Great Zimbabwe: Reclaiming a 'confiscated'past*. Routledge. DOI: 10.4324/9780367810412

Chirikure, S., Bandama, F., Chipunza, K., Mahachi, G., Matenga, E., Mupira, P., & Ndoro, W. (2017). Seen but not told: Re-mapping Great Zimbabwe using archival data, satellite imagery and geographical information systems. *Journal of Archaeological Method and Theory*, *24*(2), 489–513. DOI: 10.1007/s10816-016-9275-1

Chirikure, S., Ndoro, W., Bugarin, F. T., di Lernia, S., Ichumbaki, E. B., & Lwoga, N. B. (2021). Usable pasts forum: UNESCO and heritage tourism in Africa. *African Archaeological Review*, *38*(3), 513–533. DOI: 10.1007/s10437-021-09454-6 PMID: 34429563

Christie, M. J. (1991). Aboriginal science for the ecologically sustainable future. *Australian Science Teachers Journal*, *37*, 26–31.

Clist, B. (1987). Early Bantu settlements in west-central Africa: A review of recent research. *Current Anthropology*, *28*(3), 380–382. DOI: 10.1086/203543

Deacon, H. J., Deacon, J., Brooker, M., & Wilson, M. L. (1978). The Evidence for Herding at Boomplaas Cave in the Southern Cape, South Africa. *The South African Archaeological Bulletin, 33*(127), 39–65. JSTOR. https://doi.org/DOI: 10.2307/3888251

Denbow, J., Smith, J., Ndobochani, N. M., Atwood, K., & Miller, D. (2008). Archaeological excavations at Bosutswe, Botswana: Cultural chronology, paleo-ecology and economy. *Journal of Archaeological Science, 35*(2), 459–480. DOI: 10.1016/j.jas.2007.04.011

Denbow, J. R. (1983). *Iron Age economics: Herding, wealth, and politics along the fringes of the Kalahari Desert during the Early Iron Age*. Indiana University.

Dunne, J., Evershed, R. P., Salque, M., Cramp, L., Bruni, S., Ryan, K., Biagetti, S., & di Lernia, S. (2012). First dairying in green Saharan Africa in the fifth millennium bc. *Nature, 486*(7403), 390–394. DOI: 10.1038/nature11186 PMID: 22722200

Eerkens, J. W., & Lipo, C. P. (2007). Cultural transmission theory and the archaeological record:providing context to understanding variation and temporal changes in material culture. *Journal of Archaeological Research, 15*(3), 239–274. DOI: 10.1007/s10814-007-9013-z

Ehret, C. (1973). Patterns of Bantu and Central Sudanic settlement in central and southern Africa (ca. 1000 BC-500 AD). *Transafrican Journal of History, 3*(1/2), 1–71.

Fuller, D. Q. (2001). Dung mounds and domesticators: Early cultivation and pastoralism in Karnataka. *South Asian Archaeology, 1*, 117–127.

Gifford-Gonzalez, D. (1998). Early pastoralists in East Africa: Ecological and social dimensions. *journal of anthropological archaeology, 17*(2), 166-200.

Glickman, M. (1972). The Nuer and the Dinka: A further note. *Man, 7*(4), 586–594. DOI: 10.2307/2799950

Hagerstrand, T. (1968). *Innovation diffusion as a spatial process*. University of Chicago Press.

Hall, M. (1983). Early farming communities of southern Africa: A population discovered. *South African Historical Journal, 15*(1), 1–9. DOI: 10.1080/02582478308671573

Hays, T. R. (1975). Neolithic settlement patterns in Saharan Africa. *South African Archaeological Bulletin, 30*(117/118), 29–33. DOI: 10.2307/3888048

Hitchcock, R. K. (2002). Coping with Uncertainty: Adaptive Responses to Drought. *Sustainable livelihoods in Kalahari environments: A contribution to global debates*, 161-171.

Hodder, I. (1995). *Interpreting archaeology: Finding meaning in the past*. Psychology Press.

Hodder, I. (1995). *Interpreting archaeology: Finding meaning in the past*. Psychology Press.

Hodder, I., & Hutson, S. (2003). *Reading the past: Current approaches to interpretation in archaeology*.

Holl, A. F. C. (1998). The Dawn of African Pastoralisms: An Introductory. *Journal of Anthropological Archaeology, 17*(2), 81–96. DOI: 10.1006/jaar.1998.0318

Holl, A. F. C. (2010). The Archaeology of Africa. *ARCHEOLOGY, II*, 20–66.

Holl, A. F. C. (2022). Archaeology of African Agro-systems: A Macro-Evolutionary Perspective. *International Journal of Archaeology, 10*(1), 6–19.

Huffman, T. N. (1986). Archaeological evidence and conventional explanations of southern Bantu settlement patterns. *Africa, 56*(3), 280–298. DOI: 10.2307/1160685

Huffman, T. N., Elburg, M., & Watkeys, M. (2013). Vitrified cattle dung in the Iron Age of southern Africa. *Journal of Archaeological Science, 40*(10), 3553–3560. DOI: 10.1016/j.jas.2013.03.026

Infield, M. 2002. The values of conservation: whose values? Whose sense of beauty, integrity and equity? In Borrin-Fey- erabend, G. (Ed.), Policy Matters: Sustainable management and Co-management of natural resources. CEESP: IUCN.

Johansen, P. G. (2004). Landscape, monumental architecture, and ritual: A reconsideration of the South Indian ashmounds. *Journal of Anthropological Archaeology, 23*(3), 309–330. DOI: 10.1016/j.jaa.2004.05.003

Jones, E. O. (2024). Indigenous knowledge management practices in subsistence farming: A comprehensive evaluation. *Sustainable Technology and Entrepreneurship, 3*(2), 100058. DOI: 10.1016/j.stae.2023.100058

Kuper, A. (1975). The social structure of the Sotho-speaking peoples of Southern Africa. *Africa, 45*(1), 67–81. DOI: 10.2307/1158780

Lander, F., & Russell, T. (2018). The archaeological evidence for the appearance of pastoralism and farming in southern Africa. *PLoS One, 13*(6), e0198941. DOI: 10.1371/journal.pone.0198941 PMID: 29902271

Lane, P. (1994). The use and abuse of ethnography in the study of the southern African Iron Age. *Azania, 29*(1), 51–64. DOI: 10.1080/00672709409511661

Lane, P. (2013). Trajectories of pastoralism in northern and central Kenya: an overview of the archaeological and environmental evidence. *Pastoralism in Africa: Past, present and future*, 104-144.

Lazzari, M., Larsen, P. B., & Orlandi, F. (2024). Heritage and decoloniality: Global exchanges and unresolved questions in sedimented landscapes of injustice. *American Anthropologist*. Advance online publication. DOI: 10.1111/aman.13951

Lwoga, E. T., Ngulube, P., & Stilwell, C. (2011). Challenges of managing indigenous knowledge with other knowledge systems for agricultural growth in sub-Saharan Africa. *Libri. International Journal of Libraries & Information Services, 61*(3).

Mapesa, M. W. (2016). Traditional Management systems in Natural World Heritage Sites in Eastern Africa. In Abungu, G. H. (Ed.), *O et al (2016) Traditional Management Systems at Heritage Sites in Africa*. African World Heritage Fund.

Mapfumo, P., Mtambanengwe, F., & Chikowo, R. (2016). Building on indigenous knowledge to strengthen the capacity of smallholder farming communities to adapt to climate change and variability in southern Africa. *Climate and Development, 8*(1), 72–82. DOI: 10.1080/17565529.2014.998604

Marek-Martinez, O. V. (2021). Indigenous archaeological approaches and the refusal of colonialism in archaeology. In *Routledge Handbook of the Archaeology of Indigenous-Colonial Interaction in the Americas* (pp. 503–515). Routledge. DOI: 10.4324/9780429274251-35

Marshall, F. (1990). Origins of specialized pastoral production in East Africa. *American Anthropologist, 92*(4), 873–894. DOI: 10.1525/aa.1990.92.4.02a00020

Marshall, F., & Hildebrand, E. (2002). Cattle before crops: The beginnings of food production in Africa. *Journal of World Prehistory, 16*(2), 99–143. DOI: 10.1023/A:1019954903395

Mlay, W. F. I. (1986). Environmental implications of land-use patterns in the new villages in Tanzania.

Muchiru, A. N., Western, D., & Reid, R. S. (2009). The Impact of Abandoned Pastoral Settlements on Plant and Nutrient Succession in an African Savanna Ecosystem. *Journal of Arid Environments, 73*(3), 322–331. DOI: 10.1016/j.jaridenv.2008.09.018

Muchiru, A. N., Western, D. J., & Reid, R. S. (2008). The role of abandoned pastoral settlements in the dynamics of African large herbivore communities. *Journal of Arid Environments, 72*(6), 940–952. DOI: 10.1016/j.jaridenv.2007.11.012

Murphy, C. A. Fuller D. Q. (2016). "The Transition to Agricultural Production in India: South Asian Entanglements of Domestication." *A Companion to South Asia in the Past*, 344–57.

Musonda-Hamusonde, M., & Abungu, G. O. (2022). The Nature-Culture Divide in Heritage Classification and Management in Africa: Reality or a Convenience of Management. In *Cultural Heritage Management in Africa* (pp. 219-237). Routledge.

Ndoro, W. (2018). Drawing parallels: Authenticity in the African context–until lions learn to write, hunters will tell their history for them. *Revisiting Authenticity in the Asian Context, 2*, 37–46.

Ndoro, W., & Larsen, P. B. (2024). "Heritage is about today, it's not about what happened in the past": A conversation with Webber Ndoro, Director General of the International Centre for the Study of the Preservation and Restoration of Cultural Property. *American Anthropologist, 126*(2), 362–364. DOI: 10.1111/aman.13955

Niamir-Fuller, M. (1998). The Resilience of Pastoral Herding in Sahelian Africa. In Berkes, F., & Folke, C. (Eds.), *Linking social and ecological systems: management practices and social mechanisms for building resilience, 1* (Vol. 4, pp. 250–284). Cambridge University Press.

Paddayya, K. (1998). Evidence of Neolithic Cattle-Penning at Budihal, Gulbarga District, Karnataka. *South Asian Studies, 14*(1), 141–153. DOI: 10.1080/02666030.1998.9628556

Parson, J. (1981). Cattle, class and the state in rural Botswana. *Journal of Southern African Studies, 7*(2), 236–255. DOI: 10.1080/03057078108708027

Parsons, J. R. (1972). Archaeological settlement patterns. *Annual Review of Anthropology, 1*(1), 127–150. DOI: 10.1146/annurev.an.01.100172.001015

Pikirayi, I. (2019). Local narratives, regional histories and the demise of Great Zimbabwe. *Archaeologies of Listening*, 131–152.

Pistorius, J. C. (1994). Molokwane, a seventeenth century Tswana village. *South African Journal of Ethnology*, *17*(2), 38–53.

Porensky, L. M., & Veblen, K. E. (2015). Generation of ecosystem hotspots using short-term cattle corrals in an African savanna. *Rangeland Ecology and Management*, *68*(2), 131–141. DOI: 10.1016/j.rama.2015.01.002

Rennie, T., Light, D., Rutherford, A., Miller, M., Fisher, I., Pratchett, D., Capper, B., Buck, N., & Trail, J. (1977). Beef cattle productivity under traditional and improved management in Botswana. *Tropical Animal Health and Production*, *9*(1), 1–6. DOI: 10.1007/BF02297380 PMID: 906082

Robertshaw, P. (2021). Archaeology of early pastoralism in East Africa. In *Oxford research encyclopedia of African history*. DOI: 10.1093/acrefore/9780190277734.013.1045

Robertshaw, P., & Collett, D. (1983). A new framework for the study of early pastoral communities in East Africa. *Journal of African History*, *24*(3), 289–301. DOI: 10.1017/S0021853700022027

Rogers, E. M., & Shoemaker, F. F. (1971). *Communication of innovations: A Cross-cultural Approach*. The Free Press.

Russell, T., Silva, F., & Steele, J. (2014). Modelling the spread of farming in the Bantu-speaking regions of Africa: An archaeology-based phylogeography. *PLoS One*, *9*(1), e87854. DOI: 10.1371/journal.pone.0087854 PMID: 24498213

Sadr, K. (2008). Invisible Herders? The Archaeology of Khoekhoe Pastoralists. *Southern African Humanities*, *20*(1), 179–203.

Schapera, I. (1935). The social structure of the Tswana ward. *Bantu Studies*, *9*(1), 203–224. DOI: 10.1080/02561751.1935.9676370

Shahack-Gross, R., Marshall, F., & Weiner, S. (2003). Geo-Ethnoarchaeology of Pastoral Sites: The Identification of Livestock Enclosures in Abandoned Maasai Settlements. *Journal of Archaeological Science*, *30*(4), 439–459. DOI: 10.1006/jasc.2002.0853

Smith, J., Lee-Thorp, J., & Hall, S. (2007). Climate change and agropastoralist settlement in the Shashe-Limpopo River Basin, southern Africa: AD 880 TO 1700. *South African Archaeological Bulletin*, *62*(186), 115–125.

Stiebing, W. H. (1994). *Uncovering the past: A history of archaeology*. Oxford University Press.

Subrahmanyeswari, B., & Chander, M. (2013). Integrating indigenous knowledge of farmers for sustainable organic farming: An assessment in Uttarakhand state of India. *Indian Journal of Traditional Knowledge*, *12*(2), 259–264.

Tilley, C. (1994). Interpreting material culture. In Pearce, C. (Ed.), *Interpreting Objects and Collections* (pp. 67–75). Psychology Press.

Tilley, C. (2007). Materiality in materials. *Archaeological Dialogues*, *14*(1), 16–20. DOI: 10.1017/S1380203807002139

Wikan, G. (2004). The level of living in rural Botswana re-studied. *Norsk Geografisk Tidsskrift*, *58*(1), 1–10. DOI: 10.1080/00291950410004393

Wilson, T. H. (1982). Spatial analysis and settlement patterns on the East African coast. *Paideuma*, •••, 201–219.

Wobst, H. M. (2004). Power to the (Indigenous) past and present! Or: The theory and method behind archaeological theory and method. In *Indigenous Archaeologies* (pp. 15–29). Routledge.

Zeder, M. A. (2014). Domestication: Definition and Overview. In Smith, C. (Ed.), *Encyclopedia of Global Archaeology* (pp. 2184–2194). Springer New York., DOI: 10.1007/978-1-4419-0465-2_71

Zulu, J. (2024). Decolonizing the Mosi-oa-Tunya/Victoria Falls World Heritage Site: Engaging the Unspoken Truths. In Ripmeester, M., & Rofe, M. W. (Eds.), *Global Iconoclasm: Contesting "Official" Mnemonic Landscapes* (pp. 63–75). Springer Fachmedien Wiesbaden., DOI: 10.1007/978-3-658-43691-9_4

Chapter 16
Lending With Dharma:
How Ancient Indian Economic Thought Informs Modern Microfinance Practice

Aditi Sarawagi
http://orcid.org/0000-0002-4076-7682
Galgotias University, India

Amit Gupta
http://orcid.org/0009-0004-2525-6042
Galgotias University, India

Moirangthem Sanjoy Singh
http://orcid.org/0000-0002-5364-9263
Indira Gandhi National Tribal University, India

ABSTRACT

The study develops the Vedic-Kautilyan Microfinance Framework (VKMF), an indigenous analytical framework synthesising Vedic philosophical principle with Arthashastra governance prescriptions to examine microfinance in India. The study addresses a significant gap in microfinance scholarship: the near-exclusive reliance on Western theoretical frameworks despite India's millennia-old traditions of economic thought. Drawing upon textual analysis of classical sources and NABARD data (2000-01 to 2022-23), the chapter operationalises four Vedic principles—sangathan (collective organisation), svav-alamban (self-reliance), samajik uttardayitva (social responsibility), and rna (moral debt)—alongside Kautilya's Saam-Daam-Dand-Bhed framework for graduated stakeholder engagement. Analysis reveals that regional variations in SHG performance, unexplained by operational factors alone, invite interpretation through cultural-institutional lenses. The VKMF contributes to decolonising management knowledge while offering practical protocols for loan recovery that prioritise conciliation over coercion.

DOI: 10.4018/979-8-2600-2890-2.ch016

1. INTRODUCTION

1.1 Background and Context

The contemporary discourse on microfinance and community-based lending is characterised by a notable historical omission. Scholarly and policy literature consistently traces the origins of group-based lending to Muhammad Yunus's experiments in Bangladesh during the 1970s and the subsequent establishment of the Grameen Bank in 1983 (Yunus, 2007; Morduch, 1999). This narrative, while acknowledging a significant institutional innovation, implicitly suggests that principles of community cooperation in financial services emerged only in the latter half of the twentieth century. Such a framing overlooks the extensive documentation of collective economic organisation in ancient civilisations, particularly the sophisticated frameworks articulated in classical Indian texts.

India possesses a continuous intellectual tradition spanning over three millennia, encompassing detailed prescriptions for economic governance, community organisation, and financial transactions. The Arthashastra, attributed to Kautilya and composed approximately in the 4th century BCE, constitutes one of the world's earliest comprehensive treatises on statecraft and economic policy (Olivelle, 2013; Rangarajan, 1992). The text addresses matters ranging from taxation and trade regulation to interest rate determination and debt recovery mechanisms. Similarly, the Vedic corpus, particularly the Rigveda (circa 1500-1200 BCE), articulates foundational principles of collective action and community welfare that remain relevant to contemporary institutional design (Jamison & Brereton, 2014). The persistence of this intellectual tradition raises a fundamental question: why have modern frameworks for understanding microfinance governance largely ignored these indigenous knowledge systems?

1.2 The Contemporary Landscape

India's Self-Help Group-Bank Linkage Programme (SHG-BLP), initiated by the National Bank for Agriculture and Rural Development (NABARD) in 1992, has evolved into the world's largest microfinance initiative by outreach. As of March 2023, the programme encompasses approximately 16 crore households organised into 134 lakh savings-linked SHGs, with aggregate savings of 58,892.68 crore and loans outstanding of 1,88,078.80 crore (NABARD, 2023). Women constitute over 84 percent of SHG membership, and the programme has been credited with substantial contributions to financial inclusion, women's empowerment, and rural livelihood enhancement (Deshmukh-Ranadive, 2004; Garikipati, 2008).

Despite this scale, the sector confronts persistent challenges that conventional theoretical frameworks have struggled to adequately explain. Non-performing assets (NPAs), while declining from 5.19 percent in 2018-19 to 2.79 percent in 2022-23, exhibit significant regional variation—the Central region recorded NPAs of 9.71 percent in 2022-23 compared to 2.32 percent in the Southern region (NABARD, 2023). The Andhra Pradesh crisis of 2010, characterised by aggressive lending practices and subsequent borrower distress, exposed fundamental vulnerabilities in microfinance governance structures (Mader, 2013; Taylor, 2011). These patterns suggest that operational and regulatory factors alone may be insufficient to explain performance variations, and that deeper cultural and institutional dimensions warrant examination.

1.3 Theoretical Motivation

The dominant theoretical frameworks employed in microfinance scholarship—including group lending theory (Stiglitz, 1990; Ghatak, 1999), social capital theory (Putnam, 1993; Coleman, 1988), and institutional theory (North, 1990; Scott, 2014)—have generated valuable insights into the mechanisms underlying collective lending arrangements. However, these frameworks share a common epistemological orientation: they emerge from Western academic traditions and carry embedded assumptions about rationality, individualism, and the relationship between economic and social life that may not fully capture the cultural contexts in which microfinance operates in non-Western societies.

Scholars have increasingly called for the decolonisation of management knowledge, arguing that theories developed in North American and European contexts may exhibit limited validity in other settings and that indigenous knowledge systems merit systematic scholarly engagement (Battiste, 2002; Jack & Westwood, 2006; Nkomo, 2011). This imperative is particularly salient for India, where contemporary financial inclusion initiatives operate within communities whose cultural frameworks have been shaped by millennia of indigenous intellectual traditions. The application of concepts such as *ubuntu* in African management studies (Mangaliso, 2001) and Confucian principles in Chinese organisational research (Chen & Lee, 2008) demonstrates the scholarly value and practical utility of engaging with non-Western knowledge systems.

1.4 Research Objectives and Questions

This chapter develops an integrated analytical framework—the Vedic-Kautilyan Microfinance Framework (VKMF)—that synthesises philosophical principles from Vedic literature with operational prescriptions from the Arthashastra to examine microfinance governance in contemporary India. The framework does not seek to supplant established Western theoretical approaches but rather to complement them by offering culturally grounded interpretive categories that may illuminate dimensions of microfinance practice that remain opaque to conventional analysis.

The study is guided by three research questions:

Aim of the chapter: To build a culturally grounded, non-causal analytical framework that connects Vedic and Kautilyan ideas with contemporary microfinance governance in India, while clearly distinguishing interpretive propositions from empirically tested causal claims.

RQ1: How do Vedic principles of community cooperation—particularly those articulated in the *Samjnana Sukta* (Rigveda 10.191)—manifest in the organisational dynamics of contemporary Self-Help Groups?

RQ2: Can the Arthashastra's *Saam-Daam-Dand-Bhed* framework be operationalised as a graduated approach to loan recovery and default management in microfinance institutions?

RQ3: What insights emerge when NABARD's longitudinal microfinance data (2000-01 to 2022-23) are interpreted through the lens of the VKMF?

The corresponding objectives are: (i) to identify relevant concepts from Vedic and Kautilyan sources; (ii) to translate these concepts into microfinance governance categories; (iii) to apply the framework to secondary NABARD data descriptively; and (iv) to indicate testable propositions for future field-based research.

The significance of this research extends to policymakers, practitioners, and researchers. For NABARD and state-level rural livelihood missions, the framework offers conceptual tools for designing interventions that resonate with the cultural contexts of rural communities. For microfinance institutions, the graduated approach to borrower engagement derived from the Arthashastra may provide alternatives to coercive recovery practices that have contributed to past crises. For scholars, the study opens new avenues for research at the intersection of indigenous knowledge systems and contemporary development finance.

2. LITERATURE REVIEW

2.1 Global Evolution of Microfinance

The modern microfinance movement is conventionally traced to Muhammad Yunus's experiments in Jobra village, Bangladesh, during the mid-1970s. Observing the gap between economic theory and the lived reality of poverty, Yunus initiated small-scale lending to groups of women who were otherwise dependent on informal moneylenders charging usurious rates (Yunus, 2007). The Grameen Bank, formally established in 1983, institutionalised this approach through a model characterised by small loans, joint liability mechanisms, weekly repayment schedules, and a focus on women borrowers. By the time Yunus received the Nobel Peace Prize in 2006, the Grameen model had been replicated across numerous countries, and microfinance had become a central component of international development policy (Morduch, 1999).

The intellectual history of microfinance is inseparable from broader debates concerning market-based approaches to poverty alleviation. Proponents argued that the poor were 'bankable'—that they could be trusted to repay loans and that access to credit would enable investment in income-generating activities, consumption smoothing, and asset accumulation (Robinson, 2001). Critics, however, questioned whether microfinance constituted a genuine pathway out of poverty or merely a mechanism for managing it. Randomised controlled trials conducted during the 2010s produced sobering findings: while microfinance increased business activity and facilitated consumption smoothing, it did not generate transformative impacts on income levels or poverty reduction (Banerjee et al., 2015; Duflo et al., 2013). The field has subsequently moved towards a more nuanced understanding of microfinance as one tool among many—valuable for financial inclusion but not a panacea for poverty eradication (Cull & Morduch, 2017).

A notable feature of this global narrative is its relative silence on pre-existing traditions of community-based lending. The Grameen model is typically presented as a novel institutional innovation, a creative disruption of conventional banking practices. However, informal lending circles, rotating savings associations, and community-based credit mechanisms have existed across diverse cultures for centuries—from the *chit funds* of India to the *tontines* of West Africa to the *hui* of China (Ardener, 1964; Bouman, 1995). The question of why modern microfinance scholarship has largely overlooked these historical antecedents warrants scholarly attention.

2.2 The Indian Context: SHG-Bank Linkage Programme

India's microfinance sector has developed along a distinctive institutional trajectory. While the Grameen model influenced early microfinance institutions such as SHARE and Bandhan, the dominant model that emerged was the Self-Help Group-Bank Linkage Programme (SHG-BLP), pioneered by the NABARD in 1992. The SHG-BLP differs from the Grameen model in several important respects: it is savings-led rather than credit-led, it links community groups to formal banking institutions rather than creating parallel financial structures, and it operates through a three-tier architecture comprising SHGs, federations, and banks (Sriram, 2012; Tankha, 2012).

The scale of the SHG-BLP is substantial. As of March 2023, the programme encompasses approximately 16 crore households organised into 134 lakh savings-linked SHGs, with aggregate savings of 58,892.68 crore and loans outstanding of 1,88,078.80 crore (NABARD, 2023). This represents the largest microfinance programme globally by outreach. Women constitute over 84 percent of SHG membership, and the programme has been credited with significant contributions to financial inclusion, women's empowerment, and rural livelihood development (Deshmukh-Ranadive, 2004; Garikipati, 2008).

Despite this impressive scale, the sector has confronted persistent challenges. The Andhra Pradesh crisis of 2010, precipitated by aggressive lending practices by microfinance institutions and subsequent political intervention, exposed fundamental vulnerabilities in microfinance governance and precipitated significant regulatory reforms (Mader, 2013; Taylor, 2011). Non-performing assets (NPAs), while declining in recent years from 5.19 percent in 2018-19 to 2.79 percent in 2022-23, continue to exhibit significant regional variation. The Central region recorded NPAs of 9.71 percent in 2022-23, compared to 2.32 percent in the Southern region (NABARD, 2023). The credit gap—the proportion of savings-linked SHGs that have not accessed bank credit—remains substantial at approximately 48 percent. These patterns suggest that factors beyond standard operational and regulatory variables may influence microfinance performance, warranting examination of deeper cultural and institutional dimensions.

2.3 Dominant Theoretical Frameworks in Microfinance Scholarship

The academic literature on microfinance has drawn upon several theoretical traditions, the majority of which are rooted in Western intellectual history. A survey of these frameworks illuminates both their contributions and their limitations in explaining microfinance dynamics in non-Western contexts.

Group lending theory, developed by economists including Stiglitz (1990) and Ghatak (1999), explains the efficacy of joint liability mechanisms through concepts of peer monitoring, adverse selection, and moral hazard. The central insight is that when borrowers bear joint liability for each other's loans, they possess incentives to screen potential group members (thereby reducing adverse selection) and to monitor each other's behaviour (thereby reducing moral hazard). The group effectively performs information-gathering functions that would otherwise require costly institutional mechanisms. While this framework has been highly influential, it conceptualises groups primarily as instrumental arrangements rather than as communities with their own histories, norms, and cultural meanings.

Social capital theory, drawing from the foundational work of Putnam (1993) and Coleman (1988), provides a richer analytical account. This perspective emphasises networks of trust, norms of reciprocity, and patterns of civic engagement. Microfinance succeeds, according to this view, because it leverages existing social capital—the bonds of kinship, neighbourhood, and community that facilitate cooperation. The literature has documented how social capital facilitates group formation, enhances repayment rates, and enables collective action beyond credit transactions (Cassar et al., 2007; Karlan, 2007). However, this framework tends to treat social capital as a resource to be instrumentally exploited rather than as a set of values to be cultivated and sustained.

Institutional theory, particularly the new institutionalism associated with North (1990) and Scott (2014), has been applied to analyse how formal rules and informal norms shape microfinance governance. This perspective highlights the significance of regulatory frameworks, organisational structures, and cultural-cognitive elements in determining institutional performance. It has proven particularly useful for understanding the evolution of the Indian microfinance sector, including the roles of NABARD, the Reserve Bank of India, and state-level rural livelihood missions in shaping the institutional environment (Ghate, 2007; Harper, 2002). However, the institutional frameworks invoked in this literature are invariably Western in origin, with no systematic attempt to draw upon indigenous institutional traditions.

What unites these theoretical frameworks is their common epistemological orientation: they emerge from Western academic traditions and embody certain assumptions about rationality, individualism, and the relationship between economic and social domains that may not fully capture the cultural contexts in which microfinance operates in non-Western societies. This observation does not imply that these frameworks are invalid—they have generated substantial insights—but rather that they represent one perspective among several possible approaches.

2.4 Indigenous Knowledge Systems in Management Studies

Over the past two decades, management scholarship has increasingly recognised that the dominance of Western theoretical frameworks is both intellectually constraining and practically problematic. Scholars have called for the 'decolonisation' of management knowledge, contending that theories developed in North American and European contexts may exhibit limited validity in other settings and that indigenous knowledge systems merit systematic scholarly engagement (Battiste, 2002; Jack & Westwood, 2006; Nkomo, 2011).

This intellectual turn has generated substantial scholarly contributions. African scholars have examined the concept of *ubuntu*—an ancient Southern African philosophy centered on interconnectedness and shared humanity—as a foundation for leadership and organisational ethics (Mangaliso, 2001; Mbigi & Maree, 1995). Chinese management scholars have drawn upon Confucian, Taoist, and Legalist traditions to develop culturally grounded theories of leadership, strategy, and governance (Chen & Lee, 2008; Fang, 2012). Latin American scholars have explored indigenous concepts of *buen vivir* (living well) as alternatives to Western developmental paradigms (Gudynas, 2011).

India, with its extraordinarily rich intellectual heritage, represents fertile ground for such scholarly endeavours. A body of literature has emerged exploring Indian philosophical traditions in management contexts. Chakraborty (1991, 1995) pioneered the application of Vedantic philosophy to management, advocating for a values-based approach rooted in concepts such as *dharma* (righteous duty), *karma* (action and its consequences), and *yoga* (integration). Sharma (2007) has articulated a distinctively Indian managerial paradigm drawing on wisdom traditions. Muniapan and Satpathy (2013) have examined corporate social responsibility through the interpretive lens of the Bhagavad-Gita. While valuable, this body of work has remained somewhat peripheral to mainstream management scholarship, and its systematic application to specific domains such as microfinance governance has been limited.

2.5 Kautilya and the Arthashastra in Contemporary Scholarship

The Arthashastra, attributed to Kautilya (also known as Chanakya or Vishnugupta), constitutes a remarkable intellectual achievement. Composed approximately in the 4th century BCE, it represents one of the world's earliest comprehensive treatises on statecraft, economic policy, and governance. Unlike many ancient texts that address primarily spiritual or ethical concerns, the Arthashastra is characterised by its practical orientation—it examines taxation, trade regulation, agricultural policy, jurisprudence, diplomacy, and military strategy with a level of specificity that has led scholars to compare Kautilya to

Machiavelli, though the former preceded the latter by nearly two millennia (Boesche, 2002; Rangarajan, 1992).

Contemporary scholarship has begun to recognise the relevance of the Arthashastra to modern management and economics. Sihag (2014) has argued that Kautilya should be regarded as 'the true founder of economics,' observing that the Arthashastra anticipates numerous concepts that Western economists would develop only centuries later—including principles of public finance, monetary policy, and institutional design. Olivelle's (2013) authoritative translation has enhanced scholarly accessibility to the text, and there is growing interest in its potential contributions to contemporary governance debates.

Of particular relevance to the present study is the Arthashastra's treatment of lending and debt. Kautilya devotes considerable attention to the regulation of interest rates, the rights and obligations of creditors and debtors, and mechanisms for debt recovery. The text also articulates what has come to be known as the *Saam-Daam-Dand-Bhed* framework—a graduated approach to achieving objectives that proceeds from conciliation (*saam*) to inducement (*daam*) to punishment (*dand*) to division (*bhed*). While this framework has been analysed in the contexts of diplomacy and military strategy, its systematic application to microfinance governance—particularly loan recovery and default management—has not been explored in the existing literature.

2.6 Vedic Philosophy and Economic Thought

If the Arthashastra provides operational principles for governance, the Vedic corpus provides philosophical foundations. The Rigveda, the oldest of the Vedic texts (circa 1500-1200 BCE), contains hymns articulating fundamental principles concerning community, cooperation, and collective welfare. The *Samjnana Sukta* (Rigveda 10.191), often designated the Hymn of Unity, is particularly germane to the present inquiry. Its verses exhort people to move together (*sangacchadhvam*), speak together (*samvadadhvam*), and maintain harmony of mind (*samvo manamsi janatam*). These principles are not merely aspirational sentiments but constituted operational norms governing the organisation of ancient Indian communities (Jamison & Brereton, 2014).

Additional Vedic and post-Vedic concepts bear relevance to microfinance governance. *Svavalamban* (self-reliance) emphasises the dignity inherent in economic independence—a principle that resonates with the SHG philosophy of savings-led development. *Samajik uttardayitva* (social responsibility) locates individual welfare within the broader context of community well-being. *Rna* (debt or obligation) in Vedic thought transcends mere financial transaction to encompass moral obligation—human beings are understood to be born with debts to the divine, to sages, to ancestors, and to fellow beings, with life constituting a process of discharging these obligations (Kane, 1977). These concepts offer a philosophical vocabulary for understanding microfinance that differs substantially from the instrumental rationality characteristic of Western economic frameworks.

2.6.1 Recent Microfinance Debates: Behavioural, Gender and Political-Economy Perspectives

Recent microfinance scholarship has moved beyond the early claim that credit access alone produces poverty transformation. Behavioural studies emphasise present bias, social norms, repayment stress, financial literacy, and intra-household bargaining as factors shaping borrowing decisions and repayment behaviour. Political-economy accounts draw attention to state programmes, bank incentives, local power relations, subsidy design, and the uneven capacity of SHG-promoting institutions. Gender-focused research similarly cautions that women's participation in SHGs does not automatically translate into empowerment unless control over income, mobility, voice in household decisions, and protection from over-indebtedness are also addressed. These debates sharpen the present chapter by positioning the VKMF as an interpretive and normative framework rather than a substitute for behavioural, institutional, or political-economic analysis (Ahmad et al., 2020; Kumar et al., 2021).

2.7 Identification of the Research Gap

The foregoing review reveals a significant gap in the literature. Sophisticated theoretical frameworks exist for analysing microfinance, but these are predominantly Western in origin and orientation. A growing literature addresses indigenous knowledge systems in management, but this has not been systematically applied to microfinance governance. Scholarly work on the Arthashastra and Vedic philosophy exists, but it has not been operationalised for specific governance challenges in the financial inclusion domain.

This chapter addresses this gap by developing an integrated analytical framework—the VKMF—that synthesises the philosophical foundations of Vedic thought with the operational principles articulated in the Arthashastra. The framework does not claim superiority over Western analytical approaches; rather, it offers a complementary perspective that may resonate more deeply with the cultural contexts in which Indian microfinance operates. The following section presents this framework in detail, articulating its philosophical foundations, core principles, and operational components.

3. THEORETICAL FRAMEWORK: THE VEDIC-KAUTILYAN MICROFINANCE FRAMEWORK

3.1 Philosophical Foundations: Vedic Principles of Community and Cooperation

The Vedic corpus, comprising the Rigveda, Yajurveda, Samaveda, and Atharvaveda, represents one of humanity's oldest continuous intellectual traditions. While these texts are primarily liturgical in nature, they articulate foundational principles concerning social organisation, collective action, and the relationship between individual welfare and community well-being that bear direct relevance to contemporary institutional design. The VKMF draws upon four key Vedic principles as its philosophical foundation.

3.1.1 Sangathan: The Principle of Collective Organisation

The *Samjnana Sukta* (Rigveda 10.191) articulates the foundational principle of *sangathan* or collective organisation. The hymn's opening verse—*sangacchadhvam samvadadhvam sam vo manamsi janatam* (move together, speak together, let your minds be in harmony)—establishes unity of purpose and action as essential to community welfare (Jamison & Brereton, 2014). This principle transcends mere physical aggregation; it encompasses alignment of intention, coordination of effort, and mutual understanding among community members.

In the context of microfinance, *sangathan* provides a conceptual foundation for understanding the Self-Help Group as more than an instrumental credit mechanism. The SHG, viewed through this lens, represents a contemporary manifestation of an ancient organisational principle—a voluntary association of individuals who commit to collective action for mutual benefit. The principle suggests that group cohesion derives not merely from shared economic interests (as group lending theory would suggest) but from deeper processes of social alignment and collective identity formation.

3.1.2 Svavalamban: The Principle of Self-Reliance

Svavalamban, derived from the Sanskrit roots *sva* (self) and *avalamban* (support or dependence), denotes the principle of self-reliance. This concept appears across Vedic and post-Vedic literature as a fundamental virtue—the dignity inherent in sustaining oneself through one's own efforts rather than depending upon others (Kane, 1977). Importantly, *svavalamban* in the Indian philosophical tradition is not equivalent to Western individualism; it operates within a framework of community obligation and mutual support.

The savings-led approach of the SHG-Bank Linkage Programme resonates profoundly with this principle. Unlike credit-led models that begin with external capital injection, the SHG-BLP requires groups to first demonstrate their capacity for internal resource mobilisation through regular savings. This design embodies *svavalamban*—the group establishes its foundation through its own collective effort before seeking external support. The principle also implies that sustainable development emerges from building internal capabilities rather than creating dependencies on external resources.

3.1.3 Samajik Uttardayitva: The Principle of Social Responsibility

Samajik uttardayitva (social responsibility) locates individual welfare within the broader context of community well-being. This principle, articulated across multiple Vedic and Dharmashastra texts, posits that individuals bear responsibilities not only to themselves and their immediate families but to the wider community (Olivelle, 2005). The concept encompasses obligations to contribute to collective welfare, to refrain from actions that harm the community, and to support those in need.

Applied to microfinance, *samajik uttardayitva* provides a normative foundation for understanding member behaviour within SHGs. It suggests that repayment discipline and group participation are not merely instrumental calculations (as rational choice theory would propose) but expressions of social obligation. The principle also implies that SHGs should function as vehicles for broader community development rather than solely as mechanisms for individual credit access. This resonates with observed SHG activities in areas such as community health, education, and local governance that extend beyond their formal microfinance mandate (Deshmukh-Ranadive, 2004).

3.1.4 Rna: The Moral Dimension of Debt

The concept of *rna* in Vedic thought encompasses far more than financial indebtedness. Classical texts identify multiple forms of *rna: deva-rna* (debt to the divine), *rishi-rna* (debt to the sages who transmitted knowledge), *pitri-rna* (debt to ancestors), and *manushya-rna* (debt to fellow human beings). Human existence is conceptualised as a state of indebtedness, and virtuous life consists in progressively discharging these obligations (Kane, 1977; Olivelle, 2005).

This conceptualisation transforms the understanding of financial debt within microfinance. If debt is understood as a moral obligation rather than merely a contractual arrangement, then repayment becomes an ethical duty rather than simply an economic calculation. The concept also suggests that lending itself carries moral weight—the creditor who provides resources in time of need is performing a virtuous act, and the borrower who repays is fulfilling a sacred obligation. This framework may help explain the high repayment rates observed in SHG lending despite the absence of formal collateral requirements.

3.2 Operational Principles: The Arthashastra Framework

While the Vedic principles provide philosophical foundations, the Arthashastra contributes operational prescriptions for governance and institutional management. Kautilya's treatise is notable for its practical orientation—it addresses concrete challenges of administration, resource allocation, and conflict resolution with a level of specificity unusual in ancient texts (Rangarajan, 1992). The VKMF draws upon three key Arthashastra contributions: principles of institutional design, the graduated approach to stakeholder engagement, and prescriptions for financial governance.

3.2.1 Principles of Institutional Design

The Arthashastra articulates several principles relevant to institutional design in microfinance contexts. First, Kautilya emphasises the importance of appropriate institutional scale—administrative units should be sized to enable effective oversight while remaining responsive to local conditions (Olivelle, 2013). This resonates with the SHG model's emphasis on small group size (typically 10-20 members), which enables face-to-face interaction and mutual monitoring while maintaining administrative manageability.

Second, Kautilya advocates for layered institutional structures with clear hierarchies of authority and accountability. The three-tier architecture of the SHG-BLP—comprising primary SHGs, SHG federations, and formal banking institutions—reflects this organisational logic. Each tier performs distinct functions: primary SHGs handle savings mobilisation and internal lending, federations provide capacity building and aggregation functions, and banks supply external capital and regulatory oversight (Tankha, 2012).

Third, the Arthashastra emphasizes transparency and record-keeping as foundations of good governance. Kautilya prescribes detailed documentation of financial transactions, regular auditing, and public disclosure of accounts (Sihag, 2014). These prescriptions anticipate contemporary practices in SHG management, including the maintenance of savings registers, loan ledgers, and meeting minutes that constitute the documentary foundation of group operations.

3.2.2 Saam-Daam-Dand-Bhed: A Graduated Approach to Stakeholder Engagement

The *Saam-Daam-Dand-Bhed* framework represents one of the Arthashastra's most distinctive contributions to governance theory. Originally articulated in the context of statecraft and diplomacy, the framework prescribes a graduated sequence of approaches for achieving objectives with stakeholders:

Saam (conciliation or persuasion): The first approach involves dialogue, explanation, and appeal to shared values and interests. It seeks to achieve objectives through mutual understanding and voluntary cooperation, preserving relationships and building goodwill.

Daam (inducement or incentive): When conciliation proves insufficient, the second approach involves offering material or non-material incentives to encourage desired behaviour. This may include rewards, benefits, or opportunities contingent upon compliance.

Dand (punishment or sanction): The third approach involves the imposition of penalties for non-compliance. Kautilya emphasises that punishment should be proportionate, predictable, and applied consistently to maintain institutional credibility.

Bhed (division or differentiation): The fourth approach involves creating distinctions among stakeholders, separating cooperative actors from non-cooperative ones, and applying differentiated treatment accordingly. In extreme cases, this may involve exclusion from the community or institution.

The framework's essential logic is one of graduated escalation—beginning with the least coercive approach and proceeding to more intensive interventions only when gentler methods prove inadequate. This sequence reflects both practical wisdom (preserving relationships when possible) and ethical considerations (exhausting non-coercive options before resorting to sanctions).

3.2.3 Operationalisation for Microfinance: Loan Recovery and Default Management

The *Saam-Daam-Dand-Bhed* framework can be operationalised for loan recovery and default management in microfinance contexts. Table 1 presents a mapping of the framework to specific microfinance interventions.

Table 1. Operationalisation of Saam-Daam-Dand-Bhed for Microfinance Default Management

Stage	Principle	Microfinance Application	Specific Interventions
1	*Saam*	Dialogue and understanding; identify reasons for default; appeal to group solidarity and moral obligation	Personal counselling; group meetings; peer support; understanding household circumstances; flexible rescheduling
2	*Daam*	Positive incentives for timely repayment; benefits for compliant members	Interest rebates; priority for future loans; recognition awards; access to additional services; skill training opportunities
3	*Dand*	Proportionate penalties for continued non-compliance; graduated consequences	Late fees; suspension of new borrowing privileges; formal warnings; reduced loan eligibility; peer pressure mechanisms
4	*Bhed*	Differentiation between cooperative and non-cooperative members; ultimate separation if necessary	Formal exclusion from group (last resort); legal recovery proceedings; asset attachment; negative credit reporting

The framework's graduated structure offers a principled alternative to the coercive recovery practices that contributed to the Andhra Pradesh crisis of 2010. By mandating that institutions exhaust conciliatory and incentive-based approaches before resorting to punitive measures, the framework protects borrower welfare while maintaining institutional discipline. The explicit sequencing also provides procedural guidance for field staff, reducing discretionary variation in recovery practices.

3.2.4 Financial Governance Prescriptions

The Arthashastra contains detailed prescriptions for financial governance that anticipate contemporary regulatory principles. Kautilya addresses interest rate regulation, prescribing differential rates based on the purpose of the loan and the security offered—a practice that resonates with contemporary tiered lending structures (Sihag, 2014). He emphasises protection of vulnerable borrowers from usurious practices, prescribing limits on interest accumulation and prohibiting practices that compound borrower distress.

Kautilya also addresses the responsibilities of creditors. Lenders are expected to exercise diligence in assessing borrower capacity, to provide clear disclosure of terms, and to avoid practices that exploit borrower vulnerability. These prescriptions align with contemporary principles of responsible lending and client protection in microfinance (CGAP, 2017). The Arthashastra thus provides an indigenous normative foundation for lending practices that mainstream microfinance discourse has derived from Western regulatory traditions.

3.3 The Integrated Vedic-Kautilyan Microfinance Framework

The VKMF integrates the philosophical foundations from Vedic thought with the operational principles from the Arthashastra into a coherent analytical structure. Figure 1 presents a schematic representation of the framework.

Figure 1. The Vedic-Kautilyan Microfinance Framework (VKMF) - Schematic Representation

APPLICATION LEVEL *Contemporary SHG-BLP Practice*

Group Formation & Cohesion	Savings Mobilisation	Credit Allocation	Loan Recovery	Performance Monitoring

▲

OPERATIONAL LEVEL *Arthashastra Principles*

Institutional Design • Appropriate scale • Layered structures • Transparency	**Saam-Daam-Dand-Bhed** • Graduated escalation • Conciliation first • Proportionate response	**Financial Governance** • Interest regulation • Borrower protection • Creditor duties

▲

FOUNDATION LEVEL *Vedic Philosophical Principles*

SANGATHAN *Collective Organisation* Unity of purpose & action	**SVAVALAMBAN** *Self-Reliance* Dignity through independence	**SAMAJIK UTTARDAYITVA** *Social Responsibility* Community obligation	**RNA** *Moral Debt* Sacred obligation

Note: The framework operates bottom-up: Vedic principles provide normative foundations that inform Arthashastra operational mechanisms, which in turn guide contemporary SHG-BLP practices.

The framework operates at three interconnected levels:

Foundation Level (Vedic Principles): The four Vedic principles—*sangathan*, *svavalamban*, *samajik uttardayitva*, and *rna*—constitute the philosophical foundation. These principles provide the normative orientation that shapes institutional design and member behavior. They represent the 'why' of microfinance governance—the underlying values that justify and sustain collective financial arrangements.

Operational Level (Arthashastra Principles): The operational principles drawn from the Arthashastra—institutional design parameters, the *Saam-Daam-Dand-Bhed* framework, and financial governance prescriptions—provide the 'how' of microfinance governance. These principles translate philosophical foundations into concrete organisational practices and procedures.

Application Level (Contemporary Practice): At this level, the framework interfaces with contemporary SHG-BLP operations. The philosophical and operational principles are applied to specific governance challenges: group formation and cohesion, savings mobilisation, credit allocation, loan recovery, and performance monitoring. The framework provides interpretive categories for understanding observed practices and normative guidance for institutional design.

3.4 Theoretical Positioning: Complementarity with Existing Frameworks

The VKMF does not seek to supplant existing theoretical frameworks but rather to complement them by offering culturally grounded interpretive categories. Table 2 presents a comparative analysis of the VKMF against the dominant theoretical approaches identified in the literature review.

Table 2. Comparative Positioning of the VKMF

Dimension	Group Lending Theory	Social Capital Theory	Institutional Theory	VKMF
Core Logic	Information economics; adverse selection; moral hazard	Networks of trust; norms of reciprocity	Formal rules; informal norms; cognitive frames	Vedic values; Arthashastra governance principles
View of Groups	Instrumental arrangements for risk management	Social networks with embedded resources	Organisational fields shaped by institutional forces	Sangathan—communities with shared values and purpose
Motivation for Repayment	Rational calculation; fear of peer sanctions	Social obligation; reputation concerns	Conformity to institutional norms	Rna—moral obligation; dharmic duty
Default Management	Joint liability enforcement	Social pressure; reputation effects	Regulatory compliance; organisational routines	Saam-Daam-Dand-Bhed graduated escalation
Epistemological Origin	Western economics	Western sociology/ political science	Western organisational theory	Classical Indian texts

The VKMF's distinctive contribution lies in its cultural grounding. While existing frameworks import analytical categories from Western academic traditions, the VKMF draws upon indigenous intellectual resources that may resonate more deeply with the lived experience of SHG members. This does not render Western frameworks invalid—indeed, they may capture universal mechanisms that operate across cultural contexts. Rather, the VKMF offers an additional layer of interpretation that enriches understanding of microfinance dynamics in the Indian context.

Furthermore, the VKMF contributes to the broader project of decolonising management knowledge. By demonstrating that classical Indian texts contain sophisticated frameworks relevant to contemporary institutional challenges, the study challenges the assumption that valid management knowledge must originate from Western sources. This has implications not only for microfinance scholarship but for the wider field of management studies in non-Western contexts.

4. RESEARCH METHODOLOGY

4.1 Research Design: A Secondary Research Approach

This study employs a secondary research design, drawing upon existing texts, documents, and datasets rather than collecting primary data through surveys, interviews, or direct observation. This design choice reflects both the nature of the research questions and practical considerations. The study's primary objective—to develop and apply an analytical framework derived from classical Indian texts—is fundamentally interpretive in character. It seeks to demonstrate the relevance of ancient philosophical and governance principles to contemporary microfinance challenges, a task that requires careful textual analysis rather than primary data collection.

The research design integrates two complementary methodological approaches:

Textual Analysis: The development of the VKMF is grounded in systematic analysis of classical Indian texts. This involves identifying relevant concepts and principles in the source texts, interpreting their meaning within their original contexts, and extrapolating their potential application to contemporary microfinance governance. The approach draws upon hermeneutic traditions of textual interpretation while maintaining scholarly rigour in the treatment of sources.

Quantitative Data Analysis: The application of the VKMF to contemporary microfinance utilises quantitative data from NABARD's annual reports on the SHG-Bank Linkage Programme. This data enables examination of trends in SHG formation, savings mobilisation, credit disbursement, and portfolio quality over a 23-year period (2000-01 to 2022-23). The analysis is primarily descriptive and interpretive, seeking to identify patterns that can be understood through the lens of the VKMF rather than to establish causal relationships through inferential statistics.

The integration of these approaches reflects the study's positioning at the intersection of humanities-based scholarship (interpretation of classical texts) and social science inquiry (analysis of institutional performance data). This methodological eclecticism is appropriate given the study's objective of bridging ancient wisdom traditions and contemporary development practice.

4.1.1 Scope of Claims and Empirical Validation

The chapter does not claim causal identification. The NABARD data are used descriptively to identify broad trends and regional contrasts, and the VKMF is used to interpret these patterns as plausible propositions. Terms such as 'suggests', 'may indicate', and 'is consistent with' are used deliberately to avoid implying that cultural concepts alone explain outcomes. Future validation would require primary SHG-level data, borrower and facilitator interviews, and inferential testing that controls for income levels, bank outreach, state policy, programme maturity, infrastructure, SHG-promoting institutions, and political incentives.

4.2 Data Sources

The study draws upon two categories of data sources: classical textual sources for framework development, and contemporary institutional data for framework application. Table 3 provides an overview of the data sources utilised.

Table 3. Summary of Data Sources

Category	Source	Description	Use in Study
Classical Texts	Arthashastra (Olivelle, 2013; Rangarajan, 1992)	4th century BCE treatise on statecraft, economics, and governance attributed to Kautilya	Operational principles: institutional design, Saam-Daam-Dand-Bhed, financial governance
	Rigveda (Jamison & Brereton, 2014)	Oldest Vedic text (c. 1500-1200 BCE); collection of hymns including Samjnana Sukta (10.191)	Philosophical foundations: sangathan, collective action principles
	Dharmashastra literature (Kane, 1977; Olivelle, 2005)	Post-Vedic texts on dharma, social organization, and ethical obligations	Concepts of rna (debt/obligation), svavalamban, samajik uttardayitva
Contemporary Data	NABARD Status of Microfinance Reports (2001-2023)	Annual reports on SHG-BLP performance, including SHG numbers, savings, credit, NPAs by region	Longitudinal analysis of microfinance trends; regional variation analysis
	Secondary scholarship on Indian microfinance	Academic literature on SHG-BLP evolution, AP crisis, regional variations	Contextual understanding; triangulation of NABARD data

4.2.1 Classical Textual Sources

The development of the VKMF draws upon authoritative translations and scholarly commentaries on classical Indian texts. For the Arthashastra, the study relies primarily on two translations: Olivelle's (2013) critical edition, which represents the current scholarly standard, and Rangarajan's (1992) accessible translation with extensive commentary. These translations enable engagement with the original text while benefiting from scholarly apparatus that situates concepts within their historical and intellectual contexts.

For Vedic sources, the study draws upon Jamison and Brereton's (2014) three-volume translation of the Rigveda, which represents the most comprehensive English rendering currently available. Particular attention is given to the *Samjnana Sukta* (Rigveda 10.191), which articulates principles of collective action central to the VKMF. The study also draws upon Kane's (1977) monumental *History of Dharmaśāstra*, which provides a comprehensive analysis of concepts such as rna (debt/obligation) across the Dharmaśāstra literature.

Secondary scholarship on Kautilyan economics and Vedic thought supplements the primary textual sources. Sihag's (2014) work on Kautilya as an economic thinker provides interpretive frameworks for understanding the Arthashastra's relevance to contemporary economic governance. Scholarly commentaries by Boesche (2002) on Kautilya's political realism and by Flood (1996) on Vedic traditions provide broader intellectual context.

4.2.2 NABARD Microfinance Data

The empirical component of the study utilises data from NABARD's annual *Status of Microfinance in India* reports. These reports, published annually since 1999, constitute the most comprehensive source of data on the SHG-Bank Linkage Programme. The study analyses data from 2000-01 to 2022-23, providing a 23-year longitudinal perspective on programme evolution.

The key variables extracted from NABARD reports include:

SHG Coverage: Number of SHGs with savings linkages and credit linkages, disaggregated by region and over time. This data enables analysis of programme growth trajectories and regional penetration patterns.

Savings Mobilisation: Total savings held by SHGs with banks, disaggregated by region. This data provides insights into the savings-led character of the programme and regional variations in savings behaviour.

Credit Disbursement: Loans disbursed to SHGs during each year and loans outstanding at year-end, disaggregated by region. This data enables analysis of credit deepening and the relationship between savings and credit.

Portfolio Quality: Non-performing assets (NPAs) as a percentage of loans outstanding, disaggregated by region. NPA data is available from 2013-14 onwards with consistent regional disaggregation. This data enables analysis of repayment performance and regional variations in portfolio quality.

Regional Classification: NABARD reports classify states into six regions: Northern, North-Eastern, Eastern, Central, Western, and Southern. This regional classification enables analysis of geographical variations in programme performance and their potential relationship to cultural and institutional factors.

Table 4 presents key summary statistics from the NABARD data for selected years, illustrating the scale and evolution of the SHG-BLP.

Table 4. SHG-Bank Linkage Programme: Key Indicators (Selected Years)

Indicator	2000-01	2010-11	2015-16	2022-23	CAGR (%)
SHGs with Savings Linkage (Lakh)	2.64	74.62	79.03	134.00	19.5
Savings Amount (Crore)	292	7,016	13,691	58,893	27.6
SHGs with Credit Linkage (Lakh)	1.49	48.51	46.73	69.60	19.1
Loans Outstanding (Crore)	481	31,221	57,119	1,88,079	30.5
NPA Rate (%)	N/A	N/A	6.50	2.79	—

Source: Compiled from NABARD Status of Microfinance in India reports (various years). CAGR calculated for 2000-01 to 2022-23.

4.3 Analytical Procedures

The analysis proceeds in three stages, corresponding to the study's three research questions.

4.3.1 Stage 1: Framework Development (Addressing RQ1)

The first analytical stage involves the development of the VKMF through systematic textual analysis. The procedure involves:

(a) Identification of relevant passages in the Arthashastra and Vedic literature that address themes of collective organization, economic governance, lending and debt, and institutional design.

(b) Interpretive analysis of these passages to extract core principles and concepts, drawing upon scholarly commentaries to ensure accurate understanding within historical and intellectual contexts.

(c) Synthesis of identified principles into a coherent analytical framework, establishing logical relationships among concepts and articulating their potential relevance to contemporary microfinance.

(d) Mapping of framework components to specific microfinance governance domains (group formation, savings, credit, recovery, monitoring).

This stage addresses Research Question 1 by articulating how Vedic principles of community cooperation manifest in—and can illuminate—the organizational dynamics of contemporary Self-Help Groups.

4.3.2 Stage 2: Framework Operationalization (Addressing RQ2)

The second analytical stage involves operationalizing the *Saam-Daam-Dand-Bhed* framework for microfinance contexts, specifically loan recovery and default management. The procedure involves:

(a) Detailed analysis of the *Saam-Daam-Dand-Bhed* framework as articulated in the Arthashastra, including its original context, underlying logic, and prescribed sequence.

(b) Translation of each stage (*saam, daam, dand, bhed*) into specific microfinance interventions, drawing upon knowledge of contemporary SHG-BLP practices.

(c) Development of a graduated protocol for loan recovery that embodies the framework's principles while remaining consistent with contemporary regulatory requirements and ethical standards.

(d) Comparative analysis of the operationalized framework against recovery practices implicated in the Andhra Pradesh crisis, to demonstrate its potential contribution to more humane and sustainable approaches.

This stage addresses Research Question 2 by demonstrating how the Arthashastra's governance principles can be translated into concrete protocols for contemporary microfinance practice.

4.3.3 Stage 3: Data Interpretation through VKMF Lens (Addressing RQ3)

The third analytical stage involves interpreting NABARD's longitudinal microfinance data through the lens of the VKMF. The procedure involves:

(a) Descriptive analysis of trends in key programme indicators (SHG coverage, savings, credit, NPAs) over the 2000-01 to 2022-23 period, identifying patterns of growth, stagnation, and recovery.

(b) Regional disaggregation of performance indicators to identify geographical variations, with particular attention to differences between Southern and Central/Eastern regions.

(c) Interpretive analysis of observed patterns through VKMF concepts—for example, examining whether regional variations in NPA rates correlate with factors that might be understood through the lens of *sangathan* (collective cohesion) or *rna* (moral obligation).

(d) Generation of interpretive propositions linking VKMF principles to observed microfinance outcomes, articulated as hypotheses for future empirical testing rather than as demonstrated causal relationships.

This stage addresses Research Question 3 by demonstrating the interpretive utility of the VKMF for understanding patterns in contemporary microfinance data.

To reduce confirmation bias, the interpretation also considers rival explanations, including regional banking density, state livelihood-mission capacity, historical programme maturity, credit discipline, livelihood opportunities, and governance quality. The framework is therefore treated as one lens among several rather than as an exhaustive explanation.

5. FINDINGS

5.1 Longitudinal Trends in the SHG-Bank Linkage Programme: A VKMF Interpretation

The SHG-Bank Linkage Programme has exhibited remarkable growth over the 23-year period under analysis (2000-01 to 2022-23). This subsection presents key trends and interprets them through the lens of the VKMF's foundational principles.

5.1.1 Programme Growth and the Principle of Sangathan

The growth trajectory of the SHG-BLP demonstrates the sustained appeal of collective organisation for financial inclusion. Table 5 presents the evolution of key programme indicators across four distinct phases.

Table 5. Phases of SHG-BLP Growth (2000-01 to 2022-23)

Phase	Period	SHGs (Lakh)	Growth Rate	Characterisation
Phase 1: Foundation	2000-2005	2.64 → 22.38	53% CAGR	Rapid expansion; NABARD-led promotion; NGO partnerships
Phase 2: Expansion	2005-2010	22.38 → 69.53	25% CAGR	State livelihood missions; NRLM precursors; scale prioritised
Phase 3: Crisis & Consolidation	2010-2017	69.53 → 87.44	3% CAGR	AP crisis impact; regulatory reforms; quality focus
Phase 4: Mature Growth	2017-2023	87.44 → 134.00	7% CAGR	DAY-NRLM integration; digital innovations; sustainable growth

Source: Calculated from NABARD Status of Microfinance in India reports (various years).

Interpreted through the VKMF lens, this growth trajectory reflects the enduring appeal of *sangathan*—the principle of collective organization—in Indian society. The SHG model succeeded not merely because of its financial mechanics but because it resonated with deep-seated cultural orientations towards collective action. The Rigvedic injunction to 'move together, speak together' (*sangacchadhvam samvadadhvam*) finds contemporary expression in the SHG meeting rituals, collective decision-making processes, and mutual support mechanisms that characterize functioning groups.

The slowdown in Phase 3 (2010-2017) is particularly instructive. The Andhra Pradesh crisis of 2010 resulted not merely from operational failures but from a violation of *sangathan* principles. The aggressive lending practices of MFIs prioritized individual credit delivery over group processes, undermining the collective character that gave SHGs their distinctive strength. The subsequent recovery in Phase 4 coincided with renewed emphasis on group quality and collective functioning under the DAY-NRLM framework, suggesting that restoration of *sangathan* principles was integral to sectoral recovery.

5.1.2 The Savings-Led Character and Svavalamban

A distinctive feature of the SHG-BLP is its savings-led character—groups must demonstrate sustained savings behaviour before accessing external credit. Table 6 presents the evolution of savings mobilization and its relationship to credit access.

Table 6. Savings Mobilization and Credit Access (Selected Years)

Year	Savings-Linked SHGs (Lakh)	Credit-Linked SHGs (Lakh)	Credit Linkage Ratio (%)	Avg. Savings per SHG ( )	Credit: Savings Ratio
2005-06	22.38	10.79	48.2	15,184	2.6:1
2010-11	74.62	48.51	65.0	9,402	4.5:1
2015-16	79.03	46.73	59.1	17,323	4.2:1
2019-20	102.43	58.64	57.3	23,478	3.8:1
2022-23	134.00	69.60	51.9	43,950	3.2:1

Source: Calculated from NABARD Status of Microfinance in India reports (various years).

The savings-led design embodies the Vedic principle of *svavalamban* (self-reliance). Groups must first demonstrate their capacity for internal resource mobilization before seeking external support. The data reveals that average savings per SHG have grown nearly threefold from 15,184 in 2005-06 to 43,950 in 2022-23, indicating deepening financial capacity within groups. Simultaneously, the credit-to-savings ratio has moderated from 4.5:1 in 2010-11 (the pre-crisis peak) to 3.2:1 in 2022-23, suggesting a return to more sustainable leverage levels.

The credit linkage ratio—the proportion of savings-linked SHGs that have accessed bank credit—has fluctuated between 48% and 65%, stabilizing at approximately 52% in 2022-23. This persistent 'credit gap' might be viewed negatively from a credit-centric perspective. However, through the *svavalamban* lens, it can be interpreted more positively: nearly half of all SHGs continue to function primarily as savings and internal lending communities, demonstrating that the value of collective organization extends beyond access to external credit. These groups embody self-reliance in its fullest sense—sustaining themselves through their own resources without dependence on external capital.

5.2 Regional Variations in Performance: A VKMF Analysis

One of the most striking features of the SHG-BLP is the substantial regional variation in performance indicators. Table 7 presents the regional disaggregation of key metrics for 2022-23.

Table 7. Regional Performance Indicators (2022-23)

Region	SHGs (Lakh)	Share (%)	Loans O/S (Cr)	Avg. Loan/SHG ( )	NPA (%)	NPA Rank
Southern	50.12	37.4	1,19,794	2,39,028	2.32	1 (Best)
Eastern	39.88	29.8	38,927	97,613	3.14	2
Western	15.64	11.7	13,892	88,849	3.58	3

continued on following page

Table 7. Continued

Region	SHGs (Lakh)	Share (%)	Loans O/S (Cr)	Avg. Loan/SHG ( )	NPA (%)	NPA Rank
Northern	8.94	6.7	4,621	51,688	4.25	4
North-Eastern	7.18	5.4	4,114	57,297	5.87	5
Central	12.24	9.1	6,731	54,991	9.71	6 (Worst)
All India	**134.00**	**100.0**	**1,88,079**	**1,40,357**	**2.79**	—

Source: NABARD Status of Microfinance in India 2022-23.

The data reveals striking regional disparities. The Southern region, with 37.4% of SHGs, accounts for 63.7% of loans outstanding and exhibits the lowest NPA rate (2.32%). The Central region, by contrast, with 9.1% of SHGs, accounts for only 3.6% of loans outstanding and exhibits the highest NPA rate (9.71%)—more than four times the Southern rate. These disparities cannot be explained solely by operational factors; they invite interpretation through cultural and institutional lenses.

However, the cultural interpretation must be read alongside material and institutional factors. Southern states have longer SHG-BLP histories, stronger SHG federation structures in several districts, deeper banking penetration, more mature state rural-livelihood missions, better experience with bookkeeping, and more regular capacity-building support. Conversely, high NPA regions may reflect weaker local economies, migration shocks, uneven bank follow-up, political loan-waiver expectations, infrastructure deficits, and lower quality of SHG nurturing. The VKMF, therefore, helps frame questions about group cohesion and moral obligation, but does not by itself establish causation.

5.2.1 Interpreting Regional Variations through the VKMF

The VKMF offers several interpretive propositions for understanding regional variations in microfinance performance:

Proposition 1 (Sangathan Intensity): Regions with stronger traditions of collective organization will exhibit better microfinance performance, as measured by repayment rates and group sustainability. The Southern region's superior performance may reflect deeper historical traditions of collective action, including cooperative movements, temple-based associations, and caste-based mutual aid networks. The lower performance in Central and Northern regions may reflect weaker traditions of voluntary association outside kinship structures.

Proposition 2 (Rna Consciousness): Regions where debt is more strongly understood as a moral obligation (rather than merely a contractual arrangement) will exhibit higher repayment rates. The concept of *rna*—debt as a sacred obligation—may be differentially embedded across regions. Where this consciousness is stronger, defaulting on loans may carry not only financial but moral stigma, providing powerful non-material incentives for repayment.

Proposition 3 (Institutional Maturity): Regions with longer SHG-BLP history and more developed institutional ecosystems will exhibit better performance. The Southern region's three-decade head start (the programme originated in Andhra Pradesh and Karnataka) has enabled the development of institutional capacity, promoting organization networks, and bank familiarity with SHG lending. This resonates with the Arthashastra's emphasis on institutional development as a cumulative process requiring sustained attention.

These propositions are offered as interpretive hypotheses warranting future empirical investigation rather than as demonstrated causal relationships. They illustrate the VKMF's potential to generate culturally grounded research questions that conventional frameworks might overlook.

5.3 Portfolio Quality and the Saam-Daam-Dand-Bhed Framework

The analysis of non-performing assets (NPAs) provides an opportunity to apply the *Saam-Daam-Dand-Bhed* framework to understanding loan recovery dynamics. Table 8 presents NPA trends from 2013-14 (the first year of consistent regional data) to 2022-23.

Table 8. NPA Trends by Region (2013-14 to 2022-23)

Region	2013-14	2016-17	2018-19	2020-21	2022-23	Change
Southern	4.28	5.86	4.12	3.21	2.32	-1.96
Eastern	5.14	7.23	5.89	4.56	3.14	-2.00
Western	3.87	5.42	4.78	4.12	3.58	-0.29
Central	8.92	12.45	11.23	10.87	9.71	+0.79
All India	**5.64**	**6.50**	**5.19**	**4.23**	**2.79**	**-2.85**

Source: NABARD Status of Microfinance in India reports (various years). Northern and North-Eastern regions omitted for brevity.

The all-India NPA rate has declined substantially from 5.64% in 2013-14 to 2.79% in 2022-23, representing a significant improvement in portfolio quality. However, the Central region remains an outlier, with NPAs actually increasing from 8.92% to 9.71% over this period while other regions improved.

5.3.1 Applying Saam-Daam-Dand-Bhed to Recovery Analysis

The NPA trends can be interpreted through the *Saam-Daam-Dand-Bhed* framework. The improvement in national NPA rates post-2016 coincides with shifts in recovery approaches following the Andhra Pradesh crisis and subsequent regulatory reforms:

Strengthening of Saam (Conciliation): The post-crisis period witnessed increased emphasis on understanding borrower circumstances, livelihood support, and financial literacy—approaches that align with *Saam* principles. DAY-NRLM's focus on 'Panchasutra' (five principles, including regular meetings, regular savings, regular internal lending, regular repayment, and regular bookkeeping) institutionalizes processes of dialogue and collective problem-solving before defaults occur.

Introduction of Daam (Incentives): Several states have introduced interest subvention schemes providing rebates for timely repayment—a classic *Daam* intervention. The 'Prompt Repayment Incentive' in various state schemes offers interest rate reductions of 3-5% for groups maintaining zero NPAs, creating positive incentives for repayment rather than relying solely on penalties.

Calibration of Dand (Sanctions): The coercive recovery practices that precipitated the AP crisis represented a distorted application of *dand* without first exhausting *saam* and *daam* approaches. Post-crisis regulations (including RBI's Fair Practices Code for MFIs) have mandated graduated approaches to recovery, effectively institutionalising the *Saam-Daam-Dand-Bhed* sequence. The improvement in NPA rates suggests that this calibrated approach is more effective than coercive recovery.

The persistence of high NPAs in the Central region despite national improvements invites further investigation. One hypothesis, consistent with the VKMF, is that institutional ecosystems in these states have not yet developed the capacity for effective *saam*-based engagement—lacking adequate promoting organizations, community resource persons, and bank staff trained in relationship-based recovery approaches.

5.4 Summary of Key Findings

The analysis yields several findings that address the study's research questions:

Regarding RQ1 (Vedic principles and SHG dynamics): The growth trajectory and savings-led character of the SHG-BLP can be meaningfully interpreted through Vedic principles of *sangathan* (collective organization) and *svavalamban* (self-reliance). The programme's success appears to derive not merely from its financial mechanics but from its resonance with cultural orientations towards collective action. The post-AP-crisis recovery coincided with renewed emphasis on group processes, supporting the proposition that *sangathan* principles are integral to SHG sustainability.

Regarding RQ2 (Saam-Daam-Dand-Bhed operationalization): The *Saam-Daam-Dand-Bhed* framework provides a coherent analytical lens for understanding the evolution of loan recovery practices. The improvement in national NPA rates post-2016 correlates with policy shifts that effectively institutionalized graduated, conciliation-first approaches to recovery—consistent with the framework's prescribed sequence. The AP crisis, by contrast, can be understood as a violation of the framework's logic, with *dand* (coercive recovery) applied without prior exhaustion of *saam* and *daam* approaches.

Regarding RQ3 (NABARD data through VKMF lens): The application of the VKMF to NABARD data generates interpretive propositions linking cultural factors (traditions of collective action, moral dimensions of debt) to microfinance outcomes (regional variations in NPAs, credit linkage patterns). While these propositions require future empirical testing, they demonstrate the framework's capacity to generate culturally grounded research questions that conventional frameworks might overlook. The persistent regional disparities in performance—unexplained by operational factors alone—invite investigation through the VKMF's cultural-institutional lens.

6. DISCUSSION

6.1 Theoretical Contributions

The VKMF makes three principal contributions to microfinance scholarship. First, the concept of *sangathan* enriches the understanding of group dynamics beyond the instrumental logic of information economics. SHGs are not merely risk-management mechanisms but communities characterized by shared purpose and collective identity—resonating with the Rigvedic injunction to 'move together, speak together, let minds be in harmony.' Second, the concept of *rna* (moral debt) reframes repayment motivation. If debt carries moral weight, default represents ethical failure rather than merely contractual breach, potentially explaining high repayment rates despite the absence of collateral. Third, the *Saam-Daam-Dand-Bhed* framework provides a structured, graduated approach to stakeholder engagement absent from conventional theory—mandating that coercive measures be employed only after non-coercive approaches are exhausted.

6.2 Implications for Practice and Policy

The VKMF has several practical implications. For group formation, the *sangathan* principle suggests prioritizing collective cohesion over numerical targets, with promoting organizations investing in processes that build genuine collective identity. For savings mobilization, the *svavalamban* principle provides normative support for maintaining savings-led approaches rather than relaxing requirements to accelerate credit access. For loan recovery, the *Saam-Daam-Dand-Bhed* framework offers a structured protocol: beginning with dialogue and understanding (*saam*), proceeding to incentives (*daam*), then proportionate sanctions (*dand*), with separation (*bhed*) only as a last resort. This graduated approach offers a principled alternative to coercive practices that contributed to the Andhra Pradesh crisis.

The finding of persistent regional variations suggests the need for differentiated regional strategies. Regions with weaker traditions of collective organization may require more intensive *sangathan*-building—longer incubation periods and greater attention to collective processes before credit linkage.

6.3 Contribution to Decolonizing Management Knowledge

Beyond microfinance, this study contributes to the broader project of decolonizing management knowledge. The VKMF demonstrates that classical Indian texts contain sophisticated frameworks relevant to contemporary institutional challenges, complementing rather than displacing Western approaches. Just as African scholarship has drawn on *ubuntu* and Chinese scholarship on Confucian traditions, Indian management scholarship can productively engage with the subcontinent's rich intellectual heritage.

6.4 Critical Reflections

A further caution concerns power relations within groups and communities. Indigenous concepts such as sangathan and rna may strengthen trust and repayment discipline, but they may also be invoked by dominant members, local elites, or field-level intermediaries to pressure weaker borrowers. Any application of the VKMF must therefore include safeguards against elite capture, exclusion of vulnerable members, gendered burden-shifting, and moralized blame of borrowers facing genuine livelihood shocks. The framework is strongest when paired with rights-based client protection, transparent grievance redressal, and clear limits on peer pressure.

Several cautions warrant acknowledgment. The framework should not be essentialized as the 'Indian' approach—India's diversity defies singular characterization. Ancient texts should not be romanticized; the VKMF selectively draws upon principles that appear ethically acceptable in contemporary contexts. The framework should not be misused to justify coercive practices under indigenous cover—its logic explicitly mandates exhaustion of non-coercive approaches first. Finally, the study's propositions are interpretive hypotheses warranting empirical investigation, not demonstrated causal relationships.

7. LIMITATIONS AND FUTURE RESEARCH

7.1 Limitations of the Present Study

This study has several limitations. First, it employs a secondary research design without primary data collection; the application of the VKMF remains at the level of analytical inference rather than empirical demonstration. Second, the textual interpretation of ancient sources composed in vastly different historical contexts necessarily involves analytical extrapolation. Third, NABARD data limitations include regional aggregation that may mask intra-regional variations and NPA data availability only from 2013-14. Fourth, the analysis is descriptive rather than causal; propositions linking VKMF concepts to outcomes are interpretive hypotheses. Fifth, the study focuses only on the SHG-BLP model within India, leaving applicability to other models and contexts unexplored. Lastly, the data included in the tables from NABARD Status of Microfinance Annual Reports are from 2000-01 to 2022-23. Because of the non-availability of the latest reports, it cannot be included.

7.2 Directions for Future Research

The limitations identified above point towards several directions for future research:

7.2.1 Primary Research on VKMF Concepts

The most pressing need is for primary research examining whether and how VKMF concepts resonate with SHG members' lived experience. Qualitative research—ethnographic observation of SHG meetings, in-depth interviews with members and facilitators—could explore questions such as:

- Do SHG members articulate their group participation in terms that resonate with *sangathan* (collective purpose, harmony of mind and action)?
- How do members understand their repayment obligations—as contractual, social, or moral duties? Does the concept of *rna* resonate with their experience?
- How do groups handle default situations? Do their practices reflect something like the *Saam-Daam-Dand-Bhed* sequence?
- Are there regional or community variations in how these concepts are understood and enacted?

Such research would ground the VKMF in empirical observation, testing whether its interpretive categories align with participants' own understandings.

7.2.2 Quantitative Testing of VKMF Propositions

A feasible validation design would combine SHG-level administrative data with primary survey measures of group cohesion, repayment norms, leadership quality, bookkeeping regularity, bank interaction, and borrower vulnerability. Multivariate regressions or multilevel models could then test whether VKMF-aligned variables predict repayment performance after controlling for region, age of group, livelihood base, bank type, poverty level, state policy, and external shocks. Experimental or quasi-experimental pilots could compare a Saam-Daam-Dand-Bhed recovery protocol with standard recovery training, measuring repayment, borrower stress, grievance incidence, and group retention.

The study's interpretive propositions could be subjected to quantitative testing. This would require:

Development of measures: Operationalizing VKMF concepts as measurable variables. For example, '*sangathan* intensity' might be measured through survey items assessing collective identity, shared purpose, and harmony of group deliberation. '*Rna* consciousness' might be measured through items assessing the perceived moral (versus merely contractual) character of debt obligations.

Survey research: Administering instruments to samples of SHG members across regions, enabling statistical analysis of relationships between VKMF variables and outcome variables (repayment rates, group sustainability, member welfare).

Causal identification: Employing research designs that enable causal inference—potentially experimental or quasi-experimental designs testing the effects of interventions designed to strengthen *sangathan* or operationalize *Saam-Daam-Dand-Bhed* recovery protocols.

7.2.3 Comparative and Cross-Cultural Research

Comparative research could examine the VKMF's relevance across different contexts:

- *Intra-India comparisons:* Systematic comparison of SHG dynamics across states and regions, examining whether VKMF concepts have differential resonance in different cultural contexts within India.
- *South Asian comparisons:* Extension to Nepal, Sri Lanka, and Bangladesh, countries that share aspects of the cultural heritage from which the VKMF draws, to assess the framework's applicability beyond India.
- *Cross-cultural comparisons:* Dialogue between the VKMF and indigenous frameworks from other traditions—African *ubuntu*, East Asian Confucian approaches, Latin American *buen vivir*—to identify common principles and distinctive contributions.

7.2.4 Policy Experimentation and Evaluation

The VKMF's practical implications could be tested through policy experimentation:

- *Piloting VKMF-informed training:* Developing and piloting training curricula for community resource persons and bank staff that incorporate VKMF concepts, with evaluation of effects on group quality and performance.
- *Testing Saam-Daam-Dand-Bhed protocols:* Implementing the graduated recovery protocol (Table 9) in selected banks or regions, with rigorous evaluation comparing outcomes to conventional recovery approaches.
- *Assessing sangathan-building interventions:* Testing specific interventions designed to strengthen collective cohesion—extended incubation periods, collective visioning exercises, rituals reinforcing group bonds—and evaluating effects on subsequent performance.

7.2.5 Further Theoretical Development

The VKMF itself warrants further theoretical development:

- *Expanding the textual base:* The present study draws primarily on the Arthashastra, Rigveda, and Dharmashastra literature. Other classical texts—including the epics (Mahabharata, Ramayana), Puranas, and regional literature—may contain additional concepts relevant to microfinance governance.
- *Integration with Western frameworks:* Developing more systematic integration between VKMF concepts and established Western theories—for example, exploring how *sangathan* relates to social capital theory, or how *rna* relates to behavioural economics research on moral motivation.
- *Extension to other domains:* Applying similar methodology to develop indigenous frameworks for other management domains—corporate governance, leadership, business ethics, supply chain management—where Indian classical texts may offer distinctive insights.

These research directions would collectively advance the VKMF from its current status as a conceptual framework to a more fully developed, empirically grounded, and practically tested contribution to microfinance scholarship and practice.

REFERENCES

Ahmad, D., Mohanty, I., Irani, L., Mavalankar, D., & Niyonsenga, T. (2020). Participation in microfinance based self-help groups in India: Who becomes a member and for how long? *PLoS One*, *15*(8), e0237519. DOI: 10.1371/journal.pone.0237519 PMID: 32810162

Ardener, S. (1964). The comparative study of rotating credit associations. *Journal of the Royal Anthropological Institute*, *94*(2), 201–229.

Banerjee, A., Duflo, E., Glennerster, R., & Kinnan, C. (2015). The miracle of microfinance? Evidence from a randomized evaluation. *American Economic Journal. Applied Economics*, *7*(1), 22–53. DOI: 10.1257/app.20130533

Battiste, M. (2002). *Indigenous knowledge and pedagogy in First Nations education: A literature review with recommendations*. National Working Group on Education.

Boesche, R. (2002). *The first great political realist: Kautilya and his Arthashastra*. Lexington Books.

Bouman, F. J. A. (1995). Rotating and accumulating savings and credit associations: A development perspective. *World Development*, *23*(3), 371–384. DOI: 10.1016/0305-750X(94)00141-K

Cassar, A., Crowley, L., & Wydick, B. (2007). The effect of social capital on group loan repayment: Evidence from field experiments. *Economic Journal (London)*, *117*(517), F85–F106. DOI: 10.1111/j.1468-0297.2007.02016.x

CGAP. (2017). *Client protection in microfinance: Smart Campaign principles*. Consultative Group to Assist the Poor.

Chakraborty, S. K. (1991). *Management by values: Towards cultural congruence*. Oxford University Press.

Chakraborty, S. K. (1995). *Ethics in management: Vedantic perspectives*. Oxford University Press.

Chen, M. J., & Lee, D. (2008). The paradox of polarization: Zhongyong, harmony, and conflict management in Chinese organizations. *Management and Organization Review*, *4*(2), 273–290. DOI: 10.1111/j.1740-8784.2005.00012.x

Coleman, J. S. (1988). Social capital in the creation of human capital. *American Journal of Sociology*, *94*(Supplement), S95–S120. DOI: 10.1086/228943

Cull, R., & Morduch, J. (2017). Microfinance and economic development. In Beck, T., & Levine, R. (Eds.), *Handbook of finance and development* (pp. 550–572). Edward Elgar.

Deshmukh-Ranadive, J. (2004). Women's self-help groups in Andhra Pradesh: Participatory poverty alleviation in action. In Patel, R. (Ed.), *Reducing poverty: Is there a role for micro-credit in India?* World Bank.

Duflo, E., Banerjee, A., Glennerster, R., & Kinnan, C. (2013). *The miracle of microfinance? Evidence from a randomized evaluation* (NBER Working Paper No. 18950). National Bureau of Economic Research.

Fang, T. (2012). Yin Yang: A new perspective on culture. *Management and Organization Review*, *8*(1), 25–50. DOI: 10.1111/j.1740-8784.2011.00221.x

Flood, G. (1996). *An introduction to Hinduism*. Cambridge University Press.

Garikipati, S. (2008). The impact of lending to women on household vulnerability and women's empowerment: Evidence from India. *World Development, 36*(12), 2620–2642. DOI: 10.1016/j.worlddev.2007.11.008

Ghatak, M. (1999). Group lending, local information and peer selection. *Journal of Development Economics, 60*(1), 27–50. DOI: 10.1016/S0304-3878(99)00035-8

Ghate, P. (2007). *Indian microfinance: The challenges of rapid growth*. SAGE Publications. DOI: 10.4135/9788132101574

Gudynas, E. (2011). Buen vivir: Today's tomorrow. *Development, 54*(4), 441–447. DOI: 10.1057/dev.2011.86

Harper, M. (2002). Self-help groups and Grameen Bank groups: What are the differences? *Small Enterprise Development, 13*(3), 31–38.

Jack, G., & Westwood, R. (2006). Postcolonialism and the politics of qualitative research in international business. *MIR. Management International Review, 46*(4), 481–500. DOI: 10.1007/s11575-006-0102-x

Jamison, S. W., & Brereton, J. P. (2014). *The Rigveda: The earliest religious poetry of India* (Vols. 1–3). Oxford University Press.

Kane, P. V. (1977). *History of Dharmaśāstra* (2nd ed.). (Vols. 1–5). Bhandarkar Oriental Research Institute.

Karlan, D. (2007). Social connections and group banking. *Economic Journal (London), 117*(517), F52–F84. DOI: 10.1111/j.1468-0297.2007.02015.x

Kumar, N., Raghunathan, K., Arrieta, A., Jilani, A., & Pandey, S. (2021). The power of the collective empowers women: Evidence from self-help groups in India. *World Development, 146*, 105579. DOI: 10.1016/j.worlddev.2021.105579 PMID: 34602708

Mader, P. (2013). Rise and fall of microfinance in India: The Andhra Pradesh crisis in perspective. *Strategic Change, 22*(1–2), 47–66. DOI: 10.1002/jsc.1921

Mangaliso, M. P. (2001). Building competitive advantage from ubuntu: Management lessons from South Africa. *The Academy of Management Perspectives, 15*(3), 23–33. DOI: 10.5465/ame.2001.5229453

Mbigi, L., & Maree, J. (1995). *Ubuntu: The spirit of African transformation management*. Knowledge Resources.

Ministry of Rural Development. (2021). *Deendayal Antyodaya Yojana - National Rural Livelihoods Mission: Framework for implementation*. Government of India.

Morduch, J. (1999). The microfinance promise. *Journal of Economic Literature, 37*(4), 1569–1614. DOI: 10.1257/jel.37.4.1569

Muniapan, B., & Satpathy, B. (2013). The 'dharma' and 'karma' of CSR from the Bhagavad-Gita. *Journal of Human Values, 19*(2), 173–187. DOI: 10.1177/0971685813492265

NABARD. (2023). *Status of microfinance in India 2022-23*. National Bank for Agriculture and Rural Development.

NABARD. (2001-2023). *Status of microfinance in India* [Annual reports]. National Bank for Agriculture and Rural Development.

Nkomo, S. M. (2011). A postcolonial and anti-colonial reading of 'African' leadership and management in organization studies. *Organization, 18*(3), 365–386. DOI: 10.1177/1350508411398731

North, D. C. (1990). *Institutions, institutional change and economic performance*. Cambridge University Press. DOI: 10.1017/CBO9780511808678

Olivelle, P. (2005). *Manu's code of law: A critical edition and translation of the Mānava-Dharmaśāstra*. Oxford University Press.

Olivelle, P. (2013). *King, governance, and law in ancient India: Kau ilya's Arthaśāstra*. Oxford University Press. DOI: 10.1093/acprof:osobl/9780199891825.001.0001

Putnam, R. D. (1993). *Making democracy work: Civic traditions in modern Italy*. Princeton University Press.

Rangarajan, L. N. (1992). *Kautilya: The Arthashastra*. Penguin Books India.

Reserve Bank of India. (2015). *Master Direction - Non-Banking Financial Company - Microfinance Institutions (Reserve Bank) Directions, 2011*. RBI.

Robinson, M. S. (2001). *The microfinance revolution: Sustainable finance for the poor*. World Bank Publications. DOI: 10.1596/0-8213-4524-9

Scott, W. R. (2014). *Institutions and organizations: Ideas, interests, and identities* (4th ed.). SAGE Publications.

Sharma, S. (2007). A new managerial paradigm from Indian wisdom traditions. *Journal of Human Values, 13*(2), 99–114.

Sihag, B. S. (2014). *Kautilya: The true founder of economics*. Vitasta Publishing.

Sriram, M. S. (2012). The AP microfinance crisis 2010: Discipline or death? *Vikalpa, 37*(4), 113–128. DOI: 10.1177/0256090920120407

Stiglitz, J. E. (1990). Peer monitoring and credit markets. *The World Bank Economic Review, 4*(3), 351–366. DOI: 10.1093/wber/4.3.351

Tankha, A. (2012). *Banking on self-help groups: Twenty years on*. SAGE Publications.

Taylor, M. (2011). 'Freedom from poverty is not for free': Rural development and the microfinance crisis in Andhra Pradesh, India. *Journal of Agrarian Change, 11*(4), 484–504. DOI: 10.1111/j.1471-0366.2011.00330.x

Yunus, M. (2007). *Banker to the poor: Micro-lending and the battle against world poverty*. PublicAffairs.

ADDITIONAL READING

Banerjee, A., Breza, E., Duflo, E., & Kinnan, C. (2019). Can microfinance unlock a poverty trap for some entrepreneurs? NBER Working Paper 26346 (2019). https://doi.org/DOI: 10.3386/w26346

Cull, R., Demirgüç-Kunt, A., & Morduch, J. (2018). The Microfinance Business Model: Enduring Subsidy and Modest Profit. *The World Bank Economic Review*, *32*(2), 221–244. DOI: 10.1093/wber/lhx030

Field, E., Pande, R., Papp, J., & Rigol, N. (2013). Does the Classic Microfinance Model Discourage Entrepreneurship among the Poor? *The American Economic Review*, *103*(3), 219–223. DOI: 10.1257/aer.103.6.2196

Garikipati, S., Johnson, S., Guérin, I., & Szafarz, A. (2017). Microfinance and Gender: Issues, Challenges, and the Road Ahead. *The Journal of Development Studies*, *53*(5), 641–648. DOI: 10.1080/00220388.2016.1205736

Karlan, D., & Zinman, J. (2011). Microcredit in Theory and Practice: Using Randomized Credit Scoring for Impact Evaluation. *Science*, *332*(6035), 1278–1284. DOI: 10.1126/science.1200138 PMID: 21659596

Mader, P. (2015). *The Political Economy of Microfinance: Financializing Poverty*. Palgrave Macmillan. DOI: 10.1057/9781137364210

KEY TERMS AND DEFINITIONS

Decolonization of Management Knowledge: An academic approach that challenges the dominance of Western theories and promotes the inclusion of indigenous knowledge systems in management research.

Financial Inclusion: The process of ensuring access to appropriate financial products and services at affordable costs to all sections of society, especially vulnerable groups.

Group Lending: A microfinance mechanism where loans are provided to a group of borrowers who share collective responsibility for repayment.

Interest Subvention: A policy mechanism where part of the interest burden on loans is subsidized to encourage timely repayment and reduce financial stress on borrowers.

Microfinance: The provision of small-scale financial services—including credit, savings, and insurance—to low-income individuals who lack access to traditional banking.

Non-Performing Asset (NPA): A loan in which repayment of principal or interest has been overdue for a specified period, indicating potential default risk.

Rna (Moral Debt): A Vedic concept of debt that extends beyond financial liability to moral and social obligation, reinforcing ethical repayment behaviour.

Saam-Daam-Dand-Bhed Framework: A Kautilyan governance strategy involving four stages: • **Saam** – persuasion and dialogue • **Daam** – incentives and rewards • **Dand** – penalties or sanctions • **Bhed** – differentiation or exclusion Applied in this study as a structured approach to loan recovery.

Samajik Uttardayitva (Social Responsibility): The obligation of individuals to contribute to the welfare of the community, influencing repayment discipline and group cohesion in microfinance settings.

Sangathan (Collective Organization): A Vedic principle emphasizing unity, coordination, and shared purpose within a group, forming the foundation of collective financial behaviour in SHGs.

Self-Help Group (SHG): A voluntary association of individuals (typically 10–20 members), primarily women, who come together for savings, internal lending, and collective financial and social empowerment.

SHG-Bank Linkage Programme (SHG-BLP): A financial inclusion initiative in India that connects Self-Help Groups with formal banking institutions to facilitate savings mobilization and access to credit.

Social Capital: The networks, trust, and norms that facilitate coordination and cooperation among individuals, critical for the functioning of group lending models.

Svavalamban (Self-Reliance): The principle of economic independence achieved through one's own efforts, reflected in the savings-led approach of microfinance groups.

Vedic-Kautilyan Microfinance Framework (VKMF): An integrated analytical framework developed in this study that combines Vedic philosophical principles with Kautilya's governance mechanisms to interpret and guide microfinance practices.

Chapter 17
Visitor Motivation and Pilgrimage Experience Within Indigenous Cultural Traditions: A Study of Sabarimala, India

V. Ashwin Ranjit
http://orcid.org/0009-0006-4646-0506
Lovely Professional University, India

Pramendra Singh
http://orcid.org/0000-0002-9142-265X
Marwadi University, India

ABSTRACT

Sabarimala, one of India's most prominent pilgrimage destinations, is deeply rooted in indigenous cultural traditions and ritual practices. This study examines visitor motivation and pilgrimage experience at Sabarimala by analysing primary data collected from pilgrims. The data was collected on reasons for undertaking the pilgrimage, modes of travel, ritual participation, and perceptions of facilities and services at the site. The findings reveal why pilgrims feel associated with the pilgrimage site and how they are influenced by both the sacred atmosphere and practical aspects such as accessibility, crowd management, sanitation, and accommodation facilities. The study highlights areas requiring improvement to enhance comfort and safety, without compromising the sanctity and traditional significance of the pilgrimage. Understanding these motivational and experiential dimensions provides useful insights for policymakers, temple administration, and tourism planners in protecting the indigenous heritage of Sabarimala while enriching the visitor experience.

INTRODUCTION

Pilgrimage tourism holds an important position in the socio-economic development of the destinations due to its immense contribution in various ways (Vijayanand, 2014). With the growing popularity of factors like religious motivations, historical significance, and cultural experiences, pilgrimage tourism

DOI: 10.4018/979-8-2600-2890-2.ch017

has both challenges and opportunities (Tendero, 2023). Improved infrastructure, tourist experiences, and marketing strategies are all necessary for the sustained development of pilgrimage places (Tendero, 2023). The boundaries between pilgrimage and tourism have become blurred in the 21st century and people are finding meaning to their travel for a cause with the help of pilgrimage tourism (Abad-Galzacorta et al., 2016). Pilgrimage tourism not only promotes feelings but also communal harmony, tranquillity, and good connections between hosts and guests (Bhat, 2015). Even the cultural tourism activities are well associated with religious and pilgrimage tourism, where visitors are identified in two different categories, one is of those who seek spiritual self-discovery, and others are those who seek nature or sport experiences (Collins-Kreiner, 2010). Pilgrimage tourism is often interchangeably used with religious tourism. It has both positive and negative socio-economic impacts, and they should not be underestimated. It is also predicted that pilgrimage will re-emerge as a broad umbrella term for various segments, including heritage, spiritual, and dark tourism, all of which share a core search for meaning and deep transformation (Collins-Kreiner, 2020).

Around the world, pilgrimage sites are used for social interaction, cultural exchange, and identity building in addition to being places of prayer. These locations frequently maintain oral histories, ceremonial activities, indigenous traditions, and local customs that have been passed down through the generations. Due to the country's long-standing spiritual traditions and religious diversity, pilgrimage tourism holds a special place in India. Through the creation of jobs, transportation services, lodging options, regional handicrafts, and food-related enterprises, pilgrimage routes and sacred sites greatly boost local economies. Simultaneously, the growing pilgrimage movement puts strain on local resources and necessitates efficient management techniques to guarantee sustainability and maintain cultural authenticity.

Visitors' expectations and behavior have also changed as pilgrimage tourism has become more commercialized. Convenience, safety, cleanliness, and well-organized facilities are becoming more and more important to modern pilgrims, who also anticipate genuine spiritual experiences. Temple officials and tourism planners have been prompted to strike a balance between religious holiness and visitor satisfaction due to the evolving nature of pilgrimage. As a result, creating successful pilgrimage management methods now requires an understanding of pilgrims' motives and experiences. Travel choices, ritual participation, emotional attachment to holy sites, and general pilgrimage satisfaction are all influenced by visitor incentives. Researching these facets offers insightful information about the connection between spirituality, culture, and the growth of tourism.

The Sabarimala Sree Dharma Sastha Temple, a temple located in the Indian state of Kerala, dedicated to lord Ayyappa, is the most famous and significant among Kerala's indigenous temples (as represented in image 1). The temple is located on a hilltop called Sabarimala, which is nearly 3000 feet above sea level. This temple's year-round closure is just another distinctive feature and opens for only a few days in the year. The development of indigenous masculine identities is significantly influenced by the Sabarimala pilgrimage in Kerala, which draws on the link between renouncer and householder (Osella, 2003). Additionally, it is a transitional area that strengthens communities, especially among indigenous disadvantaged populations, but it may also be a flashpoint for conflict and contradiction (Vadakkiniyil, 2019). The temple's restrictions on women due to their menstruation have been a subject of legal debate, with a majority verdict to uplift all restrictions, but a subsequent review maintained the original verdict (Dash, 2024).

Sabarimala is well known for its distinctive pilgrimage customs and rigorous adherence to rituals performed by followers prior to their visit to the shrine, as represented in image 1. Before the pilgrimage, pilgrims typically observe a forty-one-day period of fasting, self-control, abstinence, and spiritual preparation. Devotion, sacrifice, and equality among pilgrims are symbolized by the group chanting of religious hymns, dressing in black or blue, walking barefoot through woodland paths, and carrying the holy "Irumudi." These customs set Sabarimala apart from many other holy sites in India by reflecting the pilgrimage's strong indigenous and spiritual essence. Therefore, the pilgrimage experience encompasses more than just visiting a temple; it also includes the devotees' spiritual, emotional, and physical journeys.

Pilgrimage tourism has profound impact on the natural environment and sustainability, highlighting the need for better coordination and alignment among the concerned stakeholders (Hole et al., 2019). The religious rituals and influx of pilgrims have had a notable impact on the natural environment of Sabarimala, a major pilgrimage site in Kerala (John, 2020). This has raised concerns regarding sanitation, crowd control, and the safety of water and food (Nayar, 2020). The constant flow of visitors throughout the year exerts additional pressure on the natural surroundings (Shinde, 2007). The quality of the Pamba River water in Sabarimala, which pilgrims rely on, is considerably affected during the pilgrimage season, underscoring the necessity for adequate sanitation and waste management (Jalal, 2013). In recent years, the pilgrimage journey to this indigenous shrine of Lord Ayyappa has gained popularity, attracting a diverse range of pilgrims (Sekar, 1992) and has also been a site of protests, reflecting the tensions and contradictions within Indian society (Vadakkiniyil, 2019).

In addition to environmental issues, managing large-scale pilgrimage events has grown in importance for legislators and administrators. Massive crowds arrive in Sabarimala during the busiest pilgrimage seasons, posing problems for emergency response systems, transportation, lodging, health care, and security. To protect pilgrims' comfort and safety while preserving the destination's hallowed aura, effective crowd control is crucial. Temple authorities, local government agencies, tourism departments, environmental organizations, and local communities must work together to plan pilgrimages in a sustainable manner. The long-term preservation of the pilgrimage site and its indigenous cultural legacy can be aided by the incorporation of sustainable practices like regulated tourist movement, eco-friendly trash disposal, awareness campaigns, and enhanced public facilities.

This study aims to understand the motivations and experiences of people travelling to the indigenous religious destination of Sabarimala. The primary goal is to understand what motivates these visitors to undertake this difficult journey to this sacred site. This may include spiritual seeking, fulfilment of vows, seeking blessings, or experiencing a sense of communal unity and devotion. Exploring these motives will help explain why this indigenous pilgrimage destination attracts millions of visitors annually. By using quantitative analysis, the research will try to explore the different motivations that inspire the devotees to embark on this sacred journey. This study also aims to explore the various experiences of the people visiting Sabarimala. These experiences include their accommodation, food and refreshments, crowd management and many more. Learning about these experiences, the authorities can understand what and where it is lacking. The study also aims to find out what sustainable practices can be adopted at the study location, which includes infrastructure development, environmental conservation and spreading the knowledge about the cultural significance of Sabarimala.

The study also advances knowledge of indigenous pilgrimage customs and their applicability to modern tourist research. The study looks at both experiential and motivating aspects in an effort to close the gap between pilgrims' practical realities and their spiritual hopes. Policymakers, tourist planners, and temple administrators may find the study's conclusions useful in creating visitor-focused plans that protect Sabarimala's spiritual purity and cultural distinctiveness while improving pilgrims' overall experience. In order to preserve the cultural and spiritual value of this revered indigenous location for future generations, the study also highlights the significance of striking a balance between religious customs and sustainable tourism practices.

Figure 1. AI generated image of Sabarimala temple, India (resembles the original temple)

LITERATURE REVIEW

Understanding the motivations and experiences of visitors to Sabarimala is important for several reasons. Firstly, Sharpley (2005) emphasises that recognising both spiritual and non-spiritual motivations can enhance the overall visitor experience. Many reports suggest that pilgrimage tourism in Sabarimala brings substantial economic benefits to the indigenous local community, and comprehending these motivations can help maximise these advantages. Additionally, Chand (2010) points out that distinguishing between the motivations of domestic and international tourists is crucial for effective marketing and management of travellers.

Many academics contend that because pilgrimage tourism has psychological, social, and cultural components, it cannot be fully comprehended through the prism of religion alone. Through their travels to hallowed locations, pilgrims frequently seek emotional fulfillment, mental serenity, social bonding, and a sense of identity. Pilgrimage sites are important hubs for both spiritual involvement and cultural preservation in the Indian setting since they are intricately linked to cultural customs and communal life. According to the literature, demographic factors like age, education, occupation, and previous pilgrimage experience may influence pilgrimage motivations. Some tourists may be drawn by curiosity, cultural interest, or the desire to take part in customs, while others may be driven by faith and devotion.

The pre-pilgrimage experiences of tourists visiting Sabarimala are influenced by various factors. According to Osella (2003), the pilgrimage plays a role in shaping male identities, where pilgrims engage in severe self-discipline and abstain from indulgences to connect with the deity. Jyotsna (2022) notes that the destination's popularity, safety, local culture, and spiritual atmosphere are key factors affecting the tourist experience. Pai (2016) emphasises the importance of temple management and infrastructure in determining pilgrims' overall satisfaction at sacred sites.

Because pilgrims are expected to adhere to rigorous religious rituals and disciplined lifestyles for several weeks, the preparatory period before to the Sabarimala pilgrimage is regarded as spiritually essential. Researchers have noted that devotees' emotional and psychological development is aided by this preparing phase. Pilgrims experience a sense of spiritual purity and heightened devotion when they observe fasting, celibacy, vegetarianism, and prayer routines. Regardless of caste, class, or socioeconomic status, devotees' communal engagement in these activities promotes social cohesion and equality. These encounters underscore the deeper symbolic implications associated with religious travels and set pilgrimage tourism apart from traditional tourism.

Post-pilgrimage studies reveal a range of lasting impacts and reflections. Iwaarden (2024) found that pilgrims often experience long-term psychological benefits, such as spiritual growth, after their journey. Park (2018) identified various reasons for visiting pilgrimage sites, including spiritual motives, travel, friendship, self-presentation, and status, which are also related to sharing experiences on social media. Some researchers have highlighted the significant economic benefits of the Sabarimala pilgrimage for local residents, including increased income and employment opportunities. Osella (2003) also examined how the pilgrimage influences male identities, underscoring its role in emphasizing masculinity. These studies collectively demonstrate the complex and enduring impacts of the Sabarimala pilgrimage on visitors.

According to recent research, pilgrimage experiences often go beyond just visiting the holy location. After finishing their journey, pilgrims frequently describe experiencing emotional fulfillment, a revitalized sense of purpose, reinforced faith, and personal development. Devotees frequently develop a deep emotional bond with the pilgrimage location and return there frequently as part of their spiritual routine. The way pilgrims share and preserve their experiences has also been altered by social media platforms and digital communication, enabling religious travels to be included in broader social and cultural conversations. Religious tourism places like Sabarimala have become more visible and popular as a result of this digital representation of pilgrimage experiences.

Management practices at Sabarimala are critical in shaping the tourist experience. Factors such as the site's popularity, safety, local culture, accessibility, hygiene, timing, pilgrim turnout, and temple service pricing are major determinants of the pilgrimage experience (Jyotsna 2022). These elements are shaped by management practices, which are essential for the sustainable development of religious tourism sites (Wiltshier 2020). Effective management practices, including visitor education and overall satisfaction, are key to enhancing the tourist experience (Sivakami 2023). The economic benefits of pilgrimage tourism at Sabarimala further highlight the importance of sound management practices. However, challenges such as inadequate infrastructure, damaged roads, and exclusion of certain groups persist (Khan et al., 2021). Despite these issues, the pilgrimage season positively impacts the local economy (Muraleedharan, 2009).

The significance of sustainable practices at pilgrimage sites has drawn more attention from academics studying religious tourism management. Transportation systems, sanitary facilities, waste management systems, and healthcare services are all severely impacted by the massive stream of pilgrims to Sabarimala during peak seasons. Inadequate handling of these elements can have a detrimental impact on pilgrimage site environmental conditions as well as guest happiness. Thus, integrated management strategies involving local communities, government agencies, temple boards, and tourism stakeholders are suggested by researchers. Maintaining ecological and cultural integrity while balancing infrastructure development is necessary for sustainable pilgrimage management. Improved emergency response systems, eco-friendly trash disposal techniques, improved crowd control systems, and awareness campaigns have all been shown to be crucial steps in improving the pilgrimage experience as a whole.

Research into tourist motivations and experiences at Sabarimala reveals a complex interaction of spiritual and economic factors. Sharpley (2005) suggests that spirituality in tourism exists on a continuum, with both permanent and temporary visitors seeking various experiences. Researchers support this by showing that the Sabarimala pilgrimage significantly boosts the local economy, reflecting a blend of spiritual and economic motivations. Chand (2010) further highlights the role of cultural immersion, novelty, and religious attraction in motivating visitors, noting that non-residential international (NRI) tourists have motivations similar to those of domestic tourists. Olsen (2014) adds a spatial perspective, pointing out differences in motivations and expectations among visitors to specific religious sites, themed routes, and religious areas. These studies collectively reveal the multifaceted nature of tourist motivations and experiences at Sabarimala, encompassing both spiritual and economic dimensions. Understanding the economic benefits, public health implications of mass gatherings, and the balance between tourism and religious experience is vital for managing tourism effectively at Sabarimala (Nayar 2020; Kasim 2013). While the pilgrimage generates significant economic gains for the local community (B. 2009), it also presents health and safety challenges, including high patient presentation rates and concerns about crowd management (Nayar 2020). Therefore, a balanced approach is essential to preserve the pilgrimage's spiritual significance while addressing these practical issues (Kasim, 2013).

The body of research indicates that pilgrimage tourism in Sabarimala is multifaceted, encompassing social contact, cultural engagement, religious devotion, economic contribution, and environmental concerns. However, little study has thoroughly investigated both tourist motivation and pilgrimage experience combined in the context of indigenous cultural traditions, despite a great deal of discussion on spirituality, management concerns, and socioeconomic effects. The majority of earlier research has either focused on management-related issues or religious factors independently. As a result, comprehensive study that examines the connection between pilgrims' reasons for traveling, their experiences there, and the sustainability of the pilgrimage site is still required. In an effort to close this gap, the current study looks at pilgrims' motives, experiences, and perspectives while also taking infrastructure, crowd control, cleanliness, and the preservation of indigenous cultural heritage into account.

RESEARCH METHODOLOGY

This research uses a quantitative approach to deeply understand the motivations behind the visitors travelling to the sacred and indigenous community site of Sabarimala, which is located in Kerala, India. The understanding of the motivations and experiences of these travellers will help in making the necessary changes in Sabarimala in order to improve the tourist experience and also maintain the cultural significance.

Because the quantitative research approach allows for the systematic collecting and analysis of responses from a large number of pilgrims, it was deemed suitable for this study. This method aids in determining typical behavioral patterns, driving forces, and visitor attitudes related to the pilgrimage experience. Additionally, quantitative approaches offer quantifiable insights into tourist opinions and experiences, allowing for the assessment of the overall efficacy of the destination's amenities and services. The study aims to provide an unbiased insight of the pilgrimage experiences at Sabarimala through statistical representation of replies.

The study focuses on pilgrims who have traveled to Sabarimala and encountered the customs, amenities, and administrative procedures related to the trip. Understanding tourist perceptions is crucial for enhancing destination management and sustainability strategies because pilgrimages entail both spiritual commitment and physical hardship. Therefore, the study looks at both the practical factors that affect pilgrims' overall pleasure and experience throughout the pilgrimage journey as well as their religious motivations.

Objective

The primary objective of the study is to understand the main motivations behind people visiting this indigenous pilgrimage site of Sabarimala. By using Google Forms to conduct surveys, various questions were asked to the visitors to understand their motivations, including spiritual fulfilment, seeking blessings, or adventure. The research also aims to learn about the experiences of various visitors to Sabarimala. These experiences include their accommodation, food and refreshments, crowd management and many more. Learning about these experiences, the authorities can understand what is lacking and where it is lacking.

Moreover, this study aims to contribute to the sustainable development of the destination. By understanding the motivation and experiences of the visitors, the stakeholders can come up with sustainable strategies to improve the overall pilgrimage experience, which includes infrastructure development, environmental conservation and preserving and spreading the knowledge about the cultural significance of Sabarimala.

Finding the main obstacles pilgrims encounter while visiting Sabarimala is another crucial goal of the research. Overcrowding, transportation difficulty, sanitary problems, lodging availability, and safety concerns are a few examples of these difficulties. In order to improve visitor comfort and safety while maintaining the pilgrimage destination's holiness, temple officials and tourist planners can develop efficient management strategies with the help of an understanding of these challenges.

Additionally, the study aims to investigate the connection between visitor pleasure and pilgrimage motivation. Depending on their past experiences, travel choices, cultural background, and spiritual beliefs, pilgrims visiting Sabarimala may have a variety of expectations. It is possible to gain important insights about the advantages and disadvantages of the destination's management system by analyzing these expectations and contrasting them with their actual experiences. These discoveries may help create a pilgrimage atmosphere that is more sustainable and welcoming to tourists.

Data Collection and Processing

For this study, the data were collected through surveys from pilgrims who had visited Sabarimala and a total of sixty-six responses were received. The questionnaire was self-structured based on previous research and used variables from various studies adopting a similar approach. The variables were primary focused on visitor expectations, experiences and observations at the study site. Covenient sampling method was used to collect the visitors' responses. The questionnaire included information regarding the demography, motivations and experiences of the visitors. The recorded responses were then analysed using MS Excel for descriptive analysis to show the results.

Both closed-ended and multiple-choice questions were included in the survey in order to collect comprehensive data about the pilgrims' demographics, travel habits, driving forces, and general experiences. Age, gender, level of education, occupation, and location of residence were all covered in the demographic section. Understanding the spiritual, cultural, social, and individual motivations for making the pilgrimage was the main goal of the motivation-related segment. Questions about transit options, lodging, food availability, crowd control, sanitation, safety precautions, and general pilgrimage satisfaction were all included in the experience-related segment.

Google Forms made it easy and effective to get answers from pilgrims from various geographic locations. Online surveys also made it easier for respondents to participate while cutting down on the time and expense associated with data collecting. Care was made to make sure the questions were straightforward, pertinent, and basic before distributing the questionnaire. The respondents gave their consent to participate in the study, and their answers were kept private at all times.

Because descriptive analysis enables the researcher to meaningfully summarize and present the findings using percentages, frequencies, charts, and tables, it was used to understand the data that was gathered. The responses were arranged, coded, and analyzed using Microsoft Excel. Finding common patterns in tourist motivations, pilgrimage experiences, and satisfaction levels was made easier by the data analysis. The analysis's conclusions offer legislators, tourism officials, and temple administrators useful information for enhancing pilgrimage management and encouraging sustainable tourism practices at Sabarimala.

While the study offers valuable insights into visitor experiences and motivations, it is important to recognize certain limits. Convenient sampling may limit the findings' generalizability, since the sample size was restricted to sixty-six respondents. Furthermore, the answers were predicated on pilgrims' particular perspectives and experiences, which could differ from person to person. However, the study provides important initial insights into the reasons, experiences, and management-related issues relating to the Sabarimala pilgrimage.

DATA ANALYSIS

This study aims to dive into the motivations and experiences of pilgrims visiting Sabarimala, providing a deep understanding of their demographic characteristics, reasons for visiting, overall experiences, and suggestions for improvement.

Important information about the expectations, satisfaction levels, and behavioral patterns of pilgrims visiting Sabarimala can be gleaned from the analysis of the replies gathered. The results show that practical factors like transportation, lodging, cleanliness, and crowd control have an impact on pilgrimage experiences in addition to religious commitment. The information gathered further emphasizes the connection between visitors' motivation and how they view the amenities offered at the location. Temple administrators and tourism planners can apply efficient management and sustainable development strategies with the aid of an understanding of these patterns.

Demographic Characteristics

The survey collected responses from a total of 66 visitors. The majority of respondents were male; this may be because the temple doesn't allow women from a certain age group entry into the temple. The age group 40 and above had the highest number of visitors, followed by those aged 31 and 40. This indicates that Sabarimala attracts a mature demographic. It can also be seen that the number of people visiting below the age of 18 is low.

Table 1. Demographic Characteristics of Respondents

Age Group	Male(no.)	Female(no.)
Below 18	1	0
19-30	13	2
31-40	17	3
40 and above	26	4

The study's preponderance of male responders is indicative of Sabarimala's traditional pilgrimage customs and limitations. The results also show that the majority of pilgrims are middle-aged and older, indicating that religious discipline and spiritual devotion strengthen with age. There were relatively fewer younger visitors, which could be a sign of the pilgrimage's physical demands or of younger generations' lack of interest. The respondents' demographic makeup emphasizes how crucial it is to create facilities that are appropriate for senior pilgrims, especially in the areas of lodging, transportation, and healthcare.

Frequency of visit

The respondents were asked about how often they visit Sabarimala. From the responses, it can be concluded that men travel more often than women, which may be due to the temple prohibiting female devotees. This may also be due to the rigorous trek that the visitor must go through before reaching the temple, which some women may find difficult.

Table 2. Frequency of visit

Frequency	Gender
once	Male (10) Female (5)
2-5 times	Male (21) Female (3)
6-10 times	Male (7) Female (0)
More than 10 times	Male (19) Female (1)

The results show that a sizable portion of pilgrims return to Sabarimala on a regular basis, exhibiting a deep emotional and spiritual bond with the location. Recurring visits imply that for many devotees, the pilgrimage is a continuous spiritual practice rather than just a one-time religious event. The significant cultural and religious significance of making the journey more than once is also reflected in the large proportion of male repeat travelers. Despite the physical difficulties of the trip, frequent visits further demonstrate great levels of devotion and contentment.

Motivations for Visiting Sabarimala

Understanding what motivates visitors to Sabarimala helps in getting an idea about the intentions of their journey. The main motivation for visiting Sabarimala is religious and spiritual, which can be understood due to the nature of Sabarimala as a pilgrimage site. Cultural interest also plays a significant role, showing that some visitors are interested in knowing about the heritage and traditions associated with Sabarimala. The influence of family and friends also motivates some visitors, indicating the social aspects of the pilgrimage. A smaller group of people visits for adventure or trekking, indicating that the people also see Sabarimala as more than a religious destination.

Table 3. Motivations for Visiting Sabarimala

Motivation	Number of people
Religious/spiritual reasons	51
Cultural interest	17
Adventure/trekking	3
Influence of family and friends	12

The predominance of spiritual and religious reasons attests to Sabarimala's primary role as a holy pilgrimage place rather than a traditional tourist destination. Nonetheless, the existence of social and cultural incentives suggests that travelers are also eager to experience the customs, ceremonies, and native heritage connected to the pilgrimage. The impact of friends and family emphasizes how communal and group-oriented pilgrimage travel is in India. Even while trekking and adventure were less popular reasons, they show that for some pilgrims, the actual physical travel is a significant component of the entire experience.

Reason to Visit Sabarimala

Visitors were asked how they see Sabarimala as-whether it is a pilgrimage or they see it as a form of tourism, or both. Most of the visitors perceive their visit as a pilgrimage, highlighting the religious significance of Sabarimala. Also, there were people who see it as a form of tourism indicating the potential of Sabarimala as a tourist destination.

Table 4. Reason of Visit

Perception Number of people **Pilgrimage 48** **Tourism 2** **Both 16**
The results unequivocally show that Sabarimala's spiritual identity continues to be predominant among tourists. However, the respondents' perceptions of the trip as both pilgrimage and tourism show how religious travel is evolving in contemporary society. Pilgrimage sites have steadily evolved into places that blend spirituality with travel experiences thanks to advancements in transportation, exposure to digital media, and tourism infrastructure. Because officials must maintain the site's religious holiness while simultaneously satisfying the demands of contemporary tourists, this dual view offers both opportunities and challenges for destination management.

Modes of Transportation

Most visitors hire private vehicles to reach Sabarimala, which suggests a need for big parking facilities. Public transport is also commonly used, reflecting the accessibility of the site through various means. Some of the people chose walking as a mode of transportation as it is a tradition to go on foot till Sabarimala.

Table 5. Modes of Transportation

Mode of Transportation	Number of people
Private vehicle	35
Public transport	26
Walking	3
Combination of vehicle and walking	1
Train	1

The significant reliance on private automobiles suggests that the destination's parking facilities and traffic laws need to be improved. Sabarimala's accessibility for pilgrims from many places is shown in the significant use of public transportation. For some followers, walking still has symbolic and spiritual significance because it is seen as an act of sacrifice and devotion. These results demonstrate the necessity of integrated mobility planning that guarantees environmental sustainability, ease, and safety throughout the pilgrimage season.

Facilities at Sabarimala

Respondents were also asked to provide their feedback on various facilities available at Sabarimala, including accommodation, Food and refreshments, restroom facilities, parking facilities, and souvenir availability. These facilities help in giving a better experience to the visitors, and from the given feedback, it is clear that most of them weren't satisfied with the facilities provided, and special attention should be given to them.

Table 6. Facilities at Sabarimala

Facility	Excellent	Good	Average	Poor	Very poor
Accommodation	9	9	24	10	14
Food and refreshments	10	16	21	10	9
Restroom facilities	9	8	17	15	17
Parking facilities	8	8	23	12	15
Souvenir availability	23	18	17	6	2
Medical facilities	11	16	24	6	9

The facilities analysis reveals varying degrees of visitor pleasure. While restrooms and parking facilities were scored low by several respondents, souvenir availability garnered relatively positive reviews. One of the main factors influencing tourist happiness seems to be sanitation-related difficulties. In a similar vein, complaints about lodging and parking indicate the need for infrastructure upgrades, particularly during the busiest pilgrimage seasons when tourist numbers rise dramatically. While preserving the destination's spiritual environment, improving these amenities can help improve tourist comfort, hygiene, and safety.

Overall Visitor Experience

People were asked to rate their experience at Sabarimala on a scale of 1 to 5. Most of them gave a 4- and 5-star rating, meaning people were satisfied with their experience. There were also mixed reviews, indicating that people are expecting more, and there is room for improvement.

Table 7. Overall Experience Rating

Rating	Number of people
1	5
2	5
3	19
4	21
5	16

The overall experience evaluations indicate that most pilgrims are still happy with their pilgrimage experience in spite of management and infrastructure-related issues. This could be due to the fact that for many tourists, practical inconveniences are outweighed by spiritual fulfillment and emotional affinity to the hallowed location. Nonetheless, a number of respondents' moderate ratings suggest that there is still much room for improvement in terms of visitor services and management procedures. Without sacrificing the pilgrimage's cultural and religious significance, overall satisfaction levels can be raised by improving infrastructure and resolving guest problems.

Suggestions for facility Improvement

Visitors were asked to provide suggestions on how to improve the experience at Sabarimala. Improving the restroom facilities and parking facilities were largely suggested. People also showed their dissatisfaction with crowd management and have suggested separate ques to be provided for elderly people. The suggestion for better accommodation facilities was also put forward indicating immediate attention in that area. Some people also suggested to handover the management to private agencies implicating people are not satisfied with how the management works now.

The recommendations made by the respondents emphasize how urgently Sabarimala needs visitor-centered management techniques. Among the most popular suggestions were stronger crowd control techniques, enhanced sanitation systems, and special accommodations for senior pilgrims. These recommendations show that pilgrims anticipate increased ease and comfort while maintaining the pilgrimage's spiritual authenticity. The results also show that handling the growing number of pilgrims during pilgrimage seasons requires effective administrative coordination and infrastructure planning.

Suggestions for Sustainability

Visitors also provided insights on what sustainable practices should be adopted in Sabarimala while maintaining its religious significance. Need for Proper waste management and eco-friendly accommodation were frequently mentioned suggesting the need for sustainable infrastructure. People also suggested adopting renewable energy sources and proper water conservation methods. It should be noted that many respondents didn't understand the meaning of sustainability implicating the need for spreading awareness among the visitors of Sabarimala.

The recommendations pertaining to sustainability show that pilgrims are becoming more conscious of environmental preservation and ethical travel. Among the respondents' top concerns were waste management, the use of renewable energy, water conservation, and eco-friendly lodging. Nonetheless, a number of tourists' poor comprehension of sustainability points to the necessity of awareness campaigns and educational programs. Through information boards, awareness campaigns, and ecologically conscious management systems, temple officials and tourist stakeholders can significantly contribute to the promotion of sustainable pilgrimage practices. By taking such steps, future generations will have a safer and more sustainable pilgrimage experience while also contributing to the preservation of Sabarimala's ecological and cultural significance.

CONCLUSION

The findings from the study reveal that most of the visitors are male and belong to the age group of 40 and above. The majority of respondents perceive their visit as a religious one. A large group of respondents uses private vehicles as a mode of transportation, highlighting the need for huge parking facilities. Even though most of the respondents gave positive feedback about their experience at Sabarimala, there were also people who weren't satisfied with their experience, indicating room for improvement.

It can be understood from the responses given by the visitors that most of them see their journey as a religious one. While the majority of people have given positive responses about their experience at Sabarimala, there were also people who were not satisfied with their experience, implying room for improvement, particularly in areas like accommodation, restroom facilities, crowd management and parking facilities. Addressing these issues will help in enhancing the visitor experience at Sabarimala, and by implementing the suggestions given by the visitors, the beauty of the destination can be maintained. By focusing on important areas like infrastructure, sustainability practices and crowd management, the management can ensure that Sabarimala remains a cherished destination for the visitors and also preserves its significance for future generations to come.

The report also emphasizes that for devotees, the pilgrimage to Sabarimala is an emotional, spiritual, and cultural experience in addition to a physical one. Pilgrims' reasons are closely linked to their faith, dedication, vow fulfillment, and involvement in local customs related to Lord Ayyappa. The results show that pilgrims continue to have a strong emotional attachment to and devotion to the destination in spite of a number of management and infrastructure-related issues. This illustrates Sabarimala's ongoing spiritual significance in India's religious landscape.

The study also shows that, without sacrificing the pilgrimage destination's religious atmosphere, contemporary pilgrims are demanding better facilities and well-organized administration systems. The total pilgrimage experience and tourist satisfaction are greatly impacted by amenities including lodging, cleanliness, transportation, and crowd control. For Sabarimala to flourish sustainably, it is now crucial to strike a balance between spiritual authenticity and visitor ease. The destination's sacredness and environmental sustainability can be preserved by implementing eco-friendly infrastructure, better waste management systems, better healthcare facilities, and efficient crowd control techniques.

The report also highlights how crucial it is to educate pilgrims about environmentally friendly travel and sustainable tourism. Uncontrolled visitor intake and poor waste management might have a detrimental effect on the natural environment because Sabarimala is situated within a vulnerable ecological region. Therefore, government organizations, local communities, pilgrims, and temple officials should all actively participate in sustainable pilgrimage management. The cultural, spiritual, and ecological value of Sabarimala can be preserved for future generations through such cooperative efforts.

Overall, by analyzing tourist motivations, experiences, and views within the framework of indigenous cultural traditions, the study advances knowledge of pilgrimage tourism. The results can help legislators, tourist planners, and temple administrators create visitor-focused, sustainable management plans that improve pilgrimage experiences while protecting Sabarimala's religious and cultural character.

IMPLICATIONS

The current study is conducted at a time when the communities across the globe are divided and concerned for the survival of the indigenous practices and cultural values. The study provides information about the indigenous culture and practices of people in India surrounding the fame of the Sabarimala religious and pilgrimage destination. The study not only provides information about the existing situation at Sabarimala in India but also paves the way for other indigenous pilgrimage destinations across the world to meet the expectations of the visitors, along with higher religious fulfilment. The study suggested the measures for improvement at Sabarimala by highlighting the status and expectations regarding visitor facilitation. This can also be implemented at other indigenous pilgrimage destinations.

Policymakers, temple authorities, tourism bureaus, and local communities interested in pilgrimage destination management will find the study to have significant practical ramifications. The results can assist stakeholders in creating visitor management strategies that enhance satisfaction while maintaining the destination's religious holiness by recognizing pilgrims' expectations and motives. Authorities can prioritize infrastructure upgrades and budget allocation by knowing how visitors feel about lodging, sanitation, transportation, and crowd control.

The findings also add to the expanding corpus of research on religious tourism and indigenous tourism. The study demonstrates how pilgrimage customs and visitor behavior are still influenced by indigenous cultural traditions in modern society. This knowledge can aid academics and researchers in investigating comparable pilgrimage sites where spirituality, cultural heritage, and tourism coexist. For future studies on pilgrimage motivation, tourist experience, sustainable tourism, and indigenous cultural preservation, the study is therefore a valuable resource.

The study has significant implications for sustainable tourist growth as well. The recommendations made by responders about eco-friendly infrastructure, water conservation, trash management, and renewable energy use show that pilgrims are becoming more conscious of environmental sustainability.

These results can motivate tourism administrators and planners to implement eco-friendly policies that reduce environmental harm while enhancing visitor experiences. The sustainable management techniques used at Sabarimala can also operate as a template for other pilgrimage sites dealing with comparable management and environmental issues.

The study also emphasizes the necessity of educating pilgrims about sustainability and responsible travel practices through awareness campaigns. The long-term preservation of the destination's natural and spiritual surroundings can be aided by educating tourists about cultural sensitivity, garbage disposal techniques, and environmental conservation. While guaranteeing the preservation of indigenous cultural customs, such programs can improve the connection between pilgrimage tourism and sustainable development.

LIMITATIONS

This study, Investigating Tourist Motivations and Experiences at Sabarimala, faces few limitations that may affect the findings of the study. Firstly, the sample size of the study was small, which may not have covered the diverse group of visitors. The low number of respondents limits the ability to generalise the result for the large group of visitors.

The respondents were mainly from Kerala, potentially introducing a regional bias. Since Sabarimala attracts visitors across India, the experience and motivation of visitors from other regions may vary. Being unable to represent visitors from other regions may have affected the findings. The study uses a convenience sampling method to collect data, which may have led to underrepresentation of some groups and overrepresentation of others.

The shortage of previously conducted studies on motivations and experiences of tourists at Sabarimala has made it difficult to compare the findings of this study. The lack of existing research in this area makes it difficult to validate the study's results against established benchmarks or identify trends over time.

Time and resource limitations also hindered the study's capacity to gather responses from a wider range of pilgrims over various pilgrimage seasons. Furthermore, the study mostly used a quantitative technique, which might not adequately examine the pilgrimage experience's underlying emotional, cultural, and spiritual aspects. Notwithstanding these drawbacks, the study offers insightful information about the expectations, experiences, and motives of pilgrims traveling to Sabarimala. The results add to our knowledge of pilgrimage tourism in indigenous cultural contexts and lay the groundwork for future studies on visitor happiness, sustainable pilgrimage management, and cultural heritage preservation.

FUTURE RESEARCH DIRECTIONS

Future studies on Sabarimala pilgrimage tourism can expand the scope of research by including a larger and more diverse sample of respondents from different regions of India and abroad. Since the present study primarily focuses on a limited group of pilgrims, future research may explore regional differences in visitor motivations, expectations, and experiences. Comparative studies between domestic and international pilgrims can also provide deeper insights into the changing nature of pilgrimage tourism in contemporary society.

Future researchers may adopt qualitative or mixed-method approaches to gain a more detailed understanding of pilgrims' emotional, spiritual, and cultural experiences. Methods such as in-depth interviews, focus group discussions, ethnographic observation, and case studies can help in exploring the symbolic meanings and personal transformations associated with the pilgrimage journey. Such approaches may provide richer insights into the relationship between spirituality, indigenous traditions, and tourism behaviour. Another important direction for future research involves studying the impact of digital technology and social media on pilgrimage tourism. Social media platforms increasingly influence travel decisions, destination image, and sharing of pilgrimage experiences. Future studies may examine how digital communication shapes pilgrims' expectations, perceptions, and cultural engagement with destinations like Sabarimala.

Further research can also focus on sustainability and environmental management at pilgrimage destinations. Since Sabarimala is located within an ecologically sensitive area, future studies may examine the environmental impact of mass pilgrimages, waste generation, water consumption, and ecological degradation. Researchers may additionally evaluate the effectiveness of sustainable practices such as renewable energy use, eco-friendly infrastructure, waste management systems, and awareness programmes for pilgrims. Comparative research between Sabarimala and other indigenous pilgrimage destinations in India and across the world may also contribute significantly to the literature on religious tourism and indigenous cultural preservation. Such studies can help identify best practices in visitor management, infrastructure development, and sustainability initiatives while preserving the spiritual and cultural authenticity of pilgrimage destinations.

Future studies may also investigate the economic and socio-cultural impacts of pilgrimage tourism on local communities surrounding Sabarimala. Understanding how pilgrimage tourism influences employment generation, local businesses, cultural preservation, and community development can provide valuable insights for policymakers and tourism planners aiming to promote sustainable and inclusive tourism development.

REFERENCES

Abad-Galzacorta, M., Guereño-Omil, B., Makua, A., Iriberri, J. L., & Santomà, R. (2016). Pilgrimage as tourism experience: A preliminary approach to profiling pilgrims on the Ignatian Way. *International Journal of Religious Tourism and Pilgrimage*, *4*(4), 48–66.

Bhat, I. A. (2015). Pilgrimage Tourism in Sindh Valley (Kashmir): A Case Study of Kheer Bhawani Shrine. *South Asian Journal of Tourism and Heritage*. https://www.sajth.com/journal/wp-content/uploads/2015/09/Pilgrimage-Tourism-in-Sindh-Valley-Kashmir-A-Case-Study-of-Kheer.pdf

Chand, M. (2010). A Cross-National Study of Motivational Determinants among Non-Resident Indian Visitors to Religious Centers in India. *International Journal of Hospitality & Tourism Administration*, *11*(1), 22–38. DOI: 10.1080/15256480903539412

Collins-Kreiner, N. (2010). Geographers and pilgrimages: Changing concepts in pilgrimage tourism research. *Tijdschrift voor Economische en Sociale Geografie*, *101*(4), 437–448. DOI: 10.1111/j.1467-9663.2009.00561.x

Collins-Kreiner, N. (2020). Pilgrimage tourism-past, present and future rejuvenation: A perspective article. *Tourism Review*, *75*(1), 145–148. DOI: 10.1108/TR-04-2019-0130

Dash, K. K. (2024). Strategic CSR perspectives hold key to sustainable tourism development in the Society 5.0 Perspectivas estratégicas de RSC são fundamentais para o desenvolvimento do turismo sustentável na Sociedade 5.0 Las perspectivas estratégicas de RSE son clave para el desarrollo del. *Brazilian Journal of Development*, *10*(3), 01-27.

Hole, Y., Khedkar, E. B., & Pawar, S. (2019). The significance of pilgrimage tourism to sustainable development with special reference to the Indian context. *African Journal of Hospitality, Tourism and Leisure*, *8*(3), 1–12.

Jalal, F. N., & Kumar, M. G. S. (2013). Water quality assessment of Pamba River of Kerala, India in relation to Pilgrimage season. *International Journal of Research*, *3*(1), 341–347. https://www.cabdirect.org/cabdirect/abstract/20133091512

Jyotsna, J. H., & Sai, L. P. (2022). Modelling pilgrim-tourist experience in Hindu religious destinations: An Interactive Qualitative Analysis. *Journal of Tourism and Cultural Change*, *21*(3), 364–382. DOI: 10.1080/14766825.2022.2095914

Kasim, A. (2013). Balancing tourism and religious experience: Understanding devotees' perspectives on Thaipusam in Batu Caves, Selangor, Malaysia. In *Event Tourism and Cultural Tourism* (pp. 203–218). Routledge.

Khan, M. R., Khan, H. U. R., Lim, C. K., Tan, K. L., & Ahmed, M. F. (2021). Sustainable tourism policy, destination management and sustainable tourism development: A moderated-mediation model. *Sustainability (Basel)*, *13*(21), 12156. DOI: 10.3390/su132112156

Muraleedharan, K. P. (2009). *Economic benefits of pilgrimage tourism: a case study of Sabarimala pilgrimage with special reference to Pandalam rural locality in Kerala.*

Nayar, K. R., Koya, S. F., Mohandas, K., Nair, S. S., Chitra, G. A., Abraham, M., & Lordson, J. (2020). Public health implications of Sabarimala mass gathering in India: A multi-dimensional analysis. *Travel Medicine and Infectious Disease*, *37*, 101783. DOI: 10.1016/j.tmaid.2020.101783 PMID: 32561393

Olsen, D. H. (2014). A scalar comparison of motivations and expectations of experience within the religious tourism market. *International Journal of Religious Tourism and Pilgrimage*, *1*(1), 5.

Osella, F., & Osella, C. (2003). 'Ayyappan saranam':1 masculinity and the Sabarimala pilgrimage in Kerala. *Journal of the Royal Anthropological Institute*, *9*(4), 729–754. DOI: 10.1111/j.1467-9655.2003.00171.x

P, J. S. (2020). Social ecology of the Pamba riverine at Sabarimala, the mass pilgrim centre of Kerala. *Journal of Emerging Technologies and Innovative Research, 7*(7), 1253–1257. https://www.jetir.org/papers/JETIR2007463.pdf

P, J. S. (2020b). Social ecology of the Pamba riverine at Sabarimala, the mass pilgrim centre of Kerala. *Journal of Emerging Technologies and Innovative Research, 7*(7), 1253–1257. https://www.jetir.org/papers/JETIR2007463.pdf

Pai, G. V. (2016). Producing heritage: Culture as commodity in Madurai. *International Journal of Tourism Anthropology*, *5*(1-2), 47–70.

Sekar, R. (1992). *The Sabarimalai Pilgrimage and Ayyappan Cultus*. Motilal Banarsidass.

Sharpley, R., & Sundaram, P. (2005). Tourism: a sacred journey? The case of ashram tourism, India. *International Journal of Tourism Research/˜the œInternational Journal of Tourism Research, 7*(3), 161–171. https://doi.org/DOI: 10.1002/jtr.522

Shinde, K. N. (2007). Pilgrimage and the environment: Challenges in a pilgrimage centre. *Current Issues in Tourism*, *10*(4), 343–365. DOI: 10.2167/cit259.0

Sivakami, V., Bindu, V. T., & George, B. (2023). Impact management and experience design for sustainable development of ecotourism destinations: The case of Eravikulam National Park, India. *Journal of Ecotourism*, *22*(1), 80–102. DOI: 10.1080/14724049.2022.2042542

Tendero, E. J. (2023). Pilgrimage Tourism: A case study of Fort Pilar Shrine in Zamboanga City, Philippines. *Social Science Research Network*. DOI: 10.2139/ssrn.4518755

Vadakkiniyil, D. (2019). Mahishi's rage: Communitas and protest at Sabarimala, Kerala. *Anthropology Today*, *35*(5), 16–20. DOI: 10.1111/1467-8322.12529

van Iwaarden, M., & Nawijn, J. (2024). Eudaimonic benefits of tourism: The pilgrimage experience. *Tourism Recreation Research*, *49*(1), 37–47. DOI: 10.1080/02508281.2021.1986777

Vijayanand, S. (2014). The Issues and Perspectives of Pilgrimage Tourism Development in Tanjur. *Int. J. Tour. Hosp. Rev*, *1*, 45–51.

Wiltshier, P. (2020). Health and welfare at the boundaries: Community development through tourism. *Journal of Tourism Futures*, *6*(2), 153–164. DOI: 10.1108/JTF-05-2018-0021

Yoo, C. K., Yoon, D., & Park, E. (2018). Tourist motivation: An integral approach to destination choices. *Tourism Review*, *73*(2), 169–185. DOI: 10.1108/TR-04-2017-0085

ADDITIONAL READING

Mishra, R. (2024). *Religious Tourism and Pilgrimage in India*. Bharti Publications.

Osella, F., & Osella, C. (2003). 'Ayyappan Saranam': Masculinity and the Sabarimala pilgrimage in Kerala. *Journal of the Royal Anthropological Institute*, *9*(4), 729–754. DOI: 10.1111/j.1467-9655.2003.00171.x

Ramkumar, P. (2022). *Lord Ayyappa of Sabarimala: Ayyappa Charitam*. Garuda Prakashan.

Ravindran, K. (2019). *Sabarimala: All You Wanted to Know: the Legend and the Pilgrimage*. Notion Press.

Sekar, R. (1992). *The Sabarimalai pilgrimage and Ayyappan cultus*. Motilal Banarsidass Publishing House (Mlbd).

Shinde, K. A. (2007). Case study 6: Visiting sacred sites in India: religious tourism or pilgrimage? In *CABI eBooks* (pp. 184–197). DOI: 10.1079/9781845932251.0184

Timothy, D. J., & Olsen, D. H. (Eds.). (2006). *Tourism, religion and spiritual journeys* (Vol. 4). Routledge. DOI: 10.4324/9780203001073

KEY TERMS AND DEFINITIONS

Crowd Management: The planning and control of visitor movement to ensure safety, comfort, and efficient organization at tourist or pilgrimage sites.

Cultural Heritage: The traditions, rituals, beliefs, monuments, and practices inherited from previous generations.

Cultural Preservation: The protection and continuation of traditional customs, values, and heritage for future generations.

Destination Management: The planning, development, and administration of tourism destinations to improve visitor experience and sustainability.

Eco-friendly Infrastructure: Facilities and development practices designed to minimise environmental damage and promote sustainability.

Indigenous Cultural Traditions: Customs, beliefs, rituals, and practices that are traditionally followed by native or local communities over generations.

Mass Pilgrimage: Large-scale movement of pilgrims to a sacred destination during religious seasons or festivals.

Pilgrimage Experience: The emotional, spiritual, cultural, and physical experiences encountered by pilgrims during their journey.

Pilgrimage Tourism: Travel undertaken primarily for religious, spiritual, or sacred purposes.

Religious Rituals: Sacred practices and ceremonies performed according to religious beliefs and traditions.

Religious Tourism: Tourism associated with visits to religious destinations, festivals, and sacred places for spiritual or cultural reasons.

Spiritual Fulfilment: A feeling of inner peace, devotion, or religious satisfaction achieved through spiritual activities or pilgrimage.

Sustainable Tourism: Tourism practices that balance environmental conservation, cultural preservation, and economic development for future generations.

Visitor Motivation: The internal or external factors that influence an individual's decision to visit a destination.

Visitor Satisfaction: The level of contentment experienced by visitors based on their expectations and actual experiences.

Chapter 18

Sustainable Tourism Development Through Religious Festivals: The Case of Chhath Puja in Bihar

Captain Nihal
http://orcid.org/0009-0007-6037-6447
Manipal Academy of Higher Education, India

Partho Pratim Seal
http://orcid.org/0000-0001-6867-8713
Manipal Academy of Higher Education, India

ABSTRACT

This chapter investigates the dynamic relationship between religious festivals and sustainable tourism development, with a focus on Chhath Puja in Bihar, India. Core practices of the festival centered on environmental sacredness, utilization of biodegradable offerings, and widespread community participation are examined for their capacity to foster sustainable tourism. The transformation of sacred water bodies into ritualistic spaces has been studied, highlighting how substantial economic benefits are generated for local artisans, contributing to regional socioeconomic growth while preserving cultural identity. Issues such as managing visitor numbers, addressing infrastructure deficits, disposing of waste, and preventing the commercialization of sacred traditions are also recognized. The findings reveal that Chhath Puja's foundational values and principles resonate with internationally recognized sustainability frameworks, thereby enabling socioeconomic prosperity, strengthening social cohesion, and promoting ecological conservation.

INTRODUCTION

Religious festivals consist of intangible cultural heritage across the world, working as storage of indigenous knowledge systems that have evolved over hundreds of years. These traditional events are not just done for spiritual purposes, but they also showcase complex socio-cultural phenomena that in-

DOI: 10.4018/979-8-2600-2890-2.ch018

clude environmental ethics, community cohesion, economic transactions, and transfer of knowledge to different generations (Singh, 1998). In modern discussions about sustainable development, especially in the tourism industry, there is a growing acknowledgment that the age-old customs found in religious festivals can offer important lessons for balancing economic advancement, environmental protection, and social fairness (Emmanuel, 2014).

The 21st century has brought about significant challenges, including environmental degradation, cultural identity loss, and socioeconomic inequalities. According to the United Nations World Tourism Organization (UNWTO), sustainable tourism should "take full account of its current and future economic, social and environmental impacts, addressing the needs of visitors, the industry, the environment and host communities" (UNWTO, 2018). In this context, religious festivals offer distinct opportunities to showcase how indigenous knowledge systems can meaningfully contribute to sustainable development goals while preserving cultural authenticity and spiritual importance. Chhath Puja, primarily celebrated in Bihar, Jharkhand, Eastern Uttar Pradesh, and parts of Nepal, is one of the most environmentally conscious religious festivals in the Indian subcontinent. This four-day event, dedicated to Surya (the Sun God) and Chhathi Maiya (the goddess of dawn), illustrates how ancient spiritual traditions integrate principles of environmental care, community unity, and sustainable resource use (Badikilaya, 2019). Unlike many modern festival celebrations that have increasingly embraced commercialized and environmentally damaging practices, Chhath Puja retains its fundamental ethos of natural worship using biodegradable materials, prioritising water purity, and promoting collective responsibility.

Significance of the Study

The choice of Chhath Puja as a case study for exploring sustainable tourism development through religious festivals is well-founded for several reasons.

Firstly, Bihar, known for its rich cultural heritage yet facing economic challenges, offers a unique setting to analyze how religious tourism can aid regional growth. Recent estimates suggest that Chhath Puja generates over 50,000 crore (around $6 billion USD) in economic activity annually across India, with Bihar alone contributing 15,000 crore (Organiser, 2025). This significant economic influence, along with the festival's inherent focus on environmental sustainability, makes Chhath Puja an ideal example of sustainable religious tourism.

Secondly, the festival's practices align closely with the triple bottom line framework of sustainability, which emphasizes economic viability, social equity, and environmental integrity (Tyrrell et al., 2013). Approximately 80% of the materials used during Chhath Puja are biodegradable, such as bamboo baskets, clay lamps, banana leaves, and natural offerings like fruits, vegetables, and traditional sweets (Agarwal, 2024). This environmentally friendly approach, deeply embedded in indigenous knowledge, provides practical insights for modern tourism development strategies aimed at reducing environmental impact.

Thirdly, Chhath Puja's focus on community involvement and social inclusivity offers valuable lessons on how festivals can enhance social cohesion while supporting tourism development. The festival transcends caste hierarchies, socioeconomic divisions, and gender barriers, with women taking on key roles as primary practitioners (*vratins*) and entire communities working together in preparation, celebration, and post-festival cleanup activities (Ensureias, 2025). The chapter addresses the following objectives

- To examine the historical, cultural, and spiritual foundations of Chhath Puja and its evolution as a significant religious festival in Bihar

- To analyze the environmental consciousness inherent in Chhath Puja practices and their alignment with sustainable tourism principles
- To assess the socioeconomic impacts of Chhath Puja on local artisans, vendors, and communities in Bihar
- To identify challenges and opportunities related to infrastructure, crowd management, waste disposal, and commercialization pressures
- To evaluate how Chhath Puja's indigenous knowledge systems contribute to internationally recognized sustainability frameworks
- To propose recommendations for enhancing sustainable tourism development through religious festivals while preserving cultural authenticity

This chapter undertakes an in-depth review of existing literature on sustainable tourism, religious festival tourism, and indigenous knowledge systems. It then explains the qualitative research methodology adopted for the study, detailing the case study approach, methods of data collection, and procedures used for data analysis. The findings section presents empirical insights into Chhath Puja, with particular attention to its environmental practices, economic contributions, social significance, and the infrastructural challenges associated with the festival. The discussion subsequently interprets these results by sitting them within relevant theoretical perspectives and broader debates on sustainability. The chapter concludes by highlighting the key takeaways and proposing practical recommendations for policymakers, tourism professionals, and local communities on how religious festivals can be harnessed for sustainable development without compromising their inherent spiritual and cultural essence.

REVIEW OF LITERATURE

The growth of sustainable tourism as a central framework in tourism scholarship and governance reflects mounting concerns regarding the ecological and sociocultural externalities associated with mass tourism development. The UNWTO defines sustainable tourism as tourism that "takes full account of its current and future economic, social and environmental impacts, addressing the needs of visitors, the industry, the environment and host communities" (UNWTO, 2018). This definition embodies the triple bottom line concept also known as the "three pillars of sustainability" which emphasizes the interdependence of economic viability, environmental integrity, and social equity (Dwyer & Edwards, 2013).

Tyrrell, Paris, and Biaett (2013) developed a quantified triple bottom line framework specifically for tourism assessment, arguing that sustainable tourism practices must demonstrably contribute to economic prosperity while minimizing environmental degradation and promoting social well-being. Their framework provides methodological tools for measuring tourism impacts across these three dimensions, enabling more rigorous evaluation of whether tourism development achieves sustainability goals or just employs sustainability expression.

Evolving perspectives on sustainability in tourism scholarship recognise that indigenous and traditional knowledge systems embody principles substantively aligned with contemporary sustainability goals (Mondal & Pandey, 2024). These practices, refined over generations in direct response to local environmental pressures and resource limitations, encode sophisticated understanding of ecological carrying capacity, resource seasonality, and community-based management approaches which contemporary sustainability science remains in nascent stages of theoretical development.

Religious Tourism and Festival Tourism

Religious tourism represents one of the oldest forms of travel, with pilgrimage routes and sacred sites attracting visitors for hundreds of years. Contemporary religious tourism encompasses not only traditional pilgrimage but also visits to religious festivals, heritage sites, and culturally significant locations (Pageh et al., 2022). The intersection of religion and tourism has generated considerable scholarly attention, with researchers examining motivations for religious travel, impacts on host communities, and challenges of maintaining sacred space authenticity in the face of commercialisation pressures.

Festival tourism specifically focuses on events that celebrate cultural, religious, or historical heritage through communal gatherings, performances, rituals, and celebrations (Ilieva et al., 2024). Festivals serve multiple functions: they reinforce cultural identity, transmit traditions across generations, strengthen community bonds, generate economic activity, and increasingly attract tourists seeking authentic cultural experiences. Richards and Leal Londoño (2022) identify "festival cities" as urban centers that have successfully leveraged their festival portfolios to enhance destination attractiveness, though they also highlight challenges including overtourism, commercialization, and seasonal economic fluctuations.

Research on religious festivals and sustainable development reveals mixed outcomes. Emmanuel (2014) examined the Eyo festival in Lagos, Nigeria, finding that while the festival attracted tourists and generated economic benefits, sustainability required careful management of visitor impacts, infrastructure development, and protection of authentic cultural practices. Nega (2019) studied the Ashendye festival in Ethiopia, identifying both challenges (infrastructure deficits, inadequate planning, environmental pressures) and opportunities (cultural preservation, economic multiplier effects, community engagement) in leveraging religious festivals for sustainable tourism development.

Indigenous Knowledge Systems and Environmental Conservation

Indigenous knowledge systems (IKS) comprise the understandings, skills, and philosophies developed by societies with long histories of interaction with their natural environments (Gope et al., 2017). These knowledge systems encompass resource management practices, agricultural techniques, medicinal plant usage, weather prediction methods, and spiritual beliefs that integrate environmental stewardship with cultural identity. The relationship between IKS and environmental conservation has gained prominence as researchers recognize that indigenous practices often demonstrate sustainable resource utilization strategies developed through generations of careful observation and adaptive management.

Several studies have documented how festivals and cultural practices (Seal, 2024) serve as mechanisms for preserving and transmitting indigenous environmental knowledge. Koike (2019) examined the Sumbanese cultural festival in Indonesia, demonstrating how the festival emphasizes indigenous and local knowledge (ILK) that promotes environmental preservation while contributing to Sustainable Development Goals. Skogvang (2021) analyzed Sámi festivals in Norway, showing how outdoor life activities during festivals develop cultural and environmental awareness among participants, particularly children who learn traditional skills and ecological understanding.

In the Indian context, Mondal and Pandey (2024) argued that indigenous festivals embody climate sustainability principles that deserve integration into global frameworks. They documented how festivals in tribal communities incorporate practices such as sacred grove protection, seasonal calendars aligned with agricultural cycles, and rituals that reinforce environmental responsibility. Kumar et al. (2025) studied indigenous knowledge systems in Jharkhand, highlighting how cultural narratives, festivals like Sarhul and Karma, and traditional songs create collective ecological consciousness that shapes community resource management decisions.

Chhath Puja an Overview

Academic research specifically examining Chhath Puja remains relatively limited compared to the festival's cultural significance and scale. Singh et al. (2016) conducted pioneering work on "The role of a traditional festival, Chhath Puja, in the conservation and sustainable use of tropical fruits," documenting how the festival promotes on-farm conservation of fruit species through ritual requirements for specific offerings. Their research demonstrated that religious observances create market demand that incentivizes farmers to maintain traditional crop varieties that might otherwise be abandoned in favor of commercial cultivars. The Chhath is engrained in the seasonal and sacred lunisolar cycle with cultural observations that create environmental identity supporting sustainability (Mukil et al.,2024). Chhath is now considered to be a powerful symbol of cultural retention by connecting migrant communities to their roots. By this Chhath functions not only as a spiritual practice but also as a culturally sustainable institution grounded to community life and ecological wisdom. The practices are aligned with the doctrine of Panchabhūta where there is interdependence on earth, water, fire, air, and space, which are ideal for Indian cosmology and ecological thinking. This makes Chhath an integration of the Indian traditional knowledge system and global sustainability, which is articulated in frameworks like the United Nations Sustainable Development Goals (SDGs).

Badikilaya (2019) provided a comprehensive examination of Chhath Puja from religious and cultural tourism perspectives, emphasizing its eco-friendly nature and potential for tourism promotion. Her work documented the festival's rituals, offerings, mythology, and community practices, arguing that Chhath Puja represents "the Kumbh Mela of Bihar" in terms of religious fervor and economic activity. She highlighted how the festival's simplicity worship without priest intermediation, use of natural materials, emphasis on cleanliness aligns with contemporary environmental consciousness.

Recent scholarship has expanded understanding of Chhath Puja's multidimensional impacts. Pandey and Bhart (2024) analyzed Bihar's wetlands in the context of Chhath Puja, noting that the festival brings attention to water body conservation and creates recreational and tourism opportunities. Kumar (2025) examined the tourism potential of Chhath Puja, advocating for responsible tourism practices that preserve the festival's authentic character while accommodating visitor interest.

Khadka et al. (2025) conducted a case study of Chhath in Nepal's Terai region, identifying the festival as a "culturally rich practice" that enhances religious and eco-tourism while maintaining strong connections to nature worship traditions. Their research emphasized that Chhath epitomizes a "culture of respecting and worshiping nature," offering lessons for contemporary environmental movements.

Challenges in Festival Tourism Management

Despite the opportunities religious festivals present for sustainable tourism development, do come across significant challenges which have been documented across diverse contexts. Crowd management emerges as a critical concern, particularly for large-scale religious gatherings. Kanaujiya and Tiwari (2022) examined crowd management strategies during the Prayagraj Kumbh Mela, India's largest religious gathering, highlighting difficulties in predicting crowd sizes, ensuring safety, managing traffic flows, and providing adequate infrastructure for millions of participants.

Infrastructure deficits represent another major challenge, especially in developing regions where festival locations may lack adequate transportation, accommodation, sanitation facilities, and emergency services. Nega and Tabor (2019) found that the Ashendye festival in Ethiopia suffered from inadequate infrastructure planning, limiting its potential for sustainable tourism development despite strong cultural appeal.

Waste management and environmental degradation concerns have intensified as festivals scale up and attract larger crowds. Studies of the Yamuna River during Chhath Puja in Delhi have documented severe water pollution challenges, with devotees performing rituals in water contaminated by untreated sewage and industrial effluents. While the festival's traditional practices are eco-friendly, concentration of millions of participants at limited water body sites creates waste disposal challenges that overwhelm municipal systems.

Commercialization pressures threaten to transform authentic religious festivals into commodified tourist spectacles, potentially eroding their spiritual significance and cultural authenticity. Richards and Leal Londoño (2022) noted that successful festival cities must navigate the tension between economic exploitation of festival assets and preservation of festival integrity as meaningful cultural expressions for host communities.

Gaps in Existing Literature

Despite growing research on religious tourism and sustainable development, several gaps remain. First, limited scholarship has examined how indigenous knowledge systems embedded in religious festivals can inform contemporary sustainability frameworks. While individual studies document eco-friendly practices, comprehensive analyses linking festival traditions to internationally recognized sustainability principles (such as the UN Sustainable Development Goals) remain scarce.

Second, methodologically, much festival tourism research relies on quantitative approaches emphasizing economic impact measurement, with fewer in-depth qualitative studies exploring community perspectives, cultural meanings, and social transformation processes. Phillimore and Goodson (2004) argued that qualitative research methods are particularly valuable for tourism studies seeking to understand cultural phenomena, lived experiences, and interpretive meanings, precisely the dimensions most relevant for understanding religious festivals.

Third, while challenges such as infrastructure deficits and waste management have been identified, limited research has proposed integrated solutions that balance tourism development goals with cultural preservation and environmental protection imperatives. This chapter addresses these gaps by employing qualitative case study methodology to examine Chhath Puja comprehensively, analyzing how its indigenous knowledge foundations contribute to sustainable tourism while identifying practical strategies for addressing implementation challenges.

METHODOLOGY

The research design in the study considers Chhath Puja as a primary unit of analysis while incorporating multiple embedded sub-units. These sub units are Ghat locations along the Ganges, artisan communities producing festival materials, and governing bodies managing festival infrastructure. The chapter adopts a qualitative case study method to observe sustainable tourism development through Chaat Puja in Bihar. The case study approach is best suited for understanding the well-suited and complex phenomena within real-life contexts, allowing for in-depth exploration of how multiple factors interact to achieve the necessary outcome (Yin, 2018). Researchers Beeton (2005) recommend the case study methodology for tourism research as it provides contextual understanding, which a quantitative approach may not state. The current study adopted a descriptive- explanatory case study design, which provided a detailed description of Chhath Puja's characteristics and practices and described how the festival contributes towards sustainable tourism development. The study aligns with the research study of Tasciet al. (2020), which recommends a case study research in tourism which goes beyond the analytical insights and theoretical contribution.The case study method represents a comprehensive research strategy which helps to examine a complex social phenomenon in a real- world context. This methodology is adopted to investigate the relationship between Chhath puja and sustainable tourism development in Bihar. The case study approach allowed the researchers to explore how traditional festival practices intersect with contemporary sustainability concerns while maintaining the richness and complexity of the cultural phenomena, which is deeply embedded with specific geographical, cultural and temporal context. The data collection for the study includes multiple sources to ensure triangulation to enhance the validity. The analytical approach uses thematic analysis to identify recurring patterns across different data sources and examine how the themes of sustainability, tradition, community, and commerce intersect in the context of Chhath Puja. The comparative analysis across different sites helps to show how context influences sustainability. The economic analysis tracks the financial flow from tourists to local artisans to track the equitable distribution of the economic benefits. The stakeholders analysis considers all the diverse actors involved in Chhath Puja celebrations while examining various influencers and interactions. The case study method provides a strong framework for investigating how Chhath Puja symbolises the principle of sustainable tourism development. With the capacity of apprehension of complexity, preserving context and integrating various data sources and examining the phenomena holistically, this approach helps to generate rich insights, which allows the traditional religious practices to be aligned with the sustainability goals

Justification for Qualitative Approach

The study adopted a qualitative methodology of research since to understand religious festivals requires cultural meanings, spiritual motivation and community interpretation, the dimensions which could be best measured by qualitative tools (Phillimore & Goodson, 2004). The indigenous knowledge systems of the festivals are contained and often passed to the next generation through practice and observations, which require interpretive research methods to surface the tacit knowledge (Gope et al.,2017). Considering sustainability in tourism necessitates multiple perceptions that include economic, environmental, social, and cultural, as different stakeholders define sustainability differently. The qualitative approach includes the multi-dimensional approach of researchers to obtain the varied viewpoints (Davies, 2003). For further enhancing the understanding of how the ecological and spiritual values are experienced at the community level, there is a necessity that it is complemented by lived experiences. Chhath is observed with a deep tradition shaped by memory, emotion, and engagement with nature. Chhath also affirms the ontological perspectives which are embedded in Indian traditions that consider humans to be entwined with cosmos and natural elements. Sun worship in India has been common since prehistoric times, with archaeological findings stating that sun worship dates back to Neolithic age – in the representation of sun (Saran & Pandey, 1992). Singh (1992) states about worshipping the solar deity and Chhathi Mata (mother Goddess) during the Chhath puja festivities as a site of democratic and spiritual expression

Literature Review and Document Analysis

An extensive review of academic literature, government reports, tourism policy documents, and media coverage provides basic understanding of Chhath Puja's historical development, cultural significance, and contemporary manifestations.

The document considered for analysis includes: Academic publications on Chhath Puja, religious tourism, and sustainable development, Bihar Tourism Department publications, government reports on festival management, and media reports on festival management.o,o

Data Analysis

Analysis of existing research studies, statistical databases, and archival materials does provide archival empirical evidence regarding festival impacts on census data and demographic information for Bihar, economic data on festival- related business activities, and environmental monitoring reports. To evaluate Chaath contribution to sustainable tourism development, this study employs the UNWTO sustainable tourism framework (2018), which emphasize on the Economic Sustainability: Does the festival generate economic benefits that are distributed equitably across local communities? Are economic activities viable in the long term without depleting resources?

FINDINGS

Historical and Cultural Foundations of Chhath Puja

The origins of Chhath Puja are rooted in Hindu mythology and ancient spiritual traditions. Different narratives exist about the festival, each contributing towards the symbolic significance. According to the Ramayana, Sita had performed the Chhath vrata after she returned to Ayodhya after the exile. It is believed that the observance of the rituals resulted in the birth of twin sons, Luv and Kush (Badikilaya, 2019). As per the Brahmavaivarta Purana it offers a cosmological interpretation stating how the creation involves the division of Prakriti (nature) into multiple elements, with the sixth part designated as *Shasthi* or *Chhathi*. The deity is known as Devasena, said to be the giver and protector of children, providing a maternal dimension to the festival framework (Badikilaya,2019). For this association, the Chhati Maiya, the devotees take the *vrata* (fast) for seeking the blessings for offspring or expressing gratitude for children's health.

The philosophical foundation of Chhath Puja is from the Indian Knowledge System, the doctrine of Panchabhūta or the five elements: space, air, fire, water, and earth, which establish all existence. The elements mentioned invoke observations like bathing in rivers, offering locally grown produce, and fasting in alignment with the transition of the sun. These practices are not just expressions of spiritual devotion but also vehicles for ecological perception and sustainable living (Sandahya & Kiran, 2025). Chhath is believed to promote discipline, devotion, and environmental stewardship, which values both traditional philosophy and modern ecological ethics (Jha, 2024). Considering the rate of environmental degradation and the necessity for sustainability the study of Chhath provides a unique perspective on how traditional observance does contribute towards ecological sustainability and well-being of people.

Historical Development and Regional Identity

Historical evidence from the past suggests that sun worship traditions in India can be traced back to the Bronze Age, as seen in Neolithic pottery and rock paintings. The organized sun worship was held in the 4th-5th century CE during the Gupta period. The sun temples of Bihar in Nalanda do turn up as pilgrimage centres during Chhath Puja. The Chhath, as a festival, is integral to the regional cultural identity, with migrants celebrating the festival wherever they settle, thereby creating networks of cultural connection in India.

The evolution of the festival states about the broader social transformation, along with maintaining the core ritual elements. Chhath was modified into urban contexts with artificial ponds and prepared *ghats* (river banks) to access the sites.

Ritual Practices and Environmental Consciousness

The Chhath Puja follows a structured four-day ritual that considers principles of purification, discipline, gratitude, and community solidarity:

Day 1- Nahay Khay (Bathing and Eating): The festival begins with the ritual bathing in a sacred river or any water bodies that symbolizing physical and spiritual purification. The devotees then carry the holy water home to prepare for the offerings. A Satvik meal is prepared for the day, consisting of Rice, Chana Dal and Bottle Gourd, which is cooked without usage of onion and garlic. This dietary restriction ensures that the purity required for subsequent ritual is observed (Badikilaya, 2019).

Lohanda (Kharna): On the second day of the festival the devotees do observe a complete fast—including abstinence from water for the entire day. The fast is broken only after sunset with offerings of kheer prepared from jaggery, Roti (whole wheat bread), and bananas. The entire offerings are first presented to Chhathi Maiya and then distributed to family members and the neighbours thereby initiating community sharing practices. After this meal, devotees begin a 36-hour fast without water, the most rigorous component of the observance (Utsav India, 2024).

Day 3 -Sandhya Arghya (Offerings of Evening): The third day of the festival is considered to be the most spectacular and spiritual. The entire day is spent in preparing elaborate offerings *(thekua),* whole wheat cakes with jaggery, *gujiya*, and *khaja*. In the evening, the families come together near a water body for *arghya* (water offering) to the setting sun. The offerings of seasonal fruits are all arranged in *(soops)* winnows, mostly made from bamboo.. Five stalks of sugarcane are arranged in a manner to form a canopy that resembles *Pancha Bhuta* (five elements of earth, water, fire, air, and ether), below which earthen lamps are lit. Folk songs which are dedicated to *Chhath Maiya* and the sun are played along with floating lamps, which create a sparkle, creating a profound aesthetic and spiritual atmosphere (Badikilaya, 2019).

Day 4 - Usha Arghya (Morning Offerings): Before the sun rises on the 4th day the devotees go the water body to offer *arghya* to the rising sun, thereby completing the worship cycle. The offerings express gratitude towards the sun for life- sustaining energy and look for blessings for prosperity, health and family welfare.

Biodegradable Materials and Zero-Waste Philosophy

The Chhath puja differentiates itself by its environmental consciousness, with more than 80% of material used being biodegradable (Agarwal.2024). The eco-friendly approach states that the indigenous knowledge system has understood the sustainable resource before current environmentalism emerged

Bamboo products: Winnows (*soops*), baskets (*daura*) are crafted from bamboo, a renewable resource.

Clay Items: Earthenware lamps (diyas) and cooking vessels, which are made from clay, are said to be alternatives to plastic.

Natural Fibers: The cotton threads used in rituals, banana leaves for wrapping the offerings are biodegradable materials that are sourced locally.

Organic offerings: All food offerings were seasonal, including fruits, vegetables, and sweets, which are made from traditional recipes- without synthetic additives or non-biodegradable packaging.

Traditional methods of cooking: There is a preference towards the usage of earthen ovens and mango wood as fuel, which are energy efficient and have low emissions. As compared to other festivals, which require elaborate decorations Chhath is about simplicity. There are no idols, no plastic usage and no items which pollute the air and water.

Sacredness and Purity

For Chhath Puja, water has a central importance, considering it as a medium for ritual performance and also as a sacred element. The festival happens in Karthik month between October and November, when the monsoon floods have receded, and the water bodies have been replenished, highlighting the understanding of environmental cycles. Before the festival begins, the ghats, riverbanks, and water bodies are cleaned by the government and volunteers. The collective clean-up changes the sacred space into sacred precincts, which address the environmental maintenance needs. The importance of clean water for ritual performance leads towards conservation of water bodies throughout the year. However, in the current scenario of millions of devotees converging at a water body during the festival, do create a lot of challenges. Especially in urban areas like Delhi were rituals had to be performed in the polluted waters of the Yamuna river with sewage and industrial wastage. The public health and environmental concerns has lead to the temporary creation of artificial ponds in many areas for the Chhath.

Socioeconomic Impacts

The Chhath puja has transformed itself into a major economic phenomenon, which generates an INR 50,000 Crore (approximately $6 billion USD) in economic activity across India in 2025. The state of Bihar, the heartland of Chhath accounts for around 15,000 crore, contributing significantly towards the annual income through economic events.

Agriculture and Horticulture: The festival helps to create a market for seasonal produce, which includes sugarcane, bananas, sweet limes, sweet potatoes, coconuts, ginger and turmeric. This allows the farmers to nurture traditional varieties and maintain crop diversity.

Livelihoods of Artisans: The craftsmen find an experience in demand for their products, which include bamboo and cane weavers who produce thousands of winnows and baskets, the potters create countless earthen lamps, and brass and bronze workers supply ritual vessels.

Transportation and Hospitality: Chhath Puja leads to India's largest annual migration event, where millions of people who are working in various parts of the country return to their native villages in Bihar and Eastern Uttar Pradesh. The 'reverse migration' creates demand for various transport modes, and railways operate hundreds of special trains. The accommodation facilities in Bihar, as per the reports, have full occupancy during the festivals (Swarajya, 2025).

Social Cohesion and Community Participation

The Chhath Puja's social dimensions state the capacity to strengthen the community and foster collective identity:

Although most Hindu rituals are hierarchically structured with Brahmin priests who mediate between devotees and deities, Chhath is egalitarian as the devotees perform the rituals without any intermediation of a priest. The devotees from all castes background do participate as equals, all standing in water, sharing offerings, and singing folk sings together.

Gender Dynamics: Although women are primarily practitioners (vratins), Chhath involves the entire family to support them. The male family members help in carrying the heavy basket of offerings, in preparation of ghats, and participate in arghya rituals. The collectiveness creates space for family experiences and mutual support. The Chhath also positions women as central religious actors with spiritual discipline commanding both family and community respect.

Intergenerational Transmission: The festival helps to transfer the cultural knowledge from elders to youth. The children accompany their parents to the ghat, listen to the traditional songs, and participate in food preparation. Chhath allows intergenerational continuity with the younger generation connected with the ancestral cultural practices.

Community Solidarity: The collective ghat cleaning, distribution of food offerings (*prasad*). mutual assistance during rituals, coordinated celebration activities, strengthen the village solidarity. The festival creates a social capital network of trust and reciprocity.

Identity of Migrant community- For many Bihari individuals working in various metropolitan areas, Chhath is a powerful identity marker within the community. The migrant association do arrange ghat facilities, coordinates the food preparation and cultural programs. These create emotional bonding for the homeland and provide a mutual support network for an urban environment.

Tourism Development Potential

Chhath Puja presents us mostly with the untapped potential for religious and cultural tourism development. The aesthetic dimensions created by the festival include the use of thousands of lamps sparkling on water, devotees standing for worship, music performances, and offerings, creating a spectacle. International travellers seek an immersive Indian cultural experience in Chhath as compared to other festivals like Holi or Diwali (Badikilaya, 2019). The environmental connection and water body settings of the Chhath align with the ecotourism principles. Marketing the festival as a sustainable religious practice does attract environmentally conscious travelers interested in traditional knowledge and nature based spirituality. The connectivity to Bihar has been better with the introduction of special Chhath trains. The intrastate transportation during festival is challenging with improper roads, limited public transport, and traffic congestion during the festival periods.

Tourism Information and Services: Limited tourism information does exist specifically for Chhath. The interpretive materials explain the festival's significance for non -participating visitors.

Safety Issues

Managing large crowds at water bodies provides safety challenges for both evening and early morning rituals. Although there is an increase in police deployment during Chhath a comprehensive crowd management system, which incorporates technology for monitoring the emergency response system, remains underdeveloped (Kanaujiya & Tiwari, 2022).

DISCUSSION

Chhath distributes benefits across the society with the small vendors, artisans, farmers, transporters and service providers all participating in the festival economy, which aligns with sustainable tourism principles that emphasise equity along with community benefit (Tyrrell et al., 2013).

Limitations – The economic activity is concentrated in a brief period, leading toseasonal income stability. The artisans who are dependent upon Chhath-related sales may struggle during off season months. The study has its own limitations. As the study relies only on secondary resources rather than on primary studies, it thereby limits the depth of participants' observation. In the future, the studies could be complemented by ethnographic fieldwork with the respondents during the festival. As the current study considers only the historical scope during the festival, a longitudinal study of tracking the festival's evolution over time may provide additional insights into sustainability.

FUTURE RESEARCH DIRECTIONS

The chapter states that Chhath Puja provides a valuable model for understanding the relationship between religious festivals, indigenous knowledge systems, and sustainable tourism. There is a strong need for primary study that captures the lived experiences of devotees, artisans, local residents, tourists, and policymakers. Ethnographic observation, in-depth interviews, and focus group discussions conducted during the festival period will provide richer insights into how sustainability is practised, negotiated, and interpreted at the community level. Future studies may also adopt a longitudinal approach to examine how Chhath Puja is changing over time in response to urbanization, migration, digital media, commercialisation, and environmental pressures.

Future research should move toward a more systematic measurement of sustainability indicators linked to religious festivals. Building on the triple bottom line framework, researchers may develop context-sensitive indicators for economic inclusion, waste generation, water quality, crowd safety, gender participation, and cultural continuity. There is a significant need to investigate the relationship between festival tourism and environmental governance, especially regarding river conservation, wetland protection, waste disposal systems, and public sanitation. Since Chhath Puja depends heavily on access to clean water bodies, future studies could examine how the festival may be mobilized as a platform for citizen-led conservation campaigns and public environmental awareness. .

Additionally, more interdisciplinary research is needed to connect Chhath Puja with global sustainability frameworks, especially the UN Sustainable Development Goals. Future studies may explicitly map the festival's practices to SDGs relating to sustainable communities, responsible consumption, gender inclusion, cultural preservation, and environmental protection.

CONCLUSION

Chhath Puja depicts an ancient tradition with deep mythological roots, symbolising indigenous knowledge systems that have developed over various decades. The evolution of the festival does reflect both cultural continuity and adaptive responses of changing social context which maintains the core spiritual elements while including the diverse regional manifestations

Chhath discusses environmental sustainability through the use of biodegradable materials, minimal processing, local sourcing and water purity. With 80% of festival material used being biodegradable, it demonstrates that spiritual observance and ecological responsibility can be integrated. The traditional practices include ecological knowledge regarding seasonal cycles and resource management.

Tourism Development Potential: Chhath presents a considerable, largely untapped potential for religious, cultural and eco-tourism development. The festival's characteristics include rituals, aesthetic spectacles, songs and food, eco-conscious practices which provide distinctive experiences to the tourist segments.

Sustainability: Analysis by the triple bottom line framework explains strong alignment with sustainable tourism principles. The economic benefits are distributed inclusively with environmental practices illustrating sustainable resource utilization strengthening community cohesion and cultural identity.

Practical Implications

Chhath offers a model to develop that is economically beneficial, responsible, and culturally authentic. Rather than creating artificial attractions, the agencies should facilitate respectful engagement with living cultural traditions and ensure that the benefits reach the communities.

Final Reflections

Chhath Puja, as a case, demonstrates that sustainable tourism development through religious festivals can be reinforcing when it is approached thoughtfully. Environmental consciousness, economic impacts, and social cohesion state that traditional practices developed through the indigenous knowledge system provide valuable models for addressing sustainability challenges.

Infrastructure limitations, environmental degradation, and pressures of commercialisation all constrain the sustainable development possibilities. There is a demand for an integrated approach, which involves government investment, community engagement, and coordination of policy across different domains.

The path forward necessitates balancing multiple, sometimes competing objectives and maintaining spiritual authenticity, generating economic benefits while preventing exploitative commercialization; celebrating cultural heritage while preserving traditional practices

The greatest lesson from the Chhath is about its demonstration of sustainability, which is not a novel concept requiring intervention, but also about its revival and adaptation. The indigenous knowledge system worldwide should consider rediscovering and reintegrating principles of living within the ecosystem, maintaining community solidarity, and respecting natural elements.

In the 21st century, there are unprecedented environmental, social and cultural challenges with festivals like Chhath Puja offering both tourism and economic opportunities. It is a link between ancient wisdom and contemporary needs, along with a spiritual practice that demonstrates environmental ethics, and cultural heritage, which proves continuing relevance.

REFERENCES

Agarwal, (2024). Chhath Puja: Embracing Sustainability Through Ancient Rituals. LinkedIn. Retrieved from https://www.linkedin.com/pulse/chhath-puja-embracing-sustainability-through-ancient-rituals-agarwal-81jpe

Badikilaya, G. (2019). Chhath Puja - A Study in Religious and Cultural Tourism. *Atna Journal of Tourism Studies*, *14*(2), 71–76. DOI: 10.12727/ajts.22.5

Beeton, S. (2005). The case study in tourism research: A multi-method case study approach. In Ritchie, B. W., Burns, P., & Palmer, C. (Eds.), *Tourism research methods: Integrating theory with practice* (pp. 37–48). CAB International. DOI: 10.1079/9780851999968.0037

Davies, B. (2003). The role of quantitative and qualitative research in industrial studies of tourism. *International Journal of Tourism Research*, *5*(2), 97–111. DOI: 10.1002/jtr.425

Dwyer, L., & Edwards, D. (2013). Ecotourism and the triple bottom line. In Ballantyne, R., & Packer, J. (Eds.), *International handbook on ecotourism* (pp. 245–262). Edward Elgar Publishing. DOI: 10.4337/9780857939975.00027

Emmanuel, O. K. (2014). Religious tourism and sustainable development: A study of Eyo festival in Lagos, Nigeria. *International Journal of Social Sciences and Education*, *4*(2), 474–485.

Ensureias. (2025). Chhath Puja 2025: *Cultural Significance and Social Dimensions*. Retrieved from https://www.ensureias.com/blog/current-affairs/chhath-puja-2025-cultural-significance-and-social-dimensions

Gope, L., Behera, S. K., & Roy, R. (2017). Identification of indigenous knowledge components for sustainable development among the Santhal community. *American Journal of Educational Research*, *5*(8), 883–891. DOI: 10.12691/education-5-8-8

Ilieva, L., Bozhinova, M., Todorova, L., Marinov, M., Ismailov, T., & Spasova, S. (2024). Festivals: An Opportunity for Sustainable Development of Tourism Regions. *Revista de Gestão Social e Ambiental*, *18*(3), e05234. DOI: 10.24857/rgsa.v18n11-069

Jha, J. (2024). *Beneath the sun: Equality for everyone- The spirit of Chhath Festival* (1st ed.). Self published.

Kanaujiya, A. K., & Tiwari, V. (2022). Crowd management and strategies for security and surveillance during the large mass gathering events: *The Prayagraj Kumbh Mela 2019 Experience. National Academy Science Letters*, *45*(3), 295–298. DOI: 10.1007/s40009-022-01114-w PMID: 35431383

Koike, M. (2019). Indigenous and local knowledge promoting SDGs in Indonesia: The case of the Sumbanese cultural festival. *Journal of Environmental Science and Sustainable Development*, *2*(2), 171–191. DOI: 10.7454/jessd.v2i2.1034

Kumar, A., Kumar, A., & Kumari, S. (2025). Cultural ecology and conservation: Indigenous knowledge systems in Jharkhand. *Journal of Native India & Diversity Studies*, *1*(1), 45–58.

Mondal, A., & Pandey, M. S. (2024). Indigenous festivals and climate sustainability in India: A case study of cultural practices and performances. *Rupkatha Journal on Interdisciplinary Studies in Humanities, 16*(1), 1–15. DOI: 10.21659/rupkatha.v16n1.03

Mukil, M. V., Pruthvi, M. N., & Bhavani Rao, R. (2024, September). Design for sustainability: Exploring cultural attributes in an Indian village for nature connectedness— A grounded theory approach. In 2024 *IEEE International Symposium on Technology and Society* (ISTAS) (pp. 1–8). IEEE.

Nega, D. (2019). Challenges and opportunities of managing festival tourism for sustainable tourism development: The case of the Ashendye Festival, North Wollo, Ethiopia. *African Journal of Hospitality, Tourism and Leisure, 8*(4), 1–14.

Organiser. (2025). Chhath Puja: When the sun rises on India's economy. Retrieved from https://organiser.org/2025/11/04/324170/bharat/chhath-puja-when-the-sun-rises-on-indias-economy/

Pageh, I., Rahmawati, P. I., & Pramiani, N. M. (2022). Ancient religious sites as tools for sustainable tourism development: An empirical study in the North of Bali. *International Journal of Religious Tourism and Pilgrimage, 10*(1), 78–91.

Pandey, A., & Bhart, V. (2024). Wetlands in Bihar: A comprehensive analysis of extent, socio-economic significance, ecological benefits, and associated risks. *Journal of Community Positive Practices, 24*(2), 23–45. DOI: 10.35782/JCPP.2024.1.07

Phillimore, J., & Goodson, L. (2004). *Qualitative research in tourism: Ontologies, epistemologies and methodologies*. Routledge.

Richards, G., & Leal Londoño, M. P. (2022). Festival cities and tourism: Challenges and prospects. *Journal of Policy Research in Tourism, Leisure & Events, 16*(1), 107–129. DOI: 10.1080/19407963.2022.2087664

Saran, A. B., & Pandey, G. (1992). *Sun worship in India: A Study of Deo Sun-Shrine*. Northern Book Centre.

Seal, P. P. (2024). Food Culture in China Town: Tiretti Bazar of Kolkata. In *Cultural, Gastronomy, and Adventure Tourism Development* (pp. 176-187). IGI Global

Singh, A., Nath, V., Kumar, S., & Singh, B. S. (2016). The role of a traditional festival, Chhath Puja, in the conservation and sustainable use of tropical fruits. In Normah, M. N., Chin, H. P. M., & Reed, B. M. (Eds.), *Tropical fruit tree diversity: Good practices for in situ and ex situ conservation* (pp. 217–228). Bioversity International.

Singh, K. S. (Ed.). (1998). *People of India: Rajasthan* (Vol. 38). Popular Prakashan.

Skogvang, B. O. (2021). Development of cultural and environmental awareness through Sámi outdoor life at Sámi/Indigenous festivals. *Frontiers in Sports and Active Living, 3*, 662929. DOI: 10.3389/fspor.2021.662929 PMID: 34337399

Swarajya. (2025). Bihar: How Industry And IT Hubs Can Decongest Chhath Puja Special Trains. Retrieved from https://swarajyamag.com/states/bihar-how-industry-and-it-hubs-can-decongest-chhath-puja-special-trains

Tasci, A. D. A., Wei, W., & Milman, A. (2020). Uses and misuses of the case study method. *Annals of Tourism Research, 80*, 102846. DOI: 10.1016/j.annals.2019.102815

Tyrrell, T., Paris, C. M., & Biaett, V. (2013). A quantified triple bottom line for tourism: Experimental results. *Journal of Travel Research, 52*(3), 279–293. DOI: 10.1177/0047287512465963

UNWTO. (2018). *Tourism and Culture Synergies*. United Nations World Tourism Organization., DOI: 10.18111/9789284418978

Utsav India. (2024). Chhath Puja - Events & Festivals in India. Retrieved from https://utsav.gov.in/view-event/chhath-puja-2

Yin, R. K. (2018). *Case study research and applications: Design and methods* (6th ed.). SAGE Publications.

ADDITIONAL READING

Badikilaya, G. (2019). Chhath Puja - A Study in Religious and Cultural Tourism. *Atna Journal of Tourism Studies, 14*(2), 71–76. DOI: 10.12727/ajts.22.5

Dada, M., & Kumar, S. (2026). Digital Storytelling of Chhath Puja and the Mediatization of Indigenous Ecocultural narratives. *Journal Of Emerging Trends And Novel Research, 4*(4). Advance online publication. DOI: 10.56975/jetnr.v4i4.233646

Heidl, S. E. (2025). Expanding the spiritual festival space: The online event community of a Hungarian festival. *Social Compass, 72*(2), 221–239. DOI: 10.1177/00377686251336966

Phillimore, J., & Goodson, L. (2004). *Qualitative research in tourism.* Ontologies, Epistemologies and Methodologies. Routedge.

Richards, G., & Leal Londoño, M. P. (2022). *Festival cities and tourism: Challenges and prospects.*

Singh, A., Nath, V., Singh, S. K., Sthapit, B., & Reddy, B. M. C. (2016) The role of a traditional festival, Chhath Puja, in the conservation and sustainable use of tropical fruits. In: Sthapit, B. (et al. (eds.)) *Tropical fruit tree diversity: good practices for in situ and on-farm conservation.* Abingdon, Oxon (UK): Routledge, p.217-225

Slocum, S. L. (2017). *Triple bottom line.* The SAGE International Encyclopedia of Travel and Tourism., DOI: 10.4135/9781483368924.n487

KEY TERMS AND DEFINITIONS

Arghya: A ritual offering, usually of water and prayers, made to the setting and rising sun during Chhath Puja.

Biodegradable Materials: Natural materials that decompose without causing long-term pollution, such as bamboo, clay, leaves, cotton, and organic food offerings used in Chhath rituals.

Carrying Capacity: The ability of a place or ecosystem to accommodate visitors and activities without causing unacceptable environmental damage or social strain.

Chhath Puja: A four-day Hindu festival dedicated to Surya, the Sun God, and Chhathi Maiya, celebrated mainly in Bihar, Jharkhand, Eastern Uttar Pradesh, and parts of Nepal. It is marked by fasting, ritual purity, riverbank worship, and offerings to the setting and rising sun.

Community Cohesion: The sense of solidarity, mutual support, and shared identity built through collective participation in rituals, preparation, food sharing, and cleanup activities.

Cultural Authenticity: The preservation of the original meaning, values, and lived traditions of a cultural practice without reducing it to a staged or commercial spectacle.

Environmental Consciousness: Awareness of ecological responsibility expressed through values and practices that protect natural resources, reduce waste, and promote harmony with nature.

Festival Tourism: A form of tourism centered on cultural, religious, or heritage events that attract visitors while reinforcing local identity, tradition, and community participation.

Ghat: A riverbank or stepped embankment where devotees gather to perform ritual bathing and offer prayers during the festival.

Indigenous Knowledge Systems (IKS): Bodies of knowledge, belief, and practice developed by communities over generations through close interaction with local environments, often involving sustainable resource use, cultural memory, and ecological ethics.

Panchabhūta: The Indian philosophical doctrine of the five elements—earth, water, fire, air, and space—understood as the basis of existence and reflected in ritual life and ecological thinking.

Prasad: Sanctified food that is first offered to the deity and then distributed among family, neighbors, and the wider community as a blessing.

Religious Tourism: Travel motivated wholly or partly by religious, spiritual, or sacred purposes, including pilgrimage, festival participation, and visits to religious heritage sites.

Sacred Ecology: A worldview in which elements of nature such as rivers, sun, soil, and plants are treated as sacred and therefore worthy of care, protection, and ritual respect.

Sustainable Tourism: Tourism that considers its present and future economic, social, and environmental impacts while meeting the needs of visitors, host communities, industry stakeholders, and the environment. Source

Triple Bottom Line: A sustainability framework that evaluates development in terms of three interrelated dimensions: economic viability, social equity, and environmental integrity.

Vratin: A devotee who undertakes the religious fast and observes the ritual discipline of Chhath Puja, most often women, though men may also perform the vow.

Compilation of References

Abad-Galzacorta, M., Guereño-Omil, B., Makua, A., Iriberri, J. L., & Santomà, R. (2016). Pilgrimage as tourism experience: A preliminary approach to profiling pilgrims on the Ignatian Way. *International Journal of Religious Tourism and Pilgrimage*, *4*(4), 48–66.

Abbas, H., Ullah, I., & Ullah, W. (2025). Displacement and urban alienation: An ecocritical reading of The Reluctant Fundamentalist. *ACADEMIA International Journal for Social Sciences*, *4*(1), 677–688. DOI: 10.63056/ACAD.004.01.0122

Abuarqoub, I. A. S. (2019). Language barriers to effective communication. *Utopía y Praxis Latinoamericana*, *24*(6), 64–77.

Abukutsa-Onyango, M. O. (2014). Strategic repositioning African indigenous vegetables and fruits with nutrition, economic, and climate change resilience potential. Novel plant bioresources: applications in food, medicine, and cosmetics, 361-369.

Abungu, G. H. O. (2016). Practising Archaeology in Eastern and Southern Africa: Coming of Age or the Indigenisation of a Foreign Subject? In *A future for archaeology* (pp. 143-155). Routledge.

Abungu, O. G. (2016). *Traditional Management Systems at Heritage Sites in Africa* (Abungu, O. G., Ndlovu, N., Thiombiano, F. E., Zulu, J., & Varissou, S., Eds.). African World Heritage Fund.

Acemoglu, D., Reed, T., & Robinson, J. A. (2014). Chiefs: Economic development and elite control of civil society in Sierra Leone. *Journal of Political Economy*, *122*(2), 319–368. DOI: 10.1086/674988

Adams, T. E., Ellis, C., & Jones, S. H. (2017). Autoethnography. *The international encyclopedia of communication research methods*, 1-11. DOI: 10.1002/9781118901731.iecrm0011

Adefila, A., Teixeira, R. V., Morini, L., Garcia, M. L. T., Delboni, T. M. Z. G. F., Spolander, G., & Khalil-Babatunde, M. (2022). Higher education decolonisation:# Whose voices and their geographical locations? *Globalisation, Societies and Education*, *20*(3), 262–276. DOI: 10.1080/14767724.2021.1887724

Adepoju, A., Emmanuel, A., & Audu, J. (2024). Silencing the Guns: African Traditional Proverbs and Aphorisms for Peace and Security in 21st Century Africa. *The Journal of Society and Media*, *8*(2), 537–551. DOI: 10.26740/jsm.v8n2.p537-551

Adhikary, C. D. (2023). Hindu World-Views and Social Work Practices. *Indigenising and Decolonising Social Work Education*, 124.

Africa Center for Strategic Studies. (2025). *Reflections on 60 years of strengthening military professionalism in Kenya*. Available at: https://africacenter.org/spotlight/60-years-military-professionalism-kenya/ [Accessed: 4 April 2026].

Africa Energy Chamber. (2025). Empowering Africa's Energy Sector for Growth. https://energychamber.org

African National Congress. (2025). The role of Traditional Leadership in South African Governance. Carter, E.

Agarwal, (2024). Chhath Puja: Embracing Sustainability Through Ancient Rituals. LinkedIn. Retrieved from https://www.linkedin.com/pulse/chhath-puja-embracing-sustainability-through-ancient-rituals-agarwal-81jpe

Agawu, K. (2014). *Representing African music: Postcolonial notes, queries, positions*. Routledge. DOI: 10.4324/9781315810942

Agawu, K. (2016). *African Rhythm: A Northern Perspective*. Cambridge University Press.

Agrawal, A. (1995). Dismantling the Divide Between Indigenous and Scientific Knowledge. *Development and Change*, *26*(3), 413–439. DOI: 10.1111/j.1467-7660.1995.tb00560.x

Ahmad, D., Mohanty, I., Irani, L., Mavalankar, D., & Niyonsenga, T. (2020). Participation in microfinance based self-help groups in India: Who becomes a member and for how long? *PLoS One*, *15*(8), e0237519. DOI: 10.1371/journal.pone.0237519 PMID: 32810162

Ajani, Y. A., Oladokun, B. D., Olarongbe, S. A., Amaechi, M. N., Rabiu, N., & Bashorun, M. T. (2024). Revitalizing indigenous knowledge systems via digital media technologies for sustainability of indigenous languages. *Preservation, digital technology & culture, 53*(1), 35-44.

Ajayi, V. O. (2017). Primary Sources of Data and Secondary Sources of Data. *Benue State University*, 1(1), 1-6.

Alam, M. (2025). Preserving cultural heritage and empowering indigenous communities for sustainable development in Fiji. *Social Sciences and Humanities Open, 12*(September 2024). DOI: 10.1016/j.ssaho.2025.101760

Alarima, C. I. (2018). *Factors influencing rural-urban migration of youths in Osun State*.

Alexander, L. V. (2016). Global Observed Long-Term Changes in Temperature and Precipitation Extremes: A Review of Progress and Limitations in IPCC Assessments and Beyond. *Weather and Climate Extremes*, *11*, 4–16. DOI: 10.1016/j.wace.2015.10.007

Ali, M., Fjeldstad, O. H., Jiang, B., & Shifa, A. B. (2019). Colonial legacy, state-building and the salience of ethnicity in sub-Saharan Africa. *Economic Journal (London)*, *129*(619), 1048–1081. DOI: 10.1111/ecoj.12595

Aluko, H. A., Watson, A., Islam, N., Fernandes, F. A. P., Aluko, A. O., Ayertey, S., & Baig, M. M. (2025). Rethinking Business Practices: Harnessing Indigenous Knowledge Systems and Sustainable Strategies for Resilient Entrepreneurial Success. *Business Strategy and the Environment*, •••, 1–18. DOI: 10.1002/bse.70306

Alum, E. U. (2024). The role of indigenous knowledge in advancing the therapeutic use of medicinal plants: Challenges and opportunities. *Plant Signaling & Behavior*, *19*(1), 20–23. DOI: 10.1080/15592324.2024.2439255 PMID: 39652401

Alves, F., & Vidal, D. G. (2024). Plural Nature(s): An Overview of Their Sociocultural Construction. In *Encyclopedia* (Vol. 4, Number 1, pp. 1–12). DOI: 10.3390/encyclopedia4010001

Amghar, A. (2022). Revisiting the contingency theories of leadership: Key features, meanings and lessons. *Revue Marocaine de Gestion et de Société*, *1*, 76–93.

Anderson, C. R., Bruil, J., Chappell, M. J., Kiss, C., & Pimbert, M. P. (2021). From transition to domains of transformation: Getting to sustainable and just food systems through agroecology. *Sustainability*, *13*(3), 1–24. PMID: 34676112

Andriana, W. D. (2025). The Impact of Anthropocentrism on Human–Nature Relations and Its Implications for Environmental Ethics. *Journal of Language and Literature*, *25*(2), 625–637. DOI: 10.24071/joll.v25i2.10876

Aning, K., & Edu-Afful, F. (2020). African militaries and the challenge of gender inclusion and equality. In Aning, K., & Edu-Afful, F. (Eds.), *Governing Africa's Security: The Challenges of Gender and Military Transformation*. Routledge.

Anthwal, A., Gupta, N., Sharma, A., Anthwal, S., & Kim, K. H. (2010). Conserving biodiversity through traditional beliefs in sacred groves in Uttarakhand Himalaya, India. *Resources, Conservation and Recycling*, *54*(11), 962–971. DOI: 10.1016/j.resconrec.2010.02.003

Anuja Priya, S. (2026). Negotiating science, power, and identity: The institutionalization of Ayurveda in contemporary India. *Social Sciences & Humanities Open*, *13*, 102535. DOI: 10.1016/j.ssaho.2026.102535

Apaydin, V. (2020). Heritage, memory and social justice: reclaiming space and identity. *Critical perspectives on cultural memory and heritage*, 84-97.

Ardener, S. (1964). The comparative study of rotating credit associations. *Journal of the Royal Anthropological Institute*, *94*(2), 201–229.

Argumedo, A., Song, Y., Khoury, C. K., Hunter, D., Dempewolf, H., Guarino, L., & De Haan, S. (2021). Biocultural diversity for food system transformation under global environmental change. *Frontiers in Sustainable Food Systems*, *5*, 685299. DOI: 10.3389/fsufs.2021.685299

Aristotle. (2000). Nicomachean ethics (T. Irwin, Trans., 2nd ed.). Hackett Publishing Company. (Original work published ca. 4th century BCE)

Arnold, D. (1993). *Colonizing the body: State medicine and epidemic disease in nineteenth-century India*. University of California Press.

Arsal, D. A. T. (2026, January). INTERFAITH ECOTHEOLOGY: UNIFYING RELIGIOUS PERSPECTIVES IN RESPONDING TO THE ENVIRONMENTAL CRISIS. In *International Student Conference* (Vol. 1, No. 1, pp. 8-21).

Årsheim, H. (2018). Including and Excluding Indigenous Religion through Law. *Numen*, *65*(5–6), 531–561. DOI: 10.1163/15685276-12341511

Arslan, E. N., & Dişli, G. (2025). Architectural heritage and traditional knowledge systems: Insights from the ancient settlement of Kilistra, Türkiye #. *Sustainable Communities*, *2*(1), 2477145. Advance online publication. DOI: 10.1080/29931282.2025.2477145

Asante, M. K. (2017). The philosophy of Afrocentricity. In *The Palgrave handbook of African philosophy* (pp. 231–244). Palgrave Macmillan US. DOI: 10.1057/978-1-137-59291-0_16

Ascher, R. (1961). Analogy in archaeological interpretation. *Southwestern Journal of Anthropology*, *17*(4), 317–325. DOI: 10.1086/soutjanth.17.4.3628943

Ashika, J., Shirly, S., Sharma, S., & Raymer, M. (2025). *Eco-wisdom of ancient India: Environmental conservation and sustainability through religious scriptures, Vedic teaching and Arthashastra*. In *Proceedings of the 3rd International Conference on Revitalizing Indian Knowledge Systems*. https://doi.org/ DOI: 10.2991/978-2-38476-477-8_23

Ashton, S. R. (2023). *British Policy Towards the Indian States 1905–1939*. Routledge. DOI: 10.4324/9781003368243

Attas, R., & Walker, M. E. (2019). Exploring decolonization, music, and pedagogy. *Intersections*, *39*(1), 3–20. DOI: 10.7202/1075339ar

Augustine, D. J., McNaughton, S. J., & Frank, D. A. (2003). Feedbacks between soil nutrients and large herbivores in a managed savanna ecosystem. *Ecological Applications*, *13*(5), 1325–1337. DOI: 10.1890/02-5283

Ausi, M. (2022). Lerole. Tshimo ea Tlholwa Album. URL: https://www.youtube.com/watch?v=P6ATHdMJS10&list=RDP6ATHdMJS10&start_radio=1

Ausi, M. (2022). Setimela. Tshimo ea Tlholwa Album. URL: https://www.youtube.com/watch?v=feb7Xk9xm-A&list=OLAK5uy_ncrKEvT2Kvj-keE2Hcl8GAyG2MLzFQ4vQ&index=2

Austin, G. (2018). *The Indian constitution: Cornerstone of a Nation* (Thirty-first impression). Oxford University Press.

Badikilaya, G. (2019). Chhath Puja - A Study in Religious and Cultural Tourism. *Atna Journal of Tourism Studies*, *14*(2), 71–76. DOI: 10.12727/ajts.22.5

Baena, P. A., Brunei, A., Fernández-de-Larrinoa, Y., Martinez-Cruz, T. E., Milbank, C., Way, M., & Rights, I. I. P. (2023). In brief: the white/wiphala paper on indigenous peoples' food systems. Science and innovations for food systems transformation, 229.

Bakhaya, I. (2024). *The Impact of Language Barriers in Multinational Military Operations: A Case Study of NATO Forces*. AFASES Conference Proceedings. [Accessed 20 January 2026]. https://biblioteca.afahc.ro/ro/afases/2024/Volume-AFASES2024/10-IrinaBAKHAYA.pdf

Bala, B. K., Mondol, M. R. A., Biswas, B. K., Chowdhury, B. L., & Janjai, S. (2015). Solar drying of fruits, vegetables, and spices: A review. *Renewable & Sustainable Energy Reviews*, *47*, 1–14.

Balogun, F. O., Tshabalala, N. T., & Ashafa, A. O. T. (2016). Antidiabetic medicinal plants used by the Basotho tribe of Eastern Free State: A review. *Journal of Diabetes Research, 2016*(1), 4602820. DOI: 10.1155/2016/4602820 PMID: 27437404

Banerjee, A., Duflo, E., Glennerster, R., & Kinnan, C. (2015). The miracle of microfinance? Evidence from a randomized evaluation. *American Economic Journal. Applied Economics, 7*(1), 22–53. DOI: 10.1257/app.20130533

Bannister, P. (2007). Regulating 'Tradition' South African Izangoma and the Traditional Health Practitioners Act 2004. *The Cambridge Journal of Anthropology*, 27(1), 25-61. www.jostor.org/stable/2380708

Bans-Akutey, A. (2021). *The Path-Goal Theory of Leadership*. Academia Letters., DOI: 10.20935/AL748

Baporikar, N. (2022). Strategies to manage and preserve indigenous knowledge. In *Handbook of research on protecting and managing global indigenous knowledge systems* (pp. 207–222). IGI Global Scientific Publishing. DOI: 10.4018/978-1-7998-7492-8.ch012

Barreau, A., Ibarra, J. T., Wyndham, F. S., & Kozak, R. A. (2019). Shifts in Mapuche food systems in southern Andean Forest landscapes: Historical processes and current trends of biocultural homogenization. *Mountain Research and Development, 39*(1), R12–R23. DOI: 10.1659/MRD-JOURNAL-D-18-00015.1

Barry, S. (2006). Reconciliation: The South African Truth and Reconciliation Commission's contribution to dealing with the past, reconciling and building the nation. *In die Skriflig, 40*(4), 691–714. DOI: 10.4102/ids.v40i4.365

Bates, E. (2012). The social life of musical instruments. *Ethnomusicology, 56*(3), 363–395. DOI: 10.5406/ethnomusicology.56.3.0363

Battiste, M. (2002). *Indigenous knowledge and pedagogy in First Nations education: A literature review with recommendations*. National Working Group on Education.

Battiste, M. (2013). *Decolonizing Education: Nourishing the Learning Spirit*. Purich Publishing Ltd.

Bawack, R., Roderick, S., Badhrus, A., Dennehy, D., & Corbett, J. (2025). Indigenous knowledge and information technology for sustainable development. *Information Technology for Development, 31*(2), 233–250. DOI: 10.1080/02681102.2025.2472495

Bayly, C. A. (1997). *Empire and Information: Intelligence Gathering and Social Communication in India, 1780–1870* (1st ed.). Cambridge University Press., DOI: 10.1017/CBO9780511583285

Bebey, F. (1999). *African music: A people's art*. Lawrence Hill Books.

Beeton, S. (2005). The case study in tourism research: A multi-method case study approach. In Ritchie, B. W., Burns, P., & Palmer, C. (Eds.), *Tourism research methods: Integrating theory with practice* (pp. 37–48). CAB International. DOI: 10.1079/9780851999968.0037

Bekele, W. B., & Ago, F. Y. (2022). Sample Size for Interview in Qualitative Research in Social Sciences: A Guide to Novice Researchers. *Research in Educational Policy and Management, 4*(1), 42–50. DOI: 10.46303/repam.2022.3

Ben-Eliezer, U. (2012). Operation Defensive Shield. In *Old Conflict, New War: Israel's Politics toward the Palestinians* (pp. 123–140). Palgrave Macmillan US. DOI: 10.1057/9781137027573_8

Berkes, F. (1999). *Sacred ecology: Traditional ecological knowledge and resource management.* Taylor and Francis.

Berkes, F. (2018). *Sacred ecology* (4th ed.). Routledge.

Berkes, F. (Ed.). (1989). *Common Property Resources, Ecology and Community-Based Sustainable Development.* Belhaven Press.

Berry, J. W. (2017). Theories and models of acculturation. The Oxford handbook of acculturation and health, 10, 15-28.

Bhat, I. A. (2015). Pilgrimage Tourism in Sindh Valley (Kashmir): A Case Study of Kheer Bhawani Shrine. *South Asian Journal of Tourism and Heritage.* https://www.sajth.com/journal/wp-content/uploads/2015/09/Pilgrimage-Tourism-in-Sindh-Valley-Kashmir-A-Case-Study-of-Kheer.pdf

Bhuda, M. T. (2024). Decolonising and indigenising African Indigenous research: Why Indigenous ways of knowing, being and living matter in research. In *Higher Education Transformation in Africa* (pp. 50–69). Routledge. DOI: 10.4324/9781003528319-4

Bhuda, M. T. (2024). Integrating ethnomathematical activities in the foundation phase as means of decolonisation: Insights from Indigenous Knowledge Systems. *African Journal of Development Studies, 14*(4), 401. https://hdl.handle.net/10520/ejc-aa_affrika1_v14_n4_a18

Bhuda, M. T., & Maditsi, M. E. (2023). Do Traditional Leaders still govern Indigenous communities? Insights from riots and looting in South Africa during COVID-19. *Indilinga, 22*(2), 137–153.

Bhuda, M. T., & Maditsi, M. E. (2024). Did traditional leaders succeed in restoring law and order during the 2021 riots and looting in South Africa? Insights from Indigenous scholars. *African Journal for Peace and Conflict Studies, 13*(2). Advance online publication. DOI: 10.31920/2634-3665/2024/v13n2a9

Bhuda, M. T., & Maditsi, M. E. (2025). The importance of integrating Indigenous knowledge systems in South Africa's basic education: Perspectives from Indigenous scholars. In *Resurgence and revalorization of Indigenous knowledge systems in the contemporary society* (pp. 43–68). IGI Global Scientific Publishing., DOI: 10.4018/979-8-3373-4124-8.ch003

Bhuda, M. T., Maepa, N., Matsotso, A., & Makgamatha, M. G. (2025). Can Indigenous Knowledge Foster Socio-Economic and Sustainable Development in South Africa? *African Journal of Innovation and Entrepreneurship, 4*(4), 357–382. DOI: 10.31920/2753-314X/2025/v4n4a15

Bhuda, M. T., & Marumo, P. (2022). Ubuntu Philosophy and African Indigenous Knowledge Systems: Insights from decolonization and indigenization of research. *Gender & Behaviour, 20*(1), 19133–19151.

Bhuda, M. T., Marumo, P., & Motswaledi, T. (2021). The Involvement/Non-Involvement of Traditional Leadership in Covid-19 South Africa. *Gender & Behaviour, 19*(1), 17480–17489.

Bhuda, M. T., Setshego, M. V., & Koitsiwe, M. (2025). The Role of Indigenous Participants in PAR: An Application of a Decolonial Approach. In *Enabling Indigenous Knowledge Systems in Action Research and Action Learning* (pp. 291–308). IGI Global Scientific Publishing. DOI: 10.4018/979-8-3693-3542-0.ch015

Bible Hub Online. (2025). Proverbs, 22:28. https://biblehub.com/proverbs

Blackburn, S. H. (2016). *Singing of Birth and Death: Texts in Performance*. University of Pennsylvania Press., DOI: 10.9783/9781512800579

Blacking, J. (1973). *How musical is man?* University of Washington Press.

Boesche, R. (2002). *The first great political realist: Kautilya and his Arthashastra*. Lexington Books.

Boeyens, J. C. A. (2003). The Late Iron Age Sequence in the Marico and Early Tswana History. *The South African Archaeological Bulletin, 58*(178), 63–78. JSTOR. https://doi.org/DOI: 10.2307/3889303

Bohensky, E. L., & Maru, Y. (2011). Indigenous knowledge, science, and resilience: What have we learned from a decade of international literature on "integration"? *Ecology and Society*, *16*(4), art6. DOI: 10.5751/ES-04342-160406

Boivin, N. (2004). Landscape and Cosmology in the South Indian Neolithic: New Perspectives on the Deccan Ashmounds. *Cambridge Archaeological Journal*, *14*(2), 235–257. DOI: 10.1017/S0959774304000150

Bora, A., & Kumar, A. (2025). The Role of Indian Wisdom in Addressing Climate Change and Environmental Crises. *Indian Knowledge Systems for Social Development: Traditional to Modernity Pedagogy*.

Botha, F. (2025). Market Summary: Impact of Tariffs and South Africa's Economic Challenges Us Tariffs and Reaction from Global Markets. *Investments and Markets Insights*. https://oig-invest.com/wp-content/uploads

Boughton, J. (2020). New animism: Relational epistemologies and expanding Western ontologies. *Student Anthropologist*, *7*(1), 56–59. DOI: 10.1002/j.sda2.20200701.0012

Bouman, F. J. A. (1995). Rotating and accumulating savings and credit associations: A development perspective. *World Development*, *23*(3), 371–384. DOI: 10.1016/0305-750X(94)00141-K

Brandt, M., Rasmussen, K., Hiernaux, P., Herrmann, S., Tucker, C. J., Tong, X., Tian, F., Mertz, O., Kergoat, L., Mbow, C., David, J. L., Melocik, K. A., Dendoncker, M., Vincke, C., & Fensholt, R. (2018). Reduction of tree cover in West African woodlands and promotion in semi-arid farmlands. *Nature Geoscience*, *11*(5), 328–333. DOI: 10.1038/s41561-018-0092-x PMID: 32944066

Braun, V., & Clarke, V. (2006). Using thematic analysis in psychology. *Qualitative Research in Psychology*, *3*(2), 77–101. DOI: 10.1191/1478088706qp063oa

Breidlid, A. (2009). Culture, indigenous knowledge systems and sustainable development: A critical view of education in an African context. *International Journal of Educational Development*, *29*(2), 140–148. DOI: 10.1016/j.ijedudev.2008.09.009

Breman, J. (1996). *Footloose Labour: Working in India's Informal Economy* (1st ed.). Cambridge University Press., DOI: 10.1017/CBO9781139171076

Bremner, H. A., & Southgate, D. A. T. (2007). *Food processing handbook.* Wiley-VCH.

British Broadcast Corporation. (2023). Chief Mangosuthu Buthelezi: The man who divided South Africa''. *BBC Newsletter*, www.bbc.com, 08 September 2023.

Broache, M. P. (2022). *Planning, Conducting and Analysing Interviews.* Edward Elgar Publishing.

Brockington, J. L. (1998). *The Sanskrit epics.* Brill. DOI: 10.1163/9789004492677

Brodt, S. B. (1999). Interactions of formal and informal knowledge systems in village-based tree management in central India. *Agriculture and Human Values, 16*(4), 355–363. DOI: 10.1023/A:1007537809389

Brodt, S. B. (2001). A systems perspective on the conservation and erosion of indigenous agricultural knowledge in Central India. *Human Ecology: an Interdisciplinary Journal, 29*(1), 99–120. DOI: 10.1023/A:1007147806213

Bronson, B. (1977). The earliest farming: demography as cause and consequence. In *Origins of agriculture* (pp. 23–48). Mouton Publishers. DOI: 10.1515/9783110813487.23

Brown, R. (2007). Enhancing student employability? Current practice and student experiences in HE performing arts. *Arts and Humanities in Higher Education, 6*(1), 28–49. DOI: 10.1177/1474022207072198

Bteille, A. (2017). *DEMOCRACY AND ITS INSTITUTIONS.* OXFORD University Press.

Buthelezi, S. C., Ocholla, D., & Dlamini, P. (2024). Strategies for documenting and disseminating indigenous knowledge at a South African university. *South African Journal of Information Management, 26*(1), 1648. DOI: 10.4102/SAJIM.v26i1.1648

Butler, B. (2024). *Reimagining Heritage Conservation'after Modernity'as Constellations and Cosmologies of Care and Protection.* Instituto Nacional de Antropología e Historia.

Bvenura, C., & Afolayan, A. J. (2015). The role of wild vegetables in household food security in South Africa: A review. *Food Research International, 76*, 1001–1011. DOI: 10.1016/j.foodres.2015.06.013

Campbell-Platt, G. (1994). Fermented foods: A world perspective. *Food Research International, 27*(3), 253–257. DOI: 10.1016/0963-9969(94)90093-0

Carter, S. (2019). *Lost harvests: Prairie Indian reserve farmers and government policy* (Vol. 94). McGill-Queen's Press-MQUP. DOI: 10.1515/9780773557697

Cassar, A., Crowley, L., & Wydick, B. (2007). The effect of social capital on group loan repayment: Evidence from field experiments. *Economic Journal (London), 117*(517), F85–F106. DOI: 10.1111/j.1468-0297.2007.02016.x

Castleberry, A., & Nolen, A. (2018). Thematic Analysis of Qualitative Research Data: Is it as easy as it Sounds? *Currents in Pharmacy Teaching & Learning, 10*(6), 807–815. DOI: 10.1016/j.cptl.2018.03.019 PMID: 30025784

Castro-Gómez, S. (2021). *Zero-point hubris: Science, race, and enlightenment in eighteenth-century Latin America.* Bloomsbury Publishing USA. DOI: 10.5040/9798881816308

Cavalcanti, M. L. V. D. C. (2019). The Legal Protection of the Intangible Cultural Heritage in Brazil. In *The Legal Protection of the Intangible Cultural Heritage: A Comparative Perspective* (pp. 19–35). Springer International Publishing. DOI: 10.1007/978-3-319-72983-1_2

CGAP. (2017). *Client protection in microfinance: Smart Campaign principles*. Consultative Group to Assist the Poor.

Chadha, M. (2024). Personhood in classical Indian philosophy. In Zalta, E. N., & Nodelman, U. (Eds.), *The Stanford Encyclopaedia of Philosophy* (Winter 2024 ed.)., https://plato.stanford.edu/archives/win2024/entries/personhood-india/

Chakrabarti, D. K. (2003). *Archaeology in the Third World: A history of Indian archaeology since 1947*. D.K. Printworld.

Chakrabarty, D. (2007). *Provincializing Europe: Postcolonial thought and historical difference*. Princeton university press.

Chakrabarty, S., & Bass, A. E. (2013). Encouraging Entrepreneurship: Microfinance, Knowledge support, and the costs of operating in institutional voids. *Thunderbird International Business Review*, *55*(6), 545–462. DOI: 10.1002/tie.21569

Chakraborty, S. (2002). *Managerial Effectiveness and Quality of Worklife Indian Insights*. McGraw Hill.

Chakraborty, S. K. (1991). *Management by values: Towards cultural congruence*. Oxford University Press.

Chakraborty, S. K. (1995). *Ethics in management: Vedantic perspectives*. Oxford University Press.

Chambers, L. A., Jackson, R., Worthington, C., Wilson, C. L., Tharao, W., Greenspan, N. R., Masching, R., Pierre-Pierre, V., Mbulaheni, T., Amirault, M., & Brownlee, P. (2018). Decolonizing Scoping Review Methodologies for Literature With, for, and by Indigenous Peoples and the African Diaspora: Dialoguing With the Tensions. *Qualitative Health Research*, *28*(2), 175–188. DOI: 10.1177/1049732317743237 PMID: 29182046

Chanda, T., Akpan, W. M., Mulenga, D. M., Sain, Z. H., & Gilbert, M. M. (2024). Analyzing the Interplay Between Politics and Economic Development: An African Perspective.

Chandel, P. S., Agarwal, P., Indolia, U., & Parashar, A. (2024). Reviving Ancient Indian Hospitality: Integrating Guest-Worship Traditions into Modern Homestay Tourism.

Chand, M. (2010). A Cross-National Study of Motivational Determinants among Non-Resident Indian Visitors to Religious Centers in India. *International Journal of Hospitality & Tourism Administration*, *11*(1), 22–38. DOI: 10.1080/15256480903539412

Chapple, C. K., & Tucker, M. E. (2000). *Hinduism and ecology*. Harvard Center for the Study of World Religions, Harvard University Press.

Chatterjee, P. (1993). *The nation and its fragments: Colonial and postcolonial histories*. Princeton University Press.

Chaturvedi, P., Sharma, J., & Ranjan, A. (2024). The Role of Souvenir Marketing in Sustaining Handmade Craftsmanship. *ShodhKosh: Journal of Visual and Performing Arts*, *5*(3), 278–294. DOI: 10.29121/shodhkosh.v5.i3.2024.1401

Chaudhary, N. (2025). Transcending Time and Space in the Search for Synthesis: A Commentary on Paranjpe's Understanding Yoga Psychology: Indigenous Knowledge with Global Relevance. *Integrative Psychological & Behavioral Science*, *59*(1), 15. DOI: 10.1007/s12124-024-09875-3 PMID: 39779637

Chauke, M. T. (2015). The Role of Women in Traditional Leadership with Special Reference to the Valoyi Tribe. Kamla-Raj. *Studies of Tribes and Tribals*, *13*(1), 34–39. DOI: 10.1080/0972639X.2015.11886709

Chawane, M. (2016). The development of Afrocentricity: A historical survey. *Yesterday and Today*, (16), 78–99. DOI: 10.17159/2223-0386/2016/n16a5

Chen, M. J., & Lee, D. (2008). The paradox of polarization: Zhongyong, harmony, and conflict management in Chinese organizations. *Management and Organization Review*, *4*(2), 273–290. DOI: 10.1111/j.1740-8784.2005.00012.x

Chigwada, J., Mapara, J., Ngulube, P., & Chabvutagondo, T. (2025). Influence of Indigenous data governance principles on Indigenous knowledge management: Lessons from the Southern African Intangible Cultural Heritage Project. *IFLA Journal*, •••, 1–10. DOI: 10.1177/03400352251331471

Chiimba, C. M. (2020). *Experiences of young orphans who transition into extended families in Zimbabwe* [Unpublished doctoral thesis]. Faculty of Education, University of Pretoria.

Chikoko, W., & Ruparanganda, W. (2020). *Ubuntu* or *Hunhu* perspective in understanding substance abuse and sexual behaviours of street children of Harare Central Business District, Zimbabwe. *African Journal of Social Work*, *10*(1).

Chilisa, B. (2020). *Indigenous research methodologies* (2nd ed.). Sage.

Chimbunde, P., & Kgari-Masondo, M. C. (2022). The decolonisation of African languages: Insights from Southern Africa. Language Matters. *Studies in the Languages of Southern Africa*, *53*(3), 5–24.

Chipungu, O., Mamimine, P. W., Chipotereke, C. C., Kwinje, P. C., Mumbengegwi, P., & Mirimi, K. (2005).. . *Convergence of Indigenous Knowledge Systems and Contemporary Approaches in Cultural Heritage Management.*, *2294*, 33–55.

Chirikure, S. (2020a). New Perspectives on the Political Economy of Great Zimbabwe. *Journal of Archaeological Research*, *28*(2), 139–186. DOI: 10.1007/s10814-019-09133-w

Chirikure, S. (2020b). *Great Zimbabwe: Reclaiming a 'confiscated'past.* Routledge. DOI: 10.4324/9780367810412

Chirikure, S., Bandama, F., Chipunza, K., Mahachi, G., Matenga, E., Mupira, P., & Ndoro, W. (2017). Seen but not told: Re-mapping Great Zimbabwe using archival data, satellite imagery and geographical information systems. *Journal of Archaeological Method and Theory*, *24*(2), 489–513. DOI: 10.1007/s10816-016-9275-1

Chirikure, S., Ndoro, W., Bugarin, F. T., di Lernia, S., Ichumbaki, E. B., & Lwoga, N. B. (2021). Usable pasts forum: UNESCO and heritage tourism in Africa. *African Archaeological Review*, *38*(3), 513–533. DOI: 10.1007/s10437-021-09454-6 PMID: 34429563

Chiwisa, C. C., & Qutieshate, A. (2024). A Comprehensive Literature Review of Global Leadership Styles: Assessing Universality and Impact on Workforce Motivation. *International Journal for Multidisciplinary Research (IJFMR)* 6(1), January-February 2024. www.ijfmr.com

Choi, S., Weng, S. S., Feliciano, G., Benton, A. D., & Kim, Y. (2025). Environmental Justice and Vulnerable Populations: Perspectives of Environmental Activists. *Social Work and Social Welfare*, *6*(1), 316–326. DOI: 10.25082/SWSW.2025.01.001

Chris Hani. (1991). Chirs Hani on Socialism. ENCA News Interview. www.enca.com

Christie, M. J. (1991). Aboriginal science for the ecologically sustainable future. *Australian Science Teachers Journal*, *37*, 26–31.

Churu, W. B., & Nancy, W. (2022). Afrocentricity as a basis for restoration of African identity and agency: A call to authentic education in Africa. *AKU: An African Journal of Contemporary Research*, *3*(2), 27–43.

Clapp, J., Moseley, W. G., Burlingame, B., & Termine, P. (2022). The case for a six-dimensional food security framework. *Food Policy*, *106*, 102164. DOI: 10.1016/j.foodpol.2021.102164

Clark, P. (2010). *The Gacaca courts, post-genocide justice and reconciliation in Rwanda: Justice without lawyers*. Cambridge University Press. DOI: 10.1017/CBO9780511761584

Clist, B. (1987). Early Bantu settlements in west-central Africa: A review of recent research. *Current Anthropology*, *28*(3), 380–382. DOI: 10.1086/203543

Cohn, B. S. (1996). *Colonialism and its forms of knowledge: The British in India*. Princeton University Press.

Coleman, J. S. (1988). Social capital in the creation of human capital. *American Journal of Sociology*, *94*(Supplement), S95–S120. DOI: 10.1086/228943

Collins-Kreiner, N. (2010). Geographers and pilgrimages: Changing concepts in pilgrimage tourism research. *Tijdschrift voor Economische en Sociale Geografie*, *101*(4), 437–448. DOI: 10.1111/j.1467-9663.2009.00561.x

Collins-Kreiner, N. (2020). Pilgrimage tourism-past, present and future rejuvenation: A perspective article. *Tourism Review*, *75*(1), 145–148. DOI: 10.1108/TR-04-2019-0130

Collins, T., & Esterling, S. (2019). Fluid Personality: Indigenous Rights and the Te Awa Tupua (Whanganui River Claims Settlement) Act 2017 in Aotearoa New Zealand. *Melb. J. Int'l L.*, *20*, 197.

Comaroff, J. (2002). The Empire's old clothes: Fashioning the colonial subject. In *Cross-cultural consumption* (pp. 29–48). Routledge.

Coplan, D. B. (2007). *In Township Tonight! South Africa's Black City Music and Theatre*. Ravan Press.

Coplan, D. B. (2008). *Zulu Praise Poetry and Social Identity*. Wits University Press.

Crowder, M. (2023). Senegal*: a study of French assimilation policy*. Routledge.

Crush, J., & Frayne, B. (2011). Supermarket expansion and the informal food economy in Southern African cities. *Journal of Southern African Studies, 28*(5), 781–797. DOI: 10.1080/03057070.2011.617532

Cull, R., & Morduch, J. (2017). Microfinance and economic development. In Beck, T., & Levine, R. (Eds.), *Handbook of finance and development* (pp. 550–572). Edward Elgar.

Dagnini, J. K. (2023). Afrocentrism through Afro-American Music: from the 1960's until the Early 2000's. www. revue-sociologique. org/sites/default/files/2. pdf *(Accessed 20 Jul. 2019).*

d'Ambrosio, U. (2016). Ethnomathematics: A response to the changing role of mathematics in society. DOI: 10.1108/978-1-681123-261-420251003

Damke, S. W. (2025). CONTRIBUTION OF INDIAN WOMEN IN TACKLING ENVIRONMENT PROTECTION AND CLIMATE CHANGE. *Journal of East-West Thought (JET) ISSN (O): 2168-2259 UGC CARE I, 15*(1), 659-665.

Daniels, R. (2022). Racing toward an African Inclusion: De-essentializing the Global South. *Decolonizing communication studies*, 42-56.

Das, S. M. (2019). *Scriptural importance of Hindu environmental ethics.*

Daschuk, J. W. (2013). *Clearing the plains: Disease, politics of starvation, and the loss of Aboriginal life* (Vol. 65). University of Regina Press. DOI: 10.1515/9780889772977

Dash, K. K. (2024). Strategic CSR perspectives hold key to sustainable tourism development in the Society 5.0 Perspectivas estratégicas de RSC são fundamentais para o desenvolvimento do turismo sustentável na Sociedade 5.0 Las perspectivas estratégicas de RSE son clave para el desarrollo del. *Brazilian Journal of Development, 10*(3), 01-27.

Das, S. M. (2020). *Indian environmental ethics. SSRN.* https://ssrn.com/abstract=3655033

Davies, B. (2003). The role of quantitative and qualitative research in industrial studies of tourism. *International Journal of Tourism Research, 5*(2), 97–111. DOI: 10.1002/jtr.425

Davis, S. R. (2018). *The ANC's war against apartheid: Umkhonto We Sizwe and the liberation of South Africa*. Indiana University Press. DOI: 10.2307/j.ctt2272733

Dawadi, S. (2020). Thematic Analysis Approach: A Step-by-Step Guide for ELT Research Practitioners. *Journal of NELTA, 25*(1-2), 62–71. DOI: 10.3126/nelta.v25i1-2.49731

Dawe, K. (2015). Materials matter: Towards a political ecology of musical instrument making. In *Current directions in ecomusicology* (pp. 109–121). Routledge.

De Castro, E. V. (2019). Exchanging perspectives: The transformation of objects into subjects in Amerindian ontologies. *Common Knowledge, 25*(1–3), 21–42. DOI: 10.1215/0961754X-7299066

De la Rey, A. (1990). *Suid-Afrikaanse Weermag - South African Defence Force Review*. Walker-Ramus Trading Co. SA. (PTY) LTD.

Deacon, H. J., Deacon, J., Brooker, M., & Wilson, M. L. (1978). The Evidence for Herding at Boomplaas Cave in the Southern Cape, South Africa. *The South African Archaeological Bulletin, 33*(127), 39–65. JSTOR. https://doi.org/DOI: 10.2307/3888251

Defence Web. (2022). Defence Industry Lekgotla tackles challenges facing South Africa's defence sector. DefenceWeb. Available at: https://www.defenceweb.co.za (Accessed: 28 April 2026). Dei, G. J. S. (2000). *Rethinking the role of indigenous knowledges in the academy*. International Journal of Inclusive Education, 4(2), 111–132.

Dei, G. J. S., Hall, B. L., & Rosenberg, D. G. (Eds.). (2000). *Indigenous knowledges in global contexts: Multiple readings of our world*. Published in association with University of Toronto Press.

Dei, G. J. S. (2000). Rethinking the role of indigenous knowledges in the academy. *International Journal of Inclusive Education*, *4*(2), 111–132. DOI: 10.1080/136031100284849

Dei, G. J. S. (2018). *Reframing Blackness and Black solidarities through anti-colonial and decolonial prisms*. Springer.

Delius, P. (2020). Chiefly Succession and Democracy in South Africa: Why History Matters. *Journal of Southern African Studies*, *47*(2), 209–227. DOI: 10.1080/03057070.2021.1855042

Dembedza, V. P., Chopera, P., Mapara, J., & Macheka, L. (2022). Impact of climate change-induced natural disasters on intangible cultural heritage related to food: A review. *Journal of Ethnic Foods*, *9*(1), 32. Advance online publication. DOI: 10.1186/s42779-022-00147-2

Demir, K. A., Yaman, F., & Koçak, B. (2025). Social Alienation with International Students in New York City: A Phenomenological Research. *Siyasal: Journal of Political Sciences*, *34*(1), 134–151. DOI: 10.26650/siyasal.2025.34.1478556

Denbow, J. R. (1983). *Iron Age economics: Herding, wealth, and politics along the fringes of the Kalahari Desert during the Early Iron Age*. Indiana University.

Denbow, J., Smith, J., Ndobochani, N. M., Atwood, K., & Miller, D. (2008). Archaeological excavations at Bosutswe, Botswana: Cultural chronology, paleo-ecology and economy. *Journal of Archaeological Science*, *35*(2), 459–480. DOI: 10.1016/j.jas.2007.04.011

Department of Defence. (1998). *Department of Defence Instruction: Pol and Plan No 00001/ (1998 Policy on Equal Opportunity and Affirmative Action No 00001*. Pretoria: Department of Defence.

Department of Defence. (1998). *South African Defence Review*. Department of Defence.

Department of Defence. (2020). *Bulletin Promulgation: Appointments and/or Promotions of Cols/ Capt (SAN) for the Financial Year 20/21. 22 October 2020*. Department of Defence.

Department of Defence. (2020). *Department of Defence Strategic Plan for 2020-2025, 12 March 2020*. Department of Defence.

Derrett, J. D. M. (1999). *Religion, law and the state in India* (Repr.). Oxford Univ. Press.

Descola, P. (2013). *Beyond Nature and CUlture*. The University of Chicago Press. DOI: 10.7208/chicago/9780226145006.001.0001

Deshmukh-Ranadive, J. (2004). Women's self-help groups in Andhra Pradesh: Participatory poverty alleviation in action. In Patel, R. (Ed.), *Reducing poverty: Is there a role for micro-credit in India?* World Bank.

Dharampal. (2021). *The beautiful tree: Indigenous Indian education in the eighteenth century*. Rashtrotthana Sahitya in association with Centre for Policy Studies, Chennai.

Diaz-Reyes, T. M. (2019). *Ecological Relationality and Models of Sustainability: Towards a Social, Intergenerational, Multi-Species Environmental Ethic through the Evaluation of Cultural-Spiritual Ecosystem Services*. Wake Forest University.

Dickson, N., & Meylahn, J. A. (2024). The role of storytelling as a possible trauma release for war veterans: A narrative approach. *Hervormde Teologiese Studies*, *80*(1), 9408. DOI: 10.4102/HTS.v80i1.9408

Dickson, R. K. (2023). Analysis of The Traditional Leadership Theories: A Review of Contemporary Leadership Approaches and Management Effectiveness. *Information and Knowledge Management*, *13*(5). Advance online publication. DOI: 10.7176/IKM/13-5-02

Dikakapa. (2008). *Mmamarutle*. URL: https://www.youtube.com/watch?v=neFnRNd-P0c&list=RDneFnRNd-P0c&start_radio=1

Diouf, M. (1998). The French colonial policy of assimilation and the civility of the originaires of the four communes (Senegal): A nineteenth century globalization project. *Development and Change*, *29*(4), 671–696. DOI: 10.1111/1467-7660.00095

Dirks, N. B. (2011). *Castes of Mind: Colonialism and the Making of Modern India*. Princeton University Press. DOI: 10.2307/j.ctt7rq9d

Dodd, N., Bester, P. C., & Van Der Merwe, J. (Eds.). (2020). *Contemporary Issues in South African Military Psychology*. African Sun Media. DOI: 10.18820/9781928480631

Dodo, O. (2013). Traditional Leadership Systems and Gender Recognition: Zimbabwe. *Journal for Gender and Women*, *1*(1), 29–44.

Dolan, J. M. (2016). *The restorative ecology of peace: Haudenosaunee environmental knowledge and philosophies of stewardship*. McGill University.

Donald, W. E., Ashleigh, M. J., & Baruch, Y. (2018). Students' perceptions of education and employability: Facilitating career transition from higher education into the labor market. *Career Development International*, *23*(5), 513–540. DOI: 10.1108/CDI-09-2017-0171

Dong, B. (2023). A systematic review of the Transactional Leadership Literature and Future Outlook. *Academic Journal of Management and Social Sciences*, *2*(3), 21–25. DOI: 10.54097/ajmss.v2i3.7972

Doniger, W. (2010). *The Hindus: An alternative history*. Oxford University Press.

Donkoh, S., & Mensah, J. (2023). Application of Triangulation in Qualitative Research. *Journal of Applied Biotechnology and Bioengineering*, *10*(1), 6–9. DOI: 10.15406/jabb.2023.10.00319

Drissi, S. (2022). Indigenous peoples and the nature they protect. UN Environ Programme and https://www.unep.org/new-and-stories/story/indigenous-peoples-and-nature-they- protect. Accessed 7 January 2026

Du Plessis, P. (2021). Decolonisation of education in South Africa: Challenges to decolonise the university curriculum. *South African Journal of Higher Education*, *35*(1), 54–69. DOI: 10.20853/35-1-4426

Dube, B. (2020). Go and prophesy in your own land: Foreign Prophets and popularism in South Africa. Evoking the need of Jonathanic Theology for Peaceful Resolution in Difference. *Religions*, *11*(42), 42. Advance online publication. DOI: 10.3390/rel11010042

Dube, B. (2023). *Language, power and communication in South African security institutions*. Africa Institute of South Africa.

Duflo, E., Banerjee, A., Glennerster, R., & Kinnan, C. (2013). *The miracle of microfinance? Evidence from a randomized evaluation* (NBER Working Paper No. 18950). National Bureau of Economic Research.

Dunne, J., Evershed, R. P., Salque, M., Cramp, L., Bruni, S., Ryan, K., Biagetti, S., & di Lernia, S. (2012). First dairying in green Saharan Africa in the fifth millennium bc. *Nature*, *486*(7403), 390–394. DOI: 10.1038/nature11186 PMID: 22722200

Durie, M. (2005). Indigenous knowledge within a global knowledge system. *Higher Education Policy*, *18*(3), 301–312. DOI: 10.1057/palgrave.hep.8300092

Dwivedi, O. P. (1993). Human responsibility and the environment: A Hindu perspective. *Journal of Hindu-Christian Studies*, *6*(1), 8. DOI: 10.7825/2164-6279.1077

Dwyer, L., & Edwards, D. (2013). Ecotourism and the triple bottom line. In Ballantyne, R., & Packer, J. (Eds.), *International handbook on ecotourism* (pp. 245–262). Edward Elgar Publishing. DOI: 10.4337/9780857939975.00027

Edirisinghe, A., & Suchet-Pearson, S. (2025). Legal Personhood and Rights of Nature: Bridging Relational Vedda Worldviews and Sri Lankan Law. *Transnational Environmental Law*, *14*(3), 630–658. DOI: 10.1017/S2047102525100101

Eerkens, J. W., & Lipo, C. P. (2007). Cultural transmission theory and the archaeological record:providing context to understanding variation and temporal changes in material culture. *Journal of Archaeological Research*, *15*(3), 239–274. DOI: 10.1007/s10814-007-9013-z

Ehret, C. (1973). Patterns of Bantu and Central Sudanic settlement in central and southern Africa (ca. 1000 BC-500 AD). *Transafrican Journal of History*, *3*(1/2), 1–71.

Eichler, J. (2021). Intangible cultural heritage, inequalities and participation: Who decides on heritage? *International Journal of Human Rights*, *25*(5), 793–814. DOI: 10.1080/13642987.2020.1822821

Ellis, C., Adams, T. E., & Bochner, A. P. (2011). Autoethnography: An overview. *Historical social research. Historische Sozialforschung*, •••, 273–290. https://www.jstor.org/stable/23032294

Emmanuel, O. K. (2014). Religious tourism and sustainable development: A study of Eyo festival in Lagos, Nigeria. *International Journal of Social Sciences and Education*, *4*(2), 474–485.

Ensureias. (2025). Chhath Puja 2025: *Cultural Significance and Social Dimensions*. Retrieved from https://www.ensureias.com/blog/current-affairs/chhath-puja-2025-cultural-significance-and-social-dimensions

Erbach, T. (2025). Using Cultural Heritage in Climate Adaptation: Fields of Application and Functions. *Wiley Interdisciplinary Reviews: Climate Change, 16*(4), 1–15. DOI: 10.1002/wcc.70011

Erlmann, V. (1996). The Aesthetics of the Global Imagination: Reflections on World Music in the 1990s. *Public Culture, 8*(1), 1–19. DOI: 10.1215/08992363-8-3-467

Erwin, S.I. (2005). Language barriers hinder multinational operations. *National defence, 90*(620), pp.16-17.

Erwin, S. K., & Cseh, M. (2023). Invisible and yet hypervisible: Gendered misconduct and the US military. *Industrial and Commercial Training, 55*(3), 402–413. DOI: 10.1108/ICT-10-2022-0075

Etikan, I., Musa, S. A., & Alkassim, R. S. (2016). Comparison of Convenience Sampling and Purposive Sampling. *American Journal of Theoretical and Applied Statistics, 5*(1), 1–4. DOI: 10.11648/j.ajtas.20160501.11

Euba, A. (2014). *JH Kwabena Nketia: Bridging Musicology and Composition: A Study in Creative Musicology*. MRI Press.

Fang, T. (2012). Yin Yang: A new perspective on culture. *Management and Organization Review, 8*(1), 25–50. DOI: 10.1111/j.1740-8784.2011.00221.x

FAO, Alliance of Biodiversity International and CIAT. (2021). Indigenous Peoples' Food Systems: Insights on Sustainability and Resilience in the Front Line of Climate Change.

FAO. (2018). *High Level Expert Seminar on Indigenous Food Systems*. Building on Traditional Knowledge to Achieve Zero Hunger.

Farhan, S. L., Merie, U. A. A. K., & Nasar, Z. (2024). Revitalizing historic city center a comparative methodology of current approaches and alternatives. *Journal of Cultural Heritage Management and Sustainable Development*.

Farmer, E. (2009). How do placements in kinship care compare with those in non-kin foster care: Placement patterns, progress and outcomes? *Child & Family Social Work, 14*(3), 331–342. DOI: 10.1111/j.1365-2206.2008.00600.x

Fergeus, J., Humphreys, C., Harvey, C., & Herrman, H. (2019). Supporting foster and kinship carers to promote the mental health of children. *Child & Family Social Work, 24*(1), 77–83. DOI: 10.1111/cfs.12583

Ferrer-Fons, M., Rovira-Martínez, M., & Soler-i-Martí, R. (2022). Youth empowerment through arts education: A case study of a non-formal education arts centre in Barcelona. *Social Inclusion (Lisboa), 10*(2), 85–94. DOI: 10.17645/si.v10i2.4923

Finnegan, R. (2012). *Oral literature in Africa*. Open Book Publishers. DOI: 10.11647/OBP.0025

Fitzgerald, J. L. (2004). Dharma and its translation in the *Mahābhārata. Journal of Indian Philosophy, 32*(5-6), 671–685. DOI: 10.1007/s10781-004-8640-7

Fleenor, J. W. (2000). Trait Approach to Leadership. In Rogelberg, S. (Ed.), *G* (pp. 831–833). Sage Encyclopedia of Industrial and Organizational Psychology.

Fleming, M. J., & Grace, D. M. (2016). Agents of aspiration: The (often unintended) benefits to university students working in outreach programmes. *Educational Research*, *58*(3), 300–318. DOI: 10.1080/00131881.2016.1185316

Flick, U. (2011). Methoden-Triangulation in der qualitativen Forschung. In *Triangulation: Eine Einführung* (pp. 27–49). VS Verlag für Sozialwissenschaften. DOI: 10.1007/978-3-531-92864-7_3

Flood, G. (1996). *An introduction to Hinduism*. Cambridge University Press.

Floyd, S. A. (1995). *The power of Black music: Interpreting its history from Africa to the United States*. Oxford University Press.

Folorunsho, F., & Tanga, P. T. (2021). Grandparenting and extent of fulfilling the psychosocial well-being of adolescents from single parent homes in Buffalo City Metropolitan, South Africa. *African Journal of Social Work*, *11*(2).

Fontaine, L. S. (2017). Redress for linguistics: Residential schools and assimilation in Canada. *British Journal of Canadian Studies*, *30*(2), 183–204. DOI: 10.3828/bjcs.2017.11

Food and Agriculture Organization of the United Nations. (2001). *The state of food insecurity in the world 2001*. FAO.

Food and Agriculture Organization of the United Nations. (2013). *Indigenous peoples' food systems and well-being: Interventions and policies for healthy communities*. FAO.

Food and Agriculture Organization of the United Nations. (2019). *The state of food security and nutrition in the world 2019*. FAO.

Food and Agriculture Organization of the United Nations. (2022). *The state of food security and nutrition in the world 2022*. FAO.

Fuller, D. Q. (2001). Dung mounds and domesticators: Early cultivation and pastoralism in Karnataka. *South Asian Archaeology*, *1*, 117–127.

Gadgil, M., & Guha, R. (2014). *This fissured land: An ecological history of India* (3. impr). Oxford University Press.

Ganeri, J. (2007). *Concealed Art of the Soul: Theories of the Self and Practices of Truth in Indian Ethics and Epistemology*. Oxford University Press. DOI: 10.1093/acprof:oso/9780199202416.001.0001

Ganguly, A. (2024). ECOLOGICAL WISDOM IN VEDIC LITERATURE: A COMPREHENSIVE ANALYSIS. *ENVIRONMENTAL SCIENCE*, 1.

Gape, H. (2023). *A Way of Life: Barolong Boora Tshidi. Mmabatho*. Bokone Bophirima Publishing House.

Garikipati, S. (2008). The impact of lending to women on household vulnerability and women's empowerment: Evidence from India. *World Development*, *36*(12), 2620–2642. DOI: 10.1016/j.worlddev.2007.11.008

Gauthier, P. E., Chungyalpa, D., Goldman, R. I., Davidson, R. J., & Wilson-Mendenhall, C. D. (2025). Mother Earth kinship: Centering indigenous worldviews to address the Anthropocene and rethink the ethics of human-to-nature connectedness. *Current Opinion in Psychology*, *64*, 102042. DOI: 10.1016/j.copsyc.2025.102042 PMID: 40288260

Gemeda, D. O., & Sima, A. D. (2015). The Impacts of Climate Change on African Continent and the Way Forward. *Journal of Ecology and the Natural Environment*, *7*(10), 256–262. DOI: 10.5897/JENE2015.0533

Gentles-Gibbs, N., & Zema, J. (2020). It's not about them without them: Kinship grandparents' perspectives on family empowerment in public child welfare. *Children and Youth Services Review*, *108*, 104650. DOI: 10.1016/j.childyouth.2019.104650

Georgieva, V., & Marinov, P. (2017). Intercultural interactions in a military context. *Land Forces Academy Review*, *22*(3), 153–162. DOI: 10.1515/raft-2017-0021

Ghatak, M. (1999). Group lending, local information and peer selection. *Journal of Development Economics*, *60*(1), 27–50. DOI: 10.1016/S0304-3878(99)00035-8

Ghate, P. (2007). *Indian microfinance: The challenges of rapid growth.* SAGE Publications. DOI: 10.4135/9788132101574

Gheitarani, N., El-Sayed, S., Cloutier, S., Budruk, M., Gibbons, L., & Khanian, M. (2020). Investigating the mechanism of place and community impact on quality of life of rural-urban migrants. *International Journal of Community Well-being*, *3*(1), 21–38. DOI: 10.1007/s42413-019-00052-8

Ghosh, S. C. (1975). Dalhousie, Charles Wood and the Education Despatch of 1854. *History of Education*, *4*(2), 37–47. DOI: 10.1080/0046760750040204

Gifford-Gonzalez, D. (1998). Early pastoralists in East Africa: Ecological and social dimensions. *journal of anthropological archaeology, 17*(2), 166-200.

Gita Press. (2024). *Śrīmad Bhagavad Gītā* (Sanskrit text with Hindi translation and commentary).* Gita Press Gorakhpur.

Gita Press. (2025). *Kalyan: Paryavaran ank* (99th special annual issue).* Gita Press Gorakhpur.

Gkika, E. C., Ntanos, S., & Komisopoulos, F. (2022). The impact of global competition dimensions on economic development. www.researchgate.net

Gladstone, J. W., & Brown, R. A. (2007). Grandparents' and social workers' experience with child welfare system: A case for mutual resources. *Children and Youth Services Review*, *29*(11), 1439–1453. DOI: 10.1016/j.childyouth.2007.07.002

Glickman, M. (1972). The Nuer and the Dinka: A further note. *Man*, *7*(4), 586–594. DOI: 10.2307/2799950

Godihno, T.A. (2012). The ANC's Land Reform Policy: An unsustainable and unsatisfactory compromise. *Pretoria Student Law Review,* 6.

Goji, D. S., Maikasuwa, W., & Ombogutsa, K. Y. (2025). Operation of Colonial Indirect Rule Policy and Political-Social Vices in Nigeria. *International Journal of Law, Politics and Humanities Research.*

Gomez, A. R. (2024). Religion Scholarship on Indigenous Lands: Research, Relationship, and the Potential of Engaged Scholarship. *Journal for the Study of Religion, Nature and Culture, 18*(4), 484–501. DOI: 10.1558/jsrnc.25258

Gómez-Baggethun, E. (2022). Is there a future for indigenous and local knowledge? *The Journal of Peasant Studies, 49*(6), 1139–1157. DOI: 10.1080/03066150.2021.1926994

Gondo, R. (2025). Bibliometric analysis of Indigenous knowledge systems and climate change adaptation literature, 1993–2023. *IFLA Journal*, •••, 1–14. DOI: 10.1177/03400352251317716

Goodland, R. (2001). *The Westernization of Diets: The Assessment of Impacts in Developing Countries.* World Bank.

Gope, L., Behera, S. K., & Roy, R. (2017). Identification of indigenous knowledge components for sustainable development among the Santhal community. *American Journal of Educational Research, 5*(8), 883–891. DOI: 10.12691/education-5-8-8

Gordon, G. A. (2020). *Marimbas in South African schools: gateway instruments for the indigenous African music curriculum* (Doctoral dissertation, Stellenbosch: Stellenbosch University).

Gorjestani, N. (2004). *Indigenous knowledge for development. Protecting and promoting traditional knowledge: Systems, national experiences and international dimensions.*

Gounga, M. E., Xu, S. Y., & Wang, Z. (2019). Nutritional and functional properties of fermented cereal-based products: A review. *Food Reviews International, 35*(7), 640–661.

Gudykunst, W. B. (2003). Cross-cultural and intercultural communication. *Sage (Atlanta, Ga.).*

Gudykunst, W. B. (2005). *Theorising about Intercultural Communication.* Sage.

Gudynas, E. (2011). Buen vivir: Today's tomorrow. *Development, 54*(4), 441–447. DOI: 10.1057/dev.2011.86

Guha, R. (2000). *The unquiet woods: Ecological change and peasant resistance in the Himalaya* (Expanded ed). University of California Press.

Guha, R. (2008). *India after Gandhi: The history of the world's largest democracy.* Harper Perennial.

Gumo, S., Gisege, S. O., Raballah, E., & Ouma, C. (2012). Communicating African spirituality through ecology: Challenges and prospects for the 21st century. *Religions, 3*(2), 523–543. DOI: 10.3390/rel3020523

Gupta, A. K., & Maurya, S. K. (2026). Revivalism in Indian educational philosophy: Reinterpreting ancient wisdom for modern times. *Cogent Education, 13*(1), 2584914. DOI: 10.1080/2331186X.2025.2584914

Gwerevende, S., & Mthombeni, Z. M. (2023). Safeguarding intangible cultural heritage: Exploring the synergies in the transmission of Indigenous languages, dance and music practices in Southern Africa. *International Journal of Heritage Studies, 29*(5), 398–412. DOI: 10.1080/13527258.2023.2193902

Haberman, D. L. (2006). *River of Love in an Age of Pollution: The Yamuna River of Northern India.* University of California Press.

Hagerstrand, T. (1968). *Innovation diffusion as a spatial process.* University of Chicago Press.

Hai, T. N., & Van, Q. N. (2021). Servant Leadership Styles: A Theoretical Approach. *Emerging Science Journal, 5*(2), 245–256. DOI: 10.28991/esj-2021-01273

Hall, M. (1983). Early farming communities of southern Africa: A population discovered. *South African Historical Journal, 15*(1), 1–9. DOI: 10.1080/02582478308671573

Hall, R. (2014). The legacies of the Native Land Act of 1913. Scriptura. *Journal for Contextual Hermeneutics in Southern Africa, 113*(1), 1–13.

Hameed, K., Sirwan, K., Omar, Z. O., Mohammed, E. F., Sourkan, S., Salih, A. M., & Yaqub, K. Q. (2025). Addressing the issue of poverty, economic growth, health matters, and environmental challenges. *International Journal of Scientific Research in Modern Science and Technology, 4*(3), 18–32. DOI: 10.59828/ijsrmst.v4i3.303

Hamm, M. W., & Bellows, A. C. (2003). Community food security and nutrition educators. *Journal of Nutrition Education and Behavior, 35*(1), 37–43. DOI: 10.1016/S1499-4046(06)60325-4 PMID: 12588679

Hammons, C. S. (2016). Indigenous Religion, Christianity and the State: Mobility and Nomadic Metaphysics in Siberut, Western Indonesia. *The Asia Pacific Journal of Anthropology, 17*(5), 399–418. DOI: 10.1080/14442213.2016.1208676

Haring, E. (2020). *Gender and military organisations. In the Gender and Security Agenda*. Routledge.

Harper, M. (2002). Self-help groups and Grameen Bank groups: What are the differences? *Small Enterprise Development, 13*(3), 31–38.

Harrison, D. A., Price, K. H., Gavin, J. H., & Florey, A. T. (2002). Time, teams, and task performance: Changing effects of surface-and deep-level diversity on group functioning. *Academy of Management Journal, 45*(5), 1029–1045. DOI: 10.2307/3069328

Harrison, M. (1994). *Public health in British India: Anglo-Indian preventive medicine, 1859-1914*. Cambridge University Press.

Harvey, A., & Russell-Mundine, G. (2019). Decolonising the curriculum: Using graduate qualities to embed Indigenous knowledges at the academic cultural interface. *Teaching in Higher Education, 24*(6), 789–808. DOI: 10.1080/13562517.2018.1508131

Harvey, G. (2019). Animism and ecology: Participating in the world community. *The Ecological Citizen, 3*(1), 79–84.

Hays, T. R. (1975). Neolithic settlement patterns in Saharan Africa. *South African Archaeological Bulletin, 30*(117/118), 29–33. DOI: 10.2307/3888048

Heinecken, L. (2020B). Political Reform: Integration, Representivity and Managing Diversity. South Africa's Post-Apartheid Military: Lost in Transition and Transformation, pp.71-86.

Heinecken, L. (2011). A diverse society, a representative military? The complexity of managing diversity in the South African Armed Forces. *Scientia Militaria: South African Journal of Military Studies, 37*(1), 25–49. DOI: 10.5787/37-1-58

Heinecken, L. (2017). *South Africa's post-apartheid military: Lost in transition and transformation*. Wits University Press.

Heinecken, L., & Nel, M. (2021). Leadership, organisational culture and transformation in the SANDF. *African Security Review*, *30*(3), 215–230.

Herbst, J., Mills, G., & Nicolson, G. (2003). *Botswana: A Political History*. Macmillan Botswana.

Hermanto Candra, P., Hamim, M., & Maulana Khabir, G. (2024). Water Conservation through Local Wisdom as A Manifestation of Social Justice and Ecological Justice. *BIO Web of Conferences, 96*, 04007. DOI: 10.1051/bioconf/20249604007

He, S. (2023). Liberal arts education and influence on employability: A case study of a Sino-foreign university. *Chinese Education & Society*, *56*(5-6), 350–368. DOI: 10.1080/10611932.2024.2303918

He, Y., Zeng, L., Chen, W., Gong, C., & Shen, Z. (2023). Bilateral weighted relative total variation for low-dose CT reconstruction. *Journal of Digital Imaging*, *36*(2), 458–467. DOI: 10.1007/s10278-022-00720-w PMID: 36443529

Higgins, N. (2022). indigenous Intangible Cultural Heritage. *Laws*, *11*(47), 2–15. https://uwspace.uwaterloo.ca/bitstream/handle/10012/11668/Stevens_Julia.pdf?sequence=1&isAllowed=y

Higgs, P. (2012). African philosophy and the decolonisation of education in Africa. *Educational Philosophy and Theory*, *44*(sup2, s2), 37–55. DOI: 10.1111/j.1469-5812.2011.00794.x

High Level Panel of Experts on Food Security and Nutrition. (2020). *Food security and nutrition: Building a global narrative towards 2030*. Committee on World Food Security.

Hiltebeitel, A. (2011). *Dharma: Its early history in law, religion, and narrative*. Oxford University Press., DOI: 10.1093/acprof:oso/9780195394238.001.0001

Hitchcock, R. K. (2002). Coping with Uncertainty: Adaptive Responses to Drought. *Sustainable livelihoods in Kalahari environments: A contribution to global debates*, 161-171.

Hlalele, D. J. (2019). Indigenous knowledge systems and sustainable learning in rural South Africa. *Australian and international journal of rural education, 29*(1), 88-100.

Hodder, I., & Hutson, S. (2003). *Reading the past: Current approaches to interpretation in archaeology*.

Hodder, I. (1995). *Interpreting archaeology: Finding meaning in the past*. Psychology Press.

Hole, Y., Khedkar, E. B., & Pawar, S. (2019). The significance of pilgrimage tourism to sustainable development with special reference to the Indian context. *African Journal of Hospitality, Tourism and Leisure*, *8*(3), 1–12.

Holl, A. F. C. (1998). The Dawn of African Pastoralisms: An Introductory. *Journal of Anthropological Archaeology*, *17*(2), 81–96. DOI: 10.1006/jaar.1998.0318

Holl, A. F. C. (2010). The Archaeology of Africa. *ARCHEOLOGY, II*, 20–66.

Holl, A. F. C. (2022). Archaeology of African Agro-systems: A Macro-Evolutionary Perspective. *International Journal of Archaeology*, *10*(1), 6–19.

Hoppers, C. O. (2025). Research on Indigenous Knowledge Systems: The Search for Cognitive Justice. In *Indigenous Voices and Decolonising Lifelong Education.*

Hoppers, C. A. O. (Ed.). (2002). *Indigenous Knowledge and the Integration of Knowledge Systems: Towards a philosophy of articulation.* New Africa Books.

Hoppers, C. O. (2025). Research on Indigenous knowledge systems: The search for cognitive justice. In *Indigenous Voices and Decolonising Lifelong Education* (pp. 7–24). Routledge. DOI: 10.4324/9781032627465-2

Hountondji, P. J. (1997). *Endogenous knowledge: Research trails.* CODESRIA.

Hourdequin, M. (2025). Intergenerational ethics and climate change. *Wiley Interdisciplinary Reviews: Climate Change, 16*(1), e934. DOI: 10.1002/wcc.934

Huang, Z., & Xu, T. (2022). Research on Knowledge Management of Intangible Cultural Heritage Based on Linked Data. *Mobile Information Systems, 2022*, 1–14. Advance online publication. DOI: 10.1155/2022/3384391

Hudson, R. (2006). Regions and place: Music, identity and place. *Progress in Human Geography, 30*(5), 626–634. DOI: 10.1177/0309132506070177

Huffman, T. N. (1986). Archaeological evidence and conventional explanations of southern Bantu settlement patterns. *Africa, 56*(3), 280–298. DOI: 10.2307/1160685

Huffman, T. N., Elburg, M., & Watkeys, M. (2013). Vitrified cattle dung in the Iron Age of southern Africa. *Journal of Archaeological Science, 40*(10), 3553–3560. DOI: 10.1016/j.jas.2013.03.026

Hyeonjung, C. (2021) President Biden and Climate Change: Policy and Issues. The Asan Institute for Policy Studies. https://ssrn.com/abstract=3847183, May 2021.

Ibrahim, M. (2023). The people with iron skin: Protective charms, traditional religion, and vigilante authority in Lagos, Nigeria. *Culture and Religion, 23*(3), 240–261. DOI: 10.1080/14755610.2024.2344563 PMID: 39108276

Ihara, E. S., Tompkins, C. J., & Sonethavilay, H. (2012). Culture and familism: An exploratory case study of grandparent-headed households. *Journal of Intergenerational Relationships, 10*(1), 34–47. DOI: 10.1080/15350770.2012.645737

Ilieva, L., Bozhinova, M., Todorova, L., Marinov, M., Ismailov, T., & Spasova, S. (2024). Festivals: An Opportunity for Sustainable Development of Tourism Regions. *Revista de Gestão Social e Ambiental, 18*(3), e05234. DOI: 10.24857/rgsa.v18n11-069

Inchaurrondo, A. M., Bailon, A. B., Vicente, C. P., Tio, E. V., & Bolos, A. M. (2015). Risk and protective factors associated with kinship care. *Child & Adolescent Social Work Journal, 32*(5), 417–427. DOI: 10.1007/s10560-015-0382-8

Infield, M. 2002. The values of conservation: whose values? Whose sense of beauty, integrity and equity? In Borrin-Fey- erabend, G. (Ed.), Policy Matters: Sustainable management and Co-management of natural resources. CEESP: IUCN.

Ingelaere, B. (2016). *Inside Rwanda's Gacaca courts: Seeking justice after genocide*. University of Wisconsin Press. DOI: 10.2307/jj.36032571

Inkata Freedom Party. (2018). Address by Prince Mangosuthu Buthelezi MP Inkosi of the Buthelezi Clan Traditional Prime Minister to the Zulu Monarch and Nation and President of the Inkatha Freedom Party. Lekgotla of the National House of Traditional Leaders and the Executive Committees of Provincial Houses. https://ifp.org.za

Intergovernmental Panel on Climate Change. (2022). [*Impacts, adaptation and vulnerability*. IPCC.]. *Climatic Change*, •••, 2022.

Isabirye, J. (2021). Can indigenous music learning processes inform contemporary schooling? *International Journal of Music Education*, *39*(2), 151–166. DOI: 10.1177/0255761421996373

Issa, M., Khan, H. K., Hussain, M. S., & Ali, M. A. (2023). Language erosion: An overview of the declining status of indigenous languages in Gilgit-Baltistan, Pakistan. *Language*, *25*(2).

Iticha, B., & Husen, A. (2019). Adaptation to climate change using indigenous weather forecasting systems in Borana pastoralists of southern Ethiopia. *Climate and Development*, *11*(7), 564–573. DOI: 10.1080/17565529.2018.1507896

Jack, G., & Westwood, R. (2006). Postcolonialism and the politics of qualitative research in international business. *MIR. Management International Review*, *46*(4), 481–500. DOI: 10.1007/s11575-006-0102-x

Jacoba, R. C. (2025). Sacred ties to ancestral land: Reclaiming indigenous practices in environmental stewardship in the Cordillera. *Media: Jurnal Filsafat dan Teologi, 6*(1), 1-30.

Jain, P. (2024). Sustaining Dharma, Sustainable Ecology: Dharma as Rural Environmental Ethics. In *The Routledge Companion to Indian Ethics* (pp. 223-232). Routledge India.

Jain, D., & Sharma, G. P. (2013). Solar drying of vegetables: A review. *Journal of Food Science and Technology*, *50*(2), 211–219. DOI: 10.1007/s13197-010-0161-7 PMID: 23572736

Jain, M. P. (2014). *Outlines of Indian legal and constitutional history* (7th ed.). LexisNexis.

Jakes, V. (2024). The role of traditional knowledge in sustainable development. *International Journal of Humanities and Social Science*, *3*(2), 40–55.

Jalal, F. N., & Kumar, M. G. S. (2013). Water quality assessment of Pamba River of Kerala, India in relation to Pilgrimage season. *International Journal of Research*, *3*(1), 341–347. https://www.cabdirect.org/cabdirect/abstract/20133091512

Jamison, S. W., & Brereton, J. P. (2014). *The Rigveda: The earliest religious poetry of India* (Vols. 1–3). Oxford University Press.

Jarc, J. (2015). Laissez-Faire Leadership. www.researchgate.net

JASROTIA, R., & MANHAS, A. S. (2024). Ancient Wisdom for Modern Environmental Problem. *Journal of Applied Social Science*, *11*(3&4), 137–141.

Jayanthi, V. (2001). Vegetarianism in India. *Peritoneal Dialysis International : Journal of the International Society for Peritoneal Dialysis*, *21*(3, suppl), 322–325. DOI: 10.1177/089686080102103S57 PMID: 11887845

Jessen, T. D., Ban, N. C., Claxton, N. X., & Darimont, C. T. (2022). Contributions of Indigenous Knowledge to ecological and evolutionary understanding. *Frontiers in Ecology and the Environment*, *20*(2), 93–101. DOI: 10.1002/fee.2435

Jha, G. (1999). *Manusmrti 10 Volumes (Complete Series): With the'Manubhasya'of Medhatithi*. Motilal Banarsidass.

Jha, J. (2024). *Beneath the sun: Equality for everyone- The spirit of Chhath Festival* (1st ed.). Self published.

Jindal, K., Singhi, R., & Srivastava, A. (2025). Indian Ethos for Sustainable Business Development. In *Business Management and Ethics* (pp. 171–189). Emerald Publishing Limited. DOI: 10.1108/978-1-83549-254-320251009

Jing, X., Tan, F., & Zhang, M. (2021). Digital Application of Intangible Cultural Heritage from the Perspective of Cultural Ecology. *Journal of Smart Tourism*, *1*(1), 41–52. DOI: 10.52255/smarttourism.2021.1.1.6

Johansen, P. G. (2004). Landscape, monumental architecture, and ritual: A reconsideration of the South Indian ashmounds. *Journal of Anthropological Archaeology*, *23*(3), 309–330. DOI: 10.1016/j.jaa.2004.05.003

Johnson, L. R. (2018). Basotho culture and the prayers for rain: Where climate change converges. *Southern African Journal for Folklore Studies*, *28*(2), 1–19. DOI: 10.25159/1016-8427/3799

Johnston, T. F. (1973). Aspects of Tswana Music. *Anthropos*, (H. 5./6), 889-896.

Jonas, H. (1984). *The imperative of responsibility: In search of an ethics for the technological age*. University of Chicago Press. DOI: 10.7208/chicago/9780226850337.001.0001

Jones, E. O. (2024). Indigenous knowledge management practices in subsistence farming: A comprehensive evaluation. *Sustainable Technology and Entrepreneurship*, *3*(2), 100058. DOI: 10.1016/j.stae.2023.100058

Jordan, E., Ng'weno, B., & Spencer-Oatey, H. (2020). Developing Global Leaders: Insight from African Case Studies. Palgrave: MacMillan.

Jyotsna, J. H., & Sai, L. P. (2022). Modelling pilgrim-tourist experience in Hindu religious destinations: An Interactive Qualitative Analysis. *Journal of Tourism and Cultural Change*, *21*(3), 364–382. DOI: 10.1080/14766825.2022.2095914

Kabir, G. M. (2022). Revisiting Javanese Islam: Towards a New Paradigm in Religious Studies. *Jurnal Dinamika Penelitian: Media Komunikasi Sosial Keagamaan*, *21*(2), 315–330.

Kahui, V., Armstrong, C. W., & Aanesen, M. (2024). Comparative analysis of Rights of Nature (RoN) case studies worldwide: Features of emergence and design. *Ecological Economics*, *221*, 108193. DOI: 10.1016/j.ecolecon.2024.108193

Kanaujiya, A. K., & Tiwari, V. (2022). Crowd management and strategies for security and surveillance during the large mass gathering events: *The Prayagraj Kumbh Mela 2019 Experience. National Academy Science Letters*, *45*(3), 295–298. DOI: 10.1007/s40009-022-01114-w PMID: 35431383

Kane, P. V. (1977). *History of Dharmaśāstra* (2nd ed.). (Vols. 1–5). Bhandarkar Oriental Research Institute.

Kang'ethe, S. M., Mugedya, S., & Nomngenyiya, T. (2022). Experiences of grandmothers looking after grandchildren exhibiting delinquent behaviours in Hill Crest location, Alice township, Eastern Cape: Implications for social work. *Social Work/Maatskaplike Werk, 58*(3 & 4).

Kant, I. (1998). *Groundwork of the metaphysics of morals* (Gregor, M. J., Trans.). Cambridge University Press. DOI: 10.1017/CBO9780511809590.004

Karim, Z. I. A., & Maat, S. M. (2019). Employability skills model for engineering technology students. *Journal of Technical Education and Training*, *11*(2).

Karlan, D. (2007). Social connections and group banking. *Economic Journal (London)*, *117*(517), F52–F84. DOI: 10.1111/j.1468-0297.2007.02015.x

Kashyap, A. (2025). Significance of Folk Musical Instruments in reinforcing Cultural Identity and Belonging: Discerning the dialogue between Soundscape Ecology-Ethnobiology-Ethnomusicology. *JOURNAL OF ADVANCE AND FUTURE RESEARCH*, *3*(7), 28–53.

Kasim, A. (2013). Balancing tourism and religious experience: Understanding devotees' perspectives on Thaipusam in Batu Caves, Selangor, Malaysia. In *Event Tourism and Cultural Tourism* (pp. 203–218). Routledge.

Kasiram, M., & Holscher, D. (2015). Understanding the challenges and opportunities encountered by the elderly in urban KwaZulu Natal, South Africa. *South African Family Practice : Official Journal of the South African Academy of Family Practice/Primary Care*, *57*(6), 380–385. DOI: 10.1080/20786190.2015.1078154

Katelieva, M., Muhar, A., & Penker, M. (2020). Nature-related knowledge as intangible cultural heritage: Safeguarding and tourism utilisation in Austria. *Journal of Tourism and Cultural Change*, *18*(6), 673–689. DOI: 10.1080/14766825.2019.1693581

Kaulemu, D. (2022). *The Ethics and Politics of Conceptualizing Values in Africa*. Springer International Publishing. DOI: 10.1007/978-3-031-12938-4_2

Kaya, H. O., & Seleti, Y. N. (2013). African indigenous knowledge systems and relevance of higher education in South Africa. *International Education Journal: Comparative Perspectives, 12*(1).

Keebine-Sibanda, M. J. (2021). Academics' conceptions of higher education decolonisation. *South African Journal of Higher Education*, *35*(3), 182–199.

Keegan, B. (2022). Contemporary Native American and Indigenous Religions: State of the field. *Religion Compass*, *16*(9), e12448. Advance online publication. DOI: 10.1111/rec3.12448

Keegan, T. (1985). Crisis and catharsis in the development of capitalism in South African agriculture. *African Affairs*, *84*(336), 371–398. DOI: 10.1093/oxfordjournals.afraf.a097699

Keleman Saxena, A., Cadima Fuentes, X., Gonzales Herbas, R., & Humphries, D. L. (2016). Indigenous food systems and climate change: Impacts of climatic shifts on the production and processing of native and traditional crops in the Bolivian Andes. *Frontiers in Public Health, 4*, 20. DOI: 10.3389/fpubh.2016.00020 PMID: 26973824

Kelley, S. T., Whitley, D. M., & Lamis, D. A. (2023). An intervention to improve the well-being of families in which African American grandmothers are raising grandchildren: A longitudinal mediation analysis. *Children and Youth Services Review, 151*, 107021. DOI: 10.1016/j.childyouth.2023.107021

Kennedy, D. (2023). *The Magic Mountains: Hill Stations and the British Raj* (1st ed.). University of California Press., DOI: 10.2307/jj.8085339

Kgatle, M.S. (2021). The balance between theory and praxis in South African Pentecostalism: Patmos Bible School as a model. *Theologica,* 31, 84-98. DOI: 31, 6.DOI: 10.18820/23099089/actat.sup

Khan, K. B. (2025). Driving the Indigenous Knowledge System Quest Through Music and Dance: Promoting Zimbabwe's Heritage Through Indigenous Songs and Dance. In *Harnessing Indigenous Epistemologies for Sustainable Progress* (pp. 253-272). IGI Global Scientific Publishing.

Khan, M. R., Khan, H. U. R., Lim, C. K., Tan, K. L., & Ahmed, M. F. (2021). Sustainable tourism policy, destination management and sustainable tourism development: A moderated-mediation model. *Sustainability (Basel), 13*(21), 12156. DOI: 10.3390/su132112156

Khan, R., Salim, A., Baluch, Z., Mehreen, K., & Kaleem, M. (2021). British Colonialism and the Indirect Rule: A Hierarchical Administrative Structure to Control the Unruly Tribes. *Humanities & Social Sciences Reviews, 9*(3), 1366–1372. DOI: 10.18510/hssr.2021.93136

Khatib, S. F. (2025). Social Trust, Environmental Violations, and Remedial Actions in China: Shahab, Y., Ye, Z., Liu, J., & Nadeem, M.(2024). Journal of Business Ethics, 1–18. . *Social and Environmental Accountability Journal, 45*(1), 81-83.DOI: 10.1007/s10551-024-05818-5

Khoshoo, T. N. (1999). The dharma of ecology. *Current Science, 77*(9), 1147–1153.

Khoza, R. J. (2011). *Attuned leadership: African humanism as compass.* Penguin Books.

Khunwane, M. R. (2019). *A Comparative Analysis of the influence of Folklore on the works of the following African writers: Chinua Achebe, Eskia Mphahlele, Ngungi wa Thiongo'and Andrew Nkadimeng: An Afrocentric approach* (Doctoral dissertation).

King Letsie, I. I. I. (2024). *Speech on Lesotho's Bicentenary and the 58th Anniversary of Independence* [Audio Recording]. Government of Lesotho.

King, R. D. (1998). *Nehru and the language politics of India.* Oxford Univ. Pr.

King, M. (2003). *The Penguin history of New Zealand.* Penguin Books.

Kipyegon, K. A., Wanyama, M. N., & Mapaya, M. G. (2025). Epistemics of African Music Performance: Navigating Knowledge, Tradition, and the Digital Frontier. *African Musicology Online, 14*(2), 14–24. DOI: 10.58721/amo.v14i2.1107

Kirton, G., & Greene, A. M. (2016). Understanding diversity management in the UK. In *Corporate Social Responsibility and Diversity Management* (pp. 59–73). Theoretical Approaches and Best Practices.

Klaas-Makolomakwe, N. G., & Raniga, T. (2021). A critical review of the role and functions of Traditional Leaders. *The Thinker: Opinion, 86*(8), 54–60. DOI: 10.36615/thethinker.v87i2.534

Klein, N. (2014). *This changes everything: Capitalism vs. The climate*. Simon and Schuster.

Knorr, D., & Augustin, M. A. (2025). Towards resilient food systems: Interactions with indigenous knowledge. *Trends in Food Science and Technology, 156*(December 2024), 104875. DOI: 10.1016/j.tifs.2025.104875

Knouse, S. B. (2001). Diversity and Military Effectiveness: How Diversity Enhances Performance and Cohesion in Military Units. *Armed Forces and Society*, *27*(3), 473–494.

Kobo, F. A. (2019). Spirituality trapped in androcentric celebrity cults in South Africa post- 1994. *Hervormde Teologiese Studies*, *75*(3), 1–7. DOI: 10.4102/hts.v75i3.5593

Kohn, E. (2013). *How forests think: Toward an anthropology beyond the human*. Univ of California Press.

Kohsaka, R., & Rogel, M. (2021). Traditional and local knowledge for sustainable development: Empowering the indigenous and local communities of the world. In *Partnerships for the Goals* (pp. 1261–1273). Springer International Publishing. DOI: 10.1007/978-3-319-95963-4_17

Koike, M. (2019). Indigenous and local knowledge promoting SDGs in Indonesia: The case of the Sumbanese cultural festival. *Journal of Environmental Science and Sustainable Development*, *2*(2), 171–191. DOI: 10.7454/jessd.v2i2.1034

Kolodziej-Smith, R. (2016). *A preliminary study of workgroup dynamics: A critique and reconceptualisation of Oetzel's effective intercultural workgroup communication theory*. Wayne State University.

Kone, M. T., Budeli, A. E., & Ramaano, R. (2025). Toward sustainable food system: The role of livestock farming in Matiyani village in the Vhembe Biosphere Reserve, South Africa. *Frontiers in Sustainable Food Systems*, *9*, 1637083. DOI: 10.3389/fsufs.2025.1637083

Korejan, M. M., & Shahbaz, H. (2016). An analysis of the Transformational Leadership Theory. *Revue des Sciences Fondamentales et Appliquées*, *8*(3), 452. DOI: 10.4314/jfas.v8i3s.192

Kose, L. S., Moteetee, A., & Van Vuuren, S. (2015). Ethnobotanical survey of medicinal plants used in the Maseru district of Lesotho. *Journal of Ethnopharmacology*, *170*, 184–200. DOI: 10.1016/j.jep.2015.04.047 PMID: 25957810

Kovačević, M., Dekker, T. J., & van der Velden, R. (2024). Employability development in undergraduate programmes: How different is liberal arts education? *Teaching in Higher Education*, *29*(8), 2184–2204. DOI: 10.1080/13562517.2023.2212602

Kuhnlein, H. V., Erasmus, B., & Spigelski, D. (Eds.). (2013). *Indigenous peoples' food systems: The many dimensions of culture, diversity and environment for nutrition and health*. FAO & Centre for Indigenous Peoples' Nutrition and Environment.

Kuhnlein, H. V., & Receveur, O. (1996). Dietary change and traditional food systems of indigenous peoples. *Annual Review of Nutrition, 16*(1), 417–442. DOI: 10.1146/annurev.nu.16.070196.002221 PMID: 8839933

Kumah-Abiwu, F. (2022). Urban education and academic success: The case of higher achieving Black males. *Urban Education, 57*(9), 1565–1591. DOI: 10.1177/0042085919835284

Kumar, D., Raychaudhuri, T., & Habib, I. (Eds.). (2005). The Cambridge economic history of India. *Volume II, C.1757-2003* (Édition augmentée). Orient Longman, in association with Cambridge University Press.

Kumar, A., Kumar, A., & Kumari, S. (2025). Cultural ecology and conservation: Indigenous knowledge systems in Jharkhand. *Journal of Native India & Diversity Studies, 1*(1), 45–58.

Kumar, A., & Prasad, A. (2025). Sustainable Consumption Through the Lens of Indian Tribal Wisdom. [IJSSR]. *International Journal of Social Science Research, 2*(2), 186–195.

Kumar, N., Raghunathan, K., Arrieta, A., Jilani, A., & Pandey, S. (2021). The power of the collective empowers women: Evidence from self-help groups in India. *World Development, 146*, 105579. DOI: 10.1016/j.worlddev.2021.105579 PMID: 34602708

Kuper, A. (1975). The social structure of the Sotho-speaking peoples of Southern Africa. *Africa, 45*(1), 67–81. DOI: 10.2307/1158780

Kwabena-Nketia. (1974). *The Music of Africa*. W W Norton & Company.

Labour Research Service. (2021). *Worker participation and engagement models in South Africa*. Labour Research Service.

Lagakos, D. (2020). Urban-rural gaps in the developing world: Does internal migration offer opportunities? *The Journal of Economic Perspectives, 34*(3), 174–192. DOI: 10.1257/jep.34.3.174

Lamb, G. (2018). *New Model Army: The transformation of the South African National Defence Force 1994-2004*. Stellenbosch, South Africa.

Lander, F., & Russell, T. (2018). The archaeological evidence for the appearance of pastoralism and farming in southern Africa. *PLoS One, 13*(6), e0198941. DOI: 10.1371/journal.pone.0198941 PMID: 29902271

Lane, P. (2013). Trajectories of pastoralism in northern and central Kenya: an overview of the archaeological and environmental evidence. *Pastoralism in Africa: Past, present and future*, 104-144.

Lane, P. (1994). The use and abuse of ethnography in the study of the southern African Iron Age. *Azania, 29*(1), 51–64. DOI: 10.1080/00672709409511661

Lázaro Ortiz, S., & Jiménez de Madariaga, C. (2022). The UNESCO convention for the safeguarding of the intangible cultural heritage: A critical analysis. *International Journal of Cultural Policy, 28*(3), 327–341. DOI: 10.1080/10286632.2021.1941914

Lazzari, M., Larsen, P. B., & Orlandi, F. (2024). Heritage and decoloniality: Global exchanges and unresolved questions in sedimented landscapes of injustice. *American Anthropologist*. Advance online publication. DOI: 10.1111/aman.13951

Le Grange, L. (2023). Decolonisation and anti-racism: Challenges and opportunities for (teacher) education. *Curriculum Journal*, *34*(1), 8–21. DOI: 10.1002/curj.193

Le Roux, L. (2005). *The post-apartheid South African military: Transforming with the nation. Evolutions and Revolutions: A Contemporary History of Militaries in Southern Africa*. Institute for Security Studies.

Lebaka, M. E. K. (2018). Music of politics and Religion supporting constitutional values in South Africa. *Studia Historiae Ecclesiasticae*, *44*(1), 1–15. DOI: 10.25159/2412-4265/2343

Lebaka, M. E. K. (2022). Informal learning of indigenous music and dance through observation and imitation: The case of Bapedi children's musical arts. *European Journal of Language and Literature Studies*, *8*(1), 55–67. DOI: 10.26417/658qju31

Lee, D. H. (2008). Re-imagining urban space: Mobility, connectivity, and a sense of place. *Mobile technologies: From telecommunications to media* (pp. 235-251)https://doi.org/DOI: 10.4324/9780203884317

Lema, M. W. (2025). Sustaining rural livelihoods through participatory water governance: A review of community-driven water resource management models in East and Central Africa. *Frontiers in Water*, *7*, 1704600. DOI: 10.3389/frwa.2025.1704600

Leslie, C. M. (2022). *Asian Medical Systems: A Comparative Study* (1st ed.). University of California Press.

Levstek, M., & Banerjee, R. (2021). A model of psychological mechanisms of inclusive music-making: Empowerment of marginalized young people. *Music & Science*, *4*, 20592043211059752. DOI: 10.1177/20592043211059752

Lewis, J. E., Whaanga, H., & Yolgörmez, C. (2025). Abundant intelligences: Placing AI within Indigenous knowledge frameworks. *AI & Society*, *40*(4), 2141–2157. DOI: 10.1007/s00146-024-02099-4

Liang, N., & Segalas, J. (2025). A deep and shallow sustainability intervention framework: A Taoist-inspired approach to systemic sustainability transitions. *Sustainability (Basel)*, *17*(11), 5170. DOI: 10.3390/su17115170

Lipner, J. (2010). *Hindus: Their religious beliefs and practices* (2nd ed.). Routledge.

Lithole, K. D. (2022). Afro-centred Heritage Management Approaches for Mapungubwe World Heritage Site along Vele Colliery Coal Mining in Limpopo Province, South Africa, Doctor of Philosophy in African Studies thesis, University of Venda,www.univen.ac.za

Logan-Terry, A., & Damari, R. R. (2015). Key culture-general interactional skills for military personnel. *Procedia Manufacturing*, *3*, 3990–3997. DOI: 10.1016/j.promfg.2015.07.951

Lohr, S. L. (2021). *Sampling: Design and Analysis*. Chapman and Hall/CRC. DOI: 10.1201/9780429298899

Luca, C. (2024). The Effect of Globalization on Preserving Indigenous Knowledge in Education. *Research Gate, May*.

Lukawiecki, J., Jeddore, G., Roth, R., & Moola, F. (2025). Practicing "Netukulimk": Indigenous governance of culturally significant species in Ktaqamkuk (Newfoundland, Canada). *Facets*, *10*, 1–15. DOI: 10.1139/facets-2025-0152

Lwoga, E. T., Ngulube, P., & Stilwell, C. (2011). Challenges of managing indigenous knowledge with other knowledge systems for agricultural growth in sub-Saharan Africa. *Libri. International Journal of Libraries & Information Services, 61*(3).

Maarif, S. (2019). Indigenous Religion Paradigm: Re-interpreting Religious Practices of Indigenous. *Studies in Philosophy*, 103–121. https://doi.org/http://doi.org/10.15068/00155157

Maarif, S. (2015). Ammatoan Indigenous Religion and Forest Conservation. *Worldviews: Global Religions, Culture, and Ecology, 19*(2), 144–160. DOI: 10.1163/15685357-01902005

Maarif, S. (2023). Human (Relational) Dignity: Perspectives of Followers of Indigenous Religions of Indonesia. *Religions, 14*(7), 848. DOI: 10.3390/rel14070848

Mabetha, K., DeWet-Billings, N. C., & Odimegwu, C. O. (2024). Exploring the effects of kin's caregivers' family functioning on child survival in South Africa: An application of the Family Systems Theory. *Child & Family Social Work, 29*(1), 58–77. DOI: 10.1111/cfs.13051

Macarthur, J. D. (1984). The political economy of underdevelopment A. K. Bagchi Modern Cambridge Economics, Cambridge University Press, Cambridge, 1982, 280 + viii pp. *Public Administration and Development, 4*(1), 104–105. DOI: 10.1002/pad.4230040112

Macaulay, T. B. (1862). *Macaulay's Minutes on Education in India.* (Original work published Baptist Mission Press)

Machiridza, E. R., Loveness, B., & Ottiliah, M. (2025). The Effects of Cultural Erosion on Indigenous Language Development at the Early Childhood Level in Multicultural Societies in Zimbabwe: A Case Study of Chimanimani Area. *International Journal of Research and Innovation in Social Science, 9*(3), 1431–1446. DOI: 10.47772/IJRISS.2025.90300112

MacIntyre, A. (2007). *After virtue: A study in moral theory* (3rd ed.). University of Notre Dame Press. (Original work published 1981)

Maćkowiak, A. M. (2024). Christian and Indigenous: Multiple "Religions" in Contemporary Toraja Funerals. *Religions, 15*(9), 1112. DOI: 10.3390/rel15091112

Madduma Bandara, C. M. (1985). Catchment ecosystems and village Tank Cascades in the dry zone of Sri Lanka are a time-tested system of land and water resource management. In *Strategies for river basin management: Environmental integration of land and water in a river basin* (pp. 99–113). Springer Netherlands. DOI: 10.1007/978-94-009-5458-8_11

Mader, P. (2013). Rise and fall of microfinance in India: The Andhra Pradesh crisis in perspective. *Strategic Change, 22*(1–2), 47–66. DOI: 10.1002/jsc.1921

Made, Y. A. D. N., & Somawati, A. V. (2021). THE CONCEPT OF YAJNA IN VARAHA PURANA. *Vidyottama Sanatana: International Journal of Hindu Science and Religious Studies, 5*(2), 159–168.

Maepa, N. M., Matsoso, A. N., & Muedi, H. T. (2025). Documenting and Preserving the Indigenous Foods to Ensure Food Security by the Moshana People in the Northwest Province, South Africa. In *Revaluation and Preservation of Indigenous Knowledge Systems in Modern Society* (pp. 197–214). IGI Global Scientific Publishing. DOI: 10.4018/979-8-3693-7964-6.ch011

Magadla, S. (2023A). *Guerrillas and Combative Mothers: Women and the Armed Struggle in South Africa*. University of Kwa-Zulu Natal Press. DOI: 10.4324/9781003455981

Magadla, S. (2023B). Military identity and post-apartheid integration in South Africa. *African Security Review*, *32*(2), 120–135.

Magni, G. (2017). Indigenous knowledge and implications for the sustainable development agenda. *European Journal of Education*, *52*(4), 437–447. DOI: 10.1111/ejed.12238

Magubane, B. (1996). *The making of a racist state: British imperialism and the Union of South Africa, 1875-1910*. Africa World Press.

Magubane, B. (1996). *The Making of a Racist State: British Imperialism and the Union of South Africa, 1875-1910*. Africa World Press.

Makgala, J. (2009). *History of the Bakgatla baga Kgafela in Botswana and South Africa*. Crink Publishers.

Makgamatha, M. (2024). Makgamatha-2024-analysing-the-pros-and-cons-of-party-coalitions- in-south-africa-unravelling-jacob-zuma-s-rope-a-dope. *Journal of Public Administration*, *59*(3), 531–540. DOI: 10.53973/jopa.2024.59.3.a12

Malapane, O. L., Chanza, N., & Musakwa, W. (2024). Transmission of indigenous knowledge systems under changing landscapes within the vhavenda community, South Africa. *Environmental Science & Policy*, *161*(August), 103861. DOI: 10.1016/j.envsci.2024.103861

Maletangwao Cultural Troupe. (2022). Nkwe. Thipa fa e se bogale ya lwediwa album. URL: https://www.youtube.com/watch?v=-YFDFLlZaE8&list=RD-YFDFLlZaE8&start_radio=1

Maletangwao Cultural Troupe. 2022. *Kgori Mpele Bana*. Thipa fa e se bogale ya lwediwa album. URL: https://www.youtube.com/watch?v=hOfzRhJ6w_w&list=RDhOfzRhJ6w_w&start_radio=1

Malhotra, R., & Babaji, S. D. (2020). *Sanskrit non-translatables: The importance of Sanskritizing English*. Manjul Publishing.

Malinga-Masumba, T. (2015). The nature of relationships between orphans and their kinship carers in Botswana. *Child & Family Social Work*, *20*(3), 257–266. DOI: 10.1111/cfs.12121

Mamabolo, M., Le Roux, J., & Muswede, T. (2025). *Cross-Cultural Communication Challenges in Basic Military Training: The Case of Recruits in the South African Air Force*. Southern African Journal for Folklore Studies.

Mamaleka, M. (2020). *Developing guidelines for indigenous practices: A case study of Makhuduthamaga municipality at Sekhukhune district, Limpopo province*. Limpopo.

Mamdani, M. (2018). Citizen and subject: Contemporary Africa and the legacy of late colonialism.

Manful, E., & Cudfoe, E. (2018). Is kinship failing? Views on informal support by families in contact with social services in Ghana. *Child & Family Social Work*, *23*(4), 617–624. DOI: 10.1111/cfs.12452

Mangaliso, M. P. (2001). Building competitive advantage from ubuntu: Management lessons from South Africa. *The Academy of Management Perspectives*, *15*(3), 23–33. DOI: 10.5465/ame.2001.5229453

Mani, L. (1998). *Contentious traditions: The debate on Sati in colonial India*. University of California Press. DOI: 10.1525/9780520921153

Manyike, T. V., & Shava, S. (2018). The decolonial role of African indigenous languages and indigenous knowledges in formal education processes. *Indilinga*, *17*(1), 36–52.

Mapara, J. (2009). Indigenous knowledge systems in Zimbabwe: Juxtaposing postcolonial theory. *The Journal of Pan African Studies*, *3*(1), 139–155.

Mapaya, I. (2014). *Indigenous Music and Social Pedagogy in Southern Africa*. African Sun Media.

Mapesa, M. W. (2016). Traditional Management systems in Natural World Heritage Sites in Eastern Africa. In Abungu, G. H. (Ed.), *O et al (2016) Traditional Management Systems at Heritage Sites in Africa*. African World Heritage Fund.

Mapfumo, P., Mtambanengwe, F., & Chikowo, R. (2016). Building on indigenous knowledge to strengthen the capacity of smallholder farming communities to adapt to climate change and variability in southern Africa. *Climate and Development*, *8*(1), 72–82. DOI: 10.1080/17565529.2014.998604

Marco, M. L., Heeney, D., Binda, S., Cifelli, C. J., Cotter, P. D., Foligné, B., Gänzle, M., Kort, R., Pasin, G., Pihlanto, A., Smid, E. J., & Hutkins, R. (2017). Health benefits of fermented foods: Microbiota and beyond. *Current Opinion in Biotechnology*, *44*, 94–102. DOI: 10.1016/j.copbio.2016.11.010 PMID: 27998788

Marek-Martinez, O. V. (2021). Indigenous archaeological approaches and the refusal of colonialism in archaeology. In *Routledge Handbook of the Archaeology of Indigenous-Colonial Interaction in the Americas* (pp. 503–515). Routledge. DOI: 10.4324/9780429274251-35

Marshall, F. (1990). Origins of specialized pastoral production in East Africa. *American Anthropologist*, *92*(4), 873–894. DOI: 10.1525/aa.1990.92.4.02a00020

Marshall, F., & Hildebrand, E. (2002). Cattle before crops: The beginnings of food production in Africa. *Journal of World Prehistory*, *16*(2), 99–143. DOI: 10.1023/A:1019954903395

Masamba, G., & Maguraushe, W. (2025). The Uptake of Chromatic Marimbas by Learning Institutions in Zimbabwe. *African Musicology Online*, *14*(2), 69–81. DOI: 10.58721/amo.v14i2.1208

Maseko, B., & Siziba, L. (2024). Environmental conservation and the Bulawayo CBD as a linguistic landscape construction: An ecolinguistics perspective. *Journal of Asian and African Studies*, •••, 00219096241291054.

Master, K. G. Maxy Khoisan, Makhadzi. (2020). Tshinada. URL: https://www.youtube.com/watch?v=942XC5uJoB4&list=RD942XC5uJoB4&start_radio=1

Ma, T. F., Chai, C. W., & Chao, T. W. (2022). On the Study of the Sustainable Development of Intangible Cultural Heritage of Indigenous Peoples' Diets—Take the Protection of Geographical Indications as an Example. *Sustainability (Basel)*, *14*(19), 12803. Advance online publication. DOI: 10.3390/su141912803

Mathenjwa, L. F. (1991). Zulu Literary Artist's conception of celestial bodies and associated natural phenomena. Doctor of Literature and Philosophy Thesis, University of South Africa. https://uir.unisa.ac.za

Matilal, B. K. (2002). *Ethics and epics: The collected essays of Bimal Krishna Matilal* (Ganeri, J., Ed.). Vol. 2). Oxford University Press.

Matshalaga, N. (2004). *Grandmothers and orphan care in Zimbabwe*. SAFAIDS.

Matshidze, P. E. (2013). The role of Makhadzi in traditional leadership among Venda. Doctor of Philosophy Thesis in Anthropology and Development Studies, University of Zululand. https://uzspace.unizulu.ac.za

Maurya, N. K. (2025). *Sacred Lineage and Divine Guardians: The Role of Kuldevi, Devatha, and Pitra in Material and Spiritual Growth*. Writer's Pocket.

Mawere, J., Matshidze, P., Kugara, S. L., & Madzivhandila, T. S. (2021). *The Role and Significance of Traditional Leadership in South African Local Governance. In book: Handbook of Research on Protecting and Managing Global Indigenous Knowledge Systems*. IGI Global Publisher., DOI: 10.4018/978-1-7998-7492-8.ch014

Mawere, J., & Mukonza, R. M. (2025). Empowering marginalised communities: Leveraging indigenous knowledge for sustainable energy development in South Africa. *Sustainable Development (Bradford), 33*(2), 2440–2448. DOI: 10.1002/sd.3243

Maxwell, S. (1996). Food security: A post-modern perspective. *Food Policy*, *21*(2), 155–170. DOI: 10.1016/0306-9192(95)00074-7

Mazzocchi, F., Simandan, D., Demneh, M. T., Morgan, D. R., Ghazinoory, S., Saghafi, F., & Mirzaei, M. (2018). Why 'Integrating'Western science and Indigenous knowledge is not an easy task: What lessons could be learned for the future of knowledge. *Journal of Futures Studies*, *22*(3), 19–34.

Mbigi, L. (2005). *The spirit of African leadership*. Knowres Publishing.

Mbigi, L., & Maree, J. (1995). *Ubuntu: The spirit of African transformation management*. Knowledge Resources.

Mekoa, I. (2018a). Essentialising African Indigenous Knowledge Systems in the Midst of Globalization and Modernity. *African Renaissance*, 11-28.

Mekoa, I. (2018b). Advocating for Afrocentric studies and African scholarship in the midst of the hostile Western scholarship. *African Renaissance*, *15*(3), 87–110. DOI: 10.31920/2516-5305/2018/v15n3a5

Mekoya, A., Tesfaye, A., & Bekele, D. A. (2025). Rainfall, air temperature, and reference evapotranspiration variability and trend and their implication for agroforestry development in the Lake Tana sub-basin of Ethiopia. *Discover Atmosphere*, *3*(1), 21. DOI: 10.1007/s44292-025-00045-7

Melczer, R. S. (2019). ""The Fat Which Is Begged for Does Not Make the Hair Pretty": K.T. Motsete and the Margins of Self-Determination in the Bechuanaland Protectorate". Unpublished Doctoral Dissertation: University of California, Santa Barbara.

Mendoza, R. J. (2024). Undisciplining the Museum: Indigenous Relationality as Religion. *Religions*, *15*(11), 1325. DOI: 10.3390/rel15111325

Metcalf, T. R. (1989). *An imperial vision: Indian architecture and Britain's raj*. University of California Press.

Metcalf, T. R. (1994). *Ideologies of the Raj.* Cambridge University Press.

Meyer, E. (2017). *The culture map: Breaking through the invisible boundaries of global business.* Public Affairs.

Mgbeadichie, C. F. (2015). *The critical concept of Afrocentrism in Nigerian literature.* University of Exeter.

Miao, K., & Kaewkulthorn, P. (2024). Historical Development and Learning of Marimba in Chinese Educational Institutions. *International Journal of Sociologies and Anthropologies Science Reviews, 4*(6), 553–562. DOI: 10.60027/ijsasr.2024.5129

Mieder, W. (2004). *Proverbs: A handbook.* Greenwood Press.

Mignolo, W. D., & Walsh, C. E. (2018). *On decoloniality: Concepts, analytics, praxis.* Duke University Press.

Ministry of Foreign Affairs Japan. (2025). Japan-Nigeria Foreign Minisers' meeting'. Tokyo: APO Group Distribution on Behalf of the Ministry of Foreign Affairs Japan. 20 August 2025.

Ministry of Rural Development. (2021). *Deendayal Antyodaya Yojana - National Rural Livelihoods Mission: Framework for implementation.* Government of India.

Misra, B. B. (1970). *The Administrative History of India (1834-1947).* Oxford University Press.

MiSTRA. (2019). *Traditional Leaders in a Democracy: Resources, Respect and Resistance.* Johannesburg: Mapungubwe Institute for Strategic Reflection. www.mistra.org.za

MiSTRA. (2024). *Thucydides Trap or Orderly Multipolar World? Challenges and Opportunities for Africa.* Mapungubwe Institute for Strategic Reflection. www.mistra.org.za

Mistri, S. (2026). THE USEFULNESS OF MANUSM TI IN RESOLVING THE CURRENT ENVIRONMENTAL CRISIS.

Mitra, R. (2012). The Famine in British India: Quantification Rhetoric and Colonial Disaster Management. *Journal of Creative Communications, 7*(1–2), 153–174. DOI: 10.1177/0973258613501066

Mlay, W. F. I. (1986). Environmental implications of land-use patterns in the new villages in Tanzania.

Mma Ausi & Kulenyane. (2022). *Magadigadi.* Dibeisane Album. URL: https://www.youtube.com/watch?v=ronO9ta3H8I&list=RDronO9ta3H8I&start_radio=1

MmaAusi. (2021). *Setswana.* Tshimo ea Tlholwa Album. URL: https://www.youtube.com/watch?v=2oXbSIKckfw&list=RD2oXbSIKckfw&start_radio=1

Mmalaloo. (2024). *Dumela Mmabatho.* Dithubaruba Album. URL: https://www.youtube.com/watch?v=gmGi-hiqxX8&list=RDgmGi-hiqxX8&start_radio=1

Mmausi. (2022). *Dithubaruba.* Ke Bone Molelo O Timile Album. URL: https://www.youtube.com/watch?v=Gz-JtquS6sQ&list=RDGz-JtquS6sQ&start_radio=1

Modongo, L. P. (2021). *When love kills: constructions of masculinities within the intimate femicide crisis in Botswana* (Doctoral dissertation: Stellenbosch University)

Mogapi, K. (1991). *Ngwao ya Setswana*. Longman.

Moguerane, K. (2016). Black landlords, their tenants, and the Natives Land Act of 1913. *Journal of Southern African Studies*, *42*(2), 243–266. DOI: 10.1080/03057070.2016.1148954

Mohammed, J. (2025). *The rise of African thought Leaders in Global Leadership*. African Leadership Magazine. The Rise of African Thought Leaders in Global Leadership- African Leadership Magazine.

Moher, D., Liberati, A., Tetzlaff, J., & Altman, D. G. (2009). Preferred reporting items for systematic reviews and meta-analyses: The PRISMA statement. *BMJ (Clinical Research Ed.)*, •••, 339. PMID: 19622551

Mohlamme, J. (2000). Blacks in the ex-Boer republics of the Transvaal and Orange Free State in the aftermath of the South African War of 1899-1902. *Southern Journal for Contemporary History*, *25*(2), 270–283. DOI: 10.38140/sjch.v25i2.4089

Moitra, P., & Madan, J. (2025). Stakeholder perspectives on integrating Ayurveda and Indian Indigenous Knowledge Systems into higher education: An exploratory study. *Social Sciences & Humanities Open*, *11*(March), 101453. DOI: 10.1016/j.ssaho.2025.101453

Mokotso, R. I. (2022). Considering the Basotho indigenous education and school system as resources for peace-building education in Lesotho. *The Journal for Transdisciplinary Research in Southern Africa*, *18*(1), 1213. DOI: 10.4102/td.v18i1.1213

Molato, K., & Dube, M. W. (2025). Music, Gender and Religion in Botswana: Culture Spears Traditional Group. In *Manyonganise, Molly; Gunda, Masiiwa Ragies; Naicker, Molefe, R., 2019. Traditional Music and Cultural Consciousness in Botswana*. University of Botswana Press. DOI: 10.20378/irb-106054

Molokwa, D. T., & Khazamula, X. E. (2026). Indigenous Food and Health Sovereignty Systems in South Africa: Challenges and Prospects. *The Palgrave Handbook of Religion, Health and Development in Africa*, 1-21.

Molomo, M. G. (2005). Democracy and the kgotla system in Botswana. *African Studies Quarterly*, *8*(2), 1–20.

Mondal, A., & Pandey, M. S. (2024). Indigenous festivals and climate sustainability in India: A case study of cultural practices and performances. *Rupkatha Journal on Interdisciplinary Studies in Humanities*, *16*(1), 1–15. DOI: 10.21659/rupkatha.v16n1.03

Morapedi, W. (2009). Botswana, protest and nationalism. In *Immanuel Ness. International Encyclopaedia of Protest and Revolution from the 1500 to the present* (pp. 453–463). Wiley Blackwell.

Morduch, J. (1999). The microfinance promise. *Journal of Economic Literature*, *37*(4), 1569–1614. DOI: 10.1257/jel.37.4.1569

Morojele, P. (2017). Indigenous knowledge/s of survival: Implications for lifelong learning among the Basotho herding fraternity. *Educational Research for Social Change*, *6*(1), 38–55. DOI: 10.17159/2221-4070/2017/v6i1a4

Morreira, S. (2013). *Transnational human rights and local moralities: The circulation of rights discourses in Zimbabwe and South Africa* [Unpublished doctoral thesis]. Department of Social Anthropology, University of Cape Town.

Moshesha, P. 2023. *Climate change adaptation strategies and their challenges in Lesotho*. Master's thesis. National University of Lesotho: Roma.

Motasa, K. O., & Nortjé-Meyer, L. (2021). Patriarchal Usurpation of the Modjadji Dynasty: A Gender-critical Reading of the History and Reign of the Modjadji Rain Queens. *Pharos Journal of Theology ISSN, 2414-3324 online 102.* https://orcid.org/0000-0002- 8350-8189 and Doi: https://doi.org/DOI: 10.46222/pharosjot.102.1.8

Motjotji, L., Moteetee, A., & Seleteng-Kose, L. (2023). Medicinal plants of Lesotho: A review of ethnomedicinal, pharmacological and conservation studies: Lesotho medicinal plants. *Ethnobotany Research and Applications*, *25*, 1–19. DOI: 10.32859/era.25.29.1-19

Motseki, P. D., Jojo, Z. M. M., & Gumbo, M. T. (2026). Transforming mathematics teaching through Ndebele art: An ethnomathematics perspective. *Diaspora, Indigenous, and Minority Education*, *20*(1), 1–7. DOI: 10.1080/15595692.2025.2451239

Mowakadi, J. (2025). Urbanization and Indigenous Cultural Adaptation in Lagos: A Multidisciplinary Analysis of Urban Cultural Coexistence. *Global Interdisciplinary Perspectives*, *1*(2), 20–27. DOI: 10.64229/b6jas597

Mpatlanyane, V. L., Tshikukuvhe, L. D., & Budeli, A. E. (2022). Exploring Vhavenda Indigenous Food Practices and Knowledge Contributions to Food (in) security. In *Handbook of Research on Protecting and Managing Global Indigenous Knowledge Systems* (pp. 61-88). IGI Global Scientific Publishing.

Mpeiwa, K. D. (2025). Develop a model leveraging from the shared experiences of the communities living in the heritage sites. UNESCO International Conference Paper on Cultural Heritage in Africa: A dialogue on the concept of authenticity 6th-9th May 2025. Nairobi: UNESCO.

Mpofu, B. (2020). South Africa: Re-imagining the rainbow nation. *Sociology International Journal*, *4*(3), 67–68. DOI: 10.15406/sij.2020.04.00226

Msila, V. (2015). *Ubuntu and school leadership. Journal of Social Sciences, 43(1), 67–76. Ngulube, P. (2018). Handbook of research on theoretical perspectives on indigenous knowledge systems in developing countries*. IGI Global.

Msila, V. (2017). *The Role of Indigenous Knowledge Systems in African Education*. Van Schaik Publishers.

Muchiru, A. N., Western, D. J., & Reid, R. S. (2008). The role of abandoned pastoral settlements in the dynamics of African large herbivore communities. *Journal of Arid Environments*, *72*(6), 940–952. DOI: 10.1016/j.jaridenv.2007.11.012

Muchiru, A. N., Western, D., & Reid, R. S. (2009). The Impact of Abandoned Pastoral Settlements on Plant and Nutrient Succession in an African Savanna Ecosystem. *Journal of Arid Environments*, *73*(3), 322–331. DOI: 10.1016/j.jaridenv.2008.09.018

Mueller, R., & Allan, A. (2025). *Decolonise your mind: Postcolonial theories*. IMartin. Htpps:// www .thelaosexperience.com/introduction-postcolonial-theories/ (Accessed: 20 January 2026)

Mugumbate, J., & Chereni, A. (2019). Using African Ubuntu theory in social work with children in Zimbabwe. *African Journal of Social Work*, *9*(1), 27–34.

Mugumbate, J., & Nyanguru, A. (2013). Exploring African philosophy: The value of *ubuntu* in social work. *African Journal of Social Work*, *3*(2), 82–100.

Mukharji, P. B. (2016). *Doctoring traditions: Ayurveda, small technologies, and braided sciences*. The University of Chicago Press. DOI: 10.7208/chicago/9780226381824.001.0001

Mukherjee, S. (2021). *Colonial institutions and civil war: indirect rule and Maoist insurgency in India*. Cambridge University Press. DOI: 10.1017/9781108954266

Mukil, M. V., Pruthvi, M. N., & Bhavani Rao, R. (2024, September). Design for sustainability: Exploring cultural attributes in an Indian village for nature connectedness— A grounded theory approach. In 2024 *IEEE International Symposium on Technology and Society* (ISTAS) (pp. 1–8). IEEE.

Mukil, M. V., Shinde, K., Rashed, T., & Bhavani, R. R. (2026). Religious-Cultural framework for sustaining environmental balance in rural ecosystems: Stewardship of village-temples in Tamil Nadu, India. *Local Environment*, •••, 1–27. DOI: 10.1080/13549839.2026.2631508

Muniapan, B. (2013). *Ancient Indian wisdom for sustainability and management*.

Muniapan, B. (2013). The roots of Indian corporate social responsibility (CSR) practice from a Vedantic perspective. In *Corporate social responsibility in Asia: Practice and experience* (pp. 19–33). Springer International Publishing.

Muniapan, B., & Satpathy, B. (2013). The 'dharma' and 'karma' of CSR from the Bhagavad-Gita. *Journal of Human Values*, *19*(2), 173–187. DOI: 10.1177/0971685813492265

Muraleedharan, K. P. (2009). *Economic benefits of pilgrimage tourism: a case study of Sabarimala pilgrimage with special reference to Pandalam rural locality in Kerala*.

Muralivallabhan, T. V. (2022). Protection of environment and promotion of development: The Hindu wisdom of sustainable development. *National Security*, *5*(2), 146-180.

Muriithi, G. M., & Musili, W. M. (2021). *The Effect of Science, Technology, Engineering and Mathematics Skills on Youth Employment in Kenya*. Kenya Institute for Public Research and Analysis.

Murithi, T. (2006). Practical peacemaking wisdom from Africa. *The Journal of Pan African Studies*, *1*(4), 25–34.

Murphy, C. A. Fuller D. Q. (2016). "The Transition to Agricultural Production in India: South Asian Entanglements of Domestication." *A Companion to South Asia in the Past*, 344–57.

Mushunje, M. T. (2014). Interrogating the relevance of the extended family as a social safety net for vulnerable children in Zimbabwe. *African Journal of Social Work*, *4*(2), 1–10.

Musonda-Hamusonde, M., & Abungu, G. O. (2022). The Nature-Culture Divide in Heritage Classification and Management in Africa: Reality or a Convenience of Management. In *Cultural Heritage Management in Africa* (pp. 219-237). Routledge.

NABARD. (2001-2023). *Status of microfinance in India* [Annual reports]. National Bank for Agriculture and Rural Development.

NABARD. (2023). *Status of microfinance in India 2022-23*. National Bank for Agriculture and Rural Development.

Najib, K. (2020). Government Ecology and the Indigenous Religion of the Suku Anak Dalam: Intersubjective Relations in Forest Conservation in Jambi, Indonesia. [Journal of Tropical Forest Management]. *Jurnal Manajemen Hutan Tropika, 26*(3), 303–315. DOI: 10.7226/jtfm.26.3.303

National Anthems Info. (2021). "Bophutatswana National Anthem". URL: https://www.nationalanthems.info/bpt.htm

Nayar, K. R., Koya, S. F., Mohandas, K., Nair, S. S., Chitra, G. A., Abraham, M., & Lordson, J. (2020). Public health implications of Sabarimala mass gathering in India: A multi-dimensional analysis. *Travel Medicine and Infectious Disease, 37*, 101783. DOI: 10.1016/j.tmaid.2020.101783 PMID: 32561393

Ndlela, N. (2009). *Marriage Songs and Cultural Transmission in Southern Africa*. University of KwaZulu-Natal Press.

Ndlovu-Gatsheni, S. (2018). *Epistemic freedom in Africa: Deprovincialization and decolonization*. Routledge. DOI: 10.4324/9780429492204

Ndlovu-Gatsheni, S. J. (2013). *Coloniality of power in postcolonial Africa: Myths of decolonization*. CODESRIA.

Ndlovu-Gatsheni, S. J. (2014). Global Coloniality and the challenges of creating African future. *Strategic Review for Southern Africa, 36*(2), 181–202.

Ndoro, W. (2018). Drawing parallels: Authenticity in the African context–until lions learn to write, hunters will tell their history for them. *Revisiting Authenticity in the Asian Context, 2*, 37–46.

Ndoro, W., & Larsen, P. B. (2024). "Heritage is about today, it's not about what happened in the past": A conversation with Webber Ndoro, Director General of the International Centre for the Study of the Preservation and Restoration of Cultural Property. *American Anthropologist, 126*(2), 362–364. DOI: 10.1111/aman.13955

Nega, D. (2019). Challenges and opportunities of managing festival tourism for sustainable tourism development: The case of the Ashendye Festival, North Wollo, Ethiopia. *African Journal of Hospitality, Tourism and Leisure, 8*(4), 1–14.

Nelson Mandela Foundation. (2025). Chapter Ten: Increased Repression and Continued Attack in 1995 and 1996. www.omalley.nelsonmandela.org

Nemashakwe, P. (2021). Evaluation of the State and causes of ineffective leadership in Africa. *The Fountain: Journal of Interdisciplinary Studies, 7*(1), 1–17.

Netshiluvhi, T. R. (2014). Indigenous food systems and sustainable livelihoods in rural Limpopo. *Journal of Human Ecology (Delhi, India)*, *48*(3), 279–287.

Netshivhambe, N. E. (2025). Promoting Wellness in the Aged through Indigenous Instrument. *Social Determinants of Active Ageing-Building Resilience and Empowerment: Building Resilience and Empowerment*, 21.

Netshivhambe, E. N. (2023a). Indigenous African Music Performance Re-imagined in Public Spaces. In Mpofu, P., Fadipe, I. A., & Tshabangu, T. (Eds.), *African Language Media* (1st ed., pp. 243–257). Routledge., DOI: 10.4324/9781003350194-23

Netshivhambe, E. N. (2023c). *Ignored or Invisible: The status of indigenous South African instruments In the Venda society*. UNISA Press.

Netshivhambe, E. N. (2023d). Living the heritage through Indigenous music competitions. In Kumar, S., & Lakshminarayana, M. B. (Eds.), *Indigenous People - Traditional Practices and Modern Development*. IntechOpen., DOI: 10.5772/intechopen.1003226

Netshivhambe, E. N. (2024a). De-neocolonizing the South African Music Education: Selective Transformation. In Akpan, U. S. (Ed.), *De-neocolonizing Africa: Harnessing the Digital Frontier* (pp. 393–416). Palgrave Macmillan Cham., DOI: 10.1007/978-3-031-66304-8_22

Netshivhambe, N. E. (2024b). Towards Sustainable Preservation: Revitalising Indigenous African Instruments and Craftsmanship. *E-Journal of Humanities. Arts and Social Sciences*, *5*(15), 91–107.

Newcombe, S. (2014). *Cults: History, Beliefs, Practices*, In (eds) Terry Muck, Harold A. Netland, and Gerald R. McDermott: Handbook of Religion: A Christian Engagement with Traditions, Teachings, and Practices (pp.568-572). Baker Academic

Ngcobo, K. (2025a). Trump presses ahead with 30% tariff on South Africa. Johannesburg: *British Broadcasting Corporation News*. www.BBC.com

Ngcobo, K. (2025b). ''South African president signs controversial land seizure law''. www.BBC.com

Ngcobo, A. S., Khalema, E. N., Masuku, M. M., Mlondo, N. M., Mncube, T., Ramphabana, L. B., & Matse, M. P. (2025). Role of Indigenous Knowledge Systems and Ubuntu in Addressing Gender-Based Violence Against Young Women in Rural KwaZulu-Natal. *Journal of Social Service Research*, *51*(5–6), 1203–1216. DOI: 10.1080/01488376.2025.2515420

Ngulube, P. (2002). Managing and preserving indigenous knowledge in the knowledge management era. *Information Development*, *18*(2), 95–105. DOI: 10.1177/026666602400842486

Ngwao ya Tshona. (2021). *Magosi a Tshona*. URL: https://www.youtube.com/watch?v=ADuaJpFEm08&list=RDADuaJpFEm08&start_radio=1

Ngwao, M. (2015). Makgona Ngwao. URL: https://www.youtube.com/watch?v=Jq-DGs9rfBQ&list=RDJq-DGs9rfBQ&start_radio=1

Niamir-Fuller, M. (1998). The Resilience of Pastoral Herding in Sahelian Africa. In Berkes, F., & Folke, C. (Eds.), *Linking social and ecological systems: management practices and social mechanisms for building resilience, 1* (Vol. 4, pp. 250–284). Cambridge University Press.

Niglio, O., Singh, R. P., & Gireesh Kumar, T. K. (2026). RECONNECTING WITH YOUR CULTURE (RWYC): TOWARDS CULTURAL AWAKENING, PROSPECTS AND VISION. *EDA, ESEMPI DI ARCHITETTURA, 13*(2), 317–338.

Nigro, J. (2014). Colonial Logics: *Agricultural, Commercial, & Moral Experiments in the Making of French Senega*l, 1763-1870 (Doctoral dissertation, University of Illinois at Chicago).

Nikolić, G., Kvasic, S. G., & Grbic, L. (2020). The Development of Authentic Leadership Theory. Conference: PAR International Scientific and Professional Conference 'High Impact Leadership'. https://www.researchgate.net/publication

Ninomiya, M. E. M., Burns, N., Pollock, N. J., Green, N. T., Martin, J., Linton, J., Rand, J. R., Brubacher, L. J., Keeling, A., & Latta, A. (2023). Indigenous communities and the mental health impacts of land dispossession related to industrial resource development: A systematic review. *The Lancet. Planetary Health, 7*(6), e501–e517. DOI: 10.1016/S2542-5196(23)00079-7 PMID: 37286247

Nisha, P. (2017). Effects of drying on nutritional quality of fruits and vegetables. *Critical Reviews in Food Science and Nutrition, 57*(3), 566–578. PMID: 25942011

Nkoane, M. M. (2006). The Africanisation of the university in Africa. *Alternation (Durban), 13*(1), 49–69.

Nkomo, S. M. (2011). A postcolonial and anti-colonial reading of 'African' leadership and management in organization studies. *Organization, 18*(3), 365–386. DOI: 10.1177/1350508411398731

Nobles, E. C. (2025). The ethical foundations of biodiversity metrics. *Current Opinion in Environmental Sustainability, 72*, 101503. DOI: 10.1016/j.cosust.2024.101503

North, D. C. (1990). *Institutions, institutional change and economic performance*. Cambridge University Press. DOI: 10.1017/CBO9780511808678

Northouse, P. G. (2010). *Leadership: Theory and Practice*. Sage Publications.

Ntlama, N. (2020). The Attempt to Change the Nature of Succession to the Chieftaincy in the Institution of Traditional Leadership: Mphephu v Mphephu-Ramabulana 2019 7 BCLR 862 (SCA)" PER / PELJ 2020(23) - .DOI: 10.17159/1727-3781/2020/v23i0a6804

Ntonzima, L., & Bayat, M.S. (2012). The role of Traditional Leaders in South Africa - A relic of the past, or a contemporary reality? *Arabian Journal of Business and Management Review* (OMAN Chapter) 1.

Nurit Bird-David. (1999). "Animism" Revisited: Personhood, Environment, and Relational Epistemology. *Current Anthropology Special Issue Culture—A Second Chance? 40*(S1). https://doi.org/https://doi.org/10.1086/200061

Nurullah, S., & Naik, J. P. (1951). *History of education in India (during the British period)*. Macmillan Bombay.

Nussbaum, M. C. (2001). *Upheavals of thought: The intelligence of emotions*. Cambridge University Press. DOI: 10.1017/CBO9780511840715

Nzewi, M. (2007). *A contemporary study of musical arts: Informed by African indigenous knowledge systems*. Centre for Indigenous Instrumental Music and Dance Practices.

Nzewi, M. (2007). Learning the Musical Arts in Contemporary Africa. In *Informed by Indigenous Knowledge Systems* (Vol. 1). CIIMDA Publishing.

Nzimakwe, T. I. (2014). Practising Ubuntu and leadership for good governance. *African Journal of Public Affairs*, *7*(4), 30–41.

Oberhofer, C. (2020). Decolonization and Indigenization in music education. *Canadian Music Educator*, *62*(1), 48–53.

Odora Hoppers, C. A. (2002). *Indigenous knowledge and the integration of knowledge systems*. New Africa Books.

Odora Hoppers, C. A. (2017). *Knowledge, power and transformation in higher education*. UNISA Press.

Oetzel, J. G. (2005). Effective intercultural workgroup communication theory. *Theorising about Intercultural Communication*, pp.351-371.

Oetzel, J. G. (2017). *Effective intercultural workgroup communication theory*. The International Encyclopaedia of Intercultural Communication. DOI: 10.1002/9781118783665.ieicc0025

Oetzel, J. G., McDermott, V. M., Torres, A., & Sanchez, C. (2012). The impact of individual differences and group diversity on group interaction climate and satisfaction: A test of the effective workgroup communication theory. *Journal of International and Intercultural Communication*, *5*(2), 144–167. DOI: 10.1080/17513057.2011.640754

Oguamanam, C. (2025). Information systems and digitization of traditional knowledge: Trends in cultural heritage and memory institutions and the WIPO Genetic Resources Treaty*. *The Journal of World Intellectual Property*, (August), 1–32. DOI: 10.1111/jwip.70005

Ogunleye, T. (2020). Colonial Legacy and Ethnic Imbalance in the Nigerian Armed Forces. *African Military History Review*, *4*(3), 58–75.

Ogunyemi, A. (2015). *Music and Mourning in African Societies: Case Studies from Southern Africa*. University of Lagos Press.

Oladele, O. I., & Amara, A. (2024). Farmers' use of Indigenous knowledge on climate change adaptation across farming systems and agroecological zones of Sierra Leone. *Alternative*, *20*(3), 473–483. DOI: 10.1177/11771801241251862

Oliveira, A. S., de Barros, M. D., de Carvalho Pereira, F., Gomes, C. F. S., & da Costa, H. G. (2018). Prospective scenarios: A literature review on the Scopus database. *Futures*, *100*, 20–33. DOI: 10.1016/j.futures.2018.03.005

Olivelle, P. (2005). *Manu's code of law: A critical edition and translation of the Mānava-Dharmaśāstra*. Oxford University Press.

Olivelle, P. (2013). *King, governance, and law in ancient India: Kau ilya's Arthaśāstra*. Oxford University Press. DOI: 10.1093/acprof:osobl/9780199891825.001.0001

Olsen, D. H. (2014). A scalar comparison of motivations and expectations of experience within the religious tourism market. *International Journal of Religious Tourism and Pilgrimage*, *1*(1), 5.

Ombati, M., 2017. Rainmaking rituals: Song and dance for climate change in the making of livelihoods in Africa. *International journal of modern anthropology, 1*(10), 74-96.

Ombati, M. (2021). Climate rituals: Cultural response for climate change adaptations in Africa. In *Decolonising Conflicts, Security, Peace, Gender, Environment and Development in the Anthropocene* (pp. 385–410). Springer International Publishing. DOI: 10.1007/978-3-030-62316-6_11

Omoniyi, M. B. I. (2013). The role of education in poverty alleviation and economic development: A theoretical perspective and counselling implications. British Journal of Arts and Social Sciences, 15(2), 176-185.Benešová, A., & Tupa, J. (2017). Requirements for Education and Qualification of People in Industry 4.0. *Procedia Manufacturing*, *11*, 2195–2202. DOI: 10.1016/j.promfg.2017.07.366

Organiser. (2025). Chhath Puja: When the sun rises on India's economy. Retrieved from https://organiser.org/2025/11/04/324170/bharat/chhath-puja-when-the-sun-rises-on-indias-economy/

Orlove, B., Dawson, N., Sherpa, P., Adelekan, I., Alangui, W., Carmona, R., Coen, D., Nelson, M., Reyes-García, V., Wilson, A., Rubis, J., & Sanago, G. (2022). *Intangible Cultural Heritage*. Diverse Knowledge Systems and Climate Change.

Orsini, A. (2024). Postcolonial Theory. In *Sociological Theory: From Comte to Postcolonialism* (pp. 641–660). Springer Nature Switzerland. DOI: 10.1007/978-3-031-52539-1_16

Osborne, N., & Grant-Smith, D. (2021). In-depth Interviewing. In *Methods in Urban Analysis* (pp. 105–125). Springer Singapore. DOI: 10.1007/978-981-16-1677-8_7

Osei-Tutu, A. A. (2023). African-Centered hybridity: A Reconceptualization of Africanness in this colonially guised globalized era. *Alliance for African Partnership Perspectives*, *2*(1), 61–73.

Osella, F., & Osella, C. (2003). 'Ayyappan saranam': 1 masculinity and the Sabarimala pilgrimage in Kerala. *Journal of the Royal Anthropological Institute*, *9*(4), 729–754. DOI: 10.1111/j.1467-9655.2003.00171.x

Oshame, D., & Maureen, O. E. (2023). Critical analysis of four leadership theories and principle. *World Journal of Advanced Research and Reviews*, *17*(01), 1387–1395. DOI: 10.30574/wjarr.2023.17.1.0187

Ossai, N. B. (2010). African Indigenous Knowledge Systems (AIKS). *Simbiosis, 7*(2).

Osutei, N., & Kim, J. H. (2023). Examining talent attraction and retention in small and medium-sized cities: Where do universities fit in? *Growth and Change*, *54*(3), 736–753. DOI: 10.1111/grow.12672

Owolabi, B. A., Sanuade, A. O., Fajinmolu, A. O., Atabo, L. O., Okeyode, O. S., Adeyemi, A. A., & Oganla, A. E. (2026). Understanding the Forest Decline and Habitat Loss. In *Environmental Health and Toxicology* (pp. 382–393). CRC Press. DOI: 10.1201/9781003500452-24

P, J. S. (2020). Social ecology of the Pamba riverine at Sabarimala, the mass pilgrim centre of Kerala. *Journal of Emerging Technologies and Innovative Research, 7*(7), 1253–1257. https://www.jetir.org/papers/JETIR2007463.pdf

P, J. S. (2020b). Social ecology of the Pamba riverine at Sabarimala, the mass pilgrim centre of Kerala. *Journal of Emerging Technologies and Innovative Research, 7*(7), 1253–1257. https://www.jetir.org/papers/JETIR2007463.pdf

Paddayya, K. (1998). Evidence of Neolithic Cattle-Penning at Budihal, Gulbarga District, Karnataka. *South Asian Studies*, *14*(1), 141–153. DOI: 10.1080/02666030.1998.9628556

Pageh, I., Rahmawati, P. I., & Pramiani, N. M. (2022). Ancient religious sites as tools for sustainable tourism development: An empirical study in the North of Bali. *International Journal of Religious Tourism and Pilgrimage*, *10*(1), 78–91.

Page, M. J., McKenzie, J. E., Bossuyt, P. M., Boutron, I., Hoffmann, T. C., Mulrow, C. D., Shamseer, L., Tetzlaff, J. M., Akl, E. A., Brennan, S. E., Chou, R., Glanville, J., Grimshaw, J. M., Hróbjartsson, A., Lalu, M. M., Li, T., Loder, E. W., Mayo-Wilson, E., McDonald, S., & Moher, D. (2021). The PRISMA 2020 statement: An updated guideline for reporting systematic reviews. *International Journal of Surgery*, *88*, 105906. https://doi.org/https://doi.org/10.1016/j.ijsu.2021.105906. DOI: 10.1016/j.ijsu.2021.105906 PMID: 33789826

Pai, G. V. (2016). Producing heritage: Culture as commodity in Madurai. *International Journal of Tourism Anthropology*, *5*(1-2), 47–70.

Panda, T. (2024). Sustainable living in ancient Indian teachings: Glimpses from classical Indian scriptures and traditions. *The Social Science Review: A Multidisciplinary Journal, 2*(3), 81-91.

Pandey, A., & Bhart, V. (2024). Wetlands in Bihar: A comprehensive analysis of extent, socio-economic significance, ecological benefits, and associated risks. *Journal of Community Positive Practices*, *24*(2), 23–45. DOI: 10.35782/JCPP.2024.1.07

Paraschiv, C. I. (2017). The role of education in poverty alleviation. *Theoretical & Applied Economics, 24*.Zhong, R. Y., Xu, X., Klotz, E., & Newman, S. T. (2017). Intelligent manufacturing in the context of industry 4.0: A review. *Engineering*, *3*(5), 616–630. https://doi.org/https://doi.org/10.1016/J.ENG.2017.05.015

Paris, D. (2021, July). Culturally sustaining pedagogies and our futures. In *The educational forum* (Vol. 85, No. 4, pp. 364-376). Routledge. DOI: 10.1080/00131725.2021.1957634

Parker, B., Burnett, K., Hay, T., & Skinner, K. (2019). The community food environment and food insecurity in Sioux Lookout, Ontario: Understanding the relationships between food, health, and place. *Journal of Hunger & Environmental Nutrition*, *14*(6), 762–779. DOI: 10.1080/19320248.2018.1537867

Parson, J. (1981). Cattle, class and the state in rural Botswana. *Journal of Southern African Studies*, *7*(2), 236–255. DOI: 10.1080/03057078108708027

Parsons, J. R. (1972). Archaeological settlement patterns. *Annual Review of Anthropology*, *1*(1), 127–150. DOI: 10.1146/annurev.an.01.100172.001015

Patel, R. (2009). Food sovereignty. *The Journal of Peasant Studies, 36*(3), 663–706. DOI: 10.1080/03066150903143079

Patten, E., & Parker, K. (2011). *Women in the US military: Growing share, distinctive profile*. Pew Research Center.

Pearson, J., Jackson, G., & McNamara, K. E. (2023). Climate-driven losses to knowledge systems and cultural heritage: A literature review exploring the impacts on Indigenous and local cultures. *The Anthropocene Review, 10*(2), 343–366. DOI: 10.1177/20530196211005482

Pedersen, H. D., & Gram, M. (2018). 'The brainy ones are leaving': The subtlety of (un) cool places through the eyes of rural youth. *Journal of Youth Studies, 21*(5), 620–635. DOI: 10.1080/13676261.2017.1406071

Peiris, G. H. (1981). Agrarian transformations in British Sri Lanka. *Sri Lanka Journal of Agrarian Studies*, 2.

Pellerin, M. (2012). Benefits of Afrocentricity in exploring social phenomena: Understanding Afrocentricity as a social science methodology. *The Journal of Pan African Studies, 5*(4).

Petri, D. P., & Klocek, J. (2025). External and Internal Threats to the Freedom of Religion or Belief of Indigenous Peoples in Latin America. *Religions, 16*(2), 209. DOI: 10.3390/rel16020209

Phibion, O. S. (2013). The ditlhaka music of the Balete of Botswana. *Muziki, 10*(2), 90–102. DOI: 10.1080/18125980.2013.844986

Phillimore, J., & Goodson, L. (2004). *Qualitative research in tourism: Ontologies, epistemologies and methodologies*. Routledge.

Pikirayi, I. (2019). Local narratives, regional histories and the demise of Great Zimbabwe. *Archaeologies of Listening*, 131–152.

Pinch, F. C., MacIntyre, A. T., Browne, P., & Okros, A. C. (2004). *Challenge and change in the military: Gender and diversity issues*. Wing Publishing Office.

Pistorius, J. C. (1994). Molokwane, a seventeenth century Tswana village. *South African Journal of Ethnology, 17*(2), 38–53.

Pollock, S. I. (2009). *The language of the gods in the world of men: Sanskrit, culture, and power in premodern India* (1. paperback print). Univ. of California Press.

Pollock, S. (2006). *The language of the gods in the world of men: Sanskrit, culture, and power in premodern India*. University of California Press. DOI: 10.1525/9780520932029

Popkin, B. M. (1998). The nutrition transition and its health implications in lower-income countries. *Public Health Nutrition, 1*(1), 5–21. DOI: 10.1079/PHN19980004 PMID: 10555527

Popkin, B. M., Adair, L. S., & Ng, S. W. (2012). Global nutrition transition and the pandemic of obesity. *Nutrition Reviews, 70*(1), 3–21. DOI: 10.1111/j.1753-4887.2011.00456.x PMID: 22221213

Porensky, L. M., & Veblen, K. E. (2015). Generation of ecosystem hotspots using short-term cattle corrals in an African savanna. *Rangeland Ecology and Management, 68*(2), 131–141. DOI: 10.1016/j.rama.2015.01.002

Portal do Governo Mozambique. (2025). *Japan-Malawi Foreign Ministers' meeting*. APO Portal do Governo Mozambique Distribution.

Possa, R., & Khotso, P. (2015). Naming of Basotho Medicinal Plants: Semantic Connection to their Remedies. *Southern African Journal for Folklore Studies, 25*(1), 34–47. DOI: 10.25159/1016-8427/541

Post, J. (2024). Ecosystems and Sounding Lives: Musical Instrument Makers in Greater Central Asia. *Journal of Musicological Research, 43*(2-3), 175–196. DOI: 10.1080/01411896.2024.2376352

Poteet, S., Xue, P., Patel, J., Kao, A., Giammanco, C., & Whiteley, I. (2008). Linguistic sources of coalition miscommunication. In *NATO RTO-MP-HFM-142 Adaptability in Coalition Teamwork*, p. 10.

Pradhan, M. (2026). The Cry of the Earth: Literature as a Catalyst for Sustainable Behavioral Change. In Giri, A. K., & Deshpande, R. S. (Eds.), *Quest for Planetary Well-Being*., DOI: 10.1007/978-981-95-2077-0_18

Prajapati, M. B., & Chatwani, M. J. (2025). Indigenous Knowledge Systems and Sustainable Urban Futures: Lessons from Traditional Cities and Planning Practices. In *International Conference on 'Indigenous Knowledge Systems and Sustainable Development* (Vol. 9, p. 10).

Priyadarshini, P., & Abhilash, P. C. (2019). Promoting tribal communities and indigenous knowledge as potential solutions for the sustainable development of India. *Environmental Development, 32*(September), 100459. Advance online publication. DOI: 10.1016/j.envdev.2019.100459

Puig, D. (2025). Social limits to adaptation in the context of intangible cultural heritage. *Current Opinion in Environmental Sustainability, 77*, 101569. DOI: 10.1016/j.cosust.2025.101569

Putnam, R. D. (1993). *Making democracy work: Civic traditions in modern Italy*. Princeton University Press.

Putra, M. D. P., Hastjarjo, S., & Satyawan, I. A. (2025). Environmental Communication Based on Local Wisdom in Efforts to Preserve Water Sources in Pedawa Village. *Jurnal Ilmu Sosial dan Humaniora, 14*(1), 195-209.

Qiu, Q., Zuo, Y., & Zhang, M. (2022). Intangible Cultural Heritage in Tourism: Research Review and Investigation of Future Agenda. *Land (Basel), 11*(1), 139. Advance online publication. DOI: 10.3390/land11010139

Rabie, M. (2021). One Human Race: Shared Destiny. https://www.researchgate.net/publication/350451969

Radebe, S. and Landsberg, C. (2022). Transformation, representation and leadership in South Africa's security sector, *Journal of Contemporary African Studies*, 40(4), pp. 523-538. Rasweswe, M. (2023). Indigenous governance systems and participatory democracy in South Africa. *Journal of African Governance*.

Radhakrishnan, S. (1994). The Principal Upanishads,(From Taittiriya Brahmana section).

Rahman, D., Missouri, T., & Alexopoulos, G. (2021). The social ecology of food: Where agroecology and heritage meet. *Sustainability (Basel), 13*(24), 13981. DOI: 10.3390/su132413981

Rai, G. K. (2025). Protector of the Green: Sacred Groves as Images of Ethical and Spiritual Wholeness. *International Journal for Multidisciplinary Research, 7*(3), 01-06.

Rajan, C. (2024). *Food & Decoloniality. The Decolonial Centre.* Available at: https://decolodecolonialcentre.org/2024/08/14/food-decoloniality/ (Accessed date: 13 December 2025)

Ramanujan, A. K. (1991). Three hundred Rāmāya as: Five examples and three thoughts on translation. In Richman, P. (Ed.), *Many Rāmāya as: The Diversity of a Narrative Tradition in South Asia* (pp. 22–49). University of California Press. DOI: 10.1525/9780520911758-004

Ramose, M. B. (2002). *African philosophy through Ubuntu.* Mond Books Publishers.

Ramothwala, P. (2025). State no longer backs Modjadji Coronation. www.sowetanlive.co.za

Ramoupi, N. L. L. (2025). *Robben Island Songs of Struggle: Izingoma Zomzabalazo.* Skotaville Publishers.

Ramsay, J. (2021). The Batswana-Boer War of 1852-53: How the Batswana achieved victory. *Botswana Notes and Records, 23*(1), 193–207.

Ramsay, J., Morton, F., & Morton, B. (1996). *Historical Dictionary of Botswana.* The Scarecrow Press Inc.

Rana, K. S. (2021). Exploring the Elements of Postcolonialism and its Exponents. *The Creative launcher,* 6(5), 44-52.

Rangarajan, L. N. (1992). *Kautilya: The Arthashastra.* Penguin Books India.

Ranjan, A., & Chaturvedi, P. (2023a). Digitally Sustaining: The Rural Intangible Cultural Heritage. In *Embracing Business Sustainability Through Innovation and Creativity in the Service Sector* (pp. 14–30). DOI: 10.4018/978-1-6684-6732-9.ch002

Ranjan, A., & Chaturvedi, P. (2022). Digital Technologies and the Intangible Cultural Heritage of the Rural Destination. In *Disruptive Innovation and Emerging Technologies for Business Excellence in the Service Sector* (pp. 196–218). IGI Global., DOI: 10.4018/978-1-7998-9194-9.ch010

Ranjan, A., & Chaturvedi, P. (2023b). Unleashing Digital Innovation for Sustainable Rural Intangible Cultural Heritage Marketing. In *Sustainable Development Goal Advancement Through Digital Innovation in the Service Sector* (pp. 95–113). IGI Global., DOI: 10.4018/979-8-3693-0650-5.ch007

Ranjan, A., & Chaturvedi, P. (2025). Creating Meaningful Tourism Experiences With Digital Storytelling. In *Meaningful Tourism* (pp. 127–144). Emerald Publishing Limited., DOI: 10.1108/978-1-83797-837-320251019

Ratshega, S. (2010). *Dikgafela.* Makirikiri Album. URL: https://www.youtube.com/watch?v=874NSGeNKWQ&list=RD874NSGeNKWQ&start_radio=1

Ratshega, S. (2016). *Leso la Monnamogolo.* Leso la Monnamogolo Album. URL: https://www.youtube.com/watch?v=-TrHXMQmE04&list=RD-TrHXMQmE04&start_radio=1

Ratsie, S. 2021. Perceptions of Initiation Instructors and Parents on the Role of Initiation (Bogwera Le Bojale) in Character Building of Adolescents: A Case of Mochudi Village, Kgatleng District in Botswana. Doctoral Dissertation. The University of Zambia, Zambia.

Ray, D. I. (2003). Rural Local Governance and Traditional Leadership in Africa and the Afro- Caribbean: Policy and Research Implications from Africa to the Americans and Australasia. In Ray, D. I., & Reddy, P. S. (Eds.), *Grassroots Governanc? Chiefs in Africa and the Afro-Caribbean* (pp. 1–30). University of Calgary Press. DOI: 10.2307/j.ctv6gqrhd.5

Ray, R. K. (2003). *The felt community: Commonalty and mentality before the emergence of Indian nationalism*. Oxford University Press.

Reddick, R. J., Welton, A. D., Alsandor, D. J., Denyszyn, J. L., & Platt, C. S. (2011). Stories of success: High minority, high poverty public school graduate narratives on accessing higher education. *Journal of Advanced Academics, 22*(4), 594–618. DOI: 10.1177/1932202X11414133

Ren, J. (2025). Inheritance and protection of intangible cultural heritage in drama category based on AI human–computer interaction and digital technology. *Discover Artificial Intelligence, 5*(1), 67. Advance online publication. DOI: 10.1007/s44163-025-00268-y

Rennie, T., Light, D., Rutherford, A., Miller, M., Fisher, I., Pratchett, D., Capper, B., Buck, N., & Trail, J. (1977). Beef cattle productivity under traditional and improved management in Botswana. *Tropical Animal Health and Production, 9*(1), 1–6. DOI: 10.1007/BF02297380 PMID: 906082

Republic of South Africa. (1996). *Defence in a Democracy: White Paper on National Defence for the Republic of South Africa, May 1996*. https://www.gov.za/documents/national-defence-white-paper [Accessed 10 January 2026].

Republic of South Africa. (2000). *Promotion of Equality and Prevention of Unfair Discrimination Act, No. 4 of 2000*. Government Printer.

Republic of South Africa. (2002). *Defence Act, No. 42 of 2002*. Government Printer.

Reserve Bank of India. (2015). *Master Direction - Non-Banking Financial Company - Microfinance Institutions (Reserve Bank) Directions, 2011*. RBI.

Richards, G., & Leal Londoño, M. P. (2022). Festival cities and tourism: Challenges and prospects. *Journal of Policy Research in Tourism, Leisure & Events, 16*(1), 107–129. DOI: 10.1080/19407963.2022.2087664

Richardson, M. (2025). Modelling nature connectedness within environmental systems: Human-nature relationships from 1800 to 2020 and beyond. *Earth (Basel, Switzerland), 6*(3), 82. DOI: 10.3390/earth6030082

Ricoeur, P. (1992). *Oneself as another* (Blamey, K., Trans.). University of Chicago Press.

Ringson, J., & Chereni, A. (2020). Efficacy of the extended family system in supporting orphans and vulnerable children in Zimbabwe: An indigenous knowledge perspective. *African Journal of Social Work, 10*(1).

Robertshaw, P. (2021). Archaeology of early pastoralism in East Africa. In *Oxford research encyclopedia of African history*. DOI: 10.1093/acrefore/9780190277734.013.1045

Robertshaw, P., & Collett, D. (1983). A new framework for the study of early pastoral communities in East Africa. *Journal of African History*, *24*(3), 289–301. DOI: 10.1017/S0021853700022027

Robinson, L., Helmus, T. C., Cohen, R. S., Nader, A., Radin, A., Magnuson, M., & Migacheva, K. (2018). *Modern Political Warfare: Current Practices and Possible Responses*. RAND Corporation., www.rand.org/t/RR1772 DOI: 10.7249/RR1772

Robinson, M. S. (2001). *The microfinance revolution: Sustainable finance for the poor*. World Bank Publications. DOI: 10.1596/0-8213-4524-9

Rogers, E. M., & Shoemaker, F. F. (1971). *Communication of innovations: A Cross-cultural Approach*. The Free Press.

Roy, B. (2025). The Teachings of the Pura as and Modern Science in Indian Knowledge System: A Review. [IJSSR]. *International Journal of Social Science Research*, *2*(3), 333–341.

Roy, T. (2012). *India in the world economy: From antiquity to the present*. Cambridge University Press. DOI: 10.1017/CBO9780511920516

Russell, T., Silva, F., & Steele, J. (2014). Modelling the spread of farming in the Bantu-speaking regions of Africa: An archaeology-based phylogeography. *PLoS One*, *9*(1), e87854. DOI: 10.1371/journal.pone.0087854 PMID: 24498213

S. (2025, October 11). *Sridhar Vembu's vision: How Zoho transformed Tenkasi into rural tech hub*. The Week.https://www.theweek.in/theweek/business/2025/10/11/sridhar-vembus-vision-how-zoho-transformed-tenkasi-into-rural-tech-hub.html

Sabarivasan, R., Gokul, S., & Infanta, S. C. (2025). Mapping the global scientific landscape of indigenous knowledge in agroforestry systems: A bibliometric perspective (2000–2024). *Agroforestry Systems*, *99*(8), 232. DOI: 10.1007/s10457-025-01337-y

Saberwal, V. K., & Rangarajan, M. (Eds.). (2009). *Battles over nature: Science and the politics of conservation* (2. impr). Permanent Black.

Sadiq, F. A., & Yan, B. (2018). Fermented foods and human health: Microbial contributions. *Critical Reviews in Food Science and Nutrition*, *58*(7), 1151–1166.

Sadr, K. (2008). Invisible Herders? The Archaeology of Khoekhoe Pastoralists. *Southern African Humanities*, *20*(1), 179–203.

Said, E. W. (1978). *Orientalism* (1st ed.). Pantheon Books.

Samkange, S., & Samkange, T. M. (1980). *Hunhuism or Ubuntuism: A Zimbabwe indigenous political philosophy*. Graham Publishing.

San Román, Á., & Molinero-Gerbeau, Y. (2023). Anthropocene, Capitalocene or Westernocene? On the ideological foundations of the current climate crisis. *Capitalism, Nature, Socialism*, *34*(4), 39–57. DOI: 10.1080/10455752.2023.2189131

Sansour, V., & Tartir, A. (2014). *Palestinian farmers: A last stronghold of resistance*. Al-Shabaka: The Palestinian Policy Network.

Santos, B. de S. (2016). *Epistemologies of the South: Justice against epistemicide*. Routledge., DOI: 10.4324/9781315634876

Saran, A. B., & Pandey, G. (1992). *Sun worship in India: A Study of Deo Sun-Shrine*. Northern Book Centre.

Sarkar, B. K. (2021). Giving as a Road to Fulfilment. *Bulletin of the Ramakrishna Mission Institute of Culture*.

Sarum, O. O. (2023). Current issues on leadership in Africa: The need for urgent reorientation in quest for a sustainable continent in the 21st Century Technological explosion. www.academia.edu/103775373/ 22 June 2022.

Saxena, P. (2025). *Svayambodha and Shatrubodha: Hindu View of Self and the World*. BluOne Ink Pvt. Ltd.

Schapera, I. (1935). The social structure of the Tswana ward. *Bantu Studies*, *9*(1), 203–224. DOI: 10.1080/02561751.1935.9676370

Schapera, I. (1965). *Married life in an African tribe*. Sheridon House.

Schapera, I., & Comaroff, J. L. (2015). *The Tswana*. Routledge. DOI: 10.4324/9781315684093

Scherzinger, M. (2019). *Ritual and Music in Africa: Performance, Politics, and Cultural Memory*. Oxford University Press.

Schieman, S., & Plickert, G. (2008). How knowledge is power: Education and the sense of control. *Social Forces*, *87*(1), 153–183. DOI: 10.1353/sof.0.0065

Scoppio, G. (2018). Diversity management in the military: The case of indigenous integration in the New Zealand Defence Force. *Armed Forces and Society*, *44*(2), 337–357.

Scott, J. C. (2020). *Seeing like a state: How certain schemes to improve the human condition have failed* (Veritas paperback edition). Yale University Press.

Scott, N. (2002). From the SADF to the SANDF: Safeguarding South Africa for a better life for all. Violence and transition series, 7, pp.1-88.

Scott, R. C. (2015). *The art of marimba in Tokyo: emergence in the twentieth century*. University of Toronto.

Scott, W. R. (2014). *Institutions and organizations: Ideas, interests, and identities* (4th ed.). SAGE Publications.

Seal, P. P. (2024). Food Culture in China Town: Tiretti Bazar of Kolkata. In *Cultural, Gastronomy, and Adventure Tourism Development* (pp. 176-187). IGI Global

Sekar, R. (1992). *The Sabarimalai Pilgrimage and Ayyappan Cultus*. Motilal Banarsidass.

Selwyn, J., & Nandy, S. (2014). Kinship care in the UK: Using census data to estimate the extent of formal and informal care by relatives. *Child & Family Social Work, 19*(1), 44–54. DOI: 10.1111/j.1365-2206.2012.00879.x

Sen, B. (2005). Indigenous knowledge for development: Bringing research and practice together. *The International Information & Library Review, 37*(4), 375–382. DOI: 10.1080/10572317.2005.10762695

Setiloane, G. (1976). *The image of God among the Sotho-Tswana.* Balkema.

Setlhabi, K. G. (2014). The politics of culture and the transient culture of Bojale: Bakgatla-Baga-Kgafela women's initiation in Botswana. *Journal of Southern African Studies, 40*(3), 459–477. DOI: 10.1080/03057070.2014.913424

Settee, P., & Shukla, S. (Eds.). (2020). *Indigenous food systems: Concepts, cases, and conversations.* Canadian Scholars.

Shackleton, C. M.. (2019). Livelihoods and vulnerability in rural South Africa. *Land Use Policy, 84*, 87–95.

Shahack-Gross, R., Marshall, F., & Weiner, S. (2003). Geo-Ethnoarchaeology of Pastoral Sites: The Identification of Livestock Enclosures in Abandoned Maasai Settlements. *Journal of Archaeological Science, 30*(4), 439–459. DOI: 10.1006/jasc.2002.0853

Sharma, R. S. (2006). *India's Ancient Past.*

Sharma, A. (1999). The Puru ārthas: An Axiological Exploration of Hinduism. *The Journal of Religious Ethics, 27*(2), 223–256. https://www.jstor.org/stable/40018229. DOI: 10.1111/0384-9694.00016

Sharma, K. C. (2012). *Public administration in Botswana.* Botswana Institute for Development Policy Analysis.

Sharma, S. (2007). A new managerial paradigm from Indian wisdom traditions. *Journal of Human Values, 13*(2), 99–114.

Sharma, S. K., & Biswas, D. (2024). Consciousness: Revisiting the Sacred. *Environmental Humanities in India, 25*, 33. Singh, K. Relocating Identities. *India International Centre Quarterly, 28*(3), 100–108.

Sharma, S. R., & Sharma, M. (2025). From Environmental Degradation to Social Transformation: Exploring the Role of Eco-Justice in the Struggles of Indigenous Communities. *Open Journal of Social Sciences, 13*(5), 325–357. DOI: 10.4236/jss.2025.135019

Sharma, Y., Tochhawng, L., Kumar, V., Nikam, R. J., Deka, P., & Baruah, C. J. (2024). Cultural heritage conservation in urban areas: Integrating traditional knowledge and modern planning practices. *Journal of Applied Bioanalysis, 10*, 260–265.

Sharpley, R., & Sundaram, P. (2005). Tourism: a sacred journey? The case of ashram tourism, India. *International Journal of Tourism Research/˜the œInternational Journal of Tourism Research, 7*(3), 161–171. https://doi.org/DOI: 10.1002/jtr.522

Sha, S. (2024). Environmental degradation and migration: An analysis of tribal livelihoods. *International Journal of Political Science and Governance, 6*(2), 115–118. DOI: 10.33545/26646021.2024.v6.i2b.374

Shillington, K. (1985). *The colonisation of the Southern Tswana, 1870-1900*. Routledge.

Shillington, K. T. J. (1981). *Land Loss, Labour and Dependence: The Impact of Colonialism on the Southern Tswana: c. 1870-1900*. University of London, School of Oriental and African Studies.

Shinde, K. N. (2007). Pilgrimage and the environment: Challenges in a pilgrimage centre. *Current Issues in Tourism*, *10*(4), 343–365. DOI: 10.2167/cit259.0

Shin, E., Lee, S., Kang, H., Kim, J., Kim, K., Youn, H., Jin, Y. W., Seo, S., & Youn, B. (2020). Organ-specific effects of low dose radiation exposure: A comprehensive review. *Frontiers in Genetics*, *11*, 566244. DOI: 10.3389/fgene.2020.566244 PMID: 33133150

Shiva, V. (2015). *Earth democracy: Justice, sustainability, and peace*. North Atlantic Books.

Shomolekae, T. (2024). ANC to work with traditional leaders in preparation for national dialogue. https://www.polity.org.za

Shrestha, R. K., L'Espoir Decosta, J. N. P., & Whitford, M. (2025). Indigenous knowledge systems and socio-cultural values for sustainable tourism development: Insights from Indigenous Newars of Nepal. *Journal of Sustainable Tourism*, *33*(1), 143–167. DOI: 10.1080/09669582.2024.2316298

Sihag, B. S. (2014). *Kautilya: The true founder of economics*. Vitasta Publishing.

Singh, A. (2021). An Introduction to Experimental and Exploratory Research. SSRN Electronic Journal. DOI:DOI: 10.2139/ssrn.3789360

Singh, K. S. (Ed.). (1998). *People of India: Rajasthan* (Vol. 38). Popular Prakashan.

Singh, A., Nath, V., Kumar, S., & Singh, B. S. (2016). The role of a traditional festival, Chhath Puja, in the conservation and sustainable use of tropical fruits. In Normah, M. N., Chin, H. P. M., & Reed, B. M. (Eds.), *Tropical fruit tree diversity: Good practices for in situ and ex situ conservation* (pp. 217–228). Bioversity International.

Singhal, A. K., & Sharma, U. (2025). Ethical Discourses in Ancient Indian Culture for Sustainable Protection of the Environment.

Sivakami, V., Bindu, V. T., & George, B. (2023). Impact management and experience design for sustainable development of ecotourism destinations: The case of Eravikulam National Park, India. *Journal of Ecotourism*, *22*(1), 80–102. DOI: 10.1080/14724049.2022.2042542

Sivaramakrishnan, K. (2015). Ethics of nature in Indian environmental history: A review article. *Modern Asian Studies*, *49*(4), 1261–1310. DOI: 10.1017/S0026749X14000092

Sizemore-Barber, A. (2012). The Voice of (Which?) Africa: Miriam Makeba in America. *Safundi*, *13*(3-4), 251–276. DOI: 10.1080/17533171.2012.715416

Skilleås, H. K., & Grande, B. (2024). Language Lessons Learned? NATO Requirements and Norwegian Officer Cadets. *Scandinavian Journal of Military Studies*, *7*(1). Advance online publication. DOI: 10.31374/sjms.221

Skogvang, B. O. (2021). Development of cultural and environmental awareness through Sámi outdoor life at Sámi/Indigenous festivals. *Frontiers in Sports and Active Living*, *3*, 662929. DOI: 10.3389/fspor.2021.662929 PMID: 34337399

Smart, N., & Thakur, S. C. (1993). *Ethical and political dilemmas of modern India*. No Title. DOI: 10.1007/978-1-349-23057-0

Smith, A. (2018). Reconnecting the music-making experience through musician efforts in instrument craft. *International Journal of Music Education*, *36*(4), 560–573. DOI: 10.1177/0255761418771993

Smith, J., Lee-Thorp, J., & Hall, S. (2007). Climate change and agropastoralist settlement in the Shashe-Limpopo River Basin, southern Africa: AD 880 TO 1700. *South African Archaeological Bulletin*, *62*(186), 115–125.

Smith, L. T. (1999). *Decolonizing methodologies: Research and indigenous peoples*. Zed Books.

Smith, L., & Akagawa, N. (2008). Intangible Heritage. In Smith, L., & Akagawa, N. (Eds.), *Intangible Heritage*. Routledge., DOI: 10.4324/9780203884973

Snook, J., Cunsolo, A., Borish, D., Furgal, C., Ford, J. D., Shiwak, I., Flowers, C. T. R., & Harper, S. L. (2020). "We're made criminals just to eat off the land": Colonial wildlife management and repercussions on Inuit well-being. *Sustainability (Basel)*, *12*(19), 8177. DOI: 10.3390/su12198177

Sobane, K., Lunga, W., & Setlhabane, L. (2024). Harnessing Indigenous knowledge systems for socially inclusive science communication: Working towards a 'science for us, with us' approach to science communication in the Global South. In Rasekoala, E. (Ed.), *Race and sociocultural inclusion in science communication*. Policy Press.

Soeters, J. L., & Van Der Meulen, J. (Eds.). (2007). *Cultural diversity in the armed forces: An international comparison*. Routledge. DOI: 10.4324/9780203967409

Soganga, A., & Kang'ethe, S. M. (2023). Exploring reciprocity in grandparent-grandchildren care: The case of Mdlankomo location, Libode in the Eastern Cape, South Africa. *Social Work/Maatskaplike Werk, 59*(1).

Sohad, M. K. N., & Mafrolla, E. (2025). Bridging science and society: The integration of indigenous and scientific knowledge management. *Journal of Knowledge Management*, *29*(7), 2258–2284. DOI: 10.1108/JKM-11-2024-1326

Soneji, D. (2012). *Unfinished gestures: Devadāsīs, memory, and modernity in South India*. University of Chicago Press.

South African Local Government Association. (2013). Local Briefing: Challenges regarding participation of traditional leaders in Municipal Councils. www.SALGA.org.za

Spears, C. (2006). *Jeremane*. Kulenyane Album. URL: https://www.youtube.com/watch?v=iRovbnA5lEs&list=RDiRovbnA5lEs&start_radio=1

Spears, C. (2006). Nchandinyana. Kulenyane Album. Kulenyane Studios. URL: https://www.youtube.com/watch?v=TGQa6EqO1yY&list=RDTGQa6EqO1yY&start_radio=1

Spears, C. (2006). *Phala*. Kulenyane Album. URL: https://www.youtube.com/watch?v=-iwSH6X_dwo&list=RD-iwSH6X_dwo&start_radio=1

Spears, C. (2019). Didimala Moratiwa. URL: https://www.youtube.com/watch?v=vDhvKAU3jBI&list=RDvDhvKAU3jBI&start_radio=1

Spears, C. *Lerato*. Kulenyane Album. URL: https://www.youtube.com/watch?v=taxbic-syuY&list=RDtaxbic-syuY&start_radio=1

Srinivas, M. N. (2003). *Religion and society among the Coorgs in South Asia*. Oxford University Press.

Sriram, M. S. (2012). The AP microfinance crisis 2010: Discipline or death? *Vikalpa*, *37*(4), 113–128. DOI: 10.1177/0256090920120407

Srivastava, J. (2010). *The Relevance of Gandhian Thought for Contemporary Management Systems*. Navjeevan Publications.

Srivastava, S. K. (2012). Managing indigenous and scientific knowledge for resilience building: Case studies from disaster‐prone regions of India. *Journal of Advances in Management Research*, *9*(1), 45–63. DOI: 10.1108/09727981211225644

Statistics South Africa. (2021). *General Household Survey 2020*. Statistics South Africa.

Steinkraus, K. H. (1996). *Handbook of indigenous fermented foods* (2nd ed.). Marcel Dekker.

Steinkraus, K. H. (1997). Classification of fermented foods: Worldwide review. *Food Control*, *8*(5–6), 311–317. DOI: 10.1016/S0956-7135(97)00050-9

Stiebing, W. H. (1994). *Uncovering the past: A history of archaeology*. Oxford University Press.

Stiglitz, J. E. (1990). Peer monitoring and credit markets. *The World Bank Economic Review*, *4*(3), 351–366. DOI: 10.1093/wber/4.3.351

Stiglitz, J. E. (1998). *More Instruments and Broader Goals: Moving toward the Post- Washington Consensus*. UNU World Institute for Development Economics Research.

Stokes, E. (1992). *The English utilitarians and India* (Third impression). Oxford Univ. Press.

Stratton, S. J. (2023). Population Sampling: Probability and Non-Probability Techniques. *Prehospital and Disaster Medicine*, *38*(2), 147–148. DOI: 10.1017/S1049023X23000304 PMID: 36939054

Subrahmanyeswari, B., & Chander, M. (2013). Integrating indigenous knowledge of farmers for sustainable organic farming: An assessment in Uttarakhand state of India. *Indian Journal of Traditional Knowledge*, *12*(2), 259–264.

Su, J. (2019). Understanding the changing intangible cultural heritage in tourism commodification: The music players' perspective from Lijiang, China. *Journal of Tourism and Cultural Change*, *17*(3), 247–268. DOI: 10.1080/14766825.2018.1427102

Sulaimon, M., & Olayiwola, I. (2021). Ethnic and Religious Bias in the Nigerian Armed Forces: A Critical Analysis. *African Security Studies*, *18*(4), 101–119.

Sun, M., Zhao, M., & Zhou, W. (2025). Assessment and Model Construction of Talent Attraction in the Creative Industry: A Case Study of Underdeveloped Cities and Regions in China. *Mediterranean Archaeology & Archaeometry. International Journal, 25*(1).

Sun, Y., & Shi, Y. (2024). Knowledge strategies for Indigenous studies in intercultural communication: Decentring Western epistemic hierarchies. *Social Sciences* [Special Issue]. *Role of Intercultural Communication in Multicultural or Culturally Diverse Societies, 13*(10), 1060–1072.

Suresh, V., & Sreejith, S. (2023, September). Strengthening collective action and sustainability of traditional farming organizations from an Indian philosophical perspective. In *ECKM 2023 24th European Conference on Knowledge Management Vol 2*. Academic Conferences and publishing limited.

Sustainability Directory. (2025). Displacement of Indigenous Peoples. Sustainability Directory. https://pollution.sustainability-directory.com/term/displacement-of-indigenous-people/ (Accessed: 19 January 2026)

Swanta, H.P. (2026). Taboos And Their Ethical Implications In The Formation Of Identity Among The Oegworok (Kagoro) Community Of Southern Kaduna, Nigeria. *Journal of Human, Social and Political Science Research.*

Swarajya. (2025). Bihar: How Industry And IT Hubs Can Decongest Chhath Puja Special Trains. Retrieved from https://swarajyamag.com/states/bihar-how-industry-and-it-hubs-can-decongest-chhath-puja-special-trains

Swiderska, K., & Ryan, P. (2021). *Indigenous food systems, biocultural heritage, and the SDGs: challenges, interdisciplinary research gaps, and empowering methodologies.*

Taber, N. (2018). After Deschamps: Men, masculinities, and the Canadian Armed Forces. *Journal of Military, Veteran and Family Health, 4*(1), 100–107. DOI: 10.3138/jmvfh.2017-0005

Tahseen, E., & Al-Jumaily, S. (2020). Mechanisms for reviving the intangible cultural heritage to revitalize urban spaces. *International Journal of Environment. Engineering and Education, 2*(3), 31–42.

Tak, M. (2023). The Economic De-Development of Jammu and Kashmir: Uncovering the Myth of Lagging Behind. In *The Palgrave Handbook of New Directions in Kashmir Studies* (pp. 441–457). Springer International Publishing. DOI: 10.1007/978-3-031-28520-2_27

Tak, M., Hussain, S. B., Zargar, H., & Blake, L. J. (2025). Settler colonialism de-develops food systems in protracted crises. *Nature Food, 6*(8), 751–756. DOI: 10.1038/s43016-025-01211-1 PMID: 40825868

Tankha, A. (2012). *Banking on self-help groups: Twenty years on.* SAGE Publications.

Tasci, A. D. A., Wei, W., & Milman, A. (2020). Uses and misuses of the case study method. *Annals of Tourism Research, 80*, 102846. DOI: 10.1016/j.annals.2019.102815

Taylor, E., DiFolco, S., Dupin, M., Mithen, H., Wen, L., Rose, L., & Nisbert, K. (2020). Socioeconomic deprivation and social capital in kinship carers using a helpline service. *Child & Family Social Work, 25*(4), 845–855. DOI: 10.1111/cfs.12763

Taylor, M. (2011). 'Freedom from poverty is not for free': Rural development and the microfinance crisis in Andhra Pradesh, India. *Journal of Agrarian Change*, *11*(4), 484–504. DOI: 10.1111/j.1471-0366.2011.00330.x

Teherani, A., Martimianakis, T., Stenfors-Hayes, T., Wadhwa, A., & Varpio, L. (2015). Choosing a qualitative research approach. *Journal of Graduate Medical Education*, *7*(4), 669–670. DOI: 10.4300/JGME-D-15-00414.1 PMID: 26692985

Tendero, E. J. (2023). Pilgrimage Tourism: A case study of Fort Pilar Shrine in Zamboanga City, Philippines. *Social Science Research Network*. DOI: 10.2139/ssrn.4518755

The Arab Republic of Egypt Presidency. (2025). President El-Sisi speaks with Greece Prime Minister Mitsotakis. Cairo: APO Distribution on behalf of Presidency Arab Republic of Egypt.

Tiatira, Y. C., Herwandito, S., & Krisnawati, E. (2024). Harmonising Religious Diversity: Exploring Effective Intercultural Communication in Dukuh Sampetan Village. *International Journal of Multicultural and Multireligious Understanding*, *11*(3), 108–119. DOI: 10.18415/ijmmu.v11i3.5563

Tilley, C. (1994). Interpreting material culture. In Pearce, C. (Ed.), *Interpreting Objects and Collections* (pp. 67–75). Psychology Press.

Tilley, C. (2007). Materiality in materials. *Archaeological Dialogues*, *14*(1), 16–20. DOI: 10.1017/S1380203807002139

Tirivangasi, H. M., & Kontinen, T. (2025). Reclaiming Indigenous knowledge for sustainability: Afrocentric perspectives on Indigenous intangible cultural heritage and climate adaptation in rural Zimbabwe. *International Journal of Heritage Studies*, •••, 1–28. DOI: 10.1080/13527258.2025.2595080

Tiwari, D. N. (2016). Spiritual ecology and environmental ethics. *Cultura (Iasi, Romania)*, *13*(1), 49–68. DOI: 10.5840/cultura20161313

Trinick, T., & Allen, P. (2024). Wayfinding in an indigenous initial teacher education mathematics programme. *ZDM Mathematics Education*, *56*(3), 485–495. DOI: 10.1007/s11858-024-01589-5

Tshikalange, S. (2025). Ramaphosa and King Misuzulu win SCA appeal in Zulu royal dispute. Timeslive Newspaper, https://www.timeslive.co.za/news/south-africa/2025-10-06-ramaphosa-and-king-misuzulu-win-sca-appeal-in-zulu-royal-dispute.

Tshikhudu, E. (2016). Coronation ceremony halted by court''. *Limpopo Mirror Newspaper*, www.limpopomirror.co.za. Accessed 17 October 2025.

Tshitangoni, M., & Francis, L. (2017). Relevance of Traditional Leadership in Rural Community Development amidst Democratic Institutions in Southern Africa: A Critical Review. *Studies of Tribes and Tribals*, •••, 70–83.

Tsing, A. (2023). A Multispecies Ontological Turn? In *The Routledge International Handbook of More-than-Human Studies* (pp. 116–128). Routledge. DOI: 10.4324/9781003262619-7

Turino, T. (2010). *Moving away from silence: music of the Peruvian altiplano and the experience of urban migration*. University of Chicago Press.

Tyrrell, T., Paris, C. M., & Biaett, V. (2013). A quantified triple bottom line for tourism: Experimental results. *Journal of Travel Research, 52*(3), 279–293. DOI: 10.1177/0047287512465963

Ubink, J., & Duda, T. (2021). Traditional Authority in South Africa: Reconstruction and Resistance in the Eastern Cape. *Journal of Southern African Studies, 47*(2), 191–208. DOI: 10.1080/03057070.2021.1893573

Ugwu, I. P. (2025). Environmental Protection, Rights of Nature, and Religious Beliefs in Europe. *International Journal for the Semiotics of Law - Revue Internationale de Sémiotique Juridique, 38*(3), 993–1014. DOI: 10.1007/s11196-024-10193-5

UNESCO. (2003). *Convention for the Safeguarding of the Intangible Cultural Heritage*. United Nations Educational, Scientific and Cultural Organization.

UNESCO. (2003). *Intangible Heritage - 2003 Convention*. https://ich.unesco.org/en/intangible-heritage-domains-00052

UNESCO. (2023). *What is intangible cultural heritage?* https://ich.unesco.org/en/what-is-intangible-heritage-00003

United Nations. (2023). *Women, peace and security: Report of the Secretary-General*. United Nations.

United Nations. (2025). Co-create solutions to give Africa Stronger role in Global Politics, Financial Institutions, and Debt relief. Secretary-General Antonio Gutteres Address. New York: The United Nations Tokyo Ninth International Conference. 20 August 2025/APO GROUP Distributions.

UNWTO. (2018). *Tourism and Culture Synergies*. United Nations World Tourism Organization., DOI: 10.18111/9789284418978

Urrieta, L.Jr. (2015). Learning by observing and pitching in and the connections to native and indigenous knowledge systems. []. JAI.]. *Advances in Child Development and Behavior, 49*, 357–379. DOI: 10.1016/bs.acdb.2015.08.003 PMID: 26955937

Usman, A. E., Umar, I. A., & Abubakar, H. I. 2016). *Understanding Nigerian Government and Politics*. RAFA Printing and Publishing Ltd, No 8 Keffi Street, off Makurdi Road, Lafia, Nasarawa State.

Utsav India. (2024). Chhath Puja - Events & Festivals in India. Retrieved from https://utsav.gov.in/view-event/chhath-puja-2

Uusiku, N. P., Oelofse, A., Duodu, K. G., Bester, M. J., & Faber, M. (2010). Nutritional value of leafy vegetables of sub-Saharan Africa and their potential contribution to human health: A review. *Journal of Food Composition and Analysis, 23*(6), 499–509. DOI: 10.1016/j.jfca.2010.05.002

Vadakkiniyil, D. (2019). Mahishi's rage: Communitas and protest at Sabarimala, Kerala. *Anthropology Today, 35*(5), 16–20. DOI: 10.1111/1467-8322.12529

Vallor, S. (2016). *Technology and the virtues: A philosophical guide to a future worth wanting*. Oxford University Press. DOI: 10.1093/acprof:oso/9780190498511.001.0001

Van Der Waag, I. (2015). South African defence in the age of total war, 1900-1940. *Historia (Wiesbaden, Germany), 60*(1), 129–155.

van Iwaarden, M., & Nawijn, J. (2024). Eudaimonic benefits of tourism: The pilgrimage experience. *Tourism Recreation Research*, *49*(1), 37–47. DOI: 10.1080/02508281.2021.1986777

Van Staden, M. (2025). Land loss and labour control: The intersection of land dispossession and labour law in South African history. *Industrial Law Journal*, *54*(2), 324–356. DOI: 10.1093/indlaw/dwae034

Vandendriessche, E., & Pinxten, R. (Eds.). (2022). *Indigenous knowledge and ethnomathematics*. Springer., DOI: 10.1007/978-3-030-97482-4

Vansina, J. (1985). *Oral Tradition as History*. University of Wisconsin Press. DOI: 10.2307/jj.36106057

Veldtman, S. L. (2018). Strategic Change Management in the South African National Defence Force (Doctoral dissertation, University of Cape Town).

Vijayanand, S. (2014). The Issues and Perspectives of Pilgrimage Tourism Development in Tanjur. *Int. J. Tour. Hosp. Rev*, *1*, 45–51.

Vivekananda International Foundation. (2023, April 4). *Dr. Rajendra Singh: Waterman of India*. https://www.vifindia.org/india-positive/2023/april/04/dr-rajendra-singh-waterman-of-india [page:2]

Wa Thiongo, N. (1986). *Decolonising the Mind: The Politics of Language in African Literature*. James Currey.

Wahyudi, L. (2024). *Watase Uake: Research Collaboration Tools.*

Wang, J., Lim, X. J., & Ng, S. I. (2024). Why can't I get a job? Exploring the drivers of perceived employability among fine art graduates. *Current Psychology (New Brunswick, N.J.)*, *43*(4), 3002–3020. DOI: 10.1007/s12144-023-04561-w

Washbrook, D. A. (1981). Law, State and Agrarian Society in Colonial India. *Modern Asian Studies*, *15*(3), 649–721. DOI: 10.1017/S0026749X00008714

Weatherdon, M. S. (2022). Religion, Animals, and Indigenous Traditions. *Religions*, *13*(7), 654. DOI: 10.3390/rel13070654

Weatherdon, M., & Schermerhorn, S. (2022). Movement and Indigenous Religions: A Reconsideration of Mobile Ways of Knowing and Being. *Material Religion : the Journal of Objects, Art and Belief*, *18*(1), 3–15. DOI: 10.1080/17432200.2021.2015921

Weaver, C. B. (2024). How Not to Undiscipline Religion and Science: Indigenous Ecological Knowledge, Epistemic Resistance, and the Settler Imagination. *Religions*, *15*(11), 1290. DOI: 10.3390/rel15111290

White Paper on Affirmative Action in the Public Service. (1998). *Notice 564 of 1998*. Government Printers.

Wikan, G. (2004). The level of living in rural Botswana re-studied. *Norsk Geografisk Tidsskrift*, *58*(1), 1–10. DOI: 10.1080/00291950410004393

Williams, H. L., Jr. (2022). Examining the Cross-Cultural Adaptation of Department of Defence Leaders in Joint Multinational Military Training Environments (Doctoral dissertation, Trident University International).

Williams, R. (2001). *Completing the defence transformation process: The transformation of the South African Reserve Force System*. Institute for Security Studies Papers.

Wilson, C., & Noyes, G. (2025, January). Indigenous Stewardship of Ancestral Lands Activates Land and Culture. In *Parks Stewardship Forum* (Vol. 41, No. 1, pp. 66-72). George Wright Society. DOI: 10.5070/P5.42004

Wilson, S. (2020). *Research is ceremony: Indigenous research methods*. Fernwood publishing.

Wilson, T. H. (1982). Spatial analysis and settlement patterns on the East African coast. *Paideuma*, •••, 201–219.

Wiltshier, P. (2020). Health and welfare at the boundaries: Community development through tourism. *Journal of Tourism Futures*, *6*(2), 153–164. DOI: 10.1108/JTF-05-2018-0021

Wobst, H. M. (2004). Power to the (Indigenous) past and present! Or: The theory and method behind archaeological theory and method. In *Indigenous Archaeologies* (pp. 15–29). Routledge.

Wood, E. N. (1985). Traditional Music in Botswana. *The Black Perspective in Music*, *13*(1), 13–30. DOI: 10.2307/1214791

World Bank. (2021). *Climate risk country profile: Lesotho*. Climate Change Knowledge Portal. https://climateknowledgeportal.worldbank.org/sites/default/files/2021-08/15930-WB_Lesotho%20Country%20Profile-WEB.pdf

Writer, S. (2024). A New Dawn for the Rain Queen amid succession disputes. www.mopanitimes.co.za/2024/12/a-new-dawn-for-the-rain-queen-amid- succession-dispute/accessed/01 August 2025.

WWF-India. (n.d.). Reducing India's global footprint. https://www.wwfindia.org/about_wwf/reducing_footprint/

Xavier, C. V., & Viegas, R. P. (2014). Smoking as a meat preservation technique. *Food Science & Technology International*, *20*(3), 223–232.

Xu, Y., Zhao, Q., Schuler, B. R., & Levkoff, S. E. (2022). Material hardship among custodial grandparents in COVID-19 and its association with grandchildren's physical and mental health: A latent class analysis. *Children and Youth Services Review*, *132*, 106340. DOI: 10.1016/j.childyouth.2021.106340

Yang, S., Shu, T., & Yu, T. (2022). Migration for better jobs or better living: Shifts in China. *International Journal of Environmental Research and Public Health*, *19*(21), 14576. DOI: 10.3390/ijerph192114576 PMID: 36361454

Yatim, H., Jayadi, K., Jamilah, J., Abduh, A., & Prusdianto, P. (2025). Integrating cultural heritage into education: The supportive role of schools in traditional festivals in West Sulawesi. *Discover Education*, *4*(1), 379. Advance online publication. DOI: 10.1007/s44217-025-00847-2

Yılmaz, N. (2025). Cultural Identity and Globalization: The Impact on Indigenous Traditions. *Arts and Social Sciences Journal,* Volume 16:01

Yin, R. K. (2018). *Case study research and applications: Design and methods* (6th ed.). SAGE Publications.

Yoo, C. K., Yoon, D., & Park, E. (2018). Tourist motivation: An integral approach to destination choices. *Tourism Review, 73*(2), 169–185. DOI: 10.1108/TR-04-2017-0085

Yunus, M. (2007). *Banker to the poor: Micro-lending and the battle against world poverty*. PublicAffairs.

Zabulis, X., Partarakis, N., Zidianakis, E., & Kaplanidi, D. (2025). A Critical Review of the Function of Intangible Cultural Heritage as a Driver for Social Resilience and Cohesion. *Encyclopedia*, *5*(4), 189. DOI: 10.3390/encyclopedia5040189

Zaluchu, S. E. (2025). Pray to the ancestors: A case study of Christianity and Nias indigenous religion. *Cogent Arts & Humanities*, *12*(1), 2492419. Advance online publication. DOI: 10.1080/23311983.2025.2492419

Zaman, M. Q. (2007). *The Ulama in contemporary Islam: Custodians of change* (Paperback ed.). Princeton University Press.

Zawacki-Richter, O., Kerres, M., Bedenlier, S., Bond, M., & Buntins, K. (2020). *Systematic reviews in educational research: Methodology, perspectives and application*. Springer Nature. DOI: 10.1007/978-3-658-27602-7

Zeder, M. A. (2014). Domestication: Definition and Overview. In Smith, C. (Ed.), *Encyclopedia of Global Archaeology* (pp. 2184–2194). Springer New York., DOI: 10.1007/978-1-4419-0465-2_71

Zhang, W. (2024). Teaching Design of High School Music Class: Taking Marimba piece Bamboo Forest" by Keiko Abe as Extended Example. *Arts, Culture and Language, 1*(6).

Zondo Commission. (2022). The Zondo Commission: A bite-sized summary. Public Affairs Research Institute. https://pari.org.za/zondo

Zulu, J. (2024). Decolonizing the Mosi-oa-Tunya/Victoria Falls World Heritage Site: Engaging the Unspoken Truths. In Ripmeester, M., & Rofe, M. W. (Eds.), *Global Iconoclasm: Contesting "Official" Mnemonic Landscapes* (pp. 63–75). Springer Fachmedien Wiesbaden., DOI: 10.1007/978-3-658-43691-9_4

Zweni, A. G., Faku, E. M., & Yan, B. (2025). 'The Formation of a Government of National Unity in South Africa: Issues, Trends, and Prospects'. *Journal of Nation-building &. Policy Studies*, *9*(1), 145–164. DOI: 10.31920/2516-3132/2025/v9n1a7

About the Contributors

Monicca Thulisile Bhuda is an indigenous scholar, editor, children's author, multi award winning academic, research fellow awardee and Indigenous Knowledge Systems expert. At the University of Mpumalanga, Dr. Bhuda coordinates a course on African identities and lectures on culture and heritage studies. She graduated from North-West University with a Bachelor's degree (hons) in Indigenous Knowledge Systems (with distinction), a Master's degree also in Indigenous Knowledge Systems (with distinction) and a PhD in Social Sciences with Indigenous Knowledge Systems (IKS). She has been featured on numerous media platforms such as TV, Radio, Magazines and Newspapers that were based on her interests as an indigenous scholar and expertise in the Indigenous Knowledge Systems discipline. She has also published over 40 academic works in accredited journals, including books indexed with SCOPUS. Her research is on various subjects related to Indigenous Knowledge Systems, which is multidisciplinary in nature.

Tebogo Maria Mothiba is a distinguished academic and research leader who, as of late 2025, serves as the Acting Deputy Vice-Chancellor for Research, Innovation, and Partnerships, and as Executive Dean of the Faculty of Health Sciences at the University of Limpopo. Her work centers on health service management and the management of non-communicable diseases, such as diabetes and hypertension, with extensive experience both as an academic and a registered nurse. Professor Mothiba holds a PhD in Health Service Management from the University of Johannesburg and an MCur from the University of Limpopo. She began her university career in 2006, rising to professorship in 2014, and was promoted through multiple leadership roles including Faculty Research Professor and Executive Dean. She is recognized as a South African National Research Foundation (NRF) C3-rated researcher and has published over 120 peer-reviewed articles and nine book chapters. Her research encompasses healthcare management, indigenous health practices, maternal and child health, and digital health innovations.

***

Hana An L. Alvarado is a pre-service mathematics teacher and a novice researcher with interests in mathematics education, ethnomathematics, and the integration of artificial intelligence in teaching and learning.

Priya Chaturvedi With over 3+ years of teaching experience, I'm a dedicated educator and a Senior Research Fellow specializing in Intangible Cultural Heritage. My passion for preserving and promoting cultural heritage has led her to explore various aspects of this field. I've made notable contributions to the academic community through her research and publications. I've written chapters for edited books, shedding light on the significance of intangible cultural heritage.

Furthermore, I've presented my ideas at numerous national and international conferences, fostering discussions and sharing insights with fellow researchers and scholars. As an accomplished author, I've authored several book chapters focused on intangible cultural heritage. My expertise extends to the Encyclopedia of Tourism Management and Marketing, where I've contributed manuscripts on various terms related to this field. In addition to my academic achievements, I've been actively involved in teaching various tourism-related subjects to undergraduate and postgraduate students. My dedication to imparting knowledge and nurturing the next generation of professionals in the tourism industry is commendable. My commitment to preserving cultural heritage extends beyond the classroom. I've volunteered for Bakri Chhap Agro Tourism & Natural Products Pvt. Ltd., a renowned organization dedicated to rural tangible and intangible cultural heritage preservation and rural tourism development activities in India. I served as a volunteer from January 2023 to June 2023, contributing her expertise and skills to support the organization's mission in Madhya Pradesh, India. My exemplary service has been recognized and acknowledged through a formal recommendation and experience letter provided by Bakri Chhap Agrotourism & Natural Products Pvt. Ltd. With my extensive knowledge, research contributions, and teaching experience, I continue to make valuable contributions to the field of intangible cultural heritage and tourism. My dedication to preserving cultural traditions and fostering sustainable tourism practices is an inspiration to my colleagues and students alike.

Mpeiwa Khathutshelo Donald is an emerging researcher affiliated with the University of South Africa as Postdoctoral Fellow. His interest is in Cultural heritage resources management, African Knowledge Systems, Development Studies, Anthropology and History. Ironically, he shares over 24 years of experience as a public servant until joining the academia very recently. To God the glory, Amen.

Smriti Dixit is a management educator with over four years of experience in management education, teaching and mentoring undergraduate and postgraduate students in finance, accounting, and strategy. Her academic interests lie at the intersection of Indian Knowledge Systems (IKS), leadership, and sustainability, with a focus on integrating traditional wisdom with contemporary management and sustainable business practices.

Amit Gupta is an Assistant Professor at Galgotias University, Greater Noida, specializing in Finance, Taxation, Statistics, and International Business. A Gold Medalist in M.Com from Indira Gandhi National Tribal University (IGNTU), Amarkantak, he cleared the UGC-JRF/SRF in Commerce in 2019 with an exceptional 99.74 percentile. He earned his Ph.D. from IGNTU in 2025, with his doctoral research focusing on stakeholders' perceptions towards GST in Madhya Pradesh and Chhattisgarh. With over four years of teaching and research experience, Dr. Gupta has taught subjects ranging from Income Tax Law and GST to Risk Management and Business Economics. He has published research in several peer-reviewed journals, including Scopus-indexed publications, and has co-edited two books on GST. He has presented papers at numerous national and international conferences, including the All India Commerce Conference and the All India Accounting Conference. Beyond academia, Dr. Gupta holds four design patents and is a certified tax professional trained under a Government of India initiative. He is a life member of both the Indian Accounting Association and the Indian Commerce Association, and is proficient in statistical tools such as SPSS, R, and Stata.

Snehil G. Jaiswal is a Research Scholar at the Dr. Babasaheb Ambedkar School of Law, Rashtrasant Tukadoji Maharaj Nagpur University, Maharashtra. Trained in both engineering (B.E., M.E., Ph.D. in Electronics Engineering) and law (LL.B., LL.M.), his research focuses on constitutional law, governance studies, and the historical evolution of administrative institutions in India. His work adopts an interdisciplinary approach, examining constitutional development, federal dynamics, and institutional accountability.

Pallavi Joshi, Currently serving as a Research Associate with a PhD in Management at FMS-WISDOM, Banasthali Vidyapith, with a past enriching tenure as Visiting Faculty at the School of Economics, DAVV, Indore. Earned UGC NET (Management) qualifications post the successful completion of an MBA from DAVV, Indore. Recipient of the Gold Medal for the 2015 BA program at DAVV, Indore. The pursuit of learning has been actualized by completing two Annual Refresher Programs in Teaching through SWAYAM, and several enriching courses from renowned platforms. Possessing a certification in Social and Emotional Intelligence from Socialigence, Indore, acknowledging the importance of empathy and interpersonal skills.

Gedong Maulana Kabir is a lecturer and serves as a Subject Content Specialist at the Character Building and Development Center (CBDC) at Bina Nusantara University. He is an alumnus of the Center for Religious and Cross-cultural Studies (CRCS) at Gadjah Mada University. His academic interests center on the relationship between religion, the state, and society. Specifically, his research explores issues related to indigenous religion, customary traditions (adat), and minority religions, examining their intersection with environmental, gender, cultural, and state policy aspects.

Saad Ullah Khan possesses over a decade of sustained academic and research engagement in higher education, marked by a rigorously interdisciplinary orientation that situates communication studies within its wider socio-political, cultural, and epistemic frameworks. He earned his Ph.D. from the prestigious Aligarh Muslim University, where his doctoral training foregrounded theoretically informed and methodologically robust inquiries into the social functions, power structures, and normative implications of communication practices. Dr. Khan has made substantive contributions to scholarly knowledge production, evidenced by multiple peer-reviewed publications indexed in SCOPUS and the Web of Science, alongside articles in nationally and internationally reputed journals, reflecting both disciplinary depth and cross-disciplinary relevance. His research trajectory has been supported and validated by highly competitive national fellowships, notably the Rajat Jayanti Vigyan Sancharak Fellowship (2009), awarded by the Department of Science and Technology, Government of India, in recognition of his innovative work on science communication and its societal interfaces, and the Maulana Azad National Fellowship (2011), conferred by the University Grants Commission to support his doctoral research. Dr. Khan's active participation in global scholarly dialogue is demonstrated through the presentation of over 90 research papers at national and international conferences, workshops, and symposia, underscoring his sustained intellectual engagement with evolving debates in communication and media studies. His integrated excellence in research productivity, pedagogical practice, and academic leadership has been formally acknowledged through multiple honors, including the Best Researcher Award (2020, 2023) and the Best Journalism Teacher Award (2021).

Pravina Khobragade is a Professor at the Dr. Babasaheb Ambedkar School of Law, Rashtrasant Tukadoji Maharaj Nagpur University, Maharashtra, India. Her areas of specialization include constitutional law, business law, legal history, and governance studies. She has published extensively and is actively involved in curriculum development and academic mentoring. She currently supervises eight doctoral research scholars.

Mothusiotsile Edwin Maditsi is a lecturer and Community Engagement Coordinator at the North-West University's (NWU) DSTI-NRF Centre of Excellence in Indigenous Knowledge Systems (CEIKS). He holds a BA (Hons) in Indigenous Knowledge Systems and a Master of Indigenous Knowledge Systems degree (MIKS) both from NWU. He has participated in and facilitated some community engagement (CE) activities for and on behalf of the IKS Centre at National, provincial, and local level. Some of the CE activities include the IK Act no 6 of 2019 Public awareness workshops, the Biocultural Community Protocols public awareness workshops and IK Regulations public consultations across all nine provinces of the country. Otsile lectures BIKS modules from 1st to 4th year level and he also supervises honors and master's research projects. He is a member of South African Higher Education Community Engagement Forum (SAHECEF), and he has presented at both international and national conferences. He works with the Department of Science, Technology and Innovation (DSTI) on projects such as the IK-Led smart village, IK Regulations, IKSDC project and a host of other projects at national level. In his PhD he is focusing on decolonizing university community engagement (UCE) through indigenous communities of practice (iCoPs) using the NWU as a case study. Since assuming his position as a lecturer, he has shown dedication and commitment in decolonizing the teaching, learning, research, and innovation activities within the higher education sector. He is focused on achieving scholarship of engagement which is at the peak of higher education debates. He has published several journal articles and book chapters with various publishers and some which are high index journals. For his hard work on various national projects and programmes with communities, Otsile was appointed by the Gauteng Commission on Traditional Leadership Disputes and Claims (GCTLDC) which he served for a period of 1 year. He was appointed as a technical expert to the commission. He has recently been appointed by the IKS Consortium as a deputy chairperson of the committee on IKS institutionalization in South African universities.

Mamogobo Rosinah Mamabolo is a lecturer at the University of Limpopo. She holds a PhD in Communication and specialises in culture and cross-cultural communication. Her academic interests focus on the role of cultural knowledge, intercultural interaction and communication practices in promoting understanding across diverse social and institutional contexts.

Mojuta Motlhamme is a historian of the Setswana speaking world in Southern Africa. He holds a PhD in Setswana Historical Studies from the University of the Witwatersrand. He seeks to make Setswana studies a serious academic forum and intellectual space where the curriculum will center Batswana in their various guises. So far he has published work on Onkgopotse Tiro, Diturupa cultural commemoration and he has completed a doctoral study focusing on the political life of Kgosi Lucas Manyane Mangope of Bophuthatswana. Mojuta is currently lecturing History Education at North West University Mahikeng Canpus. He is currently working on a monograph that focuses on Dikgosi tsa Basadi in the Setswana speaking world.

Reason Mupanga is a Lecturer in the Department of Social Work, University of Zimbabwe.He researches more on Indigenous Knowledge Systems, Child Protection and Climate Change.

Divhani Muvhango is a junior lecturer at the University of Venda, specializing in Indigenous Knowledge Systems and heritage studies. She is pursuing a master's degree, exploring traditional Vhavenda paintings, focusing on their cultural meaning and potential to create sustainable income opportunities for youth. Her research explores the intersections of Indigenous knowledge, heritage preservation, and community wellbeing, including traditional food systems and culturally grounded approaches to health and development. Divhani is committed to promoting African cultural knowledge, empowering youth, and integrating Indigenous practices into contemporary education and development initiatives.

Ntshengedzeni Evans Netshivhambe is a senior lecturer of music at the University of South Africa and has earned his Bachelor's, Master's, and Doctoral degrees from the University of the Witwatersrand (Wits University). He is a dedicated researcher, composer, and arranger focused on integrating African indigenous knowledge into higher education. He combines traditional African composition techniques with modern frameworks, enhancing African art music within academic curricula. His work celebrates Africa's cultural heritage, bridging the gap between tradition and contemporary scholarship. Evans' research promotes the respectful integration of African knowledge in education. His interests include music composition, African identity, formalizing indigenous African music knowledge, and professionalizing traditional African instruments.

Captain Nihal is Assistant Professor at Welcomgroup Graduate School of Hotel Administration, Manipal Academy of Higher Education, Manipal- India. His core teaching interest is in Food and Beverage Service Department. His research interests are in Hospitality Management, Education and Tourism Management.

Partho Pratim Seal is Associate Professor at Welcomgroup Graduate School of Hotel Administration, Manipal Academy of Higher Education, Manipal- India. As an avid wordsmith who believes that writing is the most influential and resourceful medium to disseminate information has authored, four books Computers in Hotels: Concepts and Application, and Food & Beverage Management from Oxford University Press- India, How to Succeed in Hotel Management Job Interviews from JAICO India, and Food Anthropology in India from Routledge, UK and authored several book chapters in Scopus Indexed books. His research interests are in Human Resource Management, Gastronomy, Education and Tourism. He has also published research papers in Scopus-Indexed journals and is a reviewer of various international journals.

Aditya Ranjan is an academic and marketing professional with experience spanning higher education, research, consulting, and rural tourism development. Currently, I serve as Assistant Professor (Marketing) at the School of Management, ITM University. I hold a Ph.D. titled *Deconstructing Rural Tourism in India* from Jamia Millia Islamia. My academic interests include consumer behavior, digital marketing, tourism, and heritage studies. I have actively contributed to curriculum development, research, and the organization of international conferences. Prior to academia, I led the Consulting Vertical at Bakri Chhap Agro Tourism and Natural Products Pvt. Ltd., where I managed rural tourism projects across

India. I have also served as Assistant Professor at Chandigarh University. My research contributions include Scopus-indexed publications, books, book chapters, and editorial assignments in the fields of marketing and tourism.

Mani Sachdev is Professor and Head of the Institute of Behavioural and Allied Sciences at Amity University Rajasthan, Jaipur. She also serves as the Head of the Amity Centre of Excellence in Indian Knowledge System at the same institution, providing academic and administrative leadership in advancing interdisciplinary scholarship, behavioural sciences, and research on the Indian Knowledge System. Her academic journey includes earning an M.Phil. in Philosophy and completing her Doctorate at the University of Rajasthan, specializing in Aesthetics, Metaphysics, and Epistemology. With an impressive tenure spanning 37 years, she has excelled as an Associate Professor of Philosophy. Dr. Sachdev's pioneering roles include founding the Department of Philosophy at Vedic Kanya P. G. College, Jaipur, and establishing the Department of Philosophy at B.N.D. Arts Government P.G. College Chimanpura, Jaipur. Her scholarly contributions and administrative leadership underscore her commitment to advancing philosophical discourse and academic excellence in higher education. Her professional portfolio encompasses the design and development of innovative curricula tailored for students, ensuring adherence to rigorous quality assurance standards within the University framework. Additionally, she contributes to the pedagogical landscape by teaching ethics to medical professionals at a distinguished medical university in Jaipur, where she serves as a distinguished member of the ethics committee. Her scholarly contributions extend to numerous international and national conferences, seminars, and scholarly journals, where she has presented research papers, authored book chapters, and contributed book reviews. Beyond academia, she has demonstrated versatility in media production, having produced, directed, and scripted corporate films for hospitals and organizations, as well as documentaries for national broadcast on Doordarshan. Furthermore, she has conducted in-depth interviews with over 50 prominent global figures, spanning governors, philosophers, academics, and renowned personalities from the film industry. Her cultural engagement includes creating, directing, writing, and narrating the acclaimed 'Light and Sound Cultural Show' celebrating Rajasthan's art, culture, and heritage, showcased at numerous tourism festivals and international conferences.

Aahana Saha is an Assistant Professor of Psychology with over nine years of teaching experience. She holds a Master's degree, is UGC-NET qualified, and completed a Ph.D. on female police personnel's job satisfaction and happiness. Her work includes developing and validating various psychological scales, publishing 20+ research papers, and securing two patents on machine-learning applications for cyber defense and forensic behaviour prediction. She has guided more than 30 postgraduate dissertations, coordinated academic programs, and launched student initiatives. Currently she is pursuing a PG certificate in forensic psychology and criminal profiling and a Master's in Data Science from Purdue University.

Aditi Sarawagi is an Assistant Professor of Finance in the School of Business at Galgotias University, Greater Noida. She holds a Ph.D. in Finance and is a UGC-JRF qualified academic in Commerce. Her teaching and research interests encompass finance, microfinance, accounting, banking, financial inclusion, sustainability, and ESG. Dr. Sarawagi has published several research papers in reputed UGC-CARE and Scopus-indexed journals and has contributed chapters to edited books on microfinance, tribal women's empowerment, and sustainable finance. She has presented papers at numerous national and international conferences and is a life member of the Indian Accounting Association and the Indian Commerce Association. In addition to her academic contributions, Dr. Sarawagi holds three design patents and is proficient in research tools such as SmartPLS, SPSS, and VOSviewer. Her research focuses on promoting inclusive and sustainable economic development through innovative financial practices.

Mamathakgalo Penelope Seakamela, she studied Bachelor of Indigenous Knowledge Systems at the Northwest University. She aspires to partake in the academic space by apply indigenous knowledge in my work and also preserve and transform the IK in the modern academia. She is interested in participating in community-based engagements and projects. Her focus area is cultural sustainability, indigenous agriculture and its practices and decolonizing the perspectives that people have about IKS

M. Sanjoy Singh, M.Com. NET and Ph.D. from Manipur University, is currently working as a Senior Assistant Professor in the Department of Commerce, Indira Gandhi National Tribal University, Amarkantak, Madhya Pradesh, for the last 9 years. He has published 7 books and 51 research papers in national/UGC-Care List journals and Scopus-indexed journals. Over the last two decades, he has attended and presented papers at 64 national and international seminars and conferences. Awarded Best Business Academic of the Year (BBAY) 2017 - Silver Medal in the 70th All India Commerce Conference of All India Commerce Association(ICA) held at IIS University, Jaipur, Rajasthan during 12th -14th October, 2017. He is a life member of the Indian Commerce Association (ICA), Indian Accounting Association (IAA), and Association of Commerce Alumni (ACA). His areas of interest are small business management, Entrepreneurship development, Microfinance, and accounting and finance. He may be contacted at joysingh357@gmail.com

Pramendra Singh is an Assistant Professor at the Faculty of Management Studies, Marwadi University, India. He has more than ten years of teaching and research experience. He has pursued two master's degrees in tourism management and another master's degree in marketing management. He completed his doctorate (Ph.D.) in tourism management. He has also done certification programs in digital marketing and continues to expand his knowledge in the domain. He has to his credit more than forty research papers and book chapters published by internationally reputed publishers, out of which thirty-two are indexed in Scopus also. He has also edited six books published by international publishers like Emerald, CABI, and IGI on topics like tourism, sustainability, education, business transformation, etc. His research interest areas are travel & tourism management, marketing, digital marketing, entrepreneurship, higher education, and sustainability.

Index